NCRP REPORT No. 58

A HANDBOOK OF RADIOACTIVITY MEASUREMENTS PROCEDURES

With Nuclear Data for Some Biologically Important Radionuclides

Recommendations of the NATIONAL COUNCIL ON RADIATION PROTECTION AND MEASUREMENTS

Issued November 1, 1978

National Council on Radiation Protection and Measurements
7910 WOODMONT AVENUE / WASHINGTON, D.C. 20014

Library of Congress Catalog Card Number 78-54276
International Standard Book Number 0-913392-41-3

Preface

In 1961 the NCRP issued NCRP Report No. 28, *A Manual of Radioactivity Procedures,* which was published as the National Bureau of Standards Handbook 80. The report came to be one of the definitive works in the area of radioactivity measurements. As the development of new techniques, procedures, and equipment made parts of the Report obsolescent, the NCRP recognized the need to update and extend the document. The report consisted essentially of three parts: Part One being "Radioactivity Standardization Procedures"; Part Two, "Measurement of Radioactivity for Clinical and Biological Purposes"; and Part Three, "Disposal of Radioactive Material". Since the disposal of radioactive material constitutes a subject in itself, the NCRP, in preparing the revision of Report No. 28, decided to concentrate on the first two parts of the Report.

Because of the projected bulk of the revised report, it was decided to divide the task of preparation into two parts. NCRP Scientific Committee 18A was given the task of drafting the material on general measurement and standardization procedures; and Scientific Committee 18B was charged with drafting the material on those measurement procedures specifically concerned with medical and biological applications. This report is the result of Committee 18A's work. As such, it follows, to a high degree, the general format of the original report. Thus, the physics of detectors is discussed in Section 2.0 and then the uses to which these detectors may be put are discussed in those sections dealing with direct and indirect measurements. At the same time, each of the sections has been developed in much more detail than previously and it is therefore inevitable that some duplication occurs. In fact, some duplication has been deliberately included so that individuals may utilize given sections of the Report without having to refer back to earlier sections where similar material may have been covered. In almost every case, it is believed that the repetition will not only emphasize the suitability of a detector for a particular application, but will also add detail to the information already given.

In a report of this length it is inevitable that the supply of unique symbols would be exhausted. Where, however, there may have been some duplications in the use of symbols, every effort has been made to keep them consistent within the short section in which they appear.

Acknowledgments are made to Dr. Ayres who acted as secretary to the Committee, and to Dr. Hoppes who, while making no specific written contribution, served invaluably by spending many hours in valuable discussion with the Chairman of the Scientific Committee during the course of editing and clarifying the original text. The assistance of Miss P. A. Mullen in the work of coordinating the references is also gratefully acknowledged.

The Council has noted the action taken by the General Conference of Weights and Measures to make available special names for some of the units of the International System (SI) used in connection with ionizing radiation. Gray (Gy) has been adopted as a special name for the SI unit, joule per kilogram, for absorbed dose, absorbed dose index, kerma, and specific energy imparted. Becquerel (Bq) has been adopted as a special name for the SI unit of activity (of a radionuclide). Since the transition from the special units currently employed—rad and curie—to the new special names is expected to take some time, the Council has determined to continue, for the time being, the use of rad and curie. To convert from one set of units to the other, the following relationships pertain:

$$1 \text{ rad} = 0.01 \text{ J kg}^{-1} = 0.01 \text{ Gy}$$
$$1 \text{ curie} = 3.7 \times 10^{10} \text{ s}^{-1} = 3.7 \times 10^{10} \text{ Bq (exactly)}.$$

Serving on Scientific Committee 18A during the preparation of this report were:

WILFRID B. MANN, *Chairman*

Members

ABRAHAM P. BAERG
J. CALVIN BRANTLEY
RUSSELL L. HEATH
BERND KAHN
HARRY H. KU
A. ALAN MOGHISSI

WILLIAM J. MACINTYRE, *Ex officio* (Chairman, Scientific Committee 18B)

Consultants

ROBERT L. AYRES
LUCY M. CAVALLO
BERT M. COURSEY
ALAN T. HIRSHFELD
DALE D. HOPPES
J. M. ROBIN HUTCHINSON
ROBERT LOEVINGER
MURRAY J. MARTIN
JANET S. MERRITT
JAMES R. NOYCE
CHIN T. PENG
FRANCIS J. SCHIMA
CARL W. SEIDEL
JOHN G. V. TAYLOR

NCRP Secretariat, CONSTANTINE J. MALETSKOS

The Council wishes to express its appreciation to the members and consultants for the time and effort they devoted to the preparation of this report.

WARREN K. SINCLAIR
President, NCRP

Bethesda, Maryland
February 15, 1978

Contents

1. Introduction

1.1 General

Radioactivity is the term used to describe those spontaneous, energy-emitting, atomic transitions that involve changes in state of the nuclei of atoms. The energy released in such exoergic transformations is emitted in the form of electromagnetic or corpuscular radiations. The radiations interact, in varying degrees, with matter as they traverse it.

Radioactivity was discovered in 1896 by A. H. Becquerel, who was investigating the fluorescence of a double sulphate of uranium and potassium, using a photographic plate. He found that the plate was affected by certain *radiations actives*, irrespective of whether or not the salt was caused to fluoresce. Marie Curie, who coined the word *radioactivité*, investigated this property in a number of minerals containing uranium, which she found to be more active, atom for atom, than uranium itself. Postulating an element that was unknown except for its radioactivity, she and Pierre Curie set about the isolation of this hypothetical element, thereby applying, for the first time, techniques that now comprise the field of radiochemistry. In 1898 they discovered polonium and, in collaboration with G. Bemont, radium. E. Rutherford showed that two types of radiation were emitted by uranium, namely the alpha rays, that were completely stopped by a thin sheet of paper, and the beta rays, that were much more penetrating. In 1900, even more highly penetrating radiations, the gamma rays, were discovered by P. Villard. Subsequently, the alpha and beta rays were shown to be ionized helium-4 atoms and electrons, respectively, and the gamma rays to be photons.

In due course, three separate generic families of naturally occurring radioactive materials were discovered, namely the uranium, thorium, and actinium series. In each of these, one radionuclide—uranium-238, thorium-232, and uranium-235 (known originally as actinium)—in undergoing radioactive change, gives rise to a series of between twelve and eighteen radioactive progeny, usually referred to as *daughters*. The atomic mass number of each member of each series is obtained by substituting n = 51, 52 . . . 59 in (4n + 2) (uranium), n = 52, 53 . . . 58 in 4n (thorium), and n = 51, 52 . . . 58 in (4n + 3) (actinium). Following the discovery of the transuranic elements, a fourth such family of radioactive elements was discovered, namely the neptunium series of which the atomic weights are given by substituting n = 52, 53 . . . 60 in (4n + 1). The common step of 4n in each series represents, of course, the loss of an alpha particle, the loss of a beta particle from the nucleus leaving the atomic mass number unchanged.

The early history of radioactivity, and the underlying principles of its detection and measurements, are treated in such textbooks as *Radiations from Radioactive Substances* by E. Rutherford, J. Chadwick, and C. D. Ellis (1930); *The Atomic Nucleus* by R. D. Evans (1955); *Nuclear and Radiochemistry* by G. Friedlander, J. W. Kennedy, and J. M. Miller (1964); and *Alpha-, Beta-, and Gamma-ray Spectroscopy*, edited by K. Siegbahn (1965). A brief and more elementary treatise, that may be useful for those involved as technicians in nuclear-medicine measurements, is *Radioactivity and Its Measurement* by W. B. Mann, R. L. Ayres, and S. B. Garfinkel (1979).

The present state of the art in radioactive metrology was the subject of an International Summer School on Radionuclide Metrology held at Herceg Novi, Yugoslavia in 1972. The proceedings, published a year later, (Herceg Novi, 1973) are a very valuable contribution to the literature of the subject.

In 1932 the neutron was discovered by James Chadwick and, in 1933, artificial radioactivity by Frederic and Irene Joliot-Curie. Thereafter, in the late 1930's and in the 1940's many artificially produced radioactive elements were discovered by many investigators using nuclear reactions produced in the cyclotron of Ernest Lawrence and the nuclear reactor developed by Enrico Fermi and his colleagues. In discussing such nuclear reactions we use the designations: n for neutron, p for proton, d for deuteron, α for alpha particle, and γ for gamma radiation; and the further convention of, for example, (n,γ), meaning capture of a neutron by a nucleus and emission of a prompt gamma ray. The reactions giving rise to artificial radioactivity in cyclotrons and other particle accelerators are many, including (p,n), (p,γ), (d,n), (d,p), (α,p), and so on. In many cases accelerated ions of higher atomic-number elements have also been used to produce nuclear reactions that result in radioactive elements. Among these, delayed proton emitters have also been observed. In reactors, in addition to the many radioactive fission products produced in the nuclear fission process, a vast quantity of radioactive materials has been produced by (n,γ), (n,2n), (n,p), (n,α), and other reactions. With high-energy accelerators producing photons of many millions of electron volts of energy, prolific photonuclear reactions occur, corresponding to ($h\nu$,xn) and ($h\nu$,xp) where $h\nu$ (the product of Planck's constant and photon frequency) represents the photon, and x (an integer representing the number of prompt neutrons and protons emitted) can be as high as ten. In the very high-energy accelerators, mesons can also be produced.

This report will be concerned with the measurement of radioactivity in general, but specifically it will deal with the vast number of different radioactive materials that have become available in the last three decades, from nuclear reactors and particle accelerators, for applications in medicine, scientific research, and industry. It will also be concerned with low-level radioactivity measurements for the monitoring of radioactivity in environmental media, such as air and water, in connection with the control of radioactive effluents associated with the production of nuclear power or the use of radionuclides.

It would be inappropriate to enlarge here upon this brief historical introduction. Some familiarity on the part of the reader with any of the textbooks referred to already, or their equivalent, must be assumed.

1.2 Radiations Emitted from Radioactive Atoms

In addition to the alpha particles, beta particles, and gamma rays already mentioned, energy from radioactive atomic transformations can be emitted as protons, neutrons, neutrinos, internal bremsstrahlung, conversion electrons, x rays, and Auger electrons. The beta particles can be either negative or positive electrons (β^- and β^+), β^- from neutron-rich and β^+ from neutron-deficient nuclei. The emission of a β^+ from the nucleus is always simultaneously accompanied by the emission of a neutrino, and that of the β^- by an anti-neutrino. An alternative to β^+ decay is capture of an atomic electron from the same atom with emission of a neutrino. The sharing of the available energy for decay between the beta particle and the neutrino accounts for the continuous beta-particle spectra; the sum of the energies of the beta particle and neutrino in a given transition is always constant, being *equivalent* to the mass difference ($E = mc^2$) between the parent and daughter atoms, less such energy as may be emitted in the form of gamma rays (or conversion electrons, x rays, or Auger electrons) by the daughter atoms in their transitions from excited energy levels to the ground level. Thus, ^{60}Co transforms (or "disintegrates" or "decays") mainly to an excited energy level of ^{60}Ni by the emission of a β^- particle and anti-neutrino with a total energy of 0.3179 million electron volts (MeV). The ^{60}Ni nucleus is left in an excited energy level, 2.5057 MeV above its ground level and decays with the prompt emission of two gamma rays, one with energy equal to 1.1732 MeV followed by another of 1.3325 MeV. Nuclear, or radioactive, transitions are conventionally depicted by means of a "decay scheme" such as those shown in Figure 1, representing the modes of decay of ^{64}Cu, ^{90}Sr-^{90}Y, ^{187}Re, and ^{210}Po. In the case of copper-64, the atomic mass of this isotope of copper is greater than the atomic masses of both its neighbors in atomic number, Z, namely nickel-64 and zinc-64, so that transitions to both these "isobars" are energetically permissible.

The case of ^{187}Re is unusual and is of interest since it illustrates the fact that the energy balance in radioactive decay involves the *whole* atom. In the transition of ^{187}Re to ^{187}Os (Fig. 1d) a total of 2.6 keV of energy is carried away by the β^- particle and anti-neutrino. Since, however, the atomic-electron binding energy of osmium is greater than that of rhenium by 15.3 keV, it follows that the mass of the rhenium nucleus is less than the mass of an osmium nucleus and an electron by the mass equivalent of 12.7 keV (i.e., 15.3-2.6 keV). A ^{187}Re nucleus stripped of all of its electrons would therefore be stable against β^- decay.

Crasemann (1973) has discussed the influence of atomic electrons on the modes of nuclear decay. In the case of ^{7}Be, an electron capturer in which the atomic binding energies are comparable with chemical binding energies, the value of the decay constant (Section 1.3) can be very slightly modified by both chemical composition and pressure (see, for example, Hensley *et al.*, 1973).

In alpha-particle decay, no simultaneous radiation, comparable to the neutrino, is emitted from the parent nucleus and hence groups of alpha particles are always homogeneous in energy. Again, gamma rays may be emitted promptly from the daughter nucleus in its decay to the ground level.

If the daughter nucleus does not decay promptly to its ground level, it may exist in a metastable state for a considerable time and exhibit radioactivity, so to speak, in its own right. Thus, ^{99}Mo decays to a 0.143-MeV level of ^{99}Tc which exists long enough, several hours, for the technetium to be chemically separated as a useful radioactive species, isolated from its parent, ^{99}Mo. The delayed transition of an excited daughter nucleus to a lower-energy level of the same nucleus is called an *isomeric transition* and such nuclear isomerism (as distinct from chemical isomerism) is denoted by the addition of the letter m (for metastable) after the atomic-mass number for the given nuclear species, e.g., ^{99m}Tc. In such cases where an isomeric species decays to a radioactive daughter, the letter g (for ground) is sometimes used to denote the ground state of the daughter nucleus, e.g., ^{81g}Kr is the radioactive daughter of ^{81m}Kr.

The energy associated with the electromagnetic decay of an excited state is not always emitted as gamma radiation; it may be transferred by a competing radiationless transition directly to a bound electron

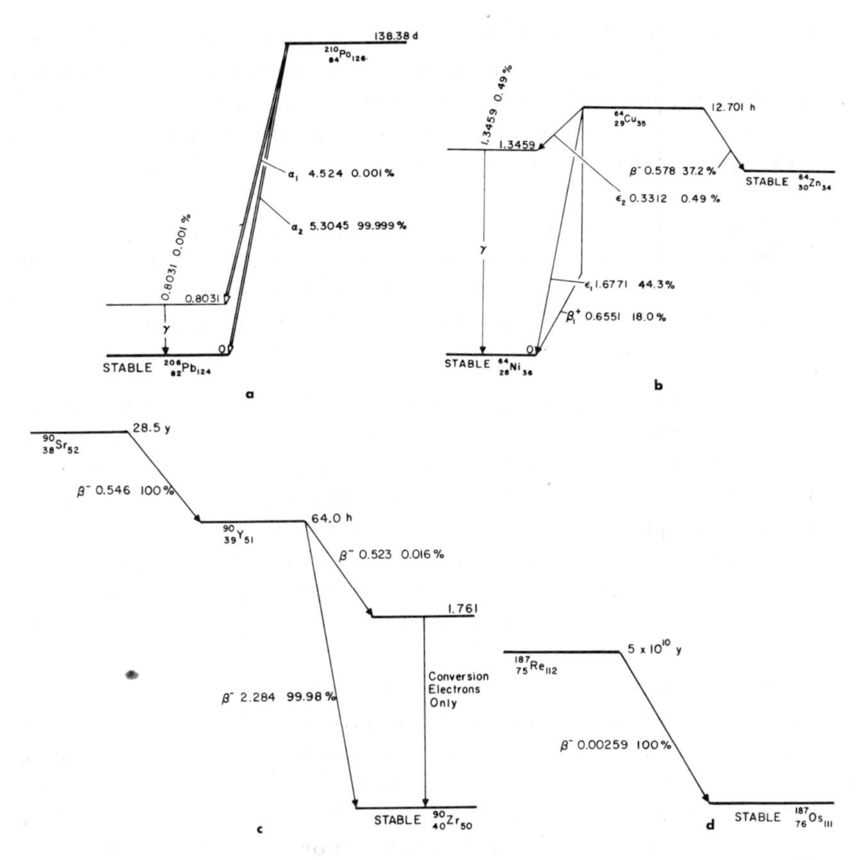

Fig. 1. Typical decay-scheme representations for (a) ^{210}Po, (b) ^{64}Cu, (c) ^{90}Sr-^{90}Y, and (d) ^{187}Re. Energies are expressed in millions of electron volts (MeV). ^{210}Po normally decays to the ground state of ^{206}Pb with the emission of a 5.3045-MeV alpha particle. Occasionally, however, the decay is to an excited level of ^{206}Pb with the emission of a 4.524-MeV alpha particle, followed by a 0.8031-MeV gamma ray. The discrepancy between the energy balances in the two branches, α_1 and α_2, is due to the difference in the energy associated with the recoil of the ^{206}Pb nucleus (see Section A.1.2.1 of Appendix A). In the case of ^{64}Cu, its atomic mass is greater than those of ^{64}Ni and ^{64}Zn by 0.0018008 and 0.0006168 atomic-mass units, respectively. From the Einstein equation, $E = mc^2$, 1 atomic-mass unit is found to be equivalent to 931 MeV. Hence the energy available for the transition of ^{64}Cu to ^{64}Ni or ^{64}Zn would be respectively 1.6765 and 0.5742 MeV. The transition to ^{64}Ni can occur either by emission of a positive electron (β+) or by electron capture (ϵ). The vertical line appearing in the β^+ branch represents the rest energy of two electrons, 1.022 MeV, because of the loss from the system, in the process of decay, of a positive electron and an excess atomic electron.

Most such decay schemes indicate gamma-ray abundances, or probabilities per decay, in percent, either before or after the gamma-ray transition energy. Thus, in Fig. 1a the 0.8031-MeV transition is shown as 0.001%. Where more than one gamma ray originates from any one energy level, all abundances would be shown.

More recently evaluated nuclear-decay data are available from the Oak Ridge Nuclear Data Project, as indicated in Section A.1.1 of Appendix A.

from a shell of the same atom. This process is known as *internal conversion,* a misnomer dating from early workers who erroneously assumed the existence of an intermediate gamma ray. The (internal) conversion electrons are ejected in monoenergetic groups having energies equal to the transition energy less the binding energies of the various electron shells.

For a given transition, the ratio of the number of decays proceeding by internal conversion to the number proceeding by gamma-ray emission is called the *internal-conversion coefficient* denoted by $\alpha(= e/\gamma)$. Similarly, coefficients for internal conversion from specified shells are written, e.g., $\alpha_K(= e_K/\gamma)$ so that $\alpha = \sum_i \alpha_i$ ($i = K, L_I, L_{II}, L_{III}, M_I \ldots$). If the transition energy is large compared to the K-shell binding energy, α_K is the largest term; if the transition energy is less than the i-shell binding energy, α_i is necessarily zero. As a rule α is small for prompt, high-energy transitions in low-Z nuclei and is larger for delayed or low-energy transitions or transitions in high-Z nuclei. Internal conversion coefficients for transitions of interest are included in compilations of decay schemes such as the *Table of Isotopes* (Lederer *et al.*, 1967). Direct conversion-electron probabilities per decay for selected radionuclides are given in Appendix A.

The emission of radiation as a conversion electron is a transition involving the *whole* atom, and this process emphasizes again how important it is to consider the interactions of the entire atom in radioactive decay. In the decay of both ^{90m}Nb and ^{99m}Tc, there are highly converted transitions of relatively low energy, of about 2.2 keV, which are again sensitive to the chemical or physical environment of the atoms. Experiments have shown that the decay constants of each of these isomers can be modified both by chemical composition (Cooper *et al.*, 1965; Crasemann, 1973) and pressure (Mazaki *et al.*, 1972).

Either an electron-capture or conversion-electron event creates a vacancy in an atomic shell of the daughter atom. The filling of this vacancy gives rise either to an x ray, characteristic of the daughter atom, or to one or more electrons (Auger electrons) from its outer shells. In the filling of atomic-shell vacancies the ratio of x rays emitted to the total number of vacancies to be filled in any given shell is called the fluorescence yield and is usually represented by ω (ω_K, ω_L, and so on). The emission of an Auger electron instead of an x ray is analogous to the emission of a conversion electron instead of a gamma ray, in that the excess energy is transferred directly to the electrons. Values for fluorescence yields and Auger-transition probabilities are available in a review article by Bambynek *et al.* (1972).

Electron capture and β^+ decay are alternate processes, the latter being possible only if the atomic mass of the parent, $M_A{}^Z$, is greater than the atomic mass of the daughter, $M_A{}^{Z-1}$, by more than two electron rest masses, $2m_e$, i.e.

$$M_A{}^Z > M_A{}^{Z-1} + 2m_e \tag{1.1}$$

The difference of $2m_e$ has nothing to do, as is sometimes asserted, with the creation of an electron-positron pair, but is simply due to the fact that the nucleus of atom $M_A{}^Z$ loses one electron mass by decay (β^+) and one atomic electron is lost because of the decrease in atomic number from Z to $Z-1$ (Mann *et al.*, 1979, Chap. 5).

If, however,

$$M_A{}^{Z-1} < M_A{}^Z < M_A{}^{Z-1} + 2m_e \tag{1.2}$$

the transition from $M_A{}^Z$ to $M_A{}^{Z-1}$ can still occur by the process of electron capture, excess energy being carried away by the neutrino. The electron captured by the transforming nucleus may originate in the K shell or outer electronic shells of the decaying atom. For each type of capture (e.g., K, L, or M), monoenergetic neutrinos are emitted.

Values for the energies and probability of emission per decay (intensity) of Auger electrons and x rays arising from K or L capture are given in Appendix A. The effects of other possible transitions within and between atomic shells are discussed by Bambynek *et al.* (1972).

It must be emphasized that the emission of gamma rays or conversion electrons from the daughter atom *follows* the transformation of the parent by intervals of time varying from fractions of a picosecond to tens of thousands of seconds. The definition of isomerism, or what time interval we choose to say has to elapse to call a transition isomeric rather than prompt, is largely a subjective judgment conditioned to some extent by our instrumental capability for measuring short intervals of time.

Finally, the parent atoms may decay either to the ground state or to excited energy levels of the daughter atoms.

For a given radioactive element, the proportions of transitions to the different nuclear energy levels of the daughter atom are given by the *branching ratios*. Such ratios for selected radionuclides, together with other decay data which are important for both their assay and use, are given in Appendix A.

For a more detailed treatment of the subject matter of this section, reference should be made to the textbooks mentioned in Section 1.1

1.3 The Half-Life, Decay Constant, and Activity of A Radioactive Element

In 1903, Rutherford and Soddy, in a paper entitled "Radioactive Change", published in the *Philosophical Magazine*, proposed "the law of radioactive change". They had observed, in every case that they had investigated, that the activity of a radionuclide decreased in geometric progression with time. This meant that if the activity decreased from any initial value to half that value in a given time, then in a further equal interval of time the activity would decrease by half again, or to one-quarter of the initial activity. In 1904, Rutherford introduced the term half-life to denote the time required for the activity of a radioactive element to decay from any given initial value to half that value. Rutherford and Soddy also stated the law that is appropriate to a geometric decrease of activity with time, namely that $N_t/N_0 = e^{-\lambda t}$, where N_0 is the number of radioactive atoms present at $t = 0$, N_t is the number after a further time t, and λ is a constant. This constant they called the "radioactive constant" but we now usually call it the *decay constant*. By differentiation they obtained $dN_t/dt = -\lambda N_t$ and concluded that, "The proportional amount of radioactive matter that changes in unit time is a constant". The fractional rate of decrease in the number of radioactive atoms is thus clearly equal to λ, which can be expressed as $(1/N_t)(dN_t/dt)$.

From the radioactive decay law, the time, $T_{1/2}$, required for a given number of radioctive atoms of a single radionuclide to decrease to half is very simply derived. Thus, if $N_t = N_0/2$ then $(N_0/2)/N_0 = e^{-\lambda T_{1/2}}$ whence the *half-life*, $T_{1/2} = \ln 2/\lambda$.

In 1905, E. von Schweidler derived the same law of radioactive decay, starting with the premise that the probability of a radioactive atom decaying in an interval of time Δt is proportional only to that interval of time. His derivation is given in Appendix B, Section B.2

For a given number of atoms N of a given radionuclide, unfed by any parent, λN is the expected number of decays per unit time at a given time, but only if we regard N as being a continuous function of t can we say that the decay rate is equal to dN/dt. But, by simple observation, it is known that the *measured* normalized decay rate $(dN/dt)/N$ varies with time over a wide range of values (see Appendix B). In other words, radioactivity is a random or stochastic phenomenon, and dN/dt, known as activity from the earliest days of radioactivity, only has meaning when we consider the average of a very large number of decays. For this reason, activity has been defined in the following rather precise manner, namely, the activity of an amount of a nuclide in a specified energy state at a given time is the expectation value, at that time, of the number of spontaneous nuclear transitions in unit time from that energy state. In this definition, zero activity would be equivalent to stability of the nuclide.

To define activity as dN/dt can be confusing as, for example, in parent-daughter relationships where dN/dt for the daughters can be positive, negative, or zero.

The dimension of activity is reciprocal time, and activity is usually expressed as the number of radioactive events per second. Such radioactive events are comprised of any atomic transitions that involve a change in nuclear state, and therefore include isomeric transitions.

A commonly used unit of activity that is accepted as a special unit in the Système International (SI) is the *curie* (Ci) (see BIPM, 1973; NBS, 1977). One curie is equal to exactly 3.7×10^{10} radioactive events per second. Multiples and sub-multiples of the curie are the megacurie (MCi), millicurie (mCi), microcurie (μCi), and so on. The becquerel (Bq) has recently been adopted as the SI unit of activity, being equal to one radioactive event per second.

As has been noted in Section 1.2, the definition of isomerism is a subjective one, and depends on whether the half-life of the excited state of the daughter nucleus is long enough for the daughter to be separated from the parent and exist as a radioactive substance in its own right. Thus, one might say that the nearly instantaneous decay of the 2.506-MeV energy level of ^{60}Ni, following ^{60}Co decay, is not considered to be radioactivity in its own right, as distinct from, say, the 0.1426-MeV level of ^{99m}Tc, decaying with a half-life of 6.02 hours. (The half-life of the 1.3325-MeV energy level of ^{60}Ni is about 0.7 picosecond). On the other hand, ^{137m}Ba, with a half-life of only 2.55 minutes, is often more closely identified with its radioactive parent, ^{137}Cs.

Care must be taken, in certifying the activity of a radioactive substance, to state very specifically what substance is under consideration. Thus, in the case of the two β^- emitters, ^{90}Sr and its daughter ^{90}Y, the two in equilibrium have approximately double the activity of freshly separated ^{90}Sr. As the ^{90}Y grows back into equilibrium after separation, the activity of the mixture increases until, in approximately 300 hours, a state of equilibrium is again approached (e.g., in 5 half-lives the ^{90}Y activity is 97 percent of the ^{90}Sr activity). It is therefore imperative, when dealing with an equilibrium mixture, to state whether or not the reported value is the *total activity* of the radionuclides in equilibrium, or the activity of the parent.

It is important to note that, in the condition of equilibrium between any radioactive parent that decays only to a short-lived daughter, the activity of the daughter is greater than the activity of the parent. It can be shown (see, for example, Mann *et al.*, 1979) that, if A represents the parent and B the daughter, and λ_A and λ_B their respective decay constants, then, at a sufficiently long time, t, after separation

$$dN_B/dt = [\lambda_B/(\lambda_B - \lambda_A)](dN_A/dt). \tag{1.3}$$

If $\lambda_B \gg \lambda_A$, then the activities of parent and daughter tend to become equal, and we have the condition of *secular* equilibrium. When, however, λ_A and λ_B are comparable in magnitude, the activity of B is significantly different from that of A, in the proportion of $\lambda_B/(\lambda_B - \lambda_A)$. This kind of equilibrium is known as *transient*.

In the case of ^{140}La ($T_{1/2}$ = 40h) in transient equilibrium with its parent ^{140}Ba ($T_{1/2}$ = 12.8d), the activity of the former is 15 percent higher than that of the latter (Figure 2).

When $\lambda_B < \lambda_A$, no state of equilibrium can exist.

1.4 The Interaction of Radiation with Matter

As mentioned in Section 1.1, the radiations emitted in radioactive transformations interact with matter, and it is in this manner, whether it be in the interaction with a photographic emulsion or in the ionization of atoms in the surrounding air, that radioactivity can be detected and the amount of radioactivity, or the quantity of radioactive material, can be measured.

The earliest measurements of radioactivity were based on the fact that radiation, as it traverses air, ionizes the gaseous molecules present. Under the influence of an electric field, the electrons and the negative and positive ions migrate respectively to and from regions of higher

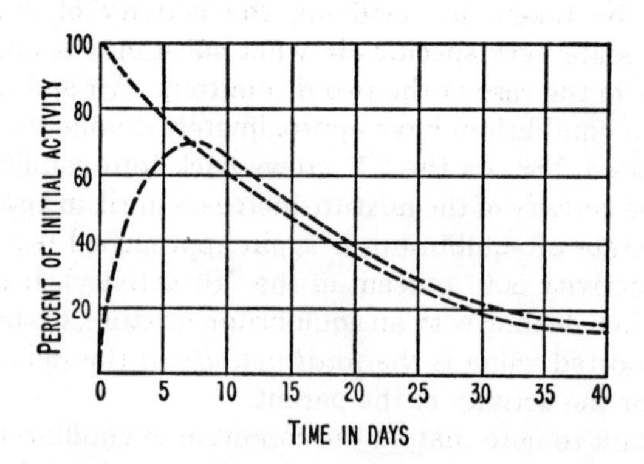

Fig. 2. Growth and decay of ^{140}La ($T_{1/2}$ = 40 h) from freshly separated ^{140}Ba ($T_{1/2}$ = 12.8 d) and the decay of the ^{140}Ba as a function of time. The ^{140}La activity is a maximum at T_m = 6.57 d. (From Mann *et al.*, 1979)

positive potential. An ionization current is thus established. In the earliest experiments, simple parallel-plate ionization chambers and gold-leaf electroscopes were used to detect and to measure radioactivity. Thus, Pierre and Marie Curie measured radioactivity in terms of units of 10^{-11} ampere of ionization current flowing across a charged parallel-plate condenser, with the plates horizontal, when powdered radioactive minerals were placed on the lower plate. For those who may be interested in the details of these earlier, yet very elegant experiments, reference is made to Mann *et al.* (1979). Reference to this and to Evans (1955), to Friedlander *et al.* (1964), and to other texts, should be made for details of the interactions of alpha and beta particles, and of photons, with matter.

1.4.1 *Interactions of Charged Particles with Matter*

When charged particles traverse air, they undergo elastic and inelastic collisions with the atoms and molecules, the elastic producing deflections and the inelastic producing positive ions and electrons, usually referred to as ion pairs. An average of nearly 35 eV of energy is expended for each ion pair formed, so that a 3.5-MeV alpha particle will, for example, produce some 10^5 ion pairs in air before expending all its energy and coming to rest. The production of ion pairs by charged particles is by Coulomb interactions. The alpha particle, being some 7000 times heavier than an electron, will "plow" its way through a gas and only rarely, when it comes close to an atomic nucleus, will it be deflected from its linear trajectory. Beta particles are, on the other hand, of the same order of mass (except at high velocity) as the atomic electrons removed by ionization. The path of a beta particle as seen, say, in a cloud chamber, is therefore meandering and circuitous by comparison with that of an alpha particle. The responses of ionization chambers, Geiger-Müller counters, and proportional counters, are all based upon the principle of gas ionization, and the physics of these detectors will be described in later sections of this report.

Charged particles also interact with photographic emulsions, with scintillating fluids and solids, such as anthracene and plastic fluors, or with semiconductor detectors. The scintillation of zinc-sulfide crystals when bombarded with alpha particles was one of the earliest methods used by Rutherford and his collaborators in the detection of these radiations. Energetic electrons, beta particles, Auger electrons, and alpha particles also cause solid and liquid fluorescent materials to emit light by imparting energy to electrons of the atoms comprising these fluors. The photons so emitted can be detected by sensitive electron-multiplier phototubes to provide quantitative information on the radioactivity present. Semiconductor detectors, to be more fully described in subsequent sections of this report, act as solid-state ionization chambers.

The energy carried by charged-particle radiations, when completely absorbed by matter, appears as an equivalent amount of heat which can be measured in a sensitive calorimeter. With the type of calorimeter described by Jordan and Blanke (1967), Jordan (1973) has achieved such sensitivity as to be able to observe, in a period of from 2 to 4 hours and with a sample of about 3 kilocuries, the 12.3-year half-life of tritium.

1.4.2 *Interactions of Energetic Photons with Matter*

The term "energetic" in this section's title is meant to refer to energies above, say, one kiloelectron volt, or in other words, to x rays and gamma rays, rather than to visible or ultraviolet photons. By way of reference, the energy of the "D line" of sodium corresponds to an approximately 2-eV transition of the atomic electrons involved.

The x and gamma rays transfer energy to matter chiefly by three extranuclear processes, namely (i) the *photoelectric effect*, which is important for energies below 1 MeV; (ii) the *Compton effect*, which is important over a wide range of energies; and (iii) *electron-positron-pair production*, which occurs at gamma-ray energies above 1.022 MeV.

In the *photoelectric effect*, the whole energy of the photon is transferred to a bound electron of another atom, which electron leaves its parent atom with an energy equal to that of the interacting photon, less the binding energy of the K, L, M, etc., shell from which it is ejected. Electrons originating in the same shell are therefore monoenergetic for incident monoenergetic photons.

In the *Compton effect*, the photon is scattered and behaves like a particle, with conservation of energy, $h\upsilon$, and momentum, $h\upsilon/c$, where h is Planck's constant, υ is the photon frequency, and c is the velocity of light. The scattering equations are derived in the textbooks referred to in Section 1.1, but, essentially, energy is imparted to a free (or loosely bound) electron with a consequent loss in frequency and energy of the scattered photon.

In *pair production*, an energetic gamma ray produces a negatron-positron pair (usually referred to as an electron-positron pair). The rest energy of the electron, derived from the Einstein relation $E = mc^2$, is equal to 0.511 MeV. A photon of energy greater than 1.022 MeV can therefore produce an electron-positron pair. To conserve energy and momentum, such a process must occur within the Coulomb field of a nucleus and any energy in excess of 1.022 MeV is imparted to the recoiling particles. The positron subsequently annihilates with an electron with the production of two 0.511-MeV photons. The processes of pair production and annihilation normally occur within the resolving time of a detector, and, for the low-energy gamma rays occurring in radioactive decay, the annihilation photons propagate essentially at 180° to each other.

The detection of gamma rays is made possible by the interactions, with matter, of the electrons produced by the three processes enumerated above. These electrons ionize gases, and consequently ionization chambers, Geiger-Müller, and proportional counters can be used to detect the photo-electrons, scattered electrons, and electron pairs. In addition, gamma rays produce such electrons within the volumes of solid detection devices, such as sodium-iodide, germanium, and silicon crystals. In the case of sodium-iodide crystals, the secondary electrons produce light pulses that can be detected by phototubes, while the germanium and silicon crystals act as solid-state ionization chambers. The physics of such detectors are dealt with in detail in Sections 2.4 and 2.5 of this report.

If gamma rays or x rays are completely absorbed by matter, the electrons produced, if also absorbed, give an equivalent amount of heat that can again be measured by sensitive calorimeters. Such measurements are more often made, however, to determine the absorbed dose rate from therapeutic gamma-ray sources and are not the concern of this report. (See, for example, ICRU, 1970).

Linear attenuation coefficients for collimated monochromatic photon beams as a function of energy are shown in Figures 3 and 4 for aluminum and lead, based on the data of Grodstein (1957), Hubbell (1969), and Hubbell *et al.* (1975). In these figures the total attenuation coefficients, together with the photoelectric, Rayleigh (coherent), Compton (incoherent), and pair-production components, are shown. The attenuation coefficient is, effectively, the probability, per unit path length, of a photon being either absorbed or scattered. Experimental observations of the removal of photons from narrow collimated beams give results that closely approximate the attenuation coefficients shown in Figures 3 and 4, and the attenuation coefficient is then synonymous with what is still sometimes called the absorption coefficient. These values, divided by the density, give the mass attenuation coefficient. The interactions of energetic photons with water are discussed in more detail in Section 4.2.2.

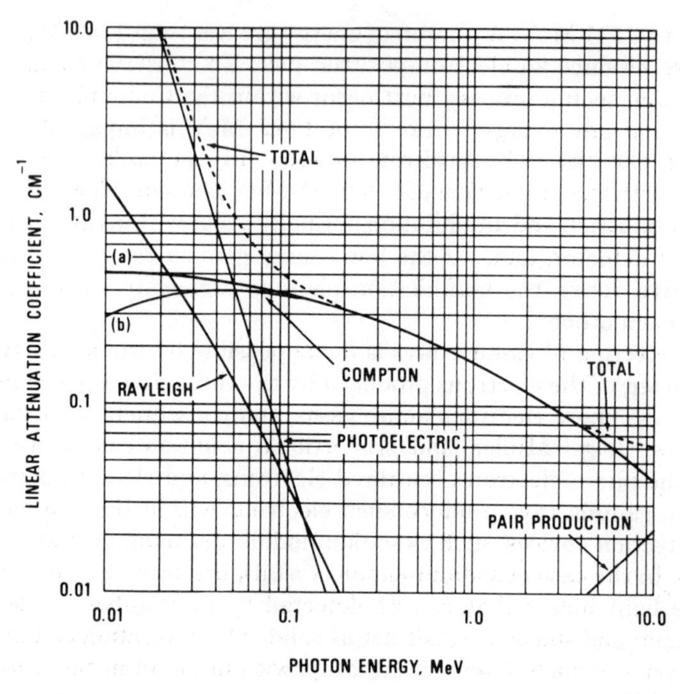

Fig. 3. Linear attenuation coefficients as a function of photon energy for a collimated, parallel beam of monochromatic photons in aluminum. The mass attenuation coefficient can be derived by dividing the ordinates by 2.70 g cm^{-3}, the density of aluminum. The "Compton (a) and (b)" curves are those pertaining to free and bound electrons, respectively. (These curves are derived from data given in Hubbell [1969] and Hubbell *et al.* [1975], kindly provided by J. H. Hubbell.)

1.4.3 *Other Interactions*

Other radiations emitted from the nucleus in radioactive decay, such as protons and neutrons, can be detected by ionization and other methods, such as, for example, by observing the etched tracks of protons and alpha particles in crystals, glass, and plastics (see Section 6.1.2.5). Neutrinos, which are particles of zero mass and zero charge and, therefore, have practically no interaction with matter, must be detected by more sophisticated means, but are of no immediate concern to the measurement of radioactivity. When electrons are accelerated or decelerated they also emit electromagnetic radiation known as *bremsstrahlung*. The beta particles or electrons, which are retarded by matter, emit photons having a range of energy from zero up to the maximum energy of the beta particles or electrons being stopped. Many of the photons from a typical x-ray tube are produced by the bremsstrahlung process. Bremsstrahlung has also been used in the

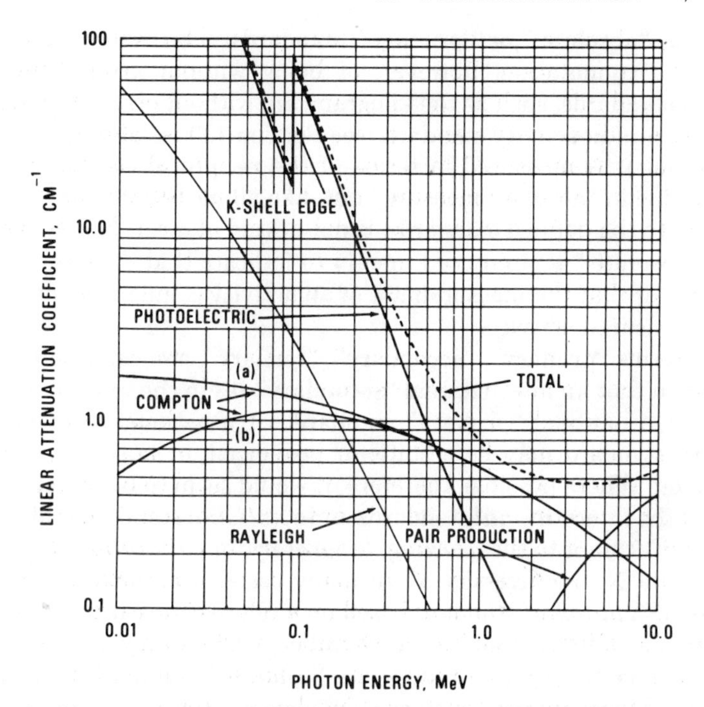

Fig. 4. Linear attenuation coefficients as a function of photon energy for a collimated, parallel beam of monochromatic photons in lead. The mass attenuation coefficient can be derived by dividing the ordinates by 11.34 g cm^{-3}, the density of lead. The "Compton (a) and (b)" curves are those pertaining to free and bound electrons, respectively. (These curves are derived from data given in Hubbell [1969] and Hubbell *et al.* [1975], kindly provided by J. H. Hubbell.)

assay of tritium (Curtis, 1972). *Internal bremsstrahlung* is emitted from the transforming nucleus, but these are rarely used to measure radioactivity, because of their very low abundance.

High-energy charged particles can also be detected by observing the Čerenkov radiation which is generated when they traverse a medium with a velocity greater than the velocity of light in that medium. This again has, however, but limited application to the measurement of radioactivity (see Sections 2.6, 4.5.1, 6.1.3.3).

1.5 Fundamental or Direct Methods of Radioactivity Standardization

1.5.1. *Introduction; Terminology*

Direct methods for the measurement of radioactivity are sometimes referred to as "absolute", and standards so produced are often designated as "absolute" or "primary" standards. The word "absolute" bears the connotation, however, of being without error. Only the defined standards, such as the kilogram, are without error. An indirect or relative radioactivity standard (see Section 1.6) is "absolute" in the sense that it is measured in terms of the reciprocal second. Such a standard is in "absolute measure" but, like all measurements made in relation to the defined standards, is not devoid of error. Nowadays the term "absolute" is often used merely to indicate that a measurement is direct, or that the measurement of an activity is independent of any other measured activity.

The terms "primary", "secondary", "tertiary", etc. can also be misleading in that an international "secondary" (as in the case of radium) could be a national "primary" standard. Or a national secondary or tertiary standard may be a state, or provincial, or hospital primary, and so on. Where possible, therefore, it would seem to be preferable to refer to the *international* standards, or to *national* standards (McNish, 1958 and 1960) or to the *working, laboratory,* or *calibration* standards of hospital X or university Y. In many cases, a national laboratory may issue a *national* standard based on a relative measurement in, for example, a calibrated ionization chamber, while a university or commercial laboratory may produce an "in-house" standard by a direct method of measurement such as coincidence counting.

It is, however, more important to describe the method of preparation and measurement, whether *direct* or *indirect* (or *relative*), and to characterize a radioactivity standard fully with regard to purity and estimated error in an accompanying statement or certificate. In the last resort, the method of production of the standard is of less importance than a realistic appraisal of the uncertainty of its value.

1.5.2 *Direct Methods of Measurement*

There are, broadly speaking, four methods of direct calibration in which the activity of a radioactive source can be determined from the count-rate data alone, without reference to any supplementary data determined from previous experiments, apart from the need to have a prior knowledge of the mode of decay of the radioactive substance itself.

The most powerful of these direct methods is probably that of coincidence counting (Geiger and Werner, 1924; Dunworth, 1940; NCRP, 1961; Herceg Novi, 1973). This method is applicable to all radionuclides decaying by the emission of two or more coincident radiations. Thus, where a radionuclide emits an alpha or beta particle, or an x ray (as a result of electron capture), followed promptly by a gamma ray, the method of coincidence counting (Sections 3.2, 3.3, and 3.4) can be used to determine the efficiency of the detectors of the two coincident radiations, and thence to determine the activity of the radionuclide in question. In the case of ^{60}Co, the activity may be determined either by beta-gamma coincidence counting of the ^{60}Co beta particles and the ^{60}Ni gamma rays, *or* by gamma-gamma coincidence counting of the two ^{60}Ni gamma rays in cascade. For more complex decay and, generally for the highest accuracy, efficiency variation techniques are now used (Section 3.2).

Modifications of the method of coincidence counting are those of anti-coincidence and correlation counting (Sections 3.2.3 and 3.2.4). These are particularly useful where a radionuclide decays to a delayed excited state of a daughter nucleus.

For charged-particle emitters the direct measurement of activity can be achieved by placing the source on a very thin source mount within a windowless detector subtending an angle of 4π steradians around the source (Section 3.5). Here the geometrical efficiency is unity, and all corrections for scattering and absorption outside the source are eliminated. For very thin, or "weightless", sources of alpha or beta particles of not too low an energy, practically 100-percent detection efficiency can be achieved in the case of 4π gas proportional counters. Internal gas counting of radionuclides in gaseous form (Section 3.6) and liquid-scintillation counting of energetic alpha- and beta-particle emitters (Sections 2.4.5 and 3.8) are also forms of 4π counting in which the total detection efficiency approaches 100 percent.

One of the oldest methods of direct standardization of activity is that of defined-solid-angle counting where the defined solid angle is 1π steradians or less. It has been used for measuring alpha-particle, beta-particle, and photon activities, but is now considered unsuitable for direct beta-particle standardization, because of the large errors introduced by scattering. Bambynek (1967) states, however, that for alpha particles with energies of several MeV, and for photons in the range of energy of 1 to 80 keV, accuracies of within ± 0.1 percent for alpha particles and ± 1 percent for photons are attainable. His criterion for the successful use of the defined-solid-angle method is that the radiations shall be "lightly scattered but heavily absorbed".

Defined-solid-angle counters have been used by Bambynek (1967) and his colleagues at the Euratom Laboratories at Geel to standardize alpha-particle and low-energy photon sources. A very successful $1\pi\alpha$ counter has been described by Robinson, H. (1960) and a modified $1\pi\alpha$ counter (based on Robinson's design) is in use at the National Bureau of Standards (Hutchinson *et al.*, 1968). Counting in a 1π geometry, normal to the surface of the thick source support, eliminates errors due to multiple alpha-particle scattering, because such scattered alpha particles are emitted at angles of less than 5° to the plane of the surface of the source support (Walker, 1965).

The third direct method is applicable to radionuclides that decay with the emission of at least two photons in coincidence with no transitions of the parent nucleus to the ground state of the daughter. This method requires a photon detector of relatively high intrinsic and geometrical efficiency. In such circumstances the two photons will sum in the detector so that not only two photopeaks corresponding to the energies of each of the photons will be produced, but also a sum-peak corresponding to the summing of the energies totally deposited by two coincident photons in the detector (Section 3.3). The count rates under these three peaks are usually measured by the analysis of the spectrum of pulses from the detector using a multichannel analyzer. From a knowledge of the total gamma-ray count rate, and of the count rates in the three photopeaks, the activity may be determined with no other measurement information, apart from certain relatively minor corrections for counting losses (Section 2.7) inherent in the nature of the electronic circuitry involved.

1.5.3 *"Near-Direct" Methods of Measurement*

In these methods the disintegration rate, or activity, is not measured directly but is calculated from a direct measurement of some property of the decay of the radionuclide in question, such as rate of energy emission, together with previously measured decay data, in this case the average energy emitted per disintegration. Some "near-direct" methods are briefly described in the subsections below. Table 1 gives conversion factors for deriving disintegration rates from the various quantities measured in different units.

1.5.3.1 *Weighing of a Radioactive Element of Known Isotopic Abundance.* Historically, weighing is the oldest method of radioactivity standardization, having first been applied in the case of radium in 1911. This radium, which was used to determine the atomic weight of radium-226, was considered to be isotopically pure. Weighing, or its equivalent, such as measuring the pressure, volume, and temperature of a gas, may also be used to determine the activities of other radioactive nuclides, provided that the isotopic abundance of the radionu-

TABLE 1—*Conversion of various measures of radioactivity into nuclear transitions per second*

Quantity of radioactive nuclide expressed as	Transitions per second Bq	Remarks
1 curie	3.7×10^{10}	Defined as exactly 3.7×10^{10} by international agreement. Symbol for the "curie" is Ci.
1 gram of pure radionuclide	$\frac{6.022 \times 10^{23}}{\text{at. wt.}} \lambda$	λ decay constant (s^{-1}), atomic weight (mass) (unified scale, i.e. carbon-12)
1 watt total rate of energy emission	$\frac{6.24 \times 10^{12}}{\bar{E}}$	$\bar{E}$ average energy emitted per disintegration (MeV).
1 roentgen/hour at 1 cm (from gamma-ray-emitting point source).	$3.7 \times 10^{7}\ \Gamma_\delta$	If exposure-rate constant Γ_δ is in R cm^2 mCi^{-1} h^{-1}. See Eq. 3.35 *et seq.*

clide in question and its decay constant are known. Where a radionuclide is mixed with inactive isotopes, such as ^{63}Ni in nickel, the relative abundances may be determined by mass spectrometry (Section 3.10). The disintegration rate, $-dN/dt$ can then be determined from the radioactive decay law $-dN/dt = \lambda N$, by using the measured value of N and the known decay constant λ.

1.5.3.2 *Calorimetry.* Particularly suitable for short-range particle emitters (such as alpha or low-energy beta) is the calorimetric method in which the rate of energy emission from the radioactive substance is measured. The conversion to activity requires a knowledge of the average energy per disintegration (Section 3.9).

1.5.3.3 *Ionization Measurements.* If the average energy expended by radiation in forming one ion pair in a gas is known, the total energy of the radiation from a source can be determined from ionization-current measurements. The determination of the activity of the source then requires a knowledge of the average radiation energy emitted per disintegration for the nuclide in question (Section 3.11).

1.5.3.4 *Loss-of-Charge Method.* For charged-particle emitters the loss-of-charge method allows the number of such particles emitted to be calculated from a measurement of the charge transfer. Since the method is rather insensitive, it is not recommended for practical use (Section 3.12).

1.5.3.5 *Activity from Production Data.* The activity of a radionuclide produced in a particular irradiation may be calculated approximately from irradiation, cross-section, and decay-scheme data (Section 3.13).

1.6 Indirect or Relative Methods of Radioactivity Standardization

In the national standardizing laboratories, the calibration data for many radionuclides (particularly a short-lived radionuclide) are usually preserved by using a standardized source to calibrate a reference instrument such as an ionization chamber. This instrument can then be used to calibrate further samples of the same radionuclide to give derived or "secondary" standards at any later date, or to check the consistency of subsequent "absolute", or direct, standardizations. The stability of the instrument is usually checked by measuring a reference source of some long-lived material, such as radium, whenever a calibration is made. One may either take the activity of the sample to be proportional to the response of the instrument (after its stability has been checked), or to the ratio of the response to that produced by the reference source (in which case variations in the sensitivity of the instrument are immaterial as long as they affect the responses to both sources equally).

Methods of relative standardization are described and discussed in Section 4 of this report.

2. Physics of Some Radiation Detectors

2.1 Ionization Chambers

An ionization chamber is an instrument in which an electric field is applied across a volume of gas. Often its geometry is cylindrical, with a cylindrical cathode enclosing the volume of gas and an axial insulated rod anode. The anode is also frequently in the form of a hollow cylinder into which a photon-emitting radioactive source can be inserted for calibration.

Charged particles moving through matter undergo inelastic collisions with atoms or molecules. In a gas, these particles produce positive ions and electrons which, in the absence of an electric field, will recombine. In some gases the electrons may become attached to neutral molecules to form negative ions. When an electric field is applied to the gas, the ions drift along the lines of force to produce an ionization current. Under usual conditions, electrons drift at speeds of the order of 10^6 cm s^{-1}. The drift velocity of ions is many orders of magnitude less. Collisions with gas molecules prevent the ions from attaining high velocities. Some values of $\overline{W}_{air}$, the average energy expended in the creation of an ion pair in air, are shown in Table 2. As can be seen, this quantity does not depend markedly on the energy, and is approximately the same for electrons and alpha particles, when measured in the same gas (see also Valentine and Curran, 1958; Klots, 1968; and ICRU, 1978).

For applications and further details of the ionization process see

TABLE 2—*Average energy expended in the production of an ion pair in air*

Radiation	$\overline{W}_{air}$ eV/ion pair	Reference
^{3}H β	33.9	Jesse and Sadauskis, 1955
^{35}S β	33.7	Bay *et al.*, 1957
^{35}S β	33.6	Gross *et al.*, 1957a
^{63}Ni β	34.0	Jesse and Sadauskis, 1955
2-MeV x rays	33.9	Weiss and Bernstein, 1955
^{210}Po α	35.5	Jesse and Sadauskis, 1953
^{210}Po α	35.0	Bortner and Hurst, 1954

Section 4.4 and Corson and Wilson (1948); Wilkinson, D. H. (1950); Staub (1953); Price (1964); Taylor, D. (1962); Curran and Wilson (1965); and Smith (1965).

2.1.1 *Current Ionization Chamber*

When an ionization chamber is in a radiation field, the measured ionization current first increases with increasing voltage and then levels off. Figure 5 shows the so-called "saturation" curve. In general, the voltage required to attain saturation current for any chamber will depend on the rate at which ionization is being produced. At saturation, the ionization current, I, is related to the number of ion pairs produced per unit time, n, by the equation $I = ne$, where e is the electronic charge. The chamber, therefore, measures the integrated effect of a large number of ionizing events. The time constant of the current-detecting device is generally made long to suppress statistical fluctuations, and whether the electrons are collected as free electrons (as, for example, in argon), or attached to slow-moving molecules as ions (as, for example, in oxygen), is of no importance. However, the probability of recombination is lower in a gas in which negative ions are not formed, and saturation can be reached with a lower field strength.

When the ionization current is sufficiently large, it can be measured with a microammeter. Generally, however, it is necessary to use more sensitive methods, for instance, by allowing the charge to collect on a capacitor and measuring the rate of change of voltage on the capacitor with an electrometer. The number of ion pairs formed per unit path

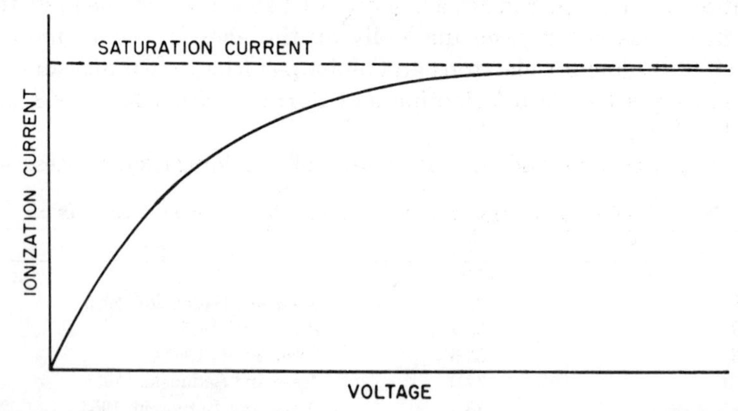

Fig. 5. Saturation curve (ion current *versus* collecting voltage) for a typical ionization chamber.

length is a function of the density of the gas in an ion chamber. Therefore, in unsealed chambers, corrections of the ionization current to standard temperature and pressure may have to be made when the range of the primary radiation is greater than the dimensions of the chamber. With the pressure ionization chambers now often in use, and which mostly contain argon at a pressure of some 20 atmospheres, this problem does not arise.

2.1.2 *Pulse Ionization Chamber*

An incident particle depositing all its energy in the gas of an ionization chamber will give rise to a charge which is proportional to the energy deposited. Suppose that the chamber, with an applied voltage, is filled with a gas in which ion pairs are formed. If the voltage drop due to the resulting ionization current is measured across a high-megohm resistor, R, and the capacity of the chamber and measuring system is C, a voltage-time dependence is found, such as is shown in the lower curve of Figure 6, which is representative of a time constant $RC = 1\ \mu s$. When the time constant is increased, the voltage rises as shown in the upper curve, for $RC = \infty$, of Figure 6. Here, a fast rise (to point a) at the beginning of the pulse, due to the rapidly moving electrons, is followed by a slower increase (to point b) due to the movement of the positive ions. When it is feasible to use a gas that allows fast electron collection, a differentiating time constant of only a few microseconds is usually chosen to prevent pulse pile-up at high counting rates. If the differentiating time constant is longer than the time for electron collection but much shorter than for collection of the positive ions, the pulse shape is determined by the movement of the electrons. The remaining positive ions induce a charge in the electron-collecting electrode and reduce the pulse amplitude, the magnitude of this charge depending on the location of the initial ionization. Therefore, in a parallel-plate chamber, the pulse amplitude will depend on

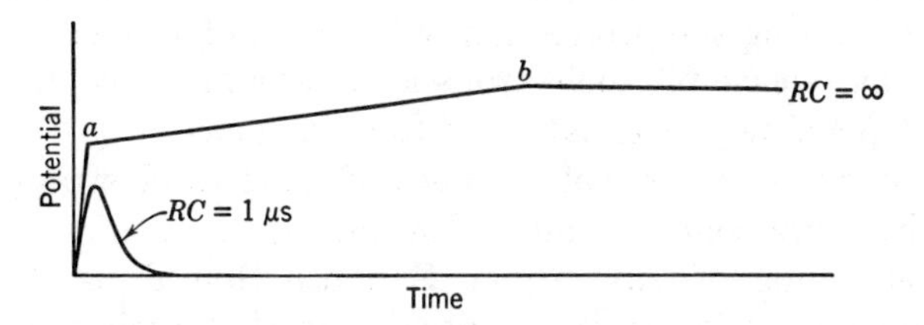

Fig. 6. Voltage pulse on collecting electrode of pulse ionization chamber. (From Halliday, 1955).

where the ionizing particle traversed the chamber. This undesirable effect can be overcome by placing a grid in the chamber (e.g., Frisch grid chamber) to screen the collecting electrode from the positive ions.

The use of a pulse ionization chamber is limited at low particle energies by the inherent noise of the associated amplifier. This noise is equivalent to the arrival on the ionization-chamber anode of about 200 electrons, which might represent an energy of about 7 keV dissipated by a particle traversing the chamber. For details, see, for instance, Price (1964).

2.2 Proportional Counters

When the electric field strength at the center electrode of a pulse ionization chamber is increased above a certain level, the size of the output pulse from the chamber starts to increase but is still proportional to the initial ionization. A device that is operated in such a fashion is called a proportional counter. The great advantage of the proportional counter is that it allows one to detect a very low initial ionization, even down to a single ion pair.

For a counter with cylindrical geometry, the field strength E_r, at radius r from the axis is given by

$$E_r = \frac{V_0}{r \ln (b/a)} \tag{2.1}$$

where a is the radius of the inner electrode, b is the radius of the outer cylinder (for $a < r < b$), and V_0 is the voltage applied to the counter. In a proportional counter the anode is usually a fine wire of diameter about 0.001 to 0.010 inch, so that the radius r can be small, and E_r can be correspondingly large. The field strength close to the wire increases rapidly and when the electrons drifting toward the wire enter the region of high-intensity electric field close to the wire, they will acquire enough energy between collisions to produce secondary ions and electrons; the latter are accelerated and produce more secondary electrons. Thus, an "avalanche" is developed. However, photons are also formed during the production of the secondary electrons. These photons may release photoelectrons anywhere in the counter volume or walls, depending on the nature of the filling gas.

The following rather simple model will serve to illustrate the operation of the proportional counter. Consider such a counter containing a pure gas filling of, say, argon. Suppose that $\bar{\text{m}}$ is the average electron multiplication in the avalanche so that each of n electrons produced in the primary ionizing event gives rise, in its acceleration toward the anode wire, to an average total of $\bar{\text{m}}$ electrons in an avalanche. As a result of the inelastic collisions between the accelerated electrons and gas atoms, photons will be emitted and these, in turn, can interact with the walls (in the case of a rare gas) to produce further photoelectrons. If P_e is the probability that such a photoelectron be formed for every electron in the avalanche, then the n electrons produced by the primary ionizing event will produce in the initial avalanche, or Townsend discharge, in the high-intensity electric field near the wire, an average of $n\bar{m}$ electrons by collision and $n\bar{m}P_e$ photoelectrons elsewhere in the counter. However, the $n\bar{m}P_e$ photoelectrons can now also be accelerated to the anode producing $n\bar{m}^2P_e$ more electrons by collision and $n\bar{m}^2P_e^2$ photoelectrons, which in turn form more generations of electrons by collision.

Thus, the final number of electrons collected at the anode wire as a result of each primary event is

$$N = n\bar{m} + n\bar{m}^2P_e + n\bar{m}^3P_e^2 + \cdots. \tag{2.2}$$

But, in a proportional counter, $\bar{m}P_e$ is much less than one, so that multiplication is essentially by electron avalanches, and the series converges giving

$$N = n\frac{\bar{m}}{1 - \bar{m}P_e}, \tag{2.3}$$

and the total number of electrons in the avalanche is proportional to the number of primary ion pairs, n. This is the essential property of a proportional counter, but it must be emphasized that m is an average number, and the actual number of electrons formed from one generation to the next will fluctuate statistically. $M = N/n$ is called the multiplication factor.

Multiplication factors have been worked out theoretically (Rose and Korff, 1941). Figure 7 shows experimental values for M as a function of applied voltage for argon. A comparison of theory and experiment has been made recently by Hendricks (1972).

2.2.1 *Shape of the Output Pulse*

In contrast to the pulse ionization chamber where the main part of the pulse is determined by the rapid movement of electrons, the pulse shape measured on the wire of a proportional counter is essentially determined by the movement of the positive ions. Since, in a proportional counter, most of the ionization develops very close to the wire, the electrons travel only short distances and their effect on the pulse rise is rather small. Figure 8 shows a typical pulse shape. A delay, of

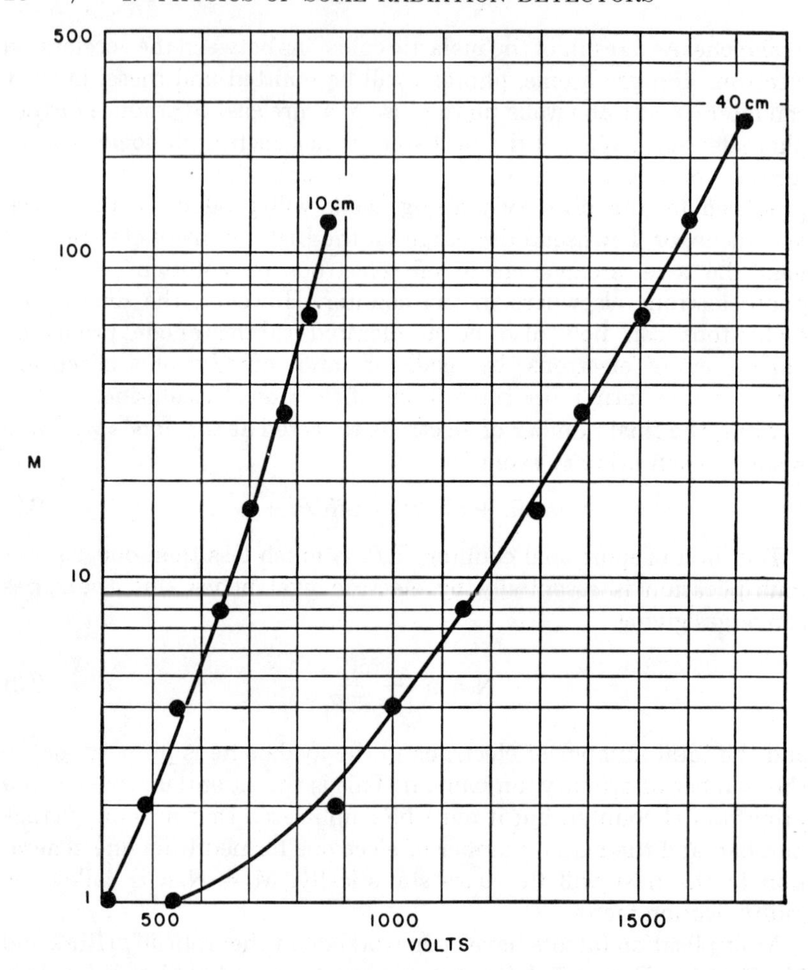

Fig. 7. Multiplication factor, M, as a function of applied voltage for argon at pressures of 10 and 40 cm Hg (1.33×10^3 Pa = 1cm Hg), radius of center wire = 0.013 cm, radius of outer electrode = 1.105 cm (Staub, 1953).

time t_1, elapses before the primary electrons reach the multiplication region. This time t_1 is determined by the location of the initial ionization in the counter and (depending on its construction) may be of the order of 1 μs. This delay time, which is variable, should not be overlooked when coincidence experiments are carried out.

Initially the pulse rises very fast (curve A), first as a result of the movement of the electrons (t_1 to t_2), then of the positive ions traveling in the strong field near the wire. As the positive ions move into the region of weaker field, the rise is slower and ceases when the ions reach the outer cylinder. This, for an average counter, takes about 200 μs.

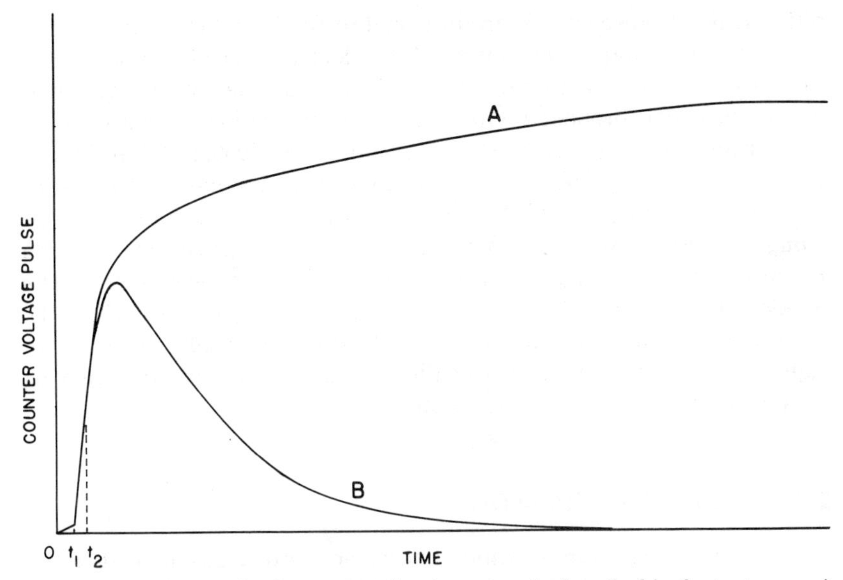

Fig. 8. Voltage pulse from proportional counter. As described in the text, curve A represents the undifferentiated and curve B the differentiated responses. (From Rossi, 1952)

Curve A shows the shape of the voltage pulse for no loss of charge from the anode wire (i.e., for an infinitely long time constant). In practice, in order to provide for time resolution, suitable *RC* differentiation is used in the output to give the type of pulse shown by curve B. This differentiation is accomplished by placing a capacitance, *C*, in series with the signal, followed by a high resistance, *R*, to ground. The charge from the detector will be exponentially dissipated by this circuit; the time required for the charge to be reduced to 1/e of its value being *RC,* known as the time constant of the circuit. When *RC* is small compared with the time between ionizing events, the charges collected during individual events (or the corresponding voltages) may be measured by means of appropriate pulse circuitry. Differentiation times as short as 0.1 μs have been used. Low-noise charge-sensitive preamplifiers are now more commonly used with proportional counters. In such preamplifiers the differentiating time constant is usually of the order of 1 ms, but double *RC* or delay-line differentiation of less than 1 μs in the main amplifier, combined with measurement of the time at which the bipolar pulse changes polarity (i.e., crosses zero voltage), can provide excellent time resolution over a wide range of pulse amplitudes.

Since all the pulses will have the same general shape (except when the ionizing event takes place very close to the wire), the height of the differentiated pulse will be proportional to the initial ionization. However, ionization and the formation of the electron cascade are statistical processes and a distribution in the height of the output pulses is observed, which broadens when the initial ionization decreases. (For further details, see Rossi, 1952; Wilkinson, D. H., 1950; and Bambynek, 1973). A further spread can also be caused by a non-uniform center wire and by end-effects, both of which cause varying field distributions along the wire. However, in most cases the shape of the outer electrode is unimportant, because multiplication takes place very close to the anode wire, where the electric field gradient is highest (Eq. 2.1). Asymmetric shapes of the outer electrode may result, however, in significant variation in electron-collection times from different regions of the sensitive volume of the counter.

2.2.2 *Choice of Counting Gas*

A proportional counter should be filled with a gas in which the negative charge is carried by electrons. However, some gases such as hydrogen, nitrogen, and the noble gases exhibit a rapid change of multiplication factor, M, with applied voltage due to the generation of photoelectrons from the walls. The addition of small proportions of gases with complex molecules that absorb photons without ionization increases the stability of the counter. Methane-argon mixtures are in common use and multiplication factors up to 10^5 are possible. Polyatomic gases, such as methane, can also be used very satisfactorily without any admixture.

2.2.3 *Characteristics of Proportional Counters*

The proportional counter can be used to determine the energy of particles and photons if they are totally absorbed in the counting gas. For low-energy x and gamma radiation, almost complete absorption can be achieved with a suitable filling gas (Figure 9), or by operating at elevated pressure (Allen, 1957; Baerg, 1973b; Legrand *et al.,* 1973; Bambynek, 1973).

Proportional counters also have wide application to particle counting without energy determination. In common practice, the source itself is located inside the counter to avoid window absorption, and the counter is operated at atmospheric pressure, generally with methane or methane-argon mixtures flowing continuously (flow counter). Counter pulses due to a beta-particle spectrum vary in size, by a factor of 1,000

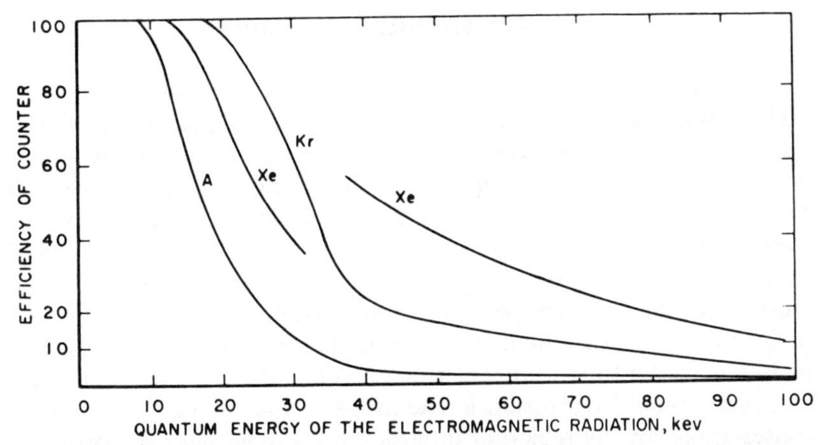

Fig. 9. Percentage of quanta absorbed in 50 mg cm^{-2} of argon, krypton, and xenon (i.e., 5-cm path at 5 atmospheres in argon, at 2.5 atmospheres in krypton, or at 1.5 atmospheres in xenon). The break in the xenon curve is at the K edge, i.e., when a photon has sufficient energy to remove an electron from the K shell. (From Curran, 1950)

or more when the counter is operated at elevated pressure. Therefore, the associated amplifiers must have good overload characteristics, which may, for example, be achieved with a low-noise charge-sensitive preamplifier and a doubly differentiating main amplifier (Baerg *et al.*, 1967).

A characteristic that is used to measure the performance of a gaseous ion-multiplication counter in the proportional region is the so-called plateau of the counter. As the applied voltage is increased, the increasing gas gain permits the triggering of the circuit discriminator by primary ionizing events, in the counter, of lower and lower energy. When the applied voltage is such that nearly all primary events are counted, a plot of counting rate against applied voltage becomes nearly horizontal, and the flatness of the curve in this region, the plateau, is a measure of the ability of the counter to detect essentially all primary ionizing events. Any further increase in voltage will cause only a small increasing in counting rate. An increase in the effective sensitive volume of the counter, or the onset of spurious pulses, can also cause the plateau to have a positive slope. The use of low gas gains, as permitted by low-noise amplifiers even for particle energies of a few keV, may be important in avoiding possible spurious pulses (Campion, 1973; BIPM, 1976), which are, however, usually negligible ($<10^{-4}$) when the counter is operated near the middle of the plateau. The applications of proportional counting to specific metrological problems are discussed in Section 3.5.6.

2.3 Geiger-Müller Counters

If the voltage on a proportional counter is increased beyond the high-voltage end of the plateau, $\overline{m}P_e$ in Eq. 2.2 approaches unity, and an electrical discharge takes place that is propagated along the anode wire. The product $\overline{m}P_e$ represents the number of photoelectrons formed by the emission and subsequent absorption of photons in the walls of the counter when the gas is a pure noble gas. In the case of, say, argon-alcohol mixtures, photons formed by the excitation and ionization of argon atoms can also produce photoelectrons from alcohol molecules in the gas filling the counter. The emission of photons will be isotropic and their number, which was less than one per secondary electron formed by collision in the proportional counter, is now very considerable. In the region of intense electric field near the wire, the Townsend avalanches caused by the electrons produced in the primary ionizing event will now produce intense photon emission with consequent photoelectric absorption throughout the counter, including the region of intense electric field around the anode wire. Under these conditions a discharge takes place, which spreads all along the wire forming a positive-ion "sheath", and gives rise to an output pulse, the size of which is independent of the amount of the initial ionization and which may amount to several volts. Less additional amplification is therefore required than for proportional counters, and the discharge can be initiated by a primary event producing as little as one ion pair. The discharge will spread along the wire at a velocity as great as 10^7 cm s^{-1}.

This mode of gas-ionization-detector operation is said to occur in the Geiger region and the detector itself is known as a Geiger-Müller counter.

Unless the discharge is "quenched" it may be self-sustaining or multiple pulses may occur, due to the neutralization of the positive ions at the cathode and consequent release of photoelectrons from the cathode after a delay of several hundred microseconds. Such quenching may be achieved externally, by electronic means, or internally, by adding a suitable polyatomic gas. A gas filling that has been frequently used is argon at a pressure of about 90 mm Hg (12 kPa) with ethyl alcohol vapor, or similar organic gas, at a pressure of about 10 mm Hg (1.3 kPa). Alcohol has a lower ionization potential than argon, so that the ions moving toward the cathode will, after a few collisions, consist only of alcohol ions. In contrast to argon ions, alcohol ions dissociate and do not produce photoelectrons when neutralized at the cathode. Therefore, multiple pulsing is avoided. However, the alcohol ions dissociate upon being neutralized and when the supply of alcohol is exhausted, after some 10^8 counts, poor plateau characteristics result. Halogen vapors also have the same quenching effect (Liebson and Friedman, 1948; Friedman, 1949). Since the halogen molecular ions do not remain dissociated, these counters have very long lives. The operating voltage is also considerably lowered. Argon, with the addition of small quantities of xenon, nitrogen, and oxygen, has also been found to be a satisfactory filling for Geiger-Müller counters (Shore, 1949; Collinson *et al.,* 1960).

After an ionizing event, the sheath of positive ions along the counter wire effectively lowers the wire potential and makes the counter inoperative until the ions have drifted to the cathode. The inoperative time is known as the dead time, and in the Geiger-Müller counter dead times are long, ranging from 10 to 500 μs. Thus, this type of counter does not allow of such high counting rates as does the proportional counter. It is also inferior because the slopes of the Geiger-Müller-counter plateau are steeper than those of proportional counters, values of less than 1 percent variation in counting rate per hundred volts being uncommon unless an artificial dead time is used to reduce spurious counts. When it is necessary to establish an actual counting rate, it is difficult to know at which point on the plateau to count. This is not so serious with a proportional counter where a plateau slope as low as 0.1 percent, counting-rate change per hundred volts, can be achieved. Therefore, for standardization work, or for any other precise counting, preference should be given to the proportional counter. However, the efficiency of either a proportional or Geiger-Müller counter may be determined, in the case of beta- and photon-emitting nuclides, by coincidence counting (Section 3.2). In the case of length-compensated internal gas counters, used for the assay of radioactive nuclides in the gaseous form, a plot of the difference in the count rates, at the same voltage, for two cylindrical counters, having the same diameters but different lengths, will give "difference plateaux" that are very flat, both in the Geiger-Müller and proportional regions (Section 3.6).

2.4 Scintillation Detectors

2.4.1 *Introduction*

It has long been known that many substances emit visible light when exposed to nuclear radiation, and this was the basis of Sir William Crookes' spinthariscope which utilized a ZnS screen to permit the observation of single alpha particles emitted in the decay of natural radioactivity. In 1944, Curran and Baker first substituted an electron-multiplier phototube for the human eye, creating the first scintillation detector as we know it today. Kallmann (1947) used naphthalene, an organic crystal, as a scintillation medium that is transparent to the fluorescent radiation emitted by the interaction of the charged particles within the scintillator. It was subsequently observed that the quantity of light emitted was proportional to the energy of the absorbed particle, and this established the scintillation process as a valid means for energy spectrometry of charged particles emitted in radioactive decay. The discovery of the scintillation properties of thallium-activated sodium-iodide crystals, NaI(Tl), by Hofstadter (1948), and the possibility of their use as gamma-ray spectrometers, led to the development of pulse-amplitude spectrometry as a method for the measurement of charged-particle and photon energies and intensities associated with radioactive decay. In order to achieve the level of sophistication required to utilize the potential of the scintillation detector as an energy-measurement device, it was necessary to develop an entire family of electronic circuitry that was unavailable at the time. The linear pulse amplifier and multichannel pulse-height analyzer were developed to utilize the energy-measurement potential of the scintillation detector. The general development of the scintillation spectrometer is outlined in the text *Alpha-, Beta-, and Gamma-Ray Spectroscopy* (Siegbahn, 1965). The characteristic energy-sensitive response of scintillators has been applied to develop techniques for gamma-ray spectrometry that are presently employed extensively for the quantitative assay of radioactivity and the identification of radionuclides.

2.4.2 *The Scintillation Process*

In order to understand the modern scintillation detector and the concepts of pulse-height analysis, let us consider the basic processes involved. The operation of a scintillation detector may be described by a sequence of four essential processes: (1) the interaction of charged particles or photons with the detector, resulting in the production of ionization; (2) the conversion of the energy of the charged particles into a proportional amount of light by the scintillator; (3) the conversion of the light emitted by the scintillator into photoelectrons at the photocathode of a phototube; and (4) the multiplication (amplification) of the initial number of photoelectrons into a measurable electrical current pulse. The result of these processes, operating in sequence, is to produce at the output of the electron-multiplier phototube an electrical pulse whose amplitude is proportional to the energy deposited in the scintillator by the incoming charged particle or photon. The fact that all of these processes occur in a very short interval of time (of the order of a microsecond) requires that the amplitude of individual electrical pulses be measured if one is to relate the output of the detector to the energy of the radiation being detected. A typical NaI(Tl) scintillation detector is shown in Figure 10 to illustrate the sequence of events following detection of a photon. In the following section, and Section 4.2, the principles of photon spectrometry will be presented together with the application of this technique to the measurement of radioactivity.

A charged particle (alpha or beta) incident upon the detector, or an electron excited in the detector following an interaction with an incoming photon, will dissipate all of its energy within the scintillator, provided that the dimensions of the scintillator are greater than the range of the particle. The scintillation process results in a portion of this energy being emitted as photons in the visible or near ultraviolet region of the electromagnetic spectrum. The number of photons, p, of average energy $h\nu$ produced by the absorption of energy E within the scintillator, may be written:

$$\mathrm{p} = \frac{E}{h\nu}\,\epsilon_\nu. \tag{2.4}$$

The quantity ϵ_ν is referred to as the intrinsic efficiency of the scintillator. It is nearly independent of the energy of the ionizing particle for most inorganic scintillators.

Many materials have been observed to emit light when excited by ionizing radiation, including both inorganic and organic chemical compounds. In organic materials such as anthracene and stilbene, this luminescence is a molecular property; thus organic compounds will fluoresce in any physical form—crystalline solid, liquid, or vapor phase. Inorganic phosphors are mainly limited to alkali halides in crystalline form. In general, the scintillation process in crystalline inorganic materials requires the presence of impurity atoms within the crystal lattice. Thallium is employed as an activator in NaI and CsI, being introduced as an impurity when the crystal is grown from the molten state.

Following the excitation of a scintillator by an individual event, the time dependence of photon emission appears to follow an exponential law of the type

$$\mathrm{p}(t) = \text{constant } \mathrm{e}^{-t/\tau}, \tag{2.5}$$

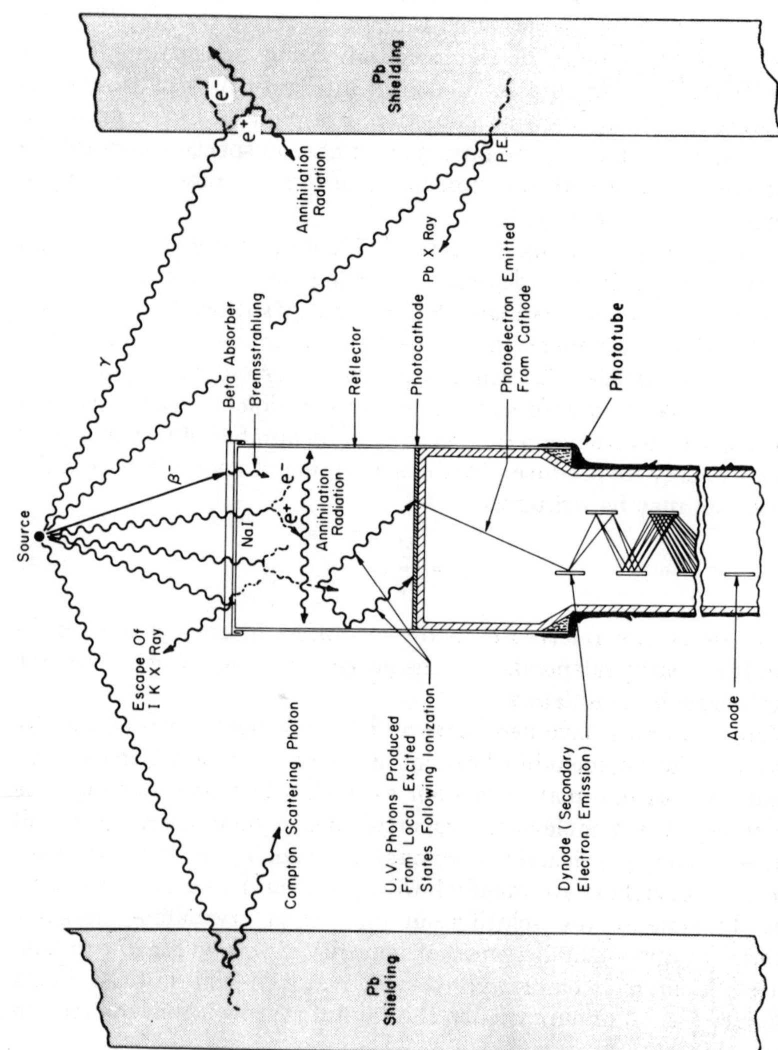

Fig. 10. Illustration of the various processes that contribute to the output from an electron-multiplier phototube by the interaction of gamma rays from a point source with a NaI(Tl) scintillation detector. (From Heath, 1964)

where p(t) is the total number of photons emitted at time t. The decay time τ is characteristic of the scintillator and determines the rise time of pulses produced at the output of the phototube. Organic scintillators are usually quite fast, with decay times of the order of 10^{-8} seconds or less. Inorganic scintillators are considerably slower, NaI(Tl) exhibiting a decay time of 2.5×10^{-7} seconds.

In addition to the time response, a major consideration in the performance of a phosphor as a scintillation detector is the absorption of the light emitted by the scintillation process in traversing the scintillator material. NaI(Tl), for example, exhibits almost no measurable attenuation of the light traversing the crystal; thus the amplitude of the light pulse emitted from a large crystal detector fabricated from this material is independent of the point of origin of the light within the volume of the scintillator. This has a significant bearing upon the performance of the device as a photon energy spectrometer. Organic phosphors, in general, exhibit appreciable attenuation. Relevant characteristics of some of the more commonly used scintillators are given in Tables 3 and 4.

2.4.3 *Conversion of Fluorescent Radiation to Electrons (Phototube)*

To increase the collection efficiency of the light emitted by the scintillator (as shown in Figure 10), it is surrounded by a thin reflector of aluminum oxide. Of the photons produced in the scintillator, the number striking the cathode surface of the phototube will depend upon the geometry and the optical properties of the system, including those of the reflector and any light guides between the scintillator and the cathode of the phototube. For good scintillators and reasonable geometries, the optical efficiency, ϵ_L, can approach unity. Photons falling on the cathode surface of the phototube are converted into electrons by the photoelectric process with an efficiency ϵ_C. Thus, the number of photoelectrons, P, produced by p photons generated in the scintillator may be expressed by the following relationship

$$P = \epsilon_L \epsilon_C p. \qquad (2.6)$$

Typically ϵ_L may be as large as 95 percent while ϵ_C usually lies between 10 and 20 percent. In order to make ϵ_C as large as possible, the phototube is designed so that the spectral response of the photocathode material closely matches the photon spectrum emitted by the scintillator material being used. It is of interest to combine the quantities presented above to obtain the number of photoelectrons to be expected for a given amount of energy expended in the scintillator, i.e.

TABLE 3—*Inorganic scintillators*

Scintillator (activator in parentheses)	Density	Emission wavelength	Refractive index	Light yield (anthracene = 1.00)	Decay time[a]	Remarks
	g cm^{-3}	nm			μs	
ZnS (Ag)	4.1	450	2.4	(2.0)[b]	(>1)[b]	Only very small crystals, transparency poor.
CdS (Ag)	4.8	760	2.5	(2.0)[b]	(>1)[b]	Only small crystals, yellow.
NaI (Tl)	3.67	410	1.7	2.0	0.25	Excellent crystals, hygroscopic.
CsI (Tl)	4.51	565	1.79	1.5	1.1	Excellent crystals, not hygroscopic.
KI (Tl)	3.13	410	1.68	0.8	>1	Excellent crystals, not hygroscopic.
NaCl (Ag)	2.17	245–385	1.54	1.15	>1	Excellent crystals.
LiI (Eu)	3.49	470	1.96	0.7	0.94	Hygroscopic.
LiF (AgCl)			1.39	0.05		
CsBr (Tl)	4.44		1.70	2.0	>1	
$CaWO_4$	6.06	430	1.92	1.0	>1	Small crystals, transparency good.

[a] Time for light pulse to fall to 1/e of initial value.

[b] The light yield and the decay time of the zinc-sulfide and cadmium-sulfide phosphors depend strongly on former treatment (quenching) and intensity of excitation. To a lesser degree this is also the case for other inorganic phosphors.

TABLE 4—*Organic scintillators*

Scintillator	Density	Emission wavelength	Refractive index	Light yield	Decay time	Remarks
	$g\ cm^{-3}$	nm			ns	
Anthracene	1.25	445	—	1.00	32	Large crystals, not quite clear.
Quaterphenyl	—	438	—	0.85	8	Pure crystals difficult to synthesize.
Stilbene	0.97	410	—	0.73	6	Good crystals readily obtainable.
Diphenyloxazole	—	—	—	0.78	—	
Diphenylbutadiene	—	461	—	0.67	8	
Diphenylanthracene	0.997	480	—	0.65	—	
Terphenyl (para)	1.23	415	—	0.55	12	Good crystals, readily obtainable.
Diphenylacetylene	—	390	—	0.26–0.92	7	
Phenanthrene	1.03	435	1.66	0.46	10	Clear crystals difficult to obtain.
Naphthalene	1.15	345	1.58	0.15	75	Good crystals easy to obtain.
Chloroanthracene	—	—	—	0.03	—	

$$\mathrm{P} = \frac{\epsilon_\nu \epsilon_\mathrm{L} \epsilon_\mathrm{C}}{h\nu} E. \tag{2.7}$$

The quantity $\epsilon_\nu \epsilon_\mathrm{L} \epsilon_\mathrm{C}/h\nu$ provides a measure of the efficiency of a given scintillation detector. For a high quality scintillation detector the energy required to release one photoelectron (E/P) from the photocathode is between 300 and 2,000 eV. This may be compared with 30 eV required to produce an ion pair in a typical gas counter.

2.4.4 *Output-Pulse Processing*

As outlined above, the electron-multiplier phototube embodies two functions that are essential to the performance of a scintillation detector: (1) the detection and conversion of the light emitted by the phosphor into a pulse of electrons; and (2) the amplification of this minute amount of current to a level that can be suitably handled in conventional electronic-amplifier circuits. The amplification of the initial number of photoelectrons emitted from the photocathode of the tube is accomplished by a process of electron multiplication. The initial photoelectrons, as illustrated in Figure 10, are accelerated toward the first of a series of metal plates which are termed *dynodes.* The dynode surface is coated with a material (typically cesium or a bimetallic compound) that has the property of emitting secondary electrons when struck by an accelerated electron, the number emitted being a function of the coating material and accelerating voltage, and typically varying from 3 to 5. These secondary electrons are then accelerated toward a second dynode and the process is repeated. With 10 or more secondary emission stages, the electron-multiplier structure can achieve a current gain of up to 10^7 or 10^8. At this level the current pulse arriving at the final collector of the phototube is sufficient in amplitude to be handled effectively by conventional electronic circuitry. It is important to note that the electron-multiplier phototube used as a current amplifier is extremely fast and linear, and has minimal influence on the speed of scintillation detectors.

At this point, an acceptable electrical output is obtained from the scintillation detector comprised of: (1) a scintillator surrounded by a suitable light collection system; and (2) a phototube that converts the fluorescent light produced by the scintillator into electrical current pulses. A typical scintillation-detector system is illustrated in Figure 11. In this figure two different configurations are presented to illustrate different applications of scintillation detectors. For simple applications, where the detector is to be employed for integral photon or particle counting, the system is comprised of the detector, phototube voltage

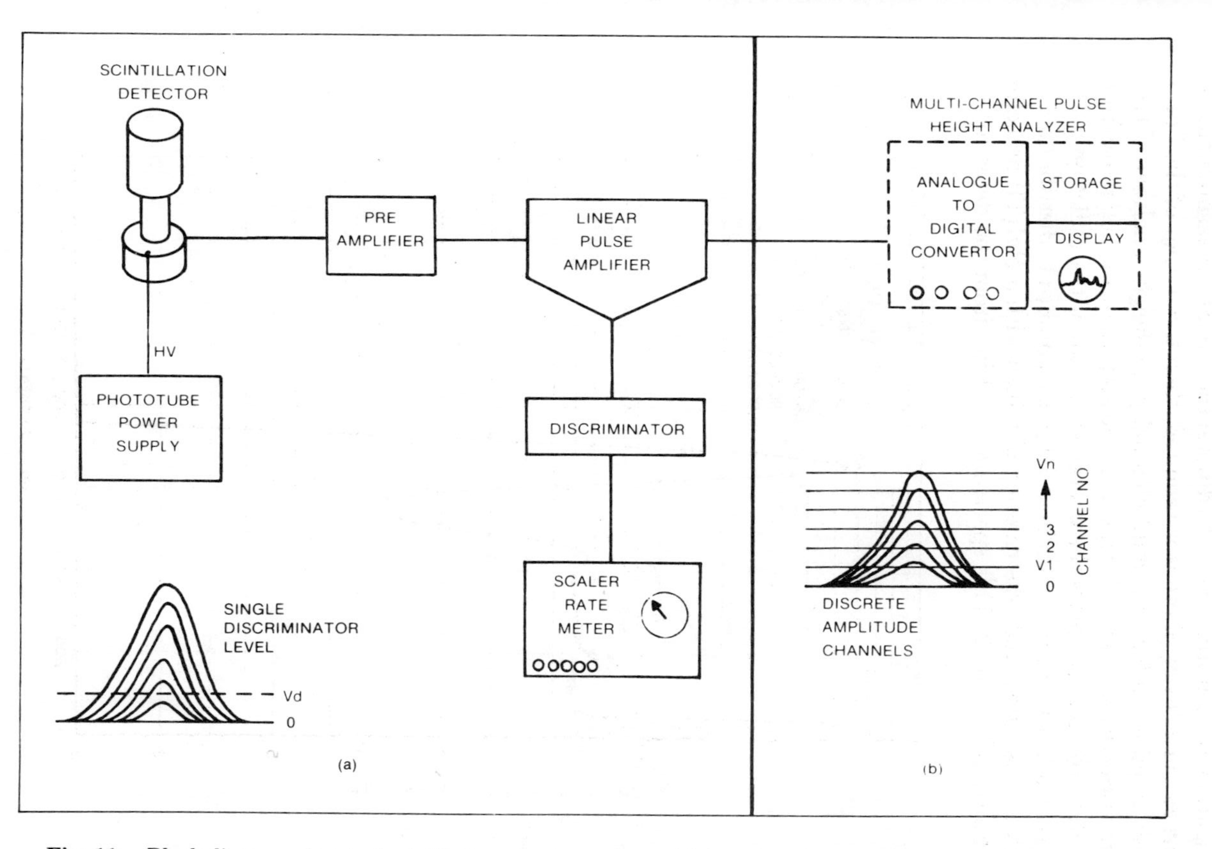

Fig. 11. Block diagram of typical scintillation-counter systems: (a) for integral count-rate measurements for photon energies above those corresponding to a discriminator voltage V_d; (b) for "pulse-height-spectrometer" measurements at different photon energies corresponding to discriminator voltage intervals $V_1, V_2 \cdots V_n$.

supply, a pulse amplifer (with preamplifier), discriminator to eliminate electronic noise from the output, and a scaler or count-rate meter for indication of counting rate. If differential energy measurement of charged particles or photons is required, then the output of the linear pulse amplifier is fed to a multichannel pulse-height analyzer (PHA) which provides a means of measuring the amplitude of each pulse from the detector, and presents the energy spectrum of the radiation being measured in the form of a differential pulse-amplitude distribution. A typical pulse-amplitude distribution obtained with a NaI(Tl) detector is shown in Figure 12. This spectrum, resulting from the detection of

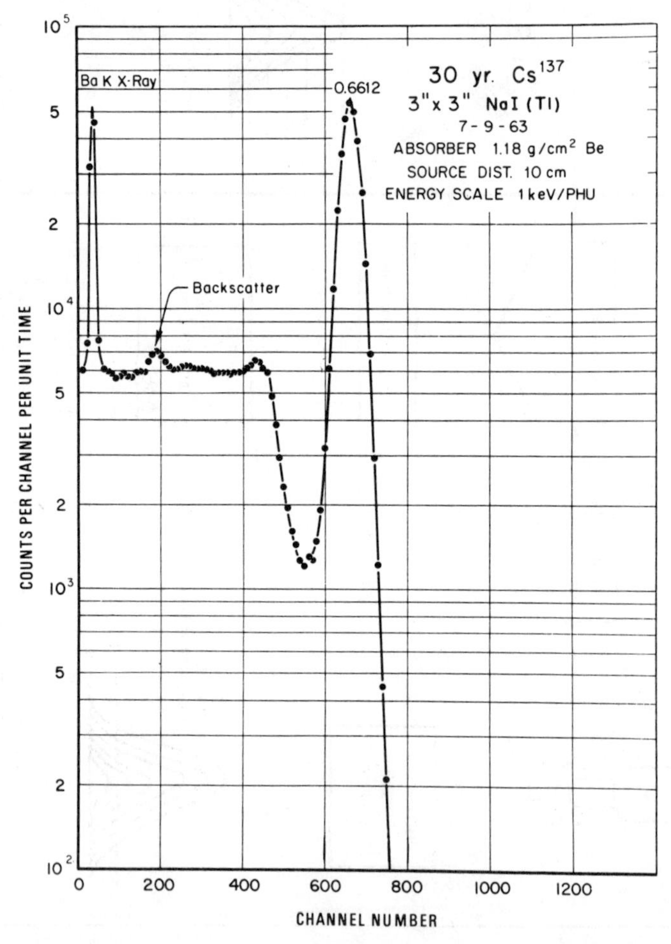

Fig. 12. A typical pulse-height spectrum obtained with a NaI(Tl) spectrometer, illustrating the energy response of inorganic scintillators. The scale of the abscissa is 1 keV per channel. (From Heath, 1964)

gamma rays emitted in the decay of ^{137}Cs, exhibits a sharp peak that results from the delayed emission from ^{137m}Ba of monoenergetic gamma rays having energies of 0.6612 MeV. This unique energy-sensitive characteristic of NaI(Tl) provides the basis for using pulse-height spectrometry to measure the energy and intensity of radiation emitted in the decay of radionuclides. A description of pulse-amplitude spectrometry and its application in the measurement of radioactivity is presented in Section 4.2.

The development of the scintillation detector in its many forms has generated a variety of detection devices that have application in the measurement and characterization of all types of radiation emitted in the decay of radioactive nuclei. Although all scintillation phosphors can be utilized for the detection of radiation, the properties of different fluorescent media may be optimized for a particular application. For example, the detection of high-energy gamma radiation with optimum efficiency requires a detector with high density and preferably with high atomic number. For this purpose, thallium-activated NaI and CsI offer the best choices. On the other hand, the measurement of low-energy beta-emitting radionuclides such as ^{3}H and ^{14}C can best be achieved by incorporating the radioactivity into a liquid scintillator. Tables 3 and 4 provide a partial listing of solid scintillators normally employed for a selection of radiation-measurements applications. The unique capabilities of the liquid-scintillation method have found many uses in biomedical applications, and have led to the development of a sophisticated technology. This specialized measurement technique is described below and in Section 4.3.

2.4.5 *Liquid-Scintillation Counting*

2.4.5.1 *Counting.* The wide application of liquid-scintillation counting derives from its particular usefulness for measurement of low-energy beta-emitting radionuclides such as ^{3}H and ^{14}C. Several monographs and reviews treat various aspects of liquid scintillation (Birks, 1964; Horrocks, 1974; Bransome, 1970; and Horrocks and Peng, 1971). Liquid scintillators are normally composed of one or more fluorescent solutes in an organic solvent. In the preparation of the sample for liquid-scintillation counting, the radioactive material is introduced into, and thoroughly mixed with, the liquid scintillator. It is generally accepted that radiation energy is expended in the ionization and excitation of the solvent. This energy is subsequently transferred to the solute (or fluor) and is re-emitted as photons in the violet and ultraviolet. These photons, having an average wavelength characteristic of the solute, can then be detected at the photocathode of one or more phototubes, although in the general application of the method a "wavelength shifter" may also be used to produce an output pulse of photons having wavelengths compatible with the optimum sensitivity of the photocathode. Following amplification in the phototube, the signal may be routed to additional amplifiers and finally recorded with a suitable counter such as a single-channel or multichannel analyzer.

The magnitude of each pulse produced in a liquid-scintillation-counter system is proportional to the energy deposited in the scintillator, but both the amplitude and the shape of the pulse will be dependent on the nature of the ionizing radiation. For example, the light output due to an alpha particle will be only about one-tenth that produced by a beta particle of the same energy. This difference arises because of the higher specific ionization along the alpha-particle path. Where beta particles produce excited and ionized solvent molecules which are relatively isolated from each other, alpha particles produce them in close proximity in regions of high-ionization density in which energy is dissipated in non-radiative transitions. This process is referred to by Birks (1964) as ionization quenching.

In order to evaluate the efficiency of the various energy-transfer steps in the liquid-scintillation process, Birks (1964, 1970) has developed an expression relating the pulse amplitude at the anode to the energy deposited in the scintillator. In an abbreviated form of his earlier equation, Birks (1975) defines a "scintillation figure of merit",

$$F = sfqm, \tag{2.8}$$

which is useful for comparing the performance of different scintillators using the same phototube and optical arrangement. The terms on the right hand side of Equation 2.8 are best defined by considering the intermediate steps in the liquid-scintillation process.

(i) In the first step, the energy of the charged particle is deposited in the solvent, and s, the relative scintillation efficiency, is a measure of the efficiency, relative, usually, to that of toluene, for the transfer of energy from the incident radiation to those excited levels of the solvent molecules from which energy can be transferred to the fluor in solution (see Table 5 and Birks (1975)).

(ii) In the second step, energy is transferred to the solute and f, the probability of this process occurring, is dependent both on the nature of the solvent and the concentration of the solute.

(iii) The third step is emission of light by the solute. The "quantum efficiency", q, of the solute is the ratio of the number of light photons emitted to the number of solute molecules in excited states.

TABLE 5. *Scintillation efficiencies relative to toluene,* s, *of solutions containing 3 g* l^{-1} *of PPO*[a]

Solvent	Chemical structure	s
p-Xylene	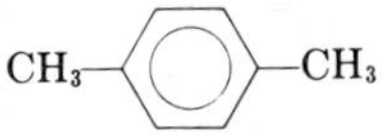 CH_3 — — CH_3	1.12
m-Xylene	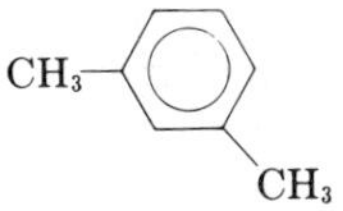 CH_3 — — CH_3	1.09
Xylene (mixture of isomers)		1.07
Phenylcyclohexane		1.02
Toluene	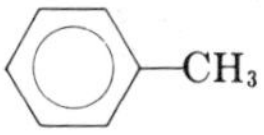 — CH_3	1.00
o-Xylene	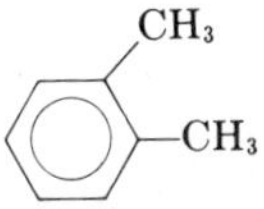 CH_3, — CH_3	0.98
Ethylbenzene	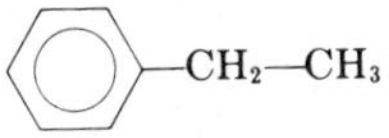 — CH_2 — CH_3	0.96
Benzene	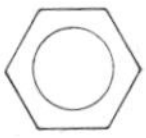	0.85
Anisole	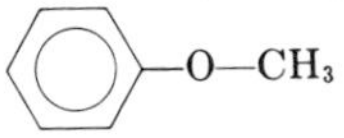 — O — CH_3	0.83
Mesitylene	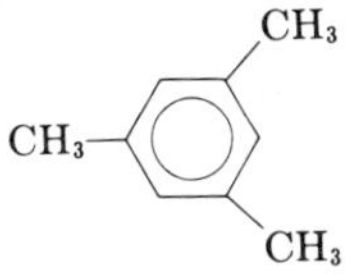 CH_3 —, CH_3, CH_3	0.82
Dioxane	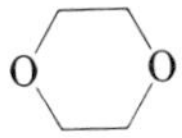 O O	0.20
Dioxane + naphthalene (~6%)	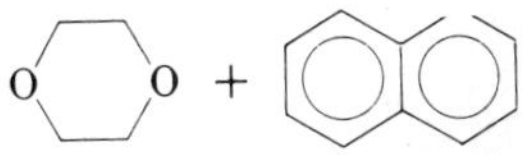 O O +	0.80

[a] From Birks (1964). For toluene, s is of the order of 3 percent.

(iv) The last term in Equation 2.8 is called the "spectral matching factor", m. This parameter, which varies from 0 to 1, indicates the degree of overlap of the emission spectrum of the solute and the absorption spectrum of the photocathode.

Thus the first three terms in F are dependent upon the choice of scintillator while m depends on the scintillator-phototube combination. Values for some of these parameters are given in Table 6. Examples of

TABLE 6. *Properties of some organic fluors (solutes) commonly used in liquid-scintillation counting: relative fluorescent quantum efficiencies (q), spectral matching factors (m), fluorescent life times (t), average fluorescent emission wavelengths (λ), and solubilities*

Solute	Chemical structure	q[a]	m[b]	t[a]	λ[a]	Solubility[c] (Toluene)
				ns	nm	$g\ l^{-1}$
TP *p*-Terphenyl		0.93	0.97	1.0	342	8.6 (20 °C)
PPO 2,5-Diphenyloxazole		1.00	0.97	1.4	375	414 (20 °C)
POPOP 1,4-Di-(2-(5-phenyloxazolyl))-benzene		0.93	0.83	1.5	415	2.2 (20 °C)
M_2-POPOP 1,4-Di-(2-(4-methyl-5-phenyloxazolyl))-benzene		0.93	0.78	1.7	427	3.9 (20 °C)
PBD 2-Phenyl-5-(4-biphenylyl)-1,3,4-oxadiazole		0.83	0.96	1.1	375	12 (20°C)
tert-Butyl-PBD 2-(4′-*t*-Butylphenyl)-5-(4″-biphenylyl)-1,3,4-oxadiazole		0.85	0.95	1.2	385	119 (20 °C)

bis-MSB *p-bis*-(*o*-Methylstyryl)-benzene	CH_3, H, C, C, H, H, C, C, H, CH_3	0.94	—	1.2	425	5.5 (25°C)

[a] Unless otherwise stated, the values shown for q, t, and λ are taken from Berlman (1971). All of these values were obtained for dilute solutions of the solutes in cyclohexane. Fluorescent quantum yields are given relative to an assumed value of 1.0 for 9, 10-diphenyl anthracene.

[b] Spectral matching factors, $m = \int_0^\infty I(\lambda)\eta(\lambda)d\lambda / \int_0^\infty I(\lambda)d\lambda$, were taken from Birks (1975). The values of $I(\lambda)$, the emission spectrum, were obtained from dilute solutions of the solutes in toluene and values for $\eta(\lambda)$, the phototube response, for bialkali phototubes.

[c] Solubilities are given here only for comparative purposes, since most data have been taken from suppliers' literature and wide variations were noted for several compounds. Suggested optimum concentrations for many of these solutes may be found in Birks (1964) and Horrocks (1974).

the use of the "scintillation figure of merit" are given in Birks (1975).

Table 5 shows solvent scintillation efficiencies relative to that of toluene. It is evident that all good solvents have an aromatic structure. Dioxane is included in the table because of its significance as a solvent for water and aqueous solutions. Dioxane alone has a low relative scintillation efficiency, but with the addition of naphthalene the efficiency becomes comparable with that of benzene.

Many scintillation liquids contain an additional component generally referred to as a secondary solute, or secondary scintillator. The secondary scintillator was originally used to match the spectral response of the phototube to that of the light emitted by the scintillator. Therefore, the term "wavelength shifter" was often used. This matching is no longer essential because many scintillators exhibit m values of 95 to 97 percent. However, secondary scintillators are still used to improve the performance of liquid-scintillation systems, particularly in quenched systems.

2.4.5.2 *Quenching.* In liquid scintillation, the production of light is sometimes inhibited, or the final light "signal" is "quenched" in the energy or light-transfer processes in the liquid. The details of the quenching processes, in particular chemical-structural effects on quenching, are not completely understood.

For practical applications three classes of quenchers may be distinguished, namely, diluting, chemical, and color quenchers. The first class consists of substances that dilute the mixture and thus reduce the probabilities of absorption of ionizing radiation by the solvent, and of the various energy transfer processes that follow the absorption of radiation. Aliphatic hydrocarbons and alcohols fall into this category.

Chemical quenchers interfere in the scintillation process by deexcitation of the solvent. If this interference is small, these quenchers are termed mild quenchers. Examples of mild quenchers are R—Cl, $R—NH_2$, R—CH = CH—R, R—SR, RCOOH (R representing an alkyl group). Compounds that strongly interfere with the process are termed strong quenchers, such as R—Br, R—SH, R—CO—R, RI, and $R—NO_2$. The distinction between these two types is, however, arbitrary. In particular, molecules with complex chemical structure often have unpredictable quenching properties. The most common quenching agent is oxygen which is often removed by bubbling an inert gas, such as nitrogen or argon, through the scintillator solution.

Color quenchers are materials that reduce the light transmission through the sample and thus lessen the probability of transfer of photons to the cathode of the phototube. Obviously, there are compounds that are both color and chemical quenchers.

2.4.6 *Gas-Scintillation Counting*

2.4.6.1 *Gas-Scintillation Counters.* Perhaps the most common effect observed due to the passage of ionizing radiation through gases is the emission of light. Light may be readily observed, for example, from electric discharges or from an intense beam of charged particles passing through a gas. As ionizing radiation passes through a gas, ion pairs and electronic excitations are produced in the molecules and atoms. The electronic excitation energy may then be given up either by non-radiative collision or by photon emission.

A detector utilizing a gas as a scintillator was first constructed by Grun and Schopper (1951). It consisted of a chamber containing argon gas at atmospheric pressure with a small nitrogen impurity. The light output was measured by a phototube attached to a quartz window. Grun and Schopper also found that the light yield from argon was about 100 times greater than that from the polyatomic molecular gases such as nitrogen, hydrogen, oxygen, and carbon dioxide. Subsequent investigations, typically by Northrup and Nobles (1956) and Nobles (1956), indicated that a noble gas of high purity made the most suitable scintillator, with xenon giving the highest light yield. It was demonstrated that the noble-gas light yield was a maximum in the visible to far-ultraviolet regions of the spectrum and that the light output was nearly proportional to the energy deposited over a wide range of stopping powers and incident radiation energies. The scintillation decay times were found to be less than 10^{-8} seconds and adjustable by pressure. The addition of even trace amounts of polyatomic-molecular gases reduced the light output and its linearity with deposited energy and lengthened the decay times. A comprehensive and detailed review of the scintillation process in gases is given by Birks (1964).

Gas-scintillation detectors have been successfully applied to alpha-particle and fission-fragment detection (Sayres and Wu, 1957). Gas-scintillation counters utilizing neutron-induced reactions have been successfully used for neutron cross-section determinations (Sayres and Wu, 1957; Engelke, 1960).

2.4.6.2 *Gas-Proportional-Scintillation Counters.* An intriguing development of the gas-scintillation counter employs a chamber resembling that of a gas proportional counter, replete with electrodes for the application of an electric field (Koch, 1958). When the electrons resulting from the passage of ionizing radiation are accelerated in the electric field, which is maintained below that required for charge multiplication, the resultant atomic excitations increase the light output by factors of 100 to 1000 (Policarpo *et al.,* 1968), and thus reduce the detection limit to a few hundred electron volts. Such a counter (Policarpo *et al.,* 1972), filled with xenon, has an energy resolution better, by almost a factor of two, than that of a conventional gas proportional counter, in which the resolution additionally involves the variance due to charge multiplication.

Studies of different gases and gas mixtures (always involving at least one noble gas) have led to a better understanding of the light production (Thiess and Miley, 1974) and of the significance of long rise times, different pressures, and wavelength shifting for better matching of the light output to the response of the phototube. The role of the electric-field distribution and counter layout was investigated (Palmer, 1975; Conde *et al.,* 1975) in attempts to increase the detector size and achieve better resolution.

The continued development of the gas-proportional-scintillation detector may lead to its general use where large window areas and good resolution are required at low energies.

2.5 Semiconductor Detectors

2.5.1 *Introduction*

The development of radiation detectors fabricated from single crystals of semiconducting material has, in recent years, revolutionized the detection of x rays and gamma rays. These devices, which are now available in reasonably large volumes, have the advantage of the high detection efficiency for gamma radiation afforded by NaI(Tl) scintillation detectors, but exhibit superior energy resolution. Pulse-amplitude-analysis systems employing detectors fabricated from single crystals of silicon and germanium are extensively utilized for elemental and radionuclidic analysis. The application to x- and gamma-ray spectrometry and the techniques of pulse-height analysis are described in Section 4.2.

The development of the semiconductor radiation detectors has paralleled the development of scintillation detectors for the detection and energy measurement of charged particles and photons. McKay (1951) reported the detection of alpha particles with a semiconductor-junction diode. This was followed by the development of p-n *diffused-junction* counters for x-ray and charged-particle energy measurement. In 1958, Walter and Dabbs (1958) obtained marked improvement in energy resolution by operating germanium surface-barrier junctions at liquid-nitrogen temperatures. Even then, however, it was only possible to collect charge carriers from very thin layers of material, and consequently the counters had very limited sensitivity to high-energy gamma radiation. Pell (1960) demonstrated that lithium ions could be diffused through large volumes of very pure silicon or germanium to compensate for the presence of residual p-type impurities and thus increase the volume of the depletion region. With this technique, lithium ions may be used to create a p—i—n structure such as is shown in Figure 13. These devices have the properties of solid-state ionization chambers. If a voltage is impressed across the detector, electrons that have been promoted to the conduction bands of the solid material by excitation are free to move and may be collected at the appropriate electrode of the device. When used as a detector of gammarays, the interaction of a gamma ray in the detector produces a primary excited electron that, in turn, excites a cascade of secondary electrons in the process of dissipating its energy within the detector volume. In this manner, the energy of the primary electron is expended in the production of electron-hole pairs which are then collected. The number of electrons collected will be proportional to the energy of the primary electron, and hence related to the energy of the detected photon.

Semiconductor crystals that are free from impurities or structural defects (which can act as traps for secondary electrons) can be used directly for the fabrication of detectors. During the early period of development of these detectors, sufficiently pure material was not available and it was necessary to compensate for the impurities by drifting lithium ions through the detector material at elevated temperature. Since the mobility of lithium ions within the crystal lattice is not negligible at room temperature, such detectors must be maintained at cryogenic temperatures throughout their lifetime. A descrip-

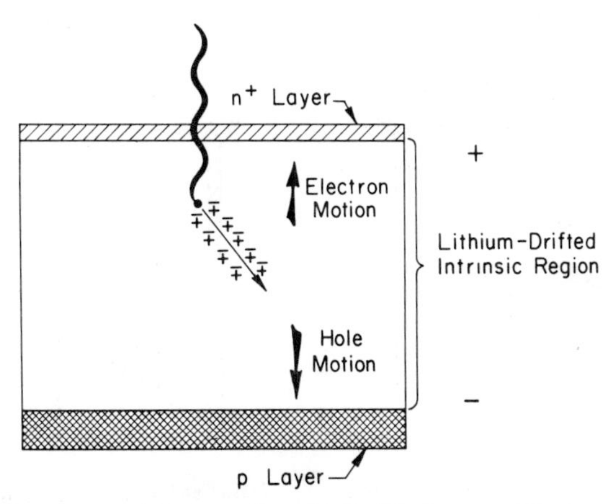

Fig. 13. Illustration of the operation of a Ge(Li) detector. The term, n^+, signifies a surfeit of electron donors. (From Heath, 1969)

tion of early work on the application of lithium-ion drifted germanium (Ge(Li)) detectors for gamma-ray-energy spectrometry is given in Webb and Williams, (1963); Tavendale, (1964); and Goulding, (1966). Due to difficulties in obtaining semiconductor material with sufficient purity, it has not been possible to achieve complete collection of charge at depths exceeding 1.5 to 2 cm. To overcome this limitation, Tavendale and Ewan (1963) proposed a coaxial geometry. Two forms of the coaxial-detector geometry are shown in Figure 14. This type of structure provides a large sensitive volume and is widely used in the fabrication of Ge(Li) detectors for gamma-ray-energy spectrometry. Detectors of this type are available in volumes up to 100 cm^3.

Single crystals of germanium are now produced with sufficient purity to fabricate high-quality detectors that require no lithium drift to compensate for residual acceptors. Both large-volume planar and coaxial detectors are now available in volumes exceeding 50 cm^3. Such detectors, which are termed "intrinsic" devices, have several advantages. Since lithium-ion mobility is not a problem, such detectors do not have to be maintained at cryogenic temperatures at all times. Although other problems exist, these devices appear to be much more stable and reliable under adverse environments. It should be stressed that these devices are operated at cryogenic temperatures to reduce thermally generated electrical noise. Both the detector and the first-stage field-effect transistor (FET) preamplifier are often operated at 100 to 140 K to achieve an optimum signal-to-noise ratio.

2.5.2 *Energy Resolution*

As previously indicated, the major application of these devices is for x- and gamma-ray-energy spectrometry. The development of low-noise

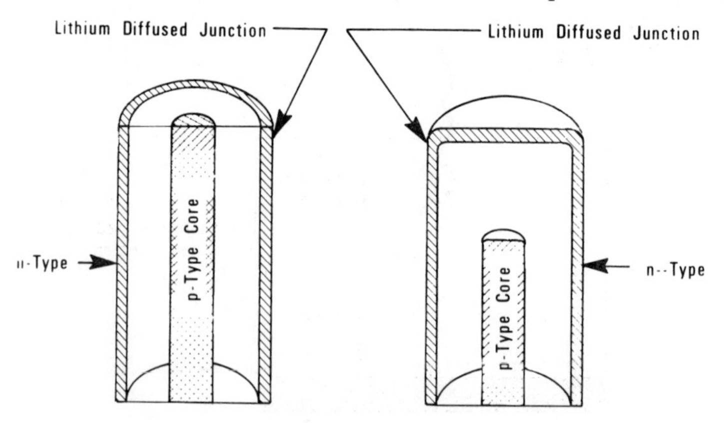

Fig. 14. Schematics of cylindrical annular and semi-annular types of Ge(Li) detector. (From Heath, 1969)

pulse amplifiers that employ field-effect transistors operated in the liquid-nitrogen-temperature range (Smith and Kline, 1966) has made it possible to achieve energy resolutions approaching the statistical limits imposed by the variance in the number of electron-hole pairs created in the detector. Figure 15 presents a pulse-height spectrum of the gamma radiation emitted in the decay of ^{205}Bi, that was obtained with a large-volume Ge(Li) detector. To indicate the superior energy-resolving capability of the semiconductor detector, the pulse-height response of a 3″ × 3″ NaI(Tl) scintillation detector is shown for comparison.

The observed experimental width of a monoenergetic electron line resulting from the interaction of a gamma ray in the intrinsic region of a semiconductor detector is determined by a number of factors. The most important of these are: (1) variations in the number of electron-hole pairs; (2) noise contribution from the detector, such as leakage current or charge collection problems; (3) electronic-noise contribution from the input stages of the pulse-amplifier system; (4) the performance of the pulse-shaping filter networks in the pulse-amplifier system; (5) the accuracy of the analogue-to-digital converter used to measure the amplitude of the detector pulse; and (6) the stability of the electronic system. Of these, the basic limitation in resolution is imposed by variations in the production of charge pairs within the detector. In germanium, on the average, 2.98 eV of energy are required to produce an electron-hole pair. This may be compared with the 30 eV of energy required to produce one ion pair in most of the proportional-counting gases. Following the production of a high-energy electron in the detector, energy is lost either by the production of electron-hole pairs or by processes that give rise to lattice vibrations (mainly optical branch). Since the sharing of energy between these two competing modes is statistical in nature we must speak of an average number of electron-hole pairs resulting from the loss of a given amount of energy in the detector. It is customary to express the experimental resolution of a detector in the following manner:

$$R = 2.355(\mathrm{F}E_\gamma \bar{E}_\mathrm{p})^{1/2}, \tag{2.9}$$

where R is the measured width of a peak in kilo-electron volts (keV) at a point equivalent to one-half the maximum amplitude (FWHM), E_γ is the energy of the gamma ray in keV, $\bar{E}_\mathrm{p}$ is the average energy required to create an electron-hole pair, and F is the so-called Fano Factor (Fano, 1947), which is related to the fractional amount of the total energy absorbed that results in the production of electron-hole pairs. To illustrate the performance of semiconductor detectors, Figure 16 shows the resolution measured as a function of photon energy of a medium-sized Ge(Li) detector used for general-purpose gamma-ray

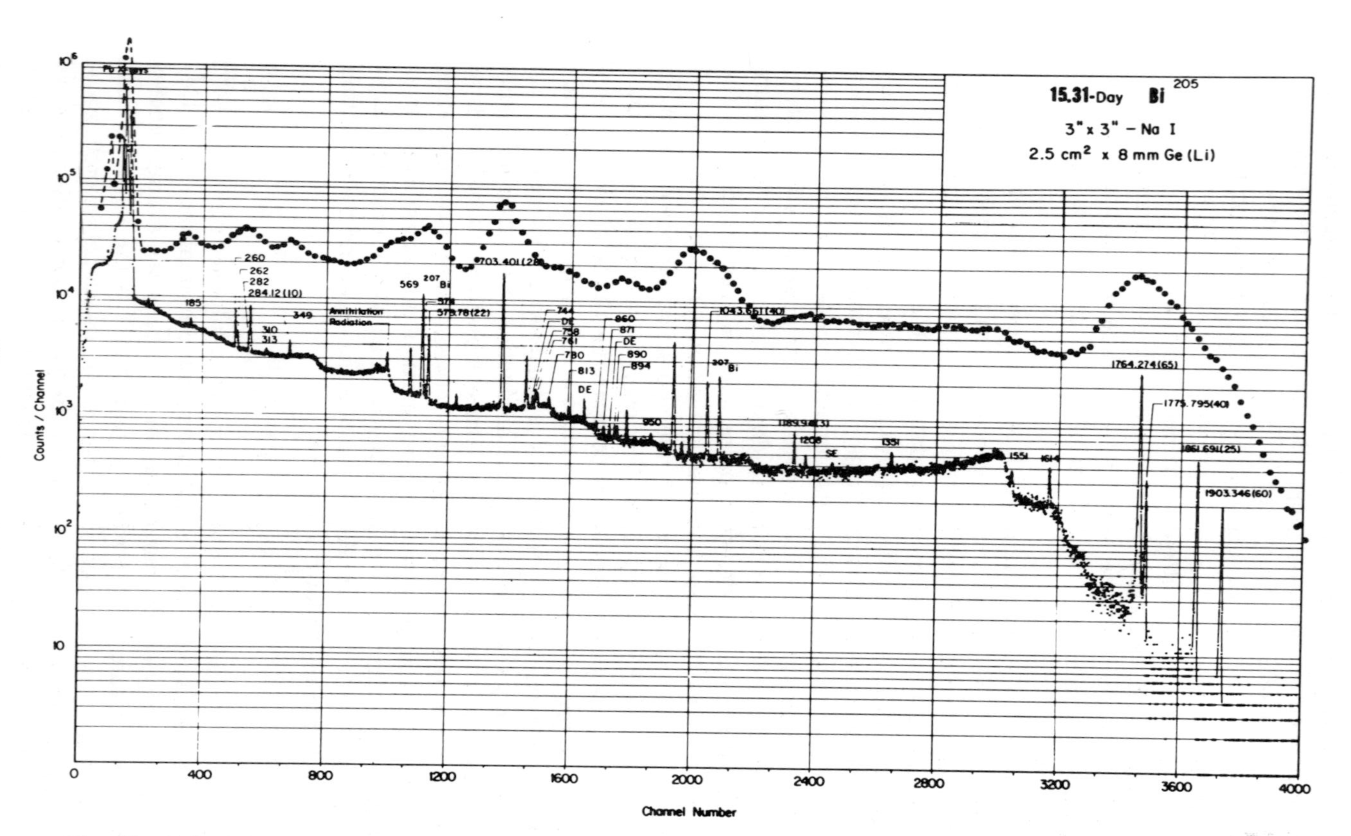

Fig. 15. Pulse-height spectrum of gamma rays emitted in the decay of ^{205}Bi obtained with a Ge(Li) detector. The response of a 3″ × 3″ cylindrical NaI(Tl) detector is shown to illustrate the improved energy resolution of semiconductor detectors. "SE" and "DE" indicate single- and double-escape peaks.

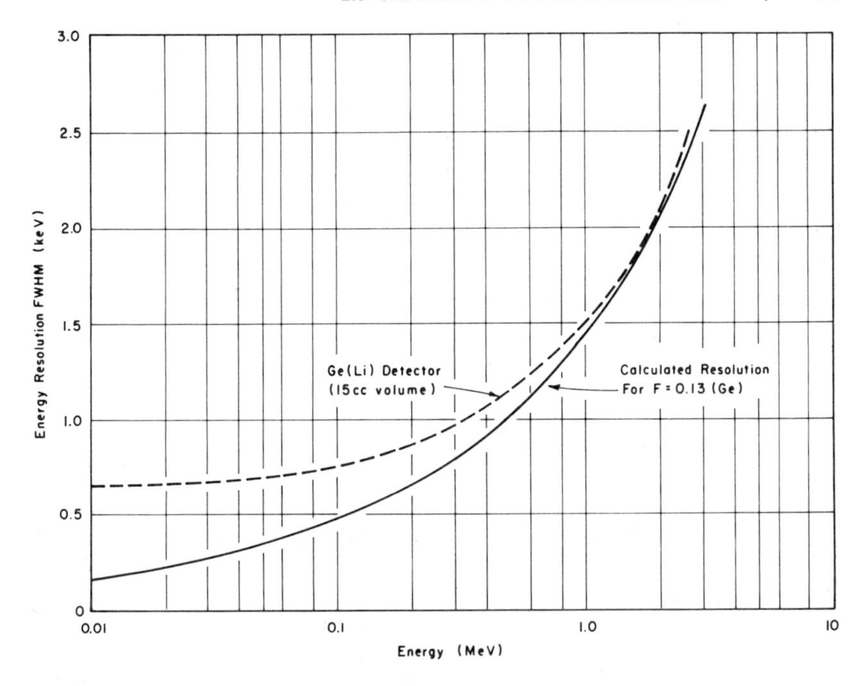

Fig. 16. The measured energy resolution, expressed as the full width at half maximum of the total-energy gamma-ray peak (FWHM), as a function of gamma-ray energy for a Ge (Li) detector. The Fano Factor is chosen so that R *vs* E_γ (Eq. 2.9) forms a lower limit to the measured curve.

spectrometry. At the low-energy end of the range, the resolution is generally limited by the noise performance of the electronic system. At higher energies, the resolution can be impaired by charge-collection problems in the detectors and instabilities of the electronic systems. For comparison, the predicted width from Eq. 2.9 is shown for a germanium device with an assumed value for the Fano Factor of 0.13.

2.6 Čerenkov Counting

When a charged particle passes through a medium, it causes local polarization along its path. When the polarized molecules return to their respective original states, they emit electromagnetic radiation. The propagation of the electromagnetic radiation occurs with a phase velocity of c/n where c is the speed of light *in vacuo* and n is the refractive index of the medium. If the velocity of the charged particle v is smaller than c/n, the electromagnetic radiation will interfere destructively within the medium. When $v > c/n$, coherent directional electromagnetic radiation results, which is called Čerenkov radiation. The threshold for production is, therefore, $v = c/n$.

Čerenkov radiation is emitted mainly in the higher-frequency range of the visible spectrum, in a narrow cone around the direction of propagation of the incident charged particle.

Table 7 shows photon yields for electrons in water, for various regions of the electromagnetic spectrum, according to Ross (1970). The threshold is about 260 keV. Figure 17 shows that the threshold electron energy decreases as the refractive index increases. Unfortunately, there are practical limitations in using high refractive index solutions, and thus water remains the most commonly used solvent for Čerenkov counting.

The use of Čerenkov radiation for the indirect measurement of radioactivity is discussed in Section 4.5.1 and in Jelley (1958).

2.7 Dead-Time, Decay, and Background Corrections

2.7.1 *Introduction*

The counting rates obtained with any of the variety of detectors available will require corrections for dead-time losses and backgrounds as well as decay during the counting interval, if this is significant, and, as a rule, decay for the elapsed time between the beginning of the counting interval and some chosen reference time. While other corrections, depending on the detector, may also be required, those just mentioned are common to all and are treated in the following subsections.

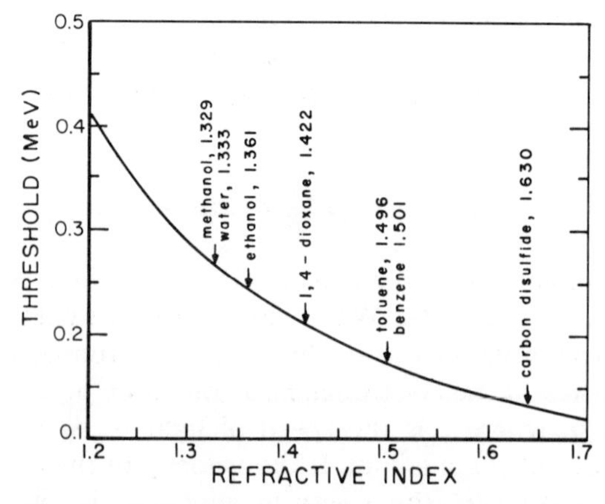

Fig. 17. Electron threshold energy for Čerenkov emission as a function of refractive index of several solvents. Refractive indices are for a wavelength of 589 nm (Na D lines) at 20 °C. (From Parker and Elrick, 1970)

TABLE 7. *Čerenkov photon yields as a function of electron energy in water for selected spectral regions*[a]

Electron Energy	Number of photons/spectral region in nm						
	250–300	300–350	350–400	400–450	450–500	500–550	550–600
keV							
275	0.0032	0.0023	0.0017	0.0013	0.0011	0.0009	0.0007
300	0.0041	0.0029	0.0022	0.0017	0.0013	0.0011	0.0009
350	0.273	0.194	0.146	0.113	0.091	0.074	0.062
400	0.723	0.513	0.385	0.300	0.240	0.196	0.163
450	1.37	0.975	0.731	0.569	0.455	0.373	0.309
500	2.22	1.58	1.18	0.922	0.737	0.604	0.501
550	3.26	2.31	1.73	1.35	1.08	0.884	0.734
600	4.45	3.16	2.37	1.85	1.48	1.21	1.00
650	5.80	4.12	3.09	2.41	1.92	1.58	1.31
700	7.59	5.39	4.04	3.15	2.51	2.06	1.71
750	8.94	6.35	4.76	3.71	2.96	2.43	2.01
800	10.7	7.59	5.69	4.43	3.54	2.90	2.41
850	12.6	8.92	6.69	5.21	4.16	3.41	2.83
900	14.6	10.3	7.75	6.04	4.83	3.95	3.28
950	16.6	11.8	8.87	6.91	5.52	4.52	3.75
1000	18.9	13.4	10.0	7.82	6.25	5.12	4.25
1200	28.5	20.3	15.2	11.8	9.46	7.75	6.43
1400	39.6	28.1	21.1	16.4	13.1	10.7	8.92
1600	51.7	36.7	27.5	21.4	17.1	14.0	11.6
1800	64.8	46.0	34.5	26.9	21.5	17.6	14.6
2000	78.8	56.0	42.0	32.7	26.1	21.4	17.8
2200	93.6	66.5	49.8	38.8	31.0	25.4	21.1
2400	109	77.6	58.2	45.3	36.2	29.7	24.6
2600	125	89.1	66.8	52.0	41.6	34.1	28.3
2800	142	101	75.8	59.0	47.2	38.7	32.1
3000	160	114	85.1	66.3	53.0	43.4	36.0
3200	178	126	94.7	73.8	58.9	48.3	40.1
3400	196	140	105	81.5	65.1	53.3	44.3
3600	215	153	115	89.3	71.4	58.5	48.6
3800	235	167	125	97.5	77.9	63.8	53.0
4000	255	181	136	106	84.5	69.3	57.5

[a] From Ross (1970).

2.7.2 *Dead-Time Corrections*

2.7.2.1 *Introduction.* For any counting system consisting of a detector with its associated electronic equipment there exists a minimum time interval, τ, by which two consecutive events must be separated for both to be recorded. Events arriving during this dead time are lost and the fraction of events lost increases with increasing counting rate. (In this section the terms dead time and dead-time losses are used instead of the alternatives, resolving time and coincidence losses, to avoid confusion with coincidence-counting terminology (see Section 3.2).)

Dead times can be of two types depending on the system response

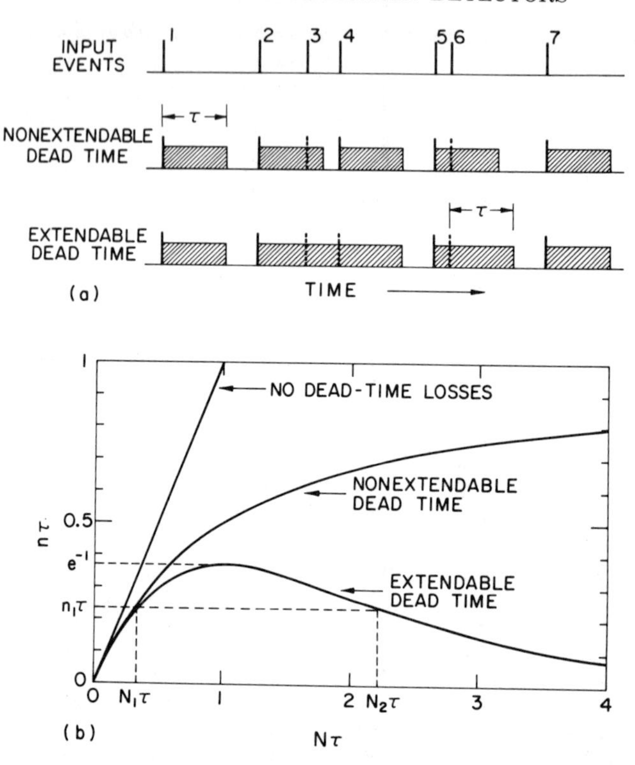

Fig. 18. Nonextendable and extendable dead times. (a) A sequence of input events (top row) is shown with the corresponding outputs from a nonextendable dead time (middle row) and an extendable dead time (bottom row). With the nonextendable dead time, events 3 and 6 are lost but 4 survives because it is separated by more than τ from 2, the last preceding event to produce an output. With the extendable dead time, event 4 is lost also because it is separated by less than τ from 3 which extended the dead time initiated by 2. Event 6 extended the dead time initiated by 5 but not by enough to block the following event. (b) The observed output rates as functions of the input rate for the two types of dead time are shown as plots of $n\tau$ *vs* $N\tau$. (This is equivalent to plotting n *vs* N both in units of τ^{-1}.) Note that with an extendable dead time, the output rate n_1 could be observed for two input rates, N_1 or the many times larger N_2.

to events arriving *during* a dead time (Figure 18a). If the dead time is increased by such input events, so that an interval τ must elapse following the last input event before another output event can be produced, it is said to be extendable (or paralyzable, cumulative, or updating). If the dead time is not increased by input events arriving while the system is dead, the dead time is said to be nonextendable (or nonparalyzable or noncumulative).[1] The responses of real systems lie

[1] The two types have also been called Type I and Type II dead times but this terminology is best avoided because different authors (e.g., Feller, 1948; Evans, 1955) do not agree as to which is which.

between these limiting cases, but, as we shall see, it is usually desirable to design systems to have as nearly nonextendable dead times as possible.

If the input events occur at random times at a mean rate N, the time intervals between events are described by the Poisson distribution (see Section 7.1.1) and the probability that no events occur in an interval τ is $e^{-N\tau}$. Then the expected output rate, n, for a system with an extendable dead time, τ, is

$$n = Ne^{-N\tau}. \tag{2.10}$$

Equation (2.10) cannot be solved for N in terms of n (i.e., in closed form), but the approximation

$$N \approx n/[1 - n\tau - (n\tau)^2/2] \tag{2.11}$$

is good to better than 0.1 percent for losses $\leq$10 percent. For somewhat larger losses, or higher accuracies, the approximate N from Eq. 2.11 may be substituted in the exponent of Eq. 2.10, and N obtained to the required accuracy by iteration. The output rate, Eq. 2.10, reaches a maximum of $(e\tau)^{-1}$ for an input rate of τ^{-1}, decreases for higher input rates, and approaches zero, with the system almost completely paralyzed, for input rates $\gg \tau^{-1}$. Because the input rate is a double-valued function of the output rate, there is danger with extendable dead times of wrongly assigning the lower input rate to an unexpectedly strong source (Figure 18b).

If a system with a nonextendable dead time τ gives an average output counting rate n, then the system is dead for a fraction $n\tau$ and "live" for a fraction $1 - n\tau$ of the counting interval, t. If the input events arrive randomly at an average rate N that does not vary significantly during t, then N (the dead-time corrected counting rate) is simply the observed rate divided by the live-time fraction, or

$$N = n/(1 - n\tau). \tag{2.12}$$

Conversely, if the input rate is known, the expected output rate given by Eq. 2.12 is

$$n = N/(1 + N\tau). \tag{2.13}$$

Equations 2.12 and 2.13 give the exact expectations for the usual case in which the beginning of the counting interval, t, is chosen at random ("equilibrium process"). Expectations given by some authors for the cases in which t commenced either with an event, i.e., at the beginning of a dead time ("ordinary process"), or commenced only if the system is "live" ("shifted" or "free-counter process"), differ, respectively, by terms of the order of τ/t or $N\tau^2/t$ (Müller, 1974) that are several orders of magnitude smaller than the statistical uncertainties of practical measurements. A system with a nonextendable dead time is never paralyzed; the output rate asymptotically approaches τ^{-1} for input rates $\gg \tau^{-1}$ (Figure 18b).

Equation 2.12 suggests a simple method for comparing dead times. If pulses from a detector are fed to two systems with dead times τ_1 and τ_2, then

$$N = n_1/(1 - n_1\tau_1) = n_2/(1 - n_2\tau_2),$$

and

$$\tau_2 = \tau_1 + (1/n_2 - 1/n_1). \tag{2.14}$$

Thus if τ_1 is known, τ_2 can be found directly, or, if τ_1 is known to be nonextendable and rate independent while τ_2 is suspect, τ_2 can be quickly checked over a range of N.

If the half life of the sample activity is so short that decay during the counting interval, t, is significant, the losses are not constant and must be integrated over t. The resultant formulae are given in Section 2.7.3.

2.7.2.2 *Statistics of Dead-Time Distorted Processes.* A property of the Poisson distribution that is applicable to radioactive events is that the variance is equal to the mean (Section 7.1.1). Therefore, for a hypothetical system with zero dead time, the estimated variance of the observed total counts in interval t is equal to the observed total counts:

$$\hat{\sigma}^2_{nt} = nt, \text{ or } \hat{\sigma}_{nt} = \sqrt{nt}. \tag{2.15}$$

This simple relation does not hold when the dead time is not zero because the intervals between observed counts no longer constitute a simple Poisson distribution; all intervals $< \tau$ have been removed. For a nonextendable dead time the standard deviation of the *observed count* is less than for a Poisson process by a factor $(1 - n\tau)$ or

$$\hat{\sigma}_{nt} = (1 - n\tau)\sqrt{nt}. \tag{2.16}$$

The standard deviation of the *corrected count* is greater than for a Poisson process by a factor $\sqrt{1 + N\tau}$, or

$$\hat{\sigma}_{Nt} = \sqrt{(1 + N\tau)Nt}. \tag{2.17}$$

These results (Feller, 1948; Vincent, 1973) should be kept in mind when assessing statistical uncertainties if the losses are large. Note, however, that the *relative* standard deviation of the *corrected rate* is still given by the usual expression

$$\hat{\sigma}_N/N = 1/\sqrt{nt}. \tag{2.18}$$

Baerg (1973a) has shown that Eq. 2.18 holds whether the data are taken by preset time, preset live time, or preset count, and, if background is negligible, even if the source decays significantly during t. For moderate losses, Eq. 2.18 also holds to a good approximation with extendable dead times (Vincent, 1973). Cohn (1966) has prepared graphs showing relative standard deviations as functions of counting losses for both types of dead time.

2.7.2.3 *Dead Times of Practical Systems.* For accurate measurements over a range of counting rates involving significant dead-time losses, one component of the system ideally should have a dead time that is stable, nonextendable, independent of pulse rate and amplitude, and longer than the maximum dead time in any other stage. If no electronic component (e.g., discriminator, single-channel analyzer, or scaler) meets these requirements, it may be necessary to insert a special dead-time-determining stage into the system. This dead time becomes the effective system dead time, the principle being that accurately determinable corrections for dead-time losses are preferred to somewhat smaller ones that are poorly defined.

Except for Geiger-Müller counters for which the intrinsic dead times range from several tens to a few hundreds of microseconds, the dead times that are imposed on systems employing proportional, scintillation, or solid-state detectors are determined by the time constants of the associated electronics and are usually a few microseconds. Some methods for measuring these are described in the following section.

Commercial liquid-scintillation-counting (LSC) systems generally have fast electronics combined with inherently fast scintillators. Because LSC is widely used to assay large numbers of relatively low-activity sources and dead times are usually $\leq 1\ \mu s$, it is common practice to neglect the dead time corrections. If, however, accuracies approaching 1 percent are required for sources giving count rates of $\geq 10^4\ s^{-1}$, rate-dependent effects should be investigated for the channel settings of interest (see Sections 3.8 and 4.3).

The methods of measuring dead times and correcting for losses applicable to single-detector systems apply to the independent singles channels of coincidence-counting arrangements, but the corrections to the observed coincidence counts are more complex and are treated in Section 3.2.

Most multichannel gamma-ray spectrometers have variable, amplitude-dependent dead times, and commercial models usually compensate the dead-time losses during each run by the use of automatic live-timing devices, as described in Section 4.2.4.4. Provided the input rate does not vary significantly over the counting interval, these are accurate to the extent that the contributions to dead-time losses from the detector and all the circuits preceding the analyzer are negligible. This may not be true for high-resolution Ge(Li) spectrometers employing long amplifier time constants, base-line restoration, and pile-up rejection circuits. The latter are essentially extendable dead-time devices and may contribute substantially to the system losses at high input rates.

2.7.2.4 *Measuring Dead Times.* The choice of method for determining dead times depends on the accuracy required, the time and equipment available, and the type of system. A careful investigation of the counting-loss characteristics of a new system can be time-consuming; routine tests to check that the dead time has not changed take only a few minutes. Some methods are outlined below.

2.7.2.4.1 *Oscilloscope Method.* High-rate output pulses from the dead-time-determining stage are viewed with a triggered oscilloscope and the minimum time between successive pulses estimated. By varying such parameters as source strength, gain, and discrimination level, any gross change of dead time with counting rate or instrument setting can be observed. Although the minimum pulse separation might be read to, say, 100 ns on the screen, the error in the dead time could be larger. For reasons discussed below (Section 2.7.2.5), it is impossible to achieve a perfectly nonextendable overall system dead time, and the average effective dead time will be larger than the minimum.

2.7.2.4.2 *Double-Pulse Method.* A train of suitably shaped pulse pairs from a double-pulse generator is fed into the preamplifier test input, the pair separation is reduced until the output rate is halved, and the separation read from an oscilloscope or from the pulser setting if it is accurately calibrated. As with the method given in Section 2.7.2.4.1, this method is rapid, but does not test for dead-time extendability. Nor does it simulate the random amplitudes or random arrival times of detector pulses. A triple pulse generator, if available, can be used to test for extendability.

2.7.2.4.3 *Two-Source Method.* Two sources, (1) and (2), of about equal activity sufficient to give several percent losses are counted in this sequence; (1), (1+2), (2) with observed rates n_1, n_{12}, and n_2, respectively, and background rate n_b. Then, if the dead time, τ, is nonextendable and independent of the counting rate,

$$\tau = \{1 - [1 - (\Delta - n_b)q/p^2]^{1/2}\}p/q, \qquad (2.19)$$

where

$$\Delta = n_1 + n_2 - n_{12},$$

$$p = n_1 n_2 - n_b n_{12},$$

and

$$q = n_1 n_2 n_{12} + n_b(n_1 n_2 - n_1 n_{12} - n_2 n_{12}).$$

Diethorn (1974) derives Eq. 2.19 and discusses various approximations that neglect the background rate. If the background can be neglected, then, for $n_b = 0$, Eq. 2.19 becomes

$$\tau = [1 - (1 - \Delta n_{12}/n_1 n_2)^{1/2}]/n_{12}. \tag{2.20}$$

The background may be included to a first order approximation by putting $\Delta - n_b$ for Δ in Eq. 2.20 (Kohman, 1945). Then Eq. 2.20 will usually be adequate for $n_b/n_1 \approx n_b/n_2 \lesssim 10^{-3}$.

The two-source method has the advantage of giving the effective overall dead time by using real pulse distributions. Several precautions must be taken to get correct results. Since Δ is a small difference between large numbers subject to statistical fluctuations the individual rates must be determined with high precision. For this reason the entire system must be extremely stable with respect, for example, to rate-dependent gain shifts. The presence of one source must not affect the input counting rate from the other source; to equalize scattering effects an identical blank should replace the absent source when each is counted alone. Rate independence is important because if $n_1 \approx n_2$, if the dead time is τ at rate n_1 (or n_2), and if the dead time is $\tau + \delta\tau$ at rate n_{12}, the method gives an apparent dead time of $\approx\tau + 2\delta\tau$. To test for rate dependence two or three pairs of sources covering the counting-rate range of interest may be used. More sophisticated multi-source methods have been described (Kohman, 1945; Thomas, 1963).

Approximations to Eq. 2.20 widely cited in textbooks, but not recommended for accurate work, are

$$\tau = \Delta/(n_{12}^2 - n_1^2 - n_2^2) \text{ or } (\Delta - n_b)/(n_{12}^2 - n_1^2 - n_2^2)$$

and (2.21)

$$\tau = \Delta/2n_1 n_2 \text{ or } (\Delta - n_b)/2n_1 n_2.$$

Of these, the last is preferred and may often be adequate for less critical work, or when only small dead-time corrections are anticipated. For an extendable dead time an exact solution for τ cannot be obtained in closed form but, neglecting background,

$$\tau \approx (n_{12}/2n_1 n_2) \ln [(n_1 + n_2)/n_{12}] \tag{2.22}$$

to better than 0.5 percent if $n_{12}\,\tau \leq 0.15$ and if $n_1 \approx n_2$ within a factor of two. (Eq. 2.22 is exact if $n_1 = n_2$ and $n_b = 0$.)

2.7.2.4.4 *Source-Pulser Method.* This is similar to the method given in Section **2.7.2.4.3** except that one source is replaced by a train of suitably shaped periodic pulses fed into the preamplifier test input. If the observed rates for the source, source plus pulser, and pulser are n_s, n_{sp}, and n_p then, if the dead time is nonextendable and rate independent,

$$\tau = [1 - (1 - \Delta/n_p)^{1/2}]/n_s, \tag{2.23}$$

where $\Delta = n_s + n_p - n_{sp}$, and the background is included in n_s. Equation 2.23 is almost exact for $n_p\tau \lesssim 0.30$, a condition that should always obtain for practical measurements. Müller (1976a, 1976b) has derived the corrections applicable to Eq. 2.23 for higher pulser rates. This method has many advantages over that described in Section 2.7.2.4.3 (Baerg, 1965), and is recommended for general use. In particular, any source may be used, the restrictions on τ with respect to extendability and rate dependence are less critical than with two sources, there are no background terms, and, since the pulser output is not subject to statistical fluctuations, a given precision can be reached in about one third the counting time. As with the two-source method, an exact solution for an extendable τ cannot be obtained in closed form but the approximation

$$\tau \approx p e^{-p}/n_s, \tag{2.24}$$

where

$$p = (q^2 + 4q - 1)^{1/2} - (q + 1)$$

and

$$q = n_p/(n_{sp} - n_s),$$

is good to better than 0.3 percent for $n_s\tau \leq 0.20$ and a wide range of pulser rates; i.e., the error in the approximation in Eq. 2.24 is independent of n_p.

2.7.2.4.5 *Two-Pulser Method.* This is another variant of the two-source method (Müller, 1973). If n_1 and n_2 are the rates of the two pulsers, then $\tau = \Delta/2n_1n_2$ exactly. The pulsers must be extremely stable, have incommensurable rates (i.e., the ratio of the rates must not be expressable as a ratio of integers), and be perfectly isolated (i.e., must never "lock-in"). The method has the disadvantage, common to the double-pulse method (Section 2.7.2.4.2), that no random pulse distributions are used. It is most useful for the fast, accurate testing of electronic dead-time circuits, especially for extendability.

2.7.2.4.6 *Decaying-Source Method.* A source of a pure activity with a conveniently short, accurately known, half life is followed (e.g., ^{128}I, $T_{1/2}$ = 24.99 m; ^{56}Mn, $T_{1/2}$ = 2.578 h). The decay plotted on semi-logarithmic paper is compared with a straight line with a slope corresponding to the known half life. Corrections are taken directly from the graph. This method is relatively time consuming and imprecise but it has several advantages. It does not matter if the dead time is extendable, nonextendable, or rate-dependent; the appropriate correction is read directly, without calculation, from data obtained under normal operating conditions. The method is particularly suitable for use with Geiger-Müller counters and may be used at higher rates for liquid-scintillation counting if the source-pulser method cannot be used. If the dead time is, in fact, nonextendable and rate-independent, τ can be found from the ratio of slope to intercept of a plot of $ne^{\lambda T}$ *vs* n (Martin, 1961), where T is the elapsed time.

2.7.2.4.7 *Proportional-Sources Method.* This method requires two or more sources for which the activity ratios are accurately known and the intrinsic (rate-independent) efficiencies are equal (Stevenson, 1966). The first condition may be met by careful quantitative source preparation and the second may obtain for, e.g., liquid-scintillation counting or integral gamma-ray counting in a well crystal, but not, as a rule, for solid-source beta-particle counting. For a nonextendable dead time, negligible background, and two sources with an activity ratio $A_1/A_2 = \mathrm{R}$,

$$\tau = (\mathrm{R} - n_1/n_2)/n_1(\mathrm{R} - 1). \qquad (2.25)$$

Typically $2 < \mathrm{R} < 10$; a series of sources may be prepared to cover a wide range of counting rates.

For an extendable dead time, the approximate solution

$$\tau \approx \ln(\mathrm{R}n_2/n_1)/n_2\,[\mathrm{R} - 1 + \ln(\mathrm{R}n_2/n_1)] \qquad (2.26)$$

is good to better than 0.5 percent for $n_2\tau \leq 0.10$ for a wide range of $\mathrm{R} > 1$: i.e., the accuracy of Eq. 2.26 depends only on the losses for the *weaker* source.

Proportional sources or a decaying source as in Section 2.7.2.4.6 may also be used to test for non-linear rate dependence in detectors operated in current-measuring or charge-integrating modes, such as ionization chambers.

2.7.2.4.8 *Reference-Peak Method.* Dead-time losses in multichannel gamma-ray spectrometers are often too complex to be calculated from measured system parameters. To monitor these losses directly, reference pulses having excellent amplitude stability and exactly the same shape as the detector pulses are injected at the preamplifier test input. The pulse amplitude is chosen so that the corresponding peak will appear in the spectrum well-removed from any gamma-ray peaks, e.g., beyond the highest energy gamma ray. The pulse rate should be comparable to the rates of the principal gamma-ray peaks but should not exceed 10 percent of the total counting rate if the pulses are periodic. This restriction is unnecessary if a random pulser is used. The total number of reference pulses presented to the preamplifier during the counting interval is recorded separately.

After computer analysis of the peak areas in the spectrum, the ratio of the recorded reference-pulse rate to the computed reference-pulse peak area is used to convert the gamma-ray peak areas to gamma-ray rates. Provided all the radionuclides in the sample have half-lives long compared to the counting interval and there are no amplitude-dependent losses, this method corrects for all rate-dependent losses, whether from analyzer dead time, pile-up (including random summing) or pile-up rejector dead time, or base-line shift, or base-line restorer dead time. It also compensates for any bias in the computer method of determining the peak areas. It does not, of course, correct for true coincidence summing, which is rate independent. For details see, e.g., Bolotin *et al.* (1970) or Wiernik (1971a, 1971b).

A similar principle has been used to monitor rate-dependent losses in medical scintillation (Anger) cameras. Here, again, the effective system dead time may be extendable and rate-dependent, effects that make accurate corrections difficult to calculate (Budinger, 1974). A reference source is placed in a peripheral position of the camera collimator outside the region of interest. The effect of all loss mechanisms can be compensated digitally, frame by frame, by having the system computer normalize all intensities to a standard value for the reference source (Freedman *et al.*, 1972; Adams *et al.*, 1974).

In systems using multichannel gamma-ray spectrometers, the reference peak method is not readily applicable to spectra distorted by amplitude-dependent (i.e., energy-dependent) losses. Such distortions are not common but may be introduced by, for example, pulsed-optical-feedback techniques in some low-noise field-effect-transistor (FET) preamplifiers because the detector pulse that initiates an FET bias restoration, or "reset", is lost, and this initiating pulse is more likely to be a large one than a small one. Losses of this type from a reference peak relative to losses from individual gamma-ray peaks depend on the peak separation and there is no direct way, when a single reference peak is used, of obtaining undistorted, dead-time-corrected spectra. If the source is a mixture of short- and long-lived activities with the shorter half-lives comparable to, or shorter than, the counting interval, there is no simple method of making accurate dead-time corrections, or even corrections that give the true relative intensities of the component activities. Electronic methods of providing either continuous automatic compensation of losses, or constant losses, have been proposed for applications involving rapidly decaying multi-component samples, e.g., activation analysis (Harms, 1967; De Bruin *et al.*, 1974). Monte Carlo studies have shown that the Harms method should give correct results automatically for a wide range of multi-component, time-dependent spectra (Masters and East, 1970).

2.7.2.5 *Dead Times In Series.* As a rule, it is impracticable to have the dead-time-determining stage immediately at the detector output before any pulse-shaping takes place. If not, the overall system dead time will always be somewhat extendable because any circuit has some dead time and it is impossible to achieve a perfectly nonextendable dead time with dead times in series, unless the first is the longest (Damon and Winters, 1954; Müller, 1973). When the longest dead time ends, a preceding stage may be dead from an event that arrived after this stage recovered, but before the end of the longer dead time (Figure 19). Thus, the average dead time of the system suffers rate-dependent extension. In this way, pulse pile-up and base-line shift in early stages may extend the effective dead time at high input rates. Therefore, rate dependence must be studied carefully before high-accuracy measurements can be made when the losses exceed 10 percent or so. At low input rates these effects are negligible. At 2-percent losses, Eq. 2.11 differs from Eq. 2.12 by only 0.02 percent, but at 10-percent losses, the difference is ≈0.5 percent. The differences for a partially extendable dead time would be much less. Practically, and sometimes even conceptually, it is difficult to distinguish between (a) rate-dependence of the dead time caused by dead-time extendability or actual dead-time variations and (b) an apparent rate-dependence caused by gain shifts, discriminator shifts, base-line shifts, pulse pile-up at high rates, and similar effects that are amplitude-dependent as well. Single-channel analyzers can show apparently anomalous effects at high rates, because pile-up may shift events both into and out of the window region. If discriminators or single-channel analyzers are set to select a small fraction of the total input events, the input rates will be many times

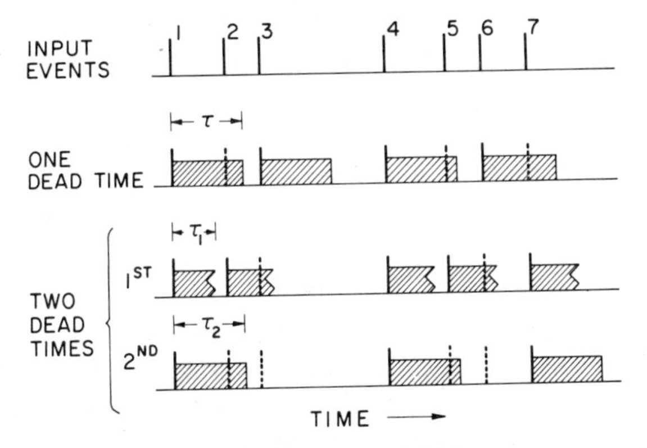

Fig. 19. Nonextendable dead times in series. A high-rate sequence of input events (top row) is shown with the corresponding output from a single dead time, τ (second row). Events 2, 5, and 7 are lost. The bottom two sequences show the effect of a preceding, shorter dead time, τ_1, followed by the principal dead time, τ_2; the irregular termination of τ_1 indicates that τ_1 may be somewhat variable, e.g., pulse-amplitude dependent. One effect of τ_1 has been to increase the losses by removing event 3 from the output of τ_2. In the sequence of events 4 to 7, however, the effect of τ_1 on the output of τ_2 has been to remove event 6 and restore event 7 for no additional net loss of counts. Thus, the effect of τ_1 is second order, complex, and nonlinearly rate dependent. Note, however, that the sequences illustrated correspond to a situation in which dead-time losses exceed 40 percent.

the observed output rates and the series-dead-time effect may be exaggerated.

2.7.2.6 *Further Reading.* More complete discussions of dead-time phenomena have been given by several authors, e.g., Evans (1955), chapter 28; Stevenson (1966), chapter VII; Vincent (1973), chapter 4; Müller (1973). A bibliography of more than 350 references dealing with the measurement of dead times, the corresponding corrections, and closely related subjects has been prepared by the International Bureau of Weights and Measures (Müller, 1975b).

2.7.3 *Background and Decay Corrections*

In the absence of decay over the counting interval, the background rate n_b is subtracted from the observed counting rate after correction for dead-time losses. For a nonextendable dead time

$$N = \frac{n}{1 - n\tau} - n_b . \tag{2.27}$$

As background count rates are normally low, dead-time corrections to the observed background count rate may nearly always be neglected. As a rule, when the ratio of source-to-background count rates is small the dead-time correction is small. Statistical considerations governing the optimum determination of the background correction are discussed in Section 7.1.2. The origins of the background, and methods of minimizing it, are discussed in Sections 6.1.1.6 and 6.1.4.2.2.

In the absence of significant background and dead time, but with significant decay during the counting interval, it can readily be shown that the counting rate N_0 at time t_0, the beginning of the counting interval, is given by

$$N_0 = n \frac{\lambda t}{1 - e^{-\lambda t}} , \tag{2.28}$$

where t is the duration of the counting interval and λ the decay constant. Alternatively, if the average counting rate is assigned to the midpoint of the counting interval, the influence of decay during that interval is much reduced. The observed rate exceeds the midpoint rate by only 0.5 percent for an interval of half a half-life or by only 0.02 percent for a counting interval of one tenth a half-life. For greater accuracy or longer counting intervals, the decay-corrected counting rate at the midpoint, N_m, is given by

$$N_m = n\lambda t e^{-\lambda t/2}/(1 - e^{-\lambda t}) = n(\lambda t/2)/\sinh(\lambda t/2) . \tag{2.29}$$

A third possibility is to specify the time, t_a, at which the rate is equal to the observed range rate, viz.,

$$t_a = t_0 + \lambda^{-1} \ln[\lambda t/(1 - e^{-\lambda t})] \tag{2.30}$$

which is earlier than the midpoint, $t_0 + t/2$, by 1.5 percent of t for t equal to half a half-life. Eq. 2.30, however, is not usually as convenient as either of the two preceding expressions if a large amount of counting data must subsequently be corrected to a common reference time.

Exact expressions treating the combined effects of dead time, decay, and background have been given by Axton and Ryves (1963). An approximation, also given by these authors, has the form,

$$N_0 = \left(\frac{n}{1 - n\tau} - n_b\right)\left\{\frac{\lambda t}{1 - e^{-\lambda t}}\right\}\left(1 + \frac{n\tau(\lambda t)^2}{12}\right). \qquad (2.31)$$

This expression is accurate to 0.1 percent for the condition that $\lambda t <$ 2 and $n\tau < 0.26$. Clearly N_m may be obtained to the same approximation by multiplying the right side of Eq. 2.31 by $e^{-\lambda t/2}$. When the combined dead-time and decay corrections are such that the last term of Eq. 2.31 can be neglected, N, the dead-time- and background-corrected rate, may be substituted for the observed rate, n, in Eqs. 2.28 and 2.29; e.g., if $\lambda t = 0.10$ and $n\tau = 0.10$, then $n\tau(\lambda t)^2/12 < 10^{-4}$. The effect of dead time, background, and decay on the variances of the corrected rates has been discussed by Baerg (1973a) and a detailed analysis including extendable dead times can be found in Vincent (1973).

Finally, the rate N_0 (or N_m) may be converted to the rate N_R at the reference time by the well-known relation derived in Section 1.3, viz.,

$$N_R = N_0 e^{-\lambda T} = N_0 e^{-(\ln 2)T/T_{1/2}}. \qquad (2.32)$$

The elapsed time, T, from the beginning of the counting interval (or, for N_m, the midpoint) to the reference time, is negative if the reference time is earlier than the time of counting.

It is important to estimate the contribution of the decay correction to the systematic error in N_R; it will depend on the uncertainty in $T_{1/2}$, the sensitivity of the detection system to any radioactive impurities in the sample, and the size of the correction. For pure samples of nuclides with accurately known half-lives, decay corrections for periods of a half-life or less may not increase the error significantly. On the other hand, corrections for long decay periods based on inaccurate half-life values or applied to impure samples can lead to large errors. Only in unusually favorable circumstances can decay corrections for more than five or six half-lives be made with any degree of confidence.

The simple decay correction, Eq. 2.32, can be applied to combined parent and daughter activities only when they are in equilibrium as discussed in Section 1.3. When the half-lives of parent and daughter are not very different, e.g., ^{95}Zr-^{95}Nb, equilibrium may not be reached during the useful life of the sample. The more complex equations describing the growth and decay of serial activities not in equilibrium were treated by Bateman (1910) and are covered in standard textbooks, e.g., Evans (1955), chapter 15, or Friedlander *et al.* (1964), chapter 3.

3. Fundamental or Direct Measurements of Activity in Radioactive Decay

3.1 General

The question arises, what is a fundamental or direct measurement of activity in radioactive decay. Possibly the simplest answer is that it is based only on count-rate measurements of the source in question. Such a measurement invokes only the standard of time, and is never based on any other radioactivity standard.

The whole subject of radioactivity is, however, not just a question of measuring so many nuclear transitions in a given period of time. Implicit in most measurement results is a fairly detailed knowledge of the modes of decay of the radionuclide under investigation. In sum peak, or gamma-gamma, and in conventional (non-efficiency-extrapolation) electron-photon coincidence counting, as examples, it is necessary to know the probability per decay of the relevant radiations, before the source disintegration rate can be deduced from the counting data, and with calorimetric methods, the average energy per decay is required. Only within the limitation that the decay-scheme data, of the kind given in Appendix A.2, are sufficiently reliable, can such measurements be regarded as direct or fundamental.

On the other hand, even seemingly direct measurements with 4π proportional counters, defined-solid-angle detectors, liquid-scintillation counters, and internal gas counters may be subject to various corrections such as those for source self-absorption, counter wall effects, detector threshold level, and the energy distribution of the radiation being measured. Where it is not possible to make these corrections to a sufficient degree of accuracy, these measurements, too, cannot properly be regarded as fundamental.

Methods using $4\pi\beta$-γ coincidence counting have been most thoroughly studied and developed to provide, for a wide range of nuclear species, measurements that are independent of the precise values of nuclear-decay-scheme parameters and free of requirements for corrections other than those for decay, dead time, resolving time, and backgrounds. These methods, based on efficiency extrapolation, must be regarded as the most reliable of the fundamental or direct methods. It should be noted, however, that no one method is applicable to all nuclear species. The choice of method is finally determined by the nature of the nuclear decay, and whether the attainable accuracy of measurement is considered to be acceptable.

In this section, methods are described for nuclear species and conditions under which the measurements are not relative to a previously determined activity of another source of the same radionuclide, or to the activities of other radionuclides with similar radiations.

3.2 Beta-Gamma Coincidence Counting

3.2.1 *General*

The measurement of activity by the coincidence method can be applied to any nuclear decay involving two or more distinguishable radiations occurring in prompt succession.[2] Two detectors are usually required, each of which responds, ideally, to one only of the two types of radiation. A third counting channel, derived from a coincidence circuit, records those events from the two detectors that occur in coincidence. Then, for simple two-stage beta-gamma decay of a source with a true disintegration rate N_0, and with beta- and gamma-detector efficiencies ϵ_β, ϵ_γ, the channel counting rates would be:

$$N_\beta = N_0\epsilon_\beta, \qquad N_\gamma = N_0\epsilon_\gamma, \text{ and } N_c = N_0\epsilon_\beta\epsilon_\gamma. \tag{3.1}$$

The value of N_0 is then simply $N_\beta N_\gamma/N_c$ and no knowledge of detector efficiencies is required, but, in this simple case, these latter clearly emerge from the data, with ϵ_β equal to N_β/N_0 or N_c/N_γ, and ϵ_γ equal to N_γ/N_0 or N_c/N_β.

Various modifications of this simple result are required to take account of (1) dead time, resolving time, and backgrounds, the effects of which are count-rate dependent and (2) decay-scheme effects that affect detection efficiencies. These modifications will be considered in the following subsections in the context of beta-gamma decay, but the treatment may be taken as equally applicable to other forms of decay.

[2] Such coincidence methods, to measure source disintegration rates, cannot be based on coincidences between primary radiation and associated secondary radiation generated outside the atom, e.g., beta particles and external bremsstrahlung or gamma rays and scintillator photons.

A $4\pi\beta$-γ-coincidence system, shown in the block schematic of Figure 20, is the most widely used in standards laboratories. Commercially available amplifiers and other electronic components may be used to assemble an acceptable system but the detector assemblies are usually of individual design, some even including automatic sample-changing facilities (e.g., Garfinkel *et al.,* 1973a; Figure 21). Two NaI(Tl) detectors (e.g., 7.6 × 7.6 cm) are commonly used for higher efficiency. The dead-time circuitry in these systems is usually of the nonextending type since it is fairly well understood theoretically and the count-rate-dependent effects are minimal. The decay-scheme effects, on the other hand, are much reduced when a $4\pi\beta$ proportional counter is used (Steyn and Haasbroek, 1958; Campion, 1959) and this type of counter also conforms well to general theoretical requirements for complex decay (Baerg, 1966).

The measurement of the counting rates for the separate beta and gamma channels is, in practice, usually made in the conventional manner appropriate for a single detector. The data are, however, usually collected simultaneously and the counting is controlled by a single timer. The determination of coincidences between events recorded in the beta and gamma channels is not so simple and several methods are described briefly in the following three sections.

3.2.2 *Direct Coincidence Counting*

The most widely used method of detecting coincident beta and gamma events depends essentially on summing the two series of shaped, square pulses. Whenever two pulses overlap, the resultant pulse is twice the amplitude of a single one and thus, by exceeding the threshold level of a trigger circuit, is recorded as a coincident event.

The resolving time, τ_r, of the coincidence circuit is equal to the width of the square input pulses, usually set equal for both beta and gamma channels. This pulse width must be long enough (~1 μs) to avoid the loss of coincidence counts due to fluctuations in the time delays in the arrivals of coincident beta and gamma pulses at the coincidence mixer circuit. This time "jitter" is due to electronic effects as well as variation of the electron collection time in proportional counters. Errors may be incurred (Gandy, 1961) unless the relative delays between beta and gamma pulses are adjusted to a net zero mean delay (Williams and Campion, 1965; Munzenmayer and Baerg, 1969).

Chance overlap of unrelated beta and gamma pulses leads to accidental coincidences. Their rate of occurrence is directly dependent on the pulse widths, which therefore should be kept to a minimum, consistent with avoiding losses due to time jitter.

When the half-life for gamma decay of the excited state of the

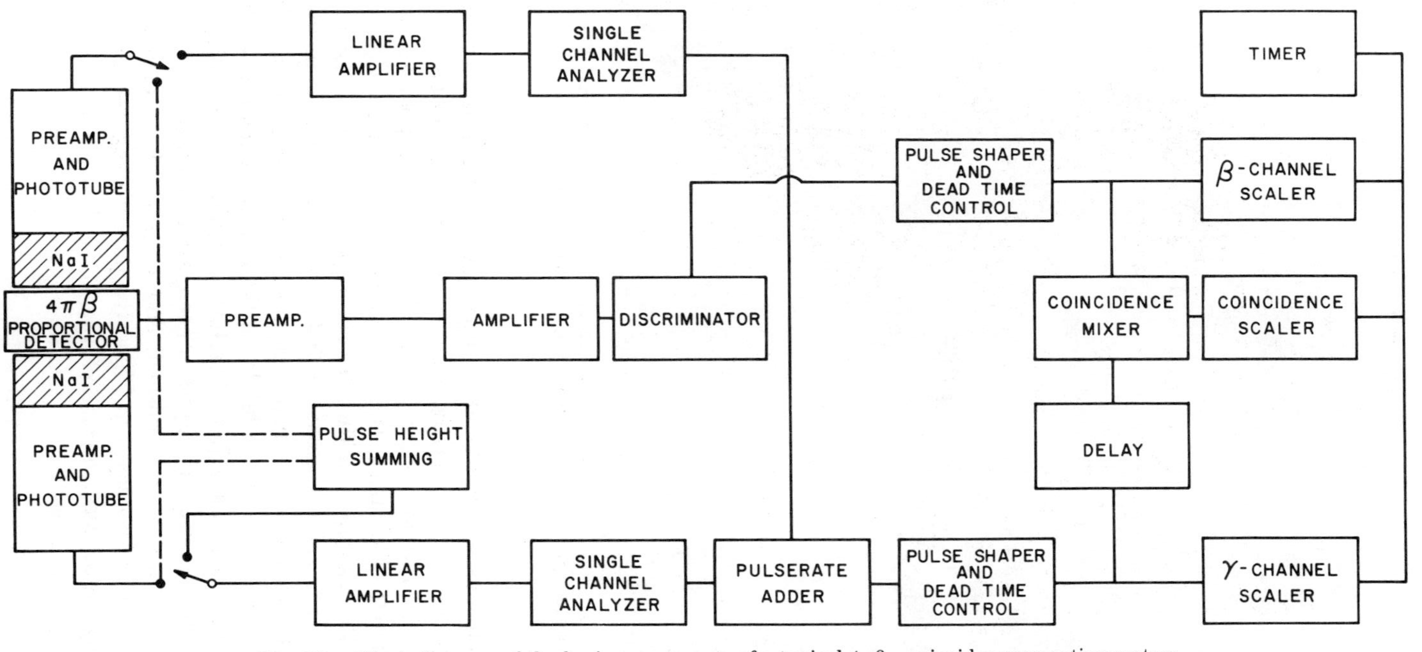

Fig. 20. Block diagram of the basic components of a typical $4\pi\beta$-γ coincidence-counting system.

Fig. 21. The $4\pi\beta$-γ coincidence-counting assembly at the National Bureau of Standards, showing the automatic sample-changing mechanism.

daughter nucleus is comparable to, or longer than the selected minimum pulse width or resolving time of about 1 μs, serious loss of coincidence events will occur. In principle, it is still possible to use direct methods of counting the coincidences if the pulse widths are increased. For example, in the electron-capture decay of ^{85}Sr to the 0.98-μs delayed, 514-keV state of ^{85}Rb, excessive loss of coincidence events could be avoided by increasing the resolving time to, say, 10 μs or more. However, the correction for accidental coincidences would be increased correspondingly. It is, therefore, considered preferable to use either anti-coincidence techniques or correlation counting methods for cases of delayed gamma-ray emission, and to reserve direct coincidence counting methods for decays in which the parent-daughter radiations occur in prompt succession ($< \sim 10^{-7}$ s).

3.2.3 *Anti-Coincidence Counting*

With anti-coincidence counting those gamma events not correlated with detected beta events are counted, and the coincidence count rate is obtained from the difference between the total and uncorrelated gamma-ray count rates (Bryant 1962, 1967; Chylinski *et al.*, 1972). As envisaged by Bryant, the advantage of this approach is that accidental coincidences are completely excluded, and the only corrections required are for dead-time losses and backgrounds. The method is therefore particularly appealing for nuclear decay involving delayed states.

In the application of the anti-coincidence method to prompt β-γ decay, only those gamma signals that precede or follow signals in the beta channel by some minimum time T, are accepted as anti-coincident gamma events. This is achieved by delaying the gamma signals by a fixed time T, and closing an anti-coincidence gamma-channel gate for a fixed time $2T$ after each beta signal.

In applying the anti-coincidence counting method to beta decay followed by delayed gamma emission (Bryant, 1967), it is only necessary that, for each beta signal, the anti-coincident gamma-channel gate remain closed for a further period τ beyond the time $2T$ required for prompt β-γ decay. By making $T+\tau$ equal to m half-lives of the delayed gamma radiation, the probability of recording unwanted "coincidence" events may be reduced to any desired practicable limit, 2^{-m}.

In the treatment given by Bryant (1962, 1967) the individual beta- and gamma-channel counting rates are corrected for dead-time losses and backgrounds in the usual way but correction for the anti-coincidence channel is not so well understood. He gives the approximate expression

$$N_A = \frac{N'_A(1 + N_\beta\tau_\beta)}{\exp\left[-N_\beta(2T + \tau)\right] - N'_A\tau_\gamma(1 + N_\beta\tau_\beta)} \tag{3.2}$$

where N_A, N_β are the dead-time corrected anti-coincident and beta-channel counting rates and where primes indicate uncorrected rates, while τ_β and τ_γ are, respectively, the beta- and gamma-channel nonextending dead times.

An approach recently introduced virtually eliminates the need to correct counting rates for instrumental effects (Baerg *et al.*, 1976). This is achieved by using extending dead-time circuitry, triggered by either beta- or gamma- channel events, and the output is applied to gate both channels as well as the live-timing clock channel. With this live-timed anti-coincidence counting method, dead-time corrections, accidental coincidences, time-jitter, and other effects can now be completely avoided. It can also be used effectively to search for and eliminate spurious double and multiple pulsing and the method is, of course, equally effective for β-γ decay with prompt or delayed gamma-ray emission.

3.2.4 *Correlation Counting*[3]

The underlying principle of correlation counting is that there is a correlation between parent-daughter decays, irrespective of whether the daughter decay is prompt or not. The method also uses the fact that, in a Poisson distribution, the variance is equal to the mean. Determination of the covariance between the two series of parent-daughter decays, measured separately in many equal and simultaneous short intervals of time by two detector systems, provides a measure of coincident or time-correlated events, without the use of a coincidence circuit and, hence, without the limitation of its resolving time. This was first pointed out in 1955 in a Russian publication (Goldansky *et al.*, 1955). The linguistic barrier was not, however, crossed until 1964 (Friedländer, 1964), and the first definitive experiments using these so-called correlation techniques were performed by Williams and Sara (1968). These experiments showed that the coincidence and correlation-counting methods gave good agreement in measuring the activity of ^{60}Co sources, although the dead-time corrections and imprecision were somewhat larger, for the same source measured for equal times, in the correlation method. The power of the correlation method, like that of anti-coincidence counting, lies in the measurement of activity where there is a significant time delay between the parent and daughter decays. The method has also been applied to the measurement of half-lives that are too long for measurement by the delayed-coincidence method (Petroff and Doggett, 1956; Foglio Para and Mandelli Bettoni, 1970), and has been used to detect the presence and number of spurious pulses in a detector consequent upon a real pulse (Campion, 1973; BIPM, 1976).

Only a brief summary of the theory of correlation counting will be given in this report, and the interested reader is referred to the literature references, given in this section, for further details.

Consider the general case of prompt intermediate states with differ-

[3] The NCRP is indebted to J. W. Müller of the Bureau International des Poids et Mesures for making available a preprint of his report (Müller, 1975a) on which much of this section is based.

ent radiations, from parent and daughter nuclides, detected by two separate counters. The series of detected events, p and d, are stochastic and the distribution of each series remains the same as time progresses. The two series are measured separately for identical time intervals and their covariance is given by

$$\mathrm{Cov}(p, d) = \mathrm{E}[(p - \langle \mathrm{p} \rangle)(d - \langle d \rangle)] = \langle pd \rangle - \langle p \rangle \langle d \rangle, \quad (3.3)$$

where E(f), or $\langle \mathrm{f} \rangle$, denotes expected value, and $\langle p \rangle$ and $\langle d \rangle$ are the expected values of p and d in a given time t, which is usually quite short. (The expected, or expectation, value is the average value of *all possible* values weighted by their respective probabilities.)

In the case of a radionuclide emitting beta particles and *prompt* gamma radiation we can say that, in the given time interval t,

$$p = b + c \text{ and } d = g + c, \quad (3.4)$$

where b and g are the number of uncorrelated pulses detected in the beta and gamma channels, respectively, and c is the number of true coincidences in each interval t. For simplicity, dead-time corrections will be ignored. Then b, g, and c are independent, and one obtains

$$\begin{aligned}\mathrm{Cov}(p, d) &= \mathrm{E}[(b + c)(g + c)] - \langle b + c \rangle \langle g + c \rangle \\ &= \langle c^2 \rangle - \langle c \rangle^2 = \mathrm{Var}(c). \quad (3.5)\end{aligned}$$

As the coincidences have a Poisson distribution, $\mathrm{Var}(c) = \langle c \rangle$ (Appendix B.3.5) and, therefore,

$$\mathrm{Cov}(p,d) = \langle c \rangle. \quad (3.6)$$

An experimental value of the Cov(p,d) can be equated to the average value $\bar{c} = \epsilon_p \epsilon_d N_0 t$, where N_0 is the activity of the source, and ϵ_p ($= \bar{p}/N_0 t$) and ϵ_d ($= \bar{d}/N_0 t$) are the respective detection efficiencies of the detectors of the parent and daughter radiations, and $\bar{p}$ and $\bar{d}$ are the experimental average values. Therefore

$$N_0 t = \bar{p}\bar{d}/\bar{c}. \quad (3.7)$$

The experimental value for the covariance is obtained from n readings (p_i, d_i), each of duration t, by using the expression

$$\mathrm{Cov}(p,d) = \frac{1}{\mathrm{n} - 1} \left[\sum_1^{\mathrm{n}} p_i d_i - \frac{1}{\mathrm{n}} \left(\sum_1^{\mathrm{n}} p_i \right) \left(\sum_1^{\mathrm{n}} d_i \right) \right]. \quad (3.8)$$

For zero dead time, it may be shown (e.g., Williams and Sara, 1968; Williams *et al.*, 1973) that

$$\frac{\mathrm{Var}(c)}{\bar{c}^2} = \frac{1}{\mathrm{n}\bar{c}} + \frac{1 + \epsilon_p \epsilon_d}{\mathrm{n} \epsilon_p \epsilon_d}. \quad (3.9)$$

It is of interest to note that the first term on the right-hand side of this equation is merely the Var $(C)/C^2$ for the ordinary coincidence method where $\bar{C}$ is equal to $n\bar{c}$, the total average number of coincidences for a total counting period nt. It is, therefore, apparent that, for equal total coincidence counts, the correlation method, because of the second term on the right-hand side of Eq. 3.9, cannot be as precise as the coincidence method.

It is also of interest to note that the second term on the right-hand side of Eq. 3.9 reduces to $(1/\epsilon_p\epsilon_d + 1)n^{-1}$. It is, therefore, important for ϵ_p and ϵ_d to be large in order to achieve higher relative precision in the determination of $\bar{c}$.

Using the relationships $\epsilon_p = \bar{p}/N_0t$, $\epsilon_d = \bar{d}/N_0t$, and $\bar{c} = \bar{p}\bar{d}/\bar{N}_0t$, Eq. 3.9 reduces to

$$\frac{\mathrm{Var}(c)}{\bar{c}^2} = \frac{1}{n}\left(\frac{\bar{p}\bar{d}}{\bar{c}^2} + 1\right) + \frac{1}{T(\bar{c}/t)}, \tag{3.10}$$

where T, equal to nt, is the total counting time and $\bar{c}/t$ is the mean coincidence count rate. As the mean coincidence count rate and the ratio of the mean count rate $\bar{p}\bar{d}/\bar{c}^2$ do not depend on the time interval t, the relative standard deviation of c, namely $\sigma(c)/\bar{c}$, is made smaller by making both n and T large. Therefore, at reasonable count rates, the duration, t, of the individual counting times is kept short, and the number, n, of runs is kept high. Thus, in practice, t may be 100 μs and n may be 10^6.

In general, the conventional method of coincidence counting for the standardization of radionuclides having long daughter lifetimes is not to be recommended because the coincidence resolving time must be increased to include a reasonable proportion of the delayed coincidences, but this increase will also increase the relative size of the correction necessary for accidental coincidences.

Accidental coincidences do not occur in the correlation method, so that, in the case of delayed daughter decays, the time interval t can be increased to include larger fractions of the daughter decays. Thus, in the measurement of the activity of ^{75}Se, which, in one branch, feeds a 17.5-ms 304-keV intermediate state in ^{75}As, Lewis *et al.* (1973) set t equal to 45ms. In such cases, the covariance equation to be identified with the coincidence count becomes

$$\mathrm{Cov}(p,d) = \epsilon_p\epsilon_d N_0 t\left(1 - \frac{1 - e^{-\lambda t}}{\lambda t}\right), \tag{3.11}$$

where λ is the decay constant of the intermediate state.

3.2.5 *Count-Rate-Dependent Corrections*

The count-rate-dependent corrections for the combined effects of dead time and resolving time in coincidence counting have been rather thoroughly studied. The corrections for dead time are not yet so well understood for correlation counting or for the earlier applications to anti-coincidence counting. (These corrections do not arise with later live-timing anti-coincidence methods.) However, because these methods are not yet widely used, it should be left to the interested reader to refer to the original literature, and we merely outline here the correction formulae used for the more familiar coincidence counting.

The count rates observed for the beta and gamma channels are corrected for non-extending dead time in the well-known manner, namely

$$N_\beta' = \frac{N_\beta''}{1 - N_\beta''\tau} \quad \text{and} \quad N_\gamma' = \frac{N_\gamma''}{1 - N_\gamma''\tau}, \tag{3.12}$$

where single primes indicate corrected rates, still including backgrounds, while double primes indicate observed rates and τ is the dead time, usually set equal for both beta and gamma channels. The corrections to induce resolving-time effects in the coincidence channel are more complicated (Campion, 1959; Gandy, 1961, 1962; Bryant, 1963). Although various formulae in current use are of comparable accuracy, the one given by Bryant is widely used and has the form,

$$N_c' = \frac{(n_c'' - 2\tau_r N_\beta'' N_\gamma'')[1 - (N_\beta'' + N_\gamma'')\tau/2]}{(1 - N_\beta''\tau)(1 - N_\gamma''\tau)[1 - (N_\beta'' + N_\gamma'')(\tau + 2\tau_r)/2 + N_c''\tau]}, \tag{3.13}$$

where τ_r is the coincidence resolving time. A recent detailed analysis by Cox and Isham (1977), for conditions which can be closely approximated in practice, offers an exact but more elaborate solution for coincidence and correlation counting. This may be needed for very high accuracy or when counting rates are very high or dead times long.

Background rates may be subtracted directly from the count-rates calculated from equations such as 3.12 and 3.13, provided source decay during the measurement is negligible. The corrections are more complicated when decay is significant, as is the decay correction itself as a result of the dead-time effect (Axton and Ryves, 1963). The decay correction for live-timed anti-coincidence counting with extending dead-time is no more complicated (Baerg *et al.*, 1976). It should be noted that dead time, decay, and backgrounds also influence the estimate of variance of the corrected counting rates (Baerg, 1973a).

3.2.6 *Efficiency-Dependent Effects*

With any practical counting system at least one of the detectors, usually the beta detector, is not exclusively sensitive to one type of radiation. Contributions to the beta-counting rate also arise from conversion electrons, and the simple expressions of Eq. 3.1 must be further modified for complex beta-gamma decay (Steyn and Haasbroek, 1958; Campion, 1959). For simple two-stage beta-gamma decay the expressions for the counting rates, already corrected for background and rate-dependent effects, become

$$\begin{aligned} N_\beta &= N_0\left[\epsilon_\beta + (1-\epsilon_\beta)\left(\frac{\alpha\epsilon_{ce}+\epsilon_{\beta\gamma}}{1+\alpha}\right)\right] \\ N_\gamma &= N_0\epsilon_\gamma/(1+\alpha) \\ N_c &= N_0[\epsilon_\beta\epsilon_\gamma/(1+\alpha) + (1-\epsilon_\beta)\epsilon_c], \end{aligned} \tag{3.14}$$

where ϵ_{ce}, $\epsilon_{\beta\gamma}$ are respectively the beta-counter detection efficiencies for conversion electrons and gamma rays and α is the conversion coefficient, while the second term in the expression for N_c represents the probability of observing additional coincidences that are not derived from coincident beta-gamma events, background, or accidental coincidences. The probability of observing such additional coincidences is the product of two probabilities, namely (i) that of the beta event being undetected; and (ii) that, ϵ_c, of simultaneous events occurring in both detectors, due, for example, in simple beta-gamma decay, to Compton scattering from one detector to another.

The considerable advantage to using a $4\pi\beta$ counter is that N_0 in Eqs. 3.14 reduces to $N_\beta N_\gamma/N_c$ when the beta-detector efficiency approaches unity. Additional terms due to complex branching are of a similar efficiency-dependent character (see next subsection) so that small corrections can sometimes be estimated to an acceptable accuracy by using available nuclear constants, and measured (Merritt and Taylor, 1967b) or calculated (Urquhart, 1967) values for $\epsilon_{\beta\gamma}$. Other effects include beta-gamma angular correlations (Hayward, 1961), non-uniform source efficiency (Putman, 1953), beta sensitivity of the gamma detector and efficiency instabilities (Campion, 1959). These may, however, usually be ignored when a $4\pi\beta$ proportional counter is used, along with some care in selecting gamma-channel energy response.

While the $4\pi\beta$ proportional counter is the most widely used, and is the only type leading to valid fundamental measurements for many radionuclides, especially those with complex decay, there are cases for which low-efficiency beta-particle detectors may be advantageously used. One example is that of ^{129}I, which decays first by beta-particle emission ($E_{\beta\ max}$ = 0.189 MeV) and then by gamma-ray (E_γ = 39 keV) or conversion-electron (E_{ce} = 34 keV) emission, the last giving rise to xenon K x rays (E_K = 29 to 34 keV). These K x rays can also be used to signal a gamma-ray transition. Mann and Hutchinson (1976) have used as the beta-particle and photon detectors, respectively, thin (0.019-inch) plastic-scintillator and thin (~1/16-inch) NaI(Tl)-crystal phototube assemblies, the latter with a 0.005-inch beryllium window that, together with the intervening air, is thick enough to absorb all beta particles. Sufficient absorber material is also placed between the source and beta-particle detector to absorb all conversion electrons. Since the gamma rays and x rays have energies well below that of $E_{\beta\ max}$, those few photons that do interact in the thin plastic scintillator are excluded by pulse-height discrimination. Under these special conditions the measurement is a simple one of determining the beta-particle, photon, and coincidence counting rates, inserting these quantities in Eqs. 3.12 and 3.13, and solving for $N_0 = N_\beta' N_\gamma' / N_c'$.

3.2.7 *Complex Beta-Gamma Decay*

When, as is often the case, the nuclear decay proceeds through two or more alternative beta branches, simple equations like 3.1 are not adequate. Instead, for coincidence measurements on a nuclide with n beta branches, each of which may be followed by prompt gamma emission, sums of equations like 3.14 are needed to describe the observed counting rates. Thus,

$$N_\beta = N_0 \sum a_r \left[\epsilon_{\beta_r} + (1 - \epsilon_{\beta_r}) \left(\frac{\alpha\epsilon_{ce} + \epsilon_{\beta\gamma}}{1 + \alpha} \right)_r \right],$$

$$N_\gamma = N_0 \sum a_r \frac{\epsilon_{\gamma_r}}{1 + \alpha_r}, \qquad (3.15)$$

$$N_c = N_0 \sum a_r \left[\frac{\epsilon_{\beta_r} \epsilon_{\gamma_r}}{1 + \alpha_r} + (1 - \epsilon_{\beta_r}) \epsilon_{c_r} \right],$$

where the summations extend over the n branches and where, for the r-th branch, a_r is the fractional-branching intensity.

The majority of quantities on the right of Eqs. 3.15 are not usually known with sufficient accuracy to extract a reliable value for N_0 directly from the observed counting rates on the left. In the general case, it is necessary to choose detectors and experimental conditions to provide additional information to solve these equations. Methods have been developed that do not require direct knowledge of nuclear decay-scheme constants or estimates of branch counting efficiencies. However, only an outline of the theoretical analyses leading to these methods will be given here. The reader interested in applying them should refer to the literature on the subject (e.g. Baerg, 1973c, and other references cited there).

The primary requirement for reducing Eqs. 3.15 to a practical formula for coincidence measurements is that the detection efficiencies for the various beta branches be functionally related and approach unit value simultaneously as efficiencies increase. Thus, as an example, for the r-th branch beta efficiency given as a function, g_r, of that for the s-th branch,

$$\epsilon_{\beta_r} = 1 - g_r(1 - \epsilon_{\beta_s}) \rightarrow 1 \text{ as } \epsilon_{\beta_s} \rightarrow 1. \quad (3.16)$$

Using expressions like this, not only for the ϵ_{β_r} but also for the ϵ_{ce}, $\epsilon_{\beta\gamma}$, ϵ_{c_r}, and ϵ_{γ_r} if these are not constant, it is then easily shown that Eqs. 3.15 reduces to the following simple coincidence formula:

$$N_\beta = N_0 f\,(N_c/N_\gamma) \rightarrow N_0 \text{ as } N_c/N_\gamma \rightarrow 1. \quad (3.17)$$

(cf. Houtermans and Miguel, 1962; Williams and Campion, 1963; Baerg, 1966). The function, $f(N_c/N_\gamma)$ is of unspecified general form, and, in practice, is defined simply as a polynomial in N_c/N_γ although in practice the functions are usually linear or of low order in this variable. The coincidence measurement then consists in taking measurements on N_β, N_γ, and N_c for various levels of beta-detection efficiency, and extrapolating the function fitting the data to unit value for N_c/N_γ.

It is particularly significant that Eq. 3.17 is an expression in observables alone and does not depend directly on decay-scheme parameters. The effects of any such dependence are now absorbed in the parameters of the "efficiency function", $f(N_c/N_\gamma)$, and these are to be determined experimentally by varying the beta-detector efficiency over a range for which expressions like Eq. 3.16 are valid.

3.2.8 *Experimental Techniques for Complex Beta-Gamma Decay*

While for simple decay schemes beta detectors of various kinds are often used, for complex beta-gamma decay it is necessary that the detector be of 4π geometric efficiency (Baerg, 1966, 1967, 1973b). In the case of the $4\pi\beta$ proportional counter, it is also useful if it can be operated at elevated pressures (~ 25 atm; 25.4 MPa) (Baerg *et al.*, 1967; Baerg, 1973b). In other respects, the counting system can be of conventional design (Section 3.2.1).

Several methods can be used to vary the $4\pi\beta$-detector efficiency to derive a function that can be reliably extrapolated to unit detection efficiency to deduce N_0. It must be emphasized, however, that careful consideration must be given to ensure that conditions defined by equations like 3.16 are fulfilled. Detailed discussions on these matters appear in the references already noted. Briefly, however, acceptable methods are restricted to variation of the physical effects intrinsically responsible for counting losses, namely source self-absorption, source-mount absorption, and beta-channel discrimination level. With any of these methods, thin sources of the highest counting efficiency should be prepared but for self-absorption variation a series of sources is usually prepared, varying in the amount of added carrier. When source-mount absorption is to be varied, a single source is used and measurements made after each of successive additions of absorber foils. With beta-channel discrimination a single source is also used, but it is necessary to operate the 4π proportional counter at elevated pressure to ensure that counting losses occur, as a result of discrimination, only from the low-energy portions of the beta spectra (Baerg *et al.*, 1967; Baerg, 1973b). Measurements are then taken after each of successive changes in discrimination level. Alternatively, multi-channel analyzers can be used to acquire the efficiency-function data in a single measurement (Smith, 1975).

In acquiring the data, the gamma-channel energy response should be chosen to provide data sets that are as nearly linear as possible. Again, the criteria to be considered in this choice are discussed in detail in the references noted above. Usually a gamma-channel window, set to respond to the photopeak of the gamma ray associated with the main beta branch, would be appropriate. Methods using multiple gamma channels and multivariate functions have also been used (Smith and Williams, 1971; Smith and Stuart, 1975). Data sets acquired with different gamma-channel responses are useful in searching for possible systematic error. As an example, results of measurements on a ^{134}Cs source are shown in Figure 22, using gamma-channel energy responses as noted. The several data sets should yield identical extrapolation values for N_0, as they do in the figure. If they do not do so, then it is likely that experimental requirements (defined by equations such as 3.16) are not properly fulfilled.

Graphical analysis of data sets and extrapolation to unit value in N_c/N_γ may often be adequate, but statistically valid extrapolation can

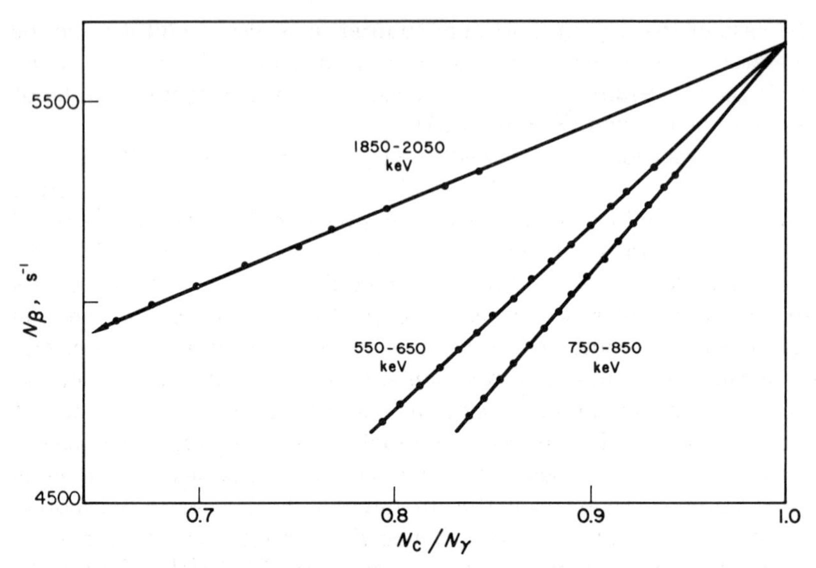

Fig. 22. Efficiency-variation plots obtained with differing gamma-channel gates for ^{134}Cs calibration, i.e., N_β, the beta-channel count rate, *vs* N_c/N_γ ($=\epsilon_\beta$), the ratio of coincidence- to gamma-channel count rates.

only be made by way of iterative least-squares data fitting (Adams and Baerg, 1967; Baerg, 1971). In the analysis, due account must be taken of errors on both the algebraically dependent and independent variables as well as correlations between them. In addition, the order of the fitted function should be determined solely by the statistical errors on the data through use of the χ^2 test for goodness of fit. The calculations are therefore rather elaborate and require the use of adequate computer facilities.

3.2.9 *Applications of Coincidence Counting*

Although the preceding discussions refer to beta-gamma decay, the coincidence methods are applicable to other modes of decay. Thus, the tracer method for pure beta emitters (Campion *et al.*, 1960; Baerg *et al.*, 1964) may be regarded as a variation on that for more complex beta-gamma decay, with the pure beta-emitter component of the mixed source considered as a ground-state branch. Electron-capture decay followed by prompt gamma-ray emission (e.g., ^{54}Mn, Weiss, 1968) presents little difficulty except for very low-energy decay (e.g., ^{7}Be, Taylor and Merritt, 1962) since the Auger-electron x-ray cascade, detected in the beta-counter, may usually be treated as single-branch beta decay in simple beta-gamma radioactivity. Alpha-gamma decay (e.g., ^{241}Am, Rytz, 1964) may be treated in like manner.

Greater care is required for mixed decay involving branch spectra differing markedly in shape. This situation arises, for example, with electron capture and low-energy conversion electrons in ^{197}Hg (Baerg *et al.*, 1967; Taylor, 1967), ^{125}I (Taylor, 1967), ^{195}Au (Adams, 1969), ^{57}Co (Garfinkel and Hutchinson, 1966, Troughton, 1966; Williams and Birdseye, 1966; Baerg, 1973a) or in mixed electron-capture-positron decay (e.g., ^{68}Ga, Smith and Williams, 1971).

Coincidence measurements on pure electron-capture decay (not accompanied by gamma-ray emission) are not yet in a well-developed state. Tracer methods and related techniques have been used (Petel and Houtermans, 1967; Steyn, 1967), as have direct coincidence measurements between x rays and Auger-photon cascades (Troughton, 1967), but rather elaborate corrections are required. Efficiency-variation methods appear to be possible (Baerg, 1973c) and coincidences with internal bremsstrahlung have also been considered (Spernol *et al.*, 1973).

More detailed procedures and reviews on coincidence measurements may be found in IAEA (1967a) and in Herceg Novi (1973).

3.3 Sum-Peak Coincidence Counting

If a radionuclide emits two photons in coincidence, with no direct transitions to the ground state, a measurement of its disintegration rate may be made in many cases by the method of sum-peak coincidence counting. So far, this method has been applied using a NaI(Tl) photon detector, with or without a well, or two NaI(Tl) crystals arranged in 4π geometry.

In the simplest application of this method (Brinkman *et al.*, 1963a; Eldridge and Crowther, 1964), one detector is used, to which the associated electronic circuitry is so connected as to record (i) the total number of photons interacting in the detector per unit time, N_t (integral counting rate), and (ii) the photopeak and sum-peak counting rates for the two photons of different energy.

Consider a radionuclide (Figure 23) that emits a photon, γ_1, in coincidence with a photon, γ_2, with photopeak efficiencies, ϵ_p' and ϵ_p'' , respectively, and total detection efficiencies, respectively, ϵ_t' and ϵ_t''. Then the count rates under the γ_1 and γ_2 peaks, and the $\gamma_1 + \gamma_2$ sum peak, A_1, A_2, and A_{12}, respectively, can be expressed as follows:

$$A_1 = N_0\epsilon_p'(1 - \epsilon_t'') \qquad (3.18)$$

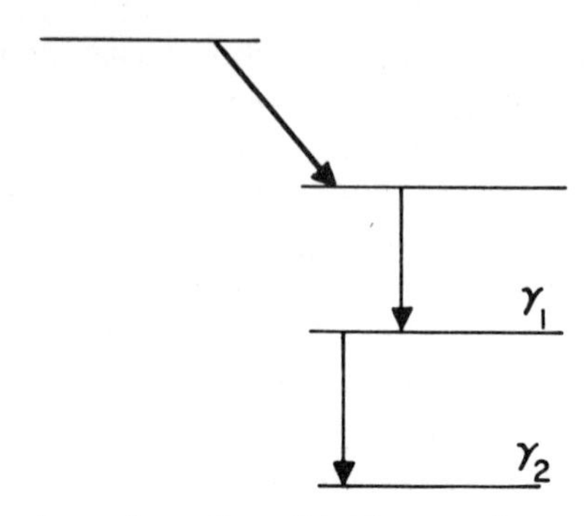

Fig. 23. Decay scheme for radionuclide that may be standardized by sum-peak coincidence counting.

and similarly,

$$A_2 = N_0 \epsilon_p''(1 - \epsilon_t') \tag{3.19}$$

and,

$$A_{12} = N_0 \epsilon_p' \epsilon_p'' \tag{3.20}$$

where N_0 is the disintegration rate.

The probability of recording neither γ_1 and γ_2 in coincidence is the product $(1\ \epsilon_t')$ $(1\ \epsilon_t'')$. Therefore,

$$N_0 - N_t = N_0(1 - \epsilon_t')(1 - \epsilon_t''), \tag{3.21}$$

whence, using Eqs. 3.18 to 3.20,

$$N_0 = N_t + \frac{A_1 A_2}{A_{12}}. \tag{3.22}$$

N_t is determined by extrapolation of the integral count rate to zero discriminator bias.

The sum-peak method, using a single NaI(Tl) detector, has also been extended by Brinkman *et al.* (1963b, 1965) and Brinkman and Aten (1963, 1965) to the measurement of pure β^+ emitters and radionuclides with complex decay schemes. Harper *et al.* (1963) and Eldridge and Crowther (1964) have applied the sum-peak method to the assay of ^{125}I. This radionuclide decays by electron capture to the 35-keV state of ^{125}Te, which then promptly decays to the ground state, either by gamma-ray emission or by electron conversion, the latter resulting in K x rays of ~29 keV. The pulse-height distribution (Figure 2, Eldridge and Crowther, 1964) from a NaI(Tl) crystal has an unresolved "singles" and a sum peak with areas I_I and I_{II}, respectively.

Eldridge and Crowther (1964) show that Eq. 3.22 can be cast in the form:

$$N_0 = \frac{P_1 P_2}{(P_1 + P_2)^2} \frac{(A_I + 2A_{II})^2}{A_{II}}, \tag{3.23}$$

where P_1 is the probability per decay of the K x ray in the electron-capture transition, and P_2 is the sum of the probabilties per decay of the 35-keV gamma ray and the K x ray arising from internal conversion. The derivation of Eq. 3.23 from Eq. 3.22 is based on the assumption that there is negligible loss of energy by Compton scattering, so that $A_1 = A_2 = A_I/2$ and $N_t = A_I + A_{II}$. Using $P_1 = 0.68$ and $P_2 = 0.75$, $P_1P_2/(p_1 + P_2)^2 = 0.249$ and is quite insensitive to possible uncertainties in P_1 and P_2.

Sutherland and Buchanan (1967) have pointed out that incorrect results for N_0 are obtained by the sum-peak method if the detection efficiency for photons emitted from different parts of the source is not a constant. For example, they demonstrate that a sum-peak measurement performed on a 25-ml sample containing ^{60}Co in a 25-mm-diameter vial placed on a 3 x 3-inch NaI(Tl) detector results in a value for the activity that is about 25 percent in error due to this effect.

Except in the case of very low-energy gamma-ray emitters (Eldridge and Crowther, 1964), the requirement for an accurate estimate of the summed Compton recoils, which must be subtracted from under the "singles" photopeaks, is the limiting factor in the attainment of high accuracies.

An alternative approach (Hutchinson *et al.*, 1973a), utilizing two large crystals in 4π geometry (Figure 87) and operated in the sum, coincidence, and anticoincidence modes, is capable of producing accuracies comparable to the coincidence method. The chief reasons for this are that (i) the total efficiencies, N_t/N_0, are extremely high, and, therefore, many subtle corrections, such as for angular correlation, are negligible; and (ii) the possibility of operating the two crystals in the various modes permits of an accurate method of separating total absorption events from otherwise indistinguishable coincident Compton events under the "singles" peaks.

The experimental arrangement for this method is shown in Figure 24. Figure 25 shows superimposed spectra of ^{94}Nb collected in the coincidence and anticoincidence modes. The three peaks correspond to 0.70- and 0.87-MeV "singles" peaks and the summed (0.70 + 0.87)-MeV peak respectively. Because each gamma ray has an equal chance of scattering from the top or bottom detector, the Compton summed continuum is the same for each mode. Thus the γ_1 and γ_2 photopeak count rates in the anticoincidence mode are obtained by means of a subtraction of the coincidence from the anticoincidence spectra. A few "singles" are recorded in the coincidence mode due to gamma rays

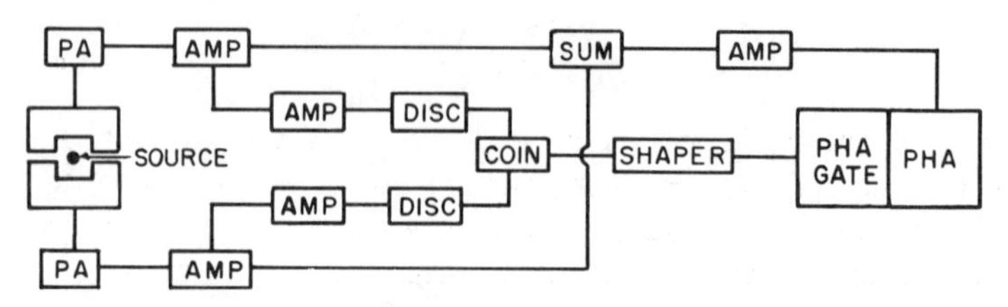

Fig. 24. Block diagram of the experimental arrangement and electronic circuitry used for sum-peak coincidence counting. The gate can be turned *off* to give the summing mode configuration, or *on* to give the coincidence or anticoincidence mode depending on whether the gating pulse is used to pass or block, respectively, the input pulse.

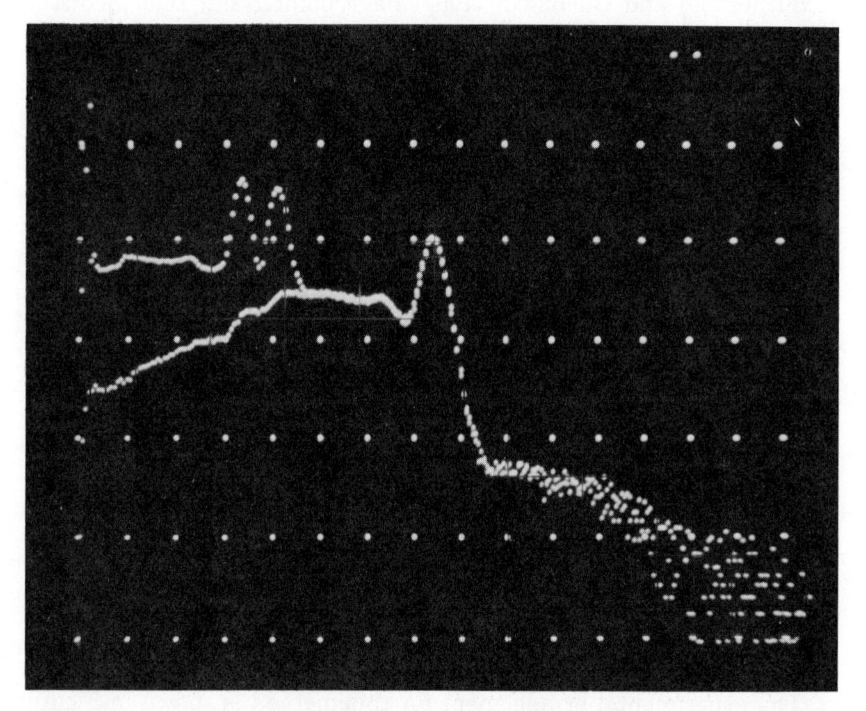

Fig. 25. Superimposed coincidence and anticoincidence spectra of ^{94}Nb, taken for the same live times.

backscattered from one detector into the other, where they lose the remainder of their energy, with the gamma rays coincident with each escaping detection. A measure of such "singles" events can be obtained from measurements of monoenergetic gamma-ray-emitting radionuclides.

To take advantage of the now accurate measurement of the photopeaks ${}^{a}A_1$ and ${}^{a}A_2$ in the anticoincidence mode it is necessary to recast

Eq. 3.22 in terms of measured quantities. It can be shown that, when this is done (Hutchinson *et al.*, 1973a), Eq. 3.22 becomes

$$N_0 = N_t + \frac{{}^aA_1{}^aA_2}{2{}^aA_{12}} \tag{3.24}$$

where N_t is still measured in the sum mode with no coincidence or anticoincidence conditions and ${}^aA_{12}$ is the count rate under the sum peak in the anticoincidence mode.

Further examples of the application of the sum-peak coincidence method with two large NaI(Tl) crystals in 4π geometry, to the assay of ^{22}Na, ^{26}Al, ^{60}Co, ^{88}Y, and ^{207}Bi are given in Hutchinson *et al.* (1973a).

3.4 Photon-Photon Coincidence Counting

Although the technique of gamma-gamma coincidence counting is more limited in application than that of beta-gamma coincidence counting, it has the advantage that sources in the form of wires or pellets can be readily standardized without the preparation of thin samples such as are required for particle counting. Also, standardizations may be made by this method on gamma-ray-emitting radionuclides in sources containing beta-ray-emitting impurities, which would make beta-gamma coincidence counting unusable.

In the gamma-gamma coincidence method there are the following important considerations: (a) the relative efficiency of each of the counters for the two gamma rays, (b) possible spurious coincidences due to the Compton scattering of a gamma ray from one counter into the other, (c) possible angular correlation of the gamma rays, and (d) the attenuation of this correlation due to the finite sizes of the source and counters. Treatments of the problem, for low detection efficiencies, have been given by Meyer *et al.* (1959), Peelle and Maienschein (1960), and Kowalski (1965). A description of x-ray-x-ray coincidence counting of ^{125}I and ^{197}Hg, which includes a careful accounting for summing corrections, has been given by Taylor (1967). The method is illustrated by reference to its use in the standardization of ^{60}Co (Hayward *et al.*, 1955). Summing corrections, that may be significant at higher efficiencies, were not discussed in this paper.

The disintegration rate in terms of observable quantities, with second-order terms neglected, for a short channel dead time, and resolving time τ_r, as given by Hayward *et al.* (1955) is

$$N_0 = \frac{(N_1 - N_{B1})(N_2 - N_{B2})}{N_c - N_{Bc}} \mathrm{F}(\phi)\mathrm{f}(\theta)\left(1 + 2\tau_r \frac{N_1N_2}{N_c}\right), \tag{3.25}$$

where N_1 and N_2 are the counting rates in each of two similar detectors, N_c is the coincidence counting rate, and $f(\theta)$ is a function depending on the geometry and on any angular correlation between the coincident gamma rays (see Feingold and Frankel, 1955). N_{B1}, N_{B2}, and N_{Bc} are the respective background counting rates, and $F(\phi)$ is a correction factor, discussed below, which arises from the change in detection efficiency with gamma-ray energy.

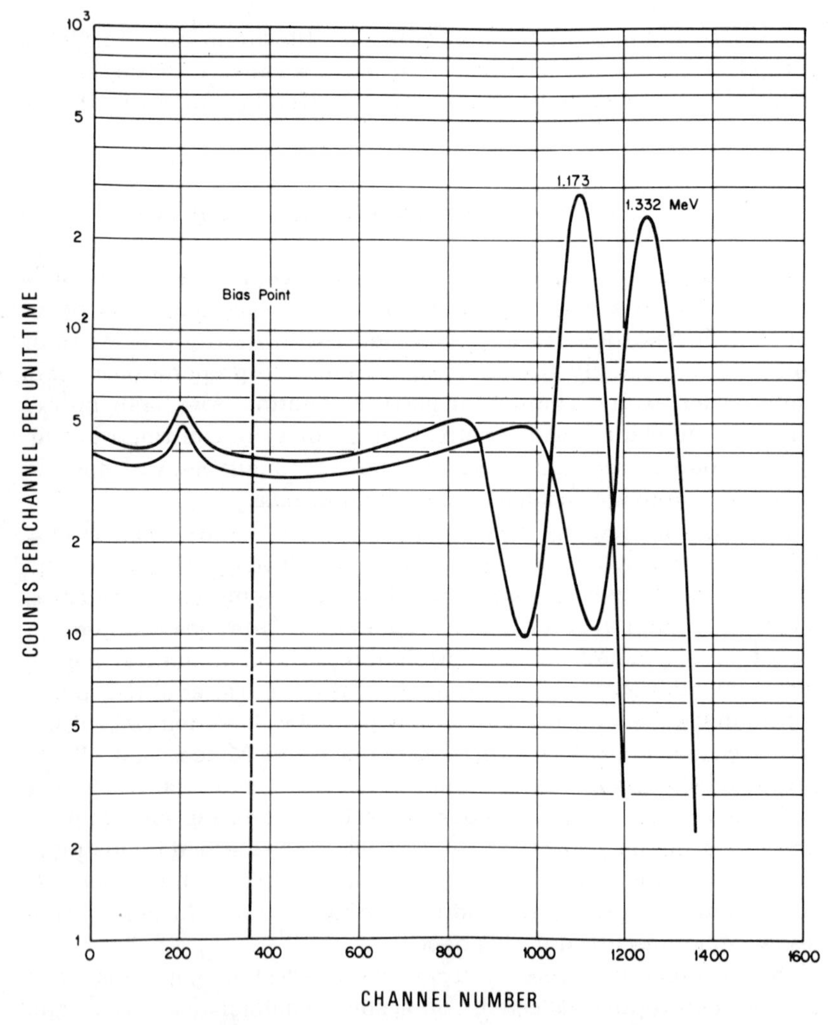

Fig. 26. Calculated pulse-height distributions of 1.173 and 1.332-MeV gamma rays for a NaI(Tl) detector system. When added, they give the distribution produced by ^{60}Co. In γ-γ coincidence counting of ^{60}Co, the integrated areas for the two spectra above the bias point are equalized by adjusting the bias voltage.

Except when two photons differ greatly in energy (e.g., x rays and higher-energy gamma rays), it is impossible to make one detector sensitive only to the first radiation and the other detector sensitive only to the second. In the case of two gamma rays of nearly equal energy, such as occurs in the decay of ^{60}Co, it is convenient, by appropriate setting of the bias in each channel, to adjust the efficiences to be nearly the same for each radiation, as determined by integration under the pulse-height-distribution curves. This can be done, as shown in Figure 26, where subtraction of the corrected pulse-height distribution from the 1.115-MeV gamma ray occurring in the decay of ^{65}Zn, has been used to separate the pulse-height distributions of the two gamma rays due to ^{60}Co. The bias point is chosen so that the areas under the pulse-height-distribution curves, for each counter, are approximately equal. If ϕ_1 is the ratio of the areas for counter 1, and ϕ_2 is the ratio for counter 2, it is easily shown that the coefficient modifying the apparent disintegration rate is

$$F(\phi) = \frac{\phi_1 + \phi_2}{(\phi_1 + 1)(\phi_2 + 1)} \tag{3.26}$$

When the ϕ's are equal to unity the coefficient is 1/2. If, for example, the ϕ's are adjusted to no better than 1.2, then the coefficient is 1/2.02; i.e., if a 20-percent error in setting the bias is made, a 1-percent error in the coefficient occurs. With reasonable care, therefore, good accuracy can be achieved.

The setting of the biases by the above criterion usually results in biases so high that spurious coincidences due to scattered gamma rays will be eliminated, except perhaps in situations where the angular separation of the two detectors with respect to the source is small. A 1-MeV gamma ray when scattered through an angle of 90° to 180° is degraded in energy to about 200 keV.

As in the beta-gamma coincidence method described in Section 3.2, the resolving time of the coincidence mixer circuit must be sufficiently long to avoid any loss of true coincidences.

3.5 4π Proportional Counting

3.5.1 *General*

Two decades ago one of the most widely used methods for the activity standardization of alpha- and beta-emitting nuclides was that of 4π counting. In its heyday the 4π counter was the preferred instrument for the assay of such radionuclides as ^{32}P, ^{35}S, ^{131}I, and ^{241}Am. When a range of values of 1 or 2 percent in international comparisons was considered to be good, many national laboratories reported results obtained by the method of 4π counting (see, for example, ICRU, 1963; Roy and Rytz, 1962, for ^{35}S; and Rytz, 1964, for ^{241}Am). Now, however, its chief use is as a high-efficiency detector in the method of $4\pi\beta$-γ coincidence counting.

As mentioned in Section 1.5.2, the 4π counter is a defined solid-angle counter, with the highest possible geometric efficiency. In the past it has been operated both in the Geiger and proportional regions, but now is used practically only in the latter (see Section 2.2) because the much lower dead times permit the use of high count rates, without large dead-time corrections.

In the method of 4π counting, a radioactive source, on a thin plastic film, is located at the center of symmetry of two detectors, each subtending a solid angle of essentially 2π steradians at the source.

The intrinsic efficiency of a well-designed 4π counter is assumed to be 100 percent for charged-particle radiation (Hawkins *et al.,* 1952; Seliger and Schwebel, 1954). Departures from 100-percent efficiency can be attributed to absorption by the source and the source mount.

Several designs of 4π counters have appeared in the literature, but two very acceptable designs are illustrated in Figures 27 and 28.

The National Bureau of Standards 4π counter, comprised of two stainless-steel hemispheres, is shown in Figure 27. The active source is deposited on the center of a metallized collodion film supported on an annular aluminum foil that divides the counter, when assembled, into two halves. Two loops of 0.001-inch-diameter wire, usually of stainless steel, form the anodes in each hemispherical section. Methane, or a mixture of argon (90 percent) and methane (10 percent), is used as a flow gas at atmospheric pressure.

Sketches of a "pill-box" type of 4π gas-flow proportional counter of a kind used at the National Research Council of Canada and other laboratories are shown in Figure 28. The source mount can be moved in and out of the counter by means of a simple slide, and the design is easily adaptable to use at pressures up to 75 atmospheres (7.6 MPa) (Baerg *et al.,* 1967; Baerg, 1973b), especially for use with $4\pi\beta$-γ coincidence counting systems (Section 3.2). The counter gas is usually a mixture of argon (90 percent) and methane (10 percent), purified by passing over heated granular calcium metal. For most beta-particle sources, the counting-gas pressure is maintained, typically, at 20 atmospheres (2.03 MPa).

As the source mount forms an intergral part of the counter cathode, it must be sufficiently conducting to provide adequate field strength at

Fig. 27. National Bureau of Standards 4π proportional counter; spherical type. (From Mann and Seliger, 1958)

the source and to prevent local accumulation of electric charge. At the same time the mount must be as thin as practical to reduce absorption losses. A common technique is the use of a thin plastic film, which can be made as thin as 1 or 2 $\mu g/cm^2$ (see, for example, Pate and Yaffe, 1955a), rendered conducting by the vacuum evaporation of a suitable metal onto the surface. Gold is commonly used since it is stable to most chemicals and can be readily evaporated; but other elements and alloys have been used. For routine work, a better film-survival rate is obtained with plastic films of 5, or even 10, $\mu g/cm^2$. Film-preparation techniques are described in Section 5.2.1.

The output pulse from proportional counters is such that external

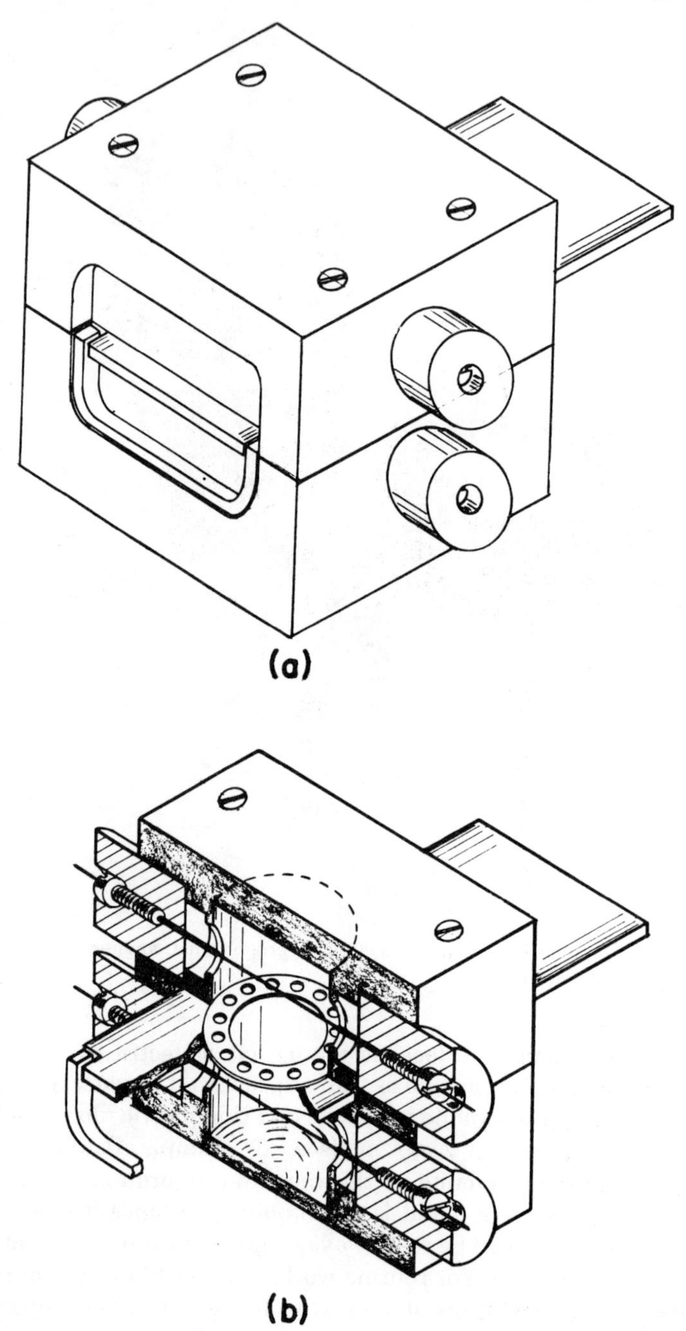

Fig. 28. Sketches of a simple slide-operated 4π counter adaptable to operation at elevated pressures.

amplification is necessary. The pulse height may vary over a considerable range, perhaps three orders of magnitude, and hence, good overload properties may be required for the circuitry. The possibility of overload may be reduced by the use of a low-noise charge-sensitive pre-amplifier which permits the use of lower gas gain. As the noise level of such a pre-amplifier is equivalent to a pulse of about 200 electrons from the detector, the gas gain need not exceed a factor of 500 to detect even single electrons. Commercially available double-differentiating (RC or delay-line) main amplifiers can provide both good overload recovery and adequately short dead times (about 1 μs). If there is a variable dead time present anywhere in the counting system, it is usually overcome by artificially introducing a constant instrumental dead time that is longer than any possibly occurring dead time. It should be noted that for disintegration-rate determinations with, say, a $4\pi\beta$-proportional counter, it is not necessary for the amplifier to be linear, since an indication of an event is all that is required.

The measurement of the activity of a source is normally made by plotting a curve of counting rate either against the potential applied to the counter, keeping the discriminator bias level constant, or against the discriminator bias for a fixed counter potential. (The first alternative must be used with Geiger-Müller counters.) The plateau of a 4π proportional counter may extend for some 200 volts of the applied potential with a beta-particle source, and its slope gives some indication of an increase in detection efficiency with voltage. The rise at the end of the plateau appears to indicate the onset of spurious events (Campion, 1973). With alpha particles the plateau begins at much lower applied voltages, because the ionization produced is usually much greater than for beta particles, but it extends to essentially the same upper limit. The slope may be of the order of 0.1 percent per 100 volts. Operation at the center of the plateau is the usually accepted practice. The counting rate so obtained must be treated for corrections described in Sections 3.5.2 to 3.5.5 below.

3.5.2 *Dead Time*

The dead times of proportional counters are of the order of 1 μs. The subject of dead-time corrections is treated in detail in Section 2.7.

3.5.3 *Background*

The background count rate is readily obtained by using a blank source mount in the detector. The background rate is then subtracted from the observed counting rate *after* corrections for dead-time losses (see Section 2.7.3). Part of the background is due to natural activity in the materials from which the counter is constructed, and a useful survey of the "cleanliness" of such materials is given by Grummitt *et al.,* (1956), DeVoe (1961), ICRU (1963), Weller (1964), and Rodriguez-Pasqués *et al.,* (1972). However, the short dead times of proportional-counter systems permit such high counting rates to be used (with small dead-time losses) that the background-rate correction can be *relatively* small, although not necessarily negligible, when low-level samples are being counted.

3.5.4 *Source-Mount Absorption*

If necessary, a correction for the absorption in the source mount should be made. The analysis of Pate and Yaffe (1955c) is recommended for the most precise work when counting beta particles.

Results obtained by the method of $4\pi\beta$-γ coincidence counting (Section 3.2) now give us insight into the efficiency of the 4π proportional counter under different conditions of source preparation with different radionuclides.

3.5.5 *Source Self-Absorption Correction*

Apart from the method of β-γ coincidence counting (Section 3.2), it is impossible to determine the source self-absorption correction with an accuracy comparable to that for other corrections in 4π counting. The technique of progressively diluting a stock solution until the observed activity expressed in terms of a given mass of the original solution reaches a constant maximum value, merely indicates that no further reduction of source self-absorption is obtainable by dilution, rather than that the correction is zero. Calculated estimates based on the size of the crystals forming the active deposits and on absorption coefficients have been made by Meyer-Schützmeister and Vincent (1952) and Seliger and Schwebel (1954), the latter by electron-shadow micrography of potassium-iodide sources. Pate and Yaffe (1956) and Yaffe and Fishman (1960) have used a vacuum-distillation method to set up self-absorption curves applicable to this source-preparation technique. Carswell and Milsted (1957) have used a spray technique to produce very thin sources, but this technique does not permit of quantitative deposition; it is possible, however, in the case of a gamma-ray emitter to derive the activity per unit mass of solution deposited on the source mount by a gamma-ray comparison with a quantitatively deposited source. Electrospraying techniques have been described by Merritt *et al.* (1959). Special methods of source preparation have also been discussed by Lowenthal and Wyllie (1973a). The preparation of uniform sources with low-self absorption is discussed further in Section 5.2.1. This question is treated in some further detail for alpha, beta, and photon emitters in Section 3.5.6.

3.5.6 *Applications*

Proportional counters with 4π geometry have been used as high-efficiency detectors of alpha particles, beta particles, and electrons, and, at elevated pressures, of x rays. Some instances of such uses are given in the following subsections.

3.5.6.1 *4$\pi\alpha$ Proportional Counting.* Provided that the effective thicknesses of the source and source mount are below the range of the alpha particles in those media, good results can be obtained in the assay of alpha-particle sources. In a 1963 international intercomparison of measurements of ^{241}Am samples, dispensed from the same master solution, 8 of 19 participating laboratories used the method of 4π proportional counting (Rytz, 1964). The range of these eight measurements was 2.5 percent, but if one outlier is removed, the range is 0.85 percent for the remaining seven measurements. The average of measurements made by 12 laboratories using 4$\pi\alpha$-γ coincidence counting was 0.4 percent higher than the average of the 4$\pi\alpha$ measurements, a result indicating $\approx$0.4-percent self-absorption in the latter measurements, and a detection efficiency of $\approx$99.6 percent.

3.5.6.2 *4$\pi\beta$ Proportional Counting.* In the assay of negatron and positron emitters the problems encountered are similar and arise chiefly from source absorption. Auger electrons can also be detected with low efficiency but are usually removed by the interposition of thin metal foils.

Twenty years ago when 4$\pi\beta$ counting was probably the most widely used method in radioactive metrology, the treatment of source self-absorption was largely empirical. In the international intercomparison of ^{35}S measurements, the National Bureau of Standards, for example, relied on preparing large numbers of sources, some 50 in number, and reporting the maximum value on the assumption that the source for this value suffered negligible absorption (Roy and Rytz, 1962). Seliger *et al.* (1958) used 121 different sources to measure the ^{35}S activity in the determinations of the average beta-particle energy in the decay of ^{35}S. An alternative method was that used by Seliger and Schwebel (1954), mentioned earlier, in which electron-shadow micrography was used to estimate the average potassium-iodide-crystal thickness to derive an absorption correction for ^{131}I sources. The state of the art at that time is well exemplified by the papers of Pate and Yaffe (1955a, b, c, and d; 1956).

With the advent of $4\pi\beta$-γ coincidence counting, a powerful tool was made available whereby the value of ϵ_β, the total efficiency for detecting beta particles, could be simply derived from the ratio of the beta-channel count rate, N_β, to the disintegration rate, N_0. Gunnink *et al.* (1959) and Merritt *et al.* (1959) used the method to measure the source self-absorption for beta particles emitted from several radionuclides. The latter authors found that, for sources having a mean superficial density of 2 μg cm^{-2}, the source self-absorption varied from about 1 percent for ^{24}Na beta-particles (β_{max} = 1.39 MeV) to between 20 and 40 percent for the 89-keV beta transition in ^{134}Cs, depending on the method of source preparation. An appreciable decrease in self-absorption was found by using spreading agents (e.g., insulin as a wetting agent, and colloidal silica as a seeding agent in the crystallization process), particularly for the less energetic beta-particle emitters. (See also Section 5.2.1.)

By incorporating two radionuclides into the same chemical compound, one a beta emitter and the other a previously standardized beta-gamma emitter, Merritt *et al.*, (1960) have used $4\pi\beta$-γ coincidence counting to determine the source self-absorption for ^{35}S. (See also Section 3.2 and Campion *et al.*, 1960.) Subsequently, this variant of the efficiency-tracing method was extended by Baerg *et al.* (1964) to two radionuclides that were not amenable to formation of a single chemical compound. The mixed efficiency-tracing method was successfully applied by Baerg and Bowes (1971), Merritt and Taylor (1971), and by Lowenthal *et al.* (1973) to intercomparative measurements of ^{63}Ni that had been standardized at the National Bureau of Standards by microcalorimetry (Barnes *et al.*, 1971).

3.5.6.3 *$4\pi x$ Proportional Counting.* A 4π-proportional counter, filled with a mixture of 90-percent argon and 10 percent-methane at pressures ranging from 1 to 3.5 atmospheres, was used by Allen (1957) to measure the emission rates of x rays from electron-capturing nuclides. By plotting count rate as a function of reciprocal pressure, and then extrapolating to zero reciprocal pressure, he was able to determine the x-ray-emission rate for infinite pressure, for which the counter was assumed to be 100-percent efficient. From the count-rate reciprocal-pressure curves the *K*-x-ray detection efficiency could then be deduced for one atmosphere. Allen measured *K*-x-ray-emission rates for a number of electron-capturing nuclides, absorbing out residual Auger electrons with about 200 μg cm^{-2} of aluminum foil, and he also measured, and corrected for, the absorption of x rays in these foils.

From a knowledge of the *K*-shell fluorescence yield (Figure 29), the *K*-capture rate can be calculated. Theoretical or experimental values of the *K/L, K/M,* etc., capture ratios can then be used to calculate the total electron-capture activity.

Garfinkel and Hutchinson (1962) have also used a pressurized 4πx counter to measure source self-absorption in the determination of *K*-x-ray-emission rates from radionuclides that decay by electron capture to the ground state. In this method, in which the count rates in the two halves of the 4π counter are measured separately, Auger electrons are used as a tracer in order to determine the *K*-x-ray absorption to ±0.3 percent for ^{55}Fe sources. Since the Auger-electron efficiency is a much more sensitive function of source thickness than *K*-x-ray efficiency, a rough measure of the fraction of absorbed Auger electrons is shown to give an accurate measure of the fraction of absorbed *K* x rays.

3.5.6.4 *4πe⁻ Proportional Counting.* The use of a 4π counter in the measurement of atomic-electron radiations is usually confined to that of internal-conversion electrons. In this case the problems of source self-absorption are not normally so great as there is no low-energy component of conversion electrons. It is often desirable, however, to obtain the complete electron spectrum, as, for example, in the case of ^{109}Cd, where the x rays and Auger electrons arising from electron capture are not in coincidence with the conversion electrons arising from the decay of ^{109m}Ag.

Legrand *et al.* (1973) have constructed a 4π proportional counter capable of operating up to a pressure of 70 atmospheres, and have used it to measure the spectra of conversion electrons arising in the decay

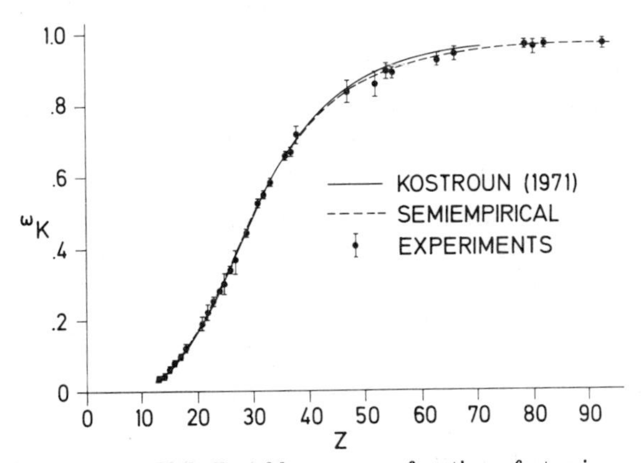

Fig. 29. Fluorescence *K*-shell yields, ω_K, as a function of atomic number, *Z*: (a) according to Kostroun *et al.* (1971); (b) a best fit to selected experimental data; and (c) critically evaluated experimental results. (From Bambynek *et al.,* 1972)

of ^{99m}Tc, ^{109}Cd, and ^{139}Ce. The sum of events under the K and L internal-conversion peaks in a given time, together with a knowledge of the K and L internal-conversion coefficients, the K: L: M ratio, and the probability of the given gamma-ray transition per decay, enables the activity of the parent radionuclide to be determined. Internal-conversion probabilities, as given in Appendix A.1.5.2, will, however, give the activity directly.

Another approach to the same problem is to construct a large 4π counter operating at atmospheric pressure. This was adopted by Troughton (1977) who constructed a cylindrical 4π counter having a diameter of 30 cm. She obtained a ^{109}Cd-^{109m}Ag spectrum, shown in Figure 30, in which the "valley" between the x-ray and K-conversion-electron peaks dropped to practically zero count rate, and was thus able to determine the number of events under the K and $L + M$, etc. peaks with no significant contribution from the x-ray peak. This slight difference from the spectrum of Legrand *et al.* (1973), where the count rate in the valley is still appreciable, may be due to some subtle difference in design, or mode of operation, or size of source mount.

3.5.7 *Further Reading*

For further information on selected topics of 4π proportional counting, reference should be made to the proceedings of the Herceg Novi (1973) conference.

3.6 Internal Gas Counting

The disintegration rate of low-energy beta-emitting, or electron-capturing, radionuclides that are in, or can be compounded into, gaseous form may be determined by the direct method of internal gas counting. In this method the radioactive gas is introduced, together with a suitable counting gas, into a counter operating in either the proportional or Geiger regions. Such counters are usually comprised of a cylindrical cathode with an anode of wire (of the order of 0.001 in. in diameter) stretched along the axis of the cylinder. (For assay of radon see Section 6.1.2.3.)

The use of such counters of known physical dimensions was first reported by Miller (1947) and they were used by W. F. Libby and his colleagues (Anderson *et al.,* 1947) for determining the activity of ^{14}C in the environment. Measurements using such counters, however, required that corrections be made for wall and end effects (see, for

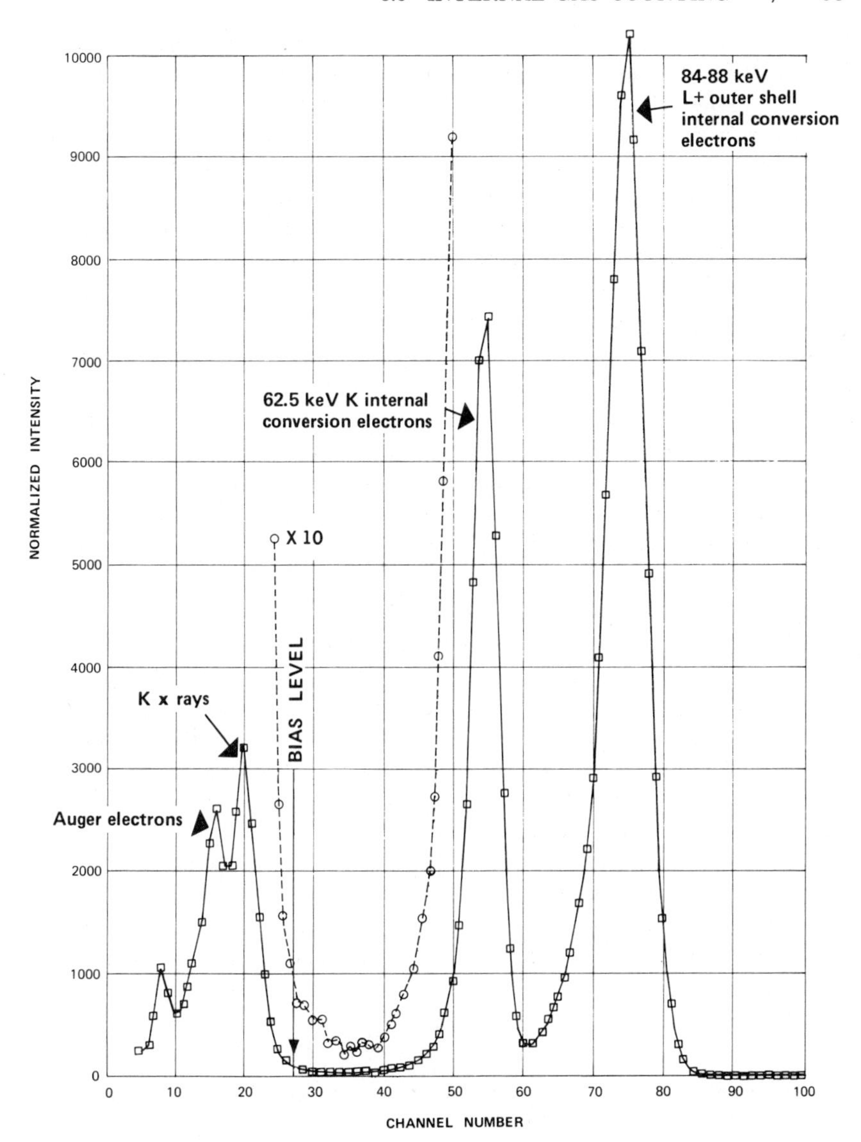

Fig. 30. ^{109}Cd-^{109m}Ag conversion-electron spectrum taken in a 30-cm-diameter cylindrical 4π gas-flow proportional counter. (From Troughton, 1977)

example, Engelkemeir and Libby, 1950), due to electrons entering the wall without the creation of an ion pair, to the decrease in electric field intensity at the ends, and, in cases where the anode was longer than the cathode, to electrons entering the counting volume from the ends. The activities of the gaseous contents could then be calculated from a knowledge of the count rate, the effective volume of the counter, and the pressure, temperature, and composition of the counting gas.

For the purpose of standardizing gases or gaseous compounds, Mann and Parkinson (1949) used a length-compensated gas-counter system. This consisted of two counters of precisely similar dimensions except for the cathode and anode lengths. To the best precision possible, the end supports for the anode wire and the cathode diameters are made the same. The difference in counting rates then gives the count rate for an ideal volume of gas contained in a cathode cyclinder equal to the difference in lengths of the two counters.

Wall effects are determined by measuring the count rates at several pressures and then extrapolating the observed count rate as a function of reciprocal pressure to zero reciprocal pressure (i.e., to infinite pressure). A correction for undetected low-energy events can be made either by a discriminator-voltage extrapolation to zero voltage or by the extrapolation of the electron, or electron plus x-ray, spectrum to zero energy. In both cases, data are taken above the level of electronic noise.

Multiwire-cathode internal gas counters with surrounding integral cylindrical ring counters in the anticoincidence mode were used by Drever *et al.* (1957) to reduce wall effects, and, recently, for the accurate measurement of electron-capture intensity ratios (Chew *et al.,* 1974a). This type of gas counter consists of a cylindrical central counter with multiwire cathode, surrounded by a ring of counters, separated from the central counter only by the multiwire cathode. The ring counters are operated in anticoincidence with the central counter, so that decay events that trigger both a ring counter and the central counter are not registered.

The pulse height spectrum recorded from such a gas counter is a much improved representation of the energy distribution between the L and K electron-capture decay events, and accurate electron-capture intensity ratios may be determined from such spectra. Corrections must, however, be made both for x rays that escape from the central counter, and for those that enter from the ring counters or end zones (Vatai, 1970; Chew *et al.,* 1974a, 1974b).

Internal gas counting is essentially a 4π method that avoids the problem of source self-absorption. A description of the method was given at the Herceg Novi International Summer School (Garfinkel *et al.,* 1973b), and has been extensively studied by Spernol and Denecke (1964a, 1964b, and 1964c). A photograph of the National Bureau of Standards length-compensated gas-counting system, consisting of two sets of three counters, one set of copper and one of stainless steel, is shown in Figure 31. The counting data from the six counters are

Fig. 31. The NBS stainless-steel and copper length-compensated internal gas counters mounted in their base with vacuum-chamber cover and one counter removed. An automatic Töppler pump for mixing the radioactive gas sample with a suitable counting gas can be seen just behind the counters. A flask containing the radioactive gas sample can be seen in the temperature equilibrating box above the steel safe, and two mercury barometers for measuring the sample pressure and counting-gas pressure are to the left. M. P. Unterweger and J. E. Harding have recently modified the system so that cryogenic mixing and transfer can be used as an alternative to mixing and transfer by means of the Töppler pump.

collected simultaneously and processed on a minicomputer, operating, when required, as a multichannel analyzer (Garfinkel *et al.*, 1973b).

Examples of NBS standards produced by this method are ^{3}H, ^{14}C (as CO_2), ^{37}Ar, ^{85}Kr, ^{131m}Xe, and ^{133}Xe.

3.7 Alpha-Particle Counting

3.7.1 *Introduction*

The use of 4π-proportional counters for alpha-particle counting has been treated in Section 3.5, pulse-ionization chambers and ZnS(Ag)-coated scintillation cells in Section 6.1.2.3, silicon surface-barrier detectors and diffused-junction detectors in Section 4.5.3, and liquid-scintillation counters in Sections 3.8, 4.3.8, and 6.1.2.2. The assay of alpha-particle-emitting sources by the methods of 2π proportional and defined-solid-angle counting is the subject of this section. Methods for the calibration of alpha-particle sources by defined-solid-angle counting have been reviewed by Brouns (1968).

3.7.2 *Standardization by 2π Proportional Counting*

The measurement of alpha-emitting radionuclides in 2π geometry requires essentially the measurement with one half of a 4π system, and all considerations in 4π proportional counting apply here equally well.

The source may, however, be electrodeposited or evaporated onto a polished-metal backing that behaves as a semi-infinite medium. Alpha particles, being more massive than electrons, are less easily scattered from the backing, so that the fraction of the alpha particles scattered into the 2π counter is far less than in the case of beta particles. Also, alpha particles are emitted in monoenergetic groups with much higher energies than beta particles, which are characteristically emitted in a spectrum of energies.

Nevertheless, to obtain accuracies comparable to those obtained in 4π counting, scattering and also source self-absorption must be evaluated. In general, such an evaluation can be readily made only for sources of uniform and known thickness.

3.7.3 *Backscattering*

Backscattering, in this section, refers specifically to the effect in which alpha particles, initially moving in the direction of the source mount, are deflected into the sensitive volume of the 2π detector. Many investigators [e.g., Deruytter (1962); Hutchinson *et al.* (1968)] have concluded that the number of backscattered alpha particles is between zero and five percent of the number emitted into 2π steradians, depending on the thickness and uniformity of the source, the atomic number of the mount and source material, and the energy of the alpha particles.

Walker (1965) and others have shown that backscattered alpha particles move predominantly into angles of less than 5° from the plane of the source mount, as a result of *multiple* small-angle scattering primarily due to the nuclear electrostatic field. Large-angle Rutherford scattering occurs in the order of a few times in 10^5 emitted alpha particles, and is, therefore, negligible for almost all 2π counting.

Backscattering increases with increasing atomic number, *Z*, of the scattering material, as demonstrated in Figure 32, for which 2 μg cm^{-2} collodion films impregnated with ^{210}Po were placed on polished source mounts of a number of materials. The backscattered fraction, f_b, of the total for "weightless" sources is given by:

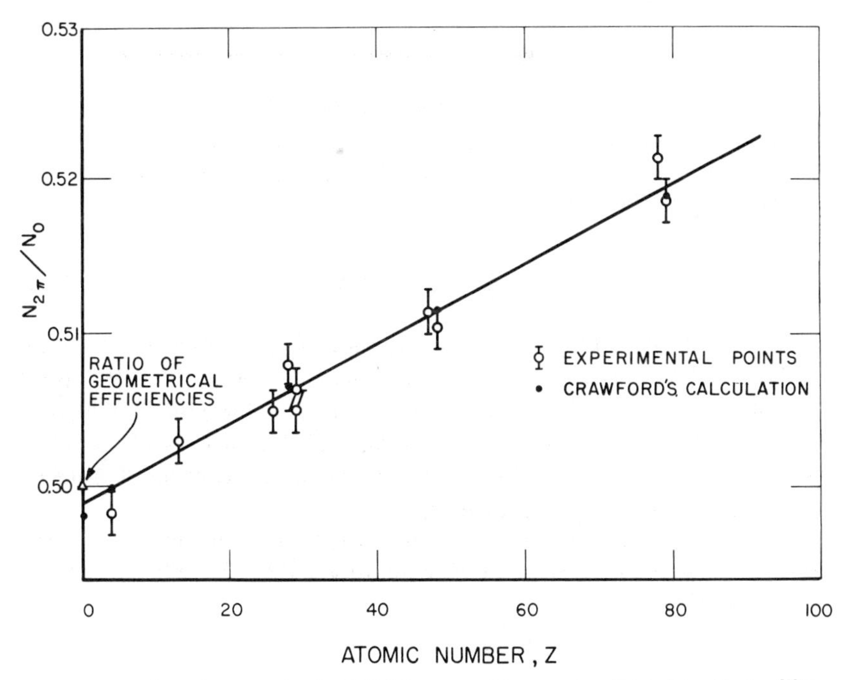

Fig. 32. Counting rate in the NBS 2π α-particle counter divided by N_0 for ^{210}Po sources impregnated in collodion on polished mounts plotted against *Z*, the atomic number of the backing material. The error bars represent plus and minus one standard error. The solid points are obtained from a relation developed by Crawford (1949). (From Hutchinson *et al.*, 1968)

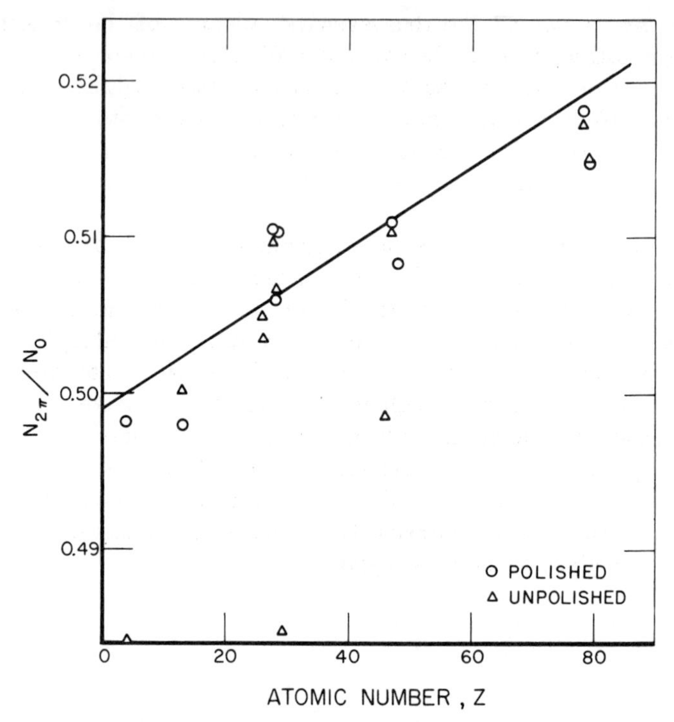

Fig. 33. Counting rate in the NBS 2π α-particle counter divided by N_0, for ^{210}Po sources adsorbed onto polished and unpolished mounts plotted against Z, the atomic number of the backing material.

$$f_b = \frac{N_{2\pi}}{N_0} - 0.5 \tag{3.27}$$

where $N_{2\pi}$ is the alpha-particle count rate in 2π geometry, and N_0 is the total alpha-particle-emission rate. Figure 33 shows the experimental values of $N_{2\pi}/N_0$ plotted as a function of Z for sources adsorbed onto polished and unpolished mounts, and the large scatter of the points demonstrates the often great effect of even minor surface irregularities on the backscattering.

It has also been shown (Crawford, 1949; Hutchinson *et al.*, 1968) that backscattering varies approximately as $E^{-1/2}$, where E is the alpha-particle energy.

3.7.4 *Source Self-Absorption*

The absorption of alpha particles by the source is primarily a function of the thickness, d_s, of the source and the range, R_s, of the alpha particles in the source material. Thus, several authors [Robinson, H. (1960); Friedlander *et al.* (1964); Gold *et al.* (1968); White (1970)] have shown, both theoretically and experimentally, that for a uniformly thick source, the fraction of alpha particles absorbed by the source increases proportionately to $d_s/2R_s$ for $d_s < R_s$. Figure 34 demonstrates this dependence, which can be calculated from simple geometrical considerations and the assumptions that alpha particles have a well defined range and are not deflected from their initial direction in the source material.

3.7.5 *Calculation of Scattering and Source Self-Absorption*

These last assumptions (Section 3.7.4) represent only first order approximations, and scattering would not occur if they were strictly true.

An expression for $N_{2\pi}/N_0$, for the source configuration shown in Figure 35, is given in Eq. 3.28. A non-radioactive layer of absorber a, thickness d_a, covers an alpha-particle emitting source s, thickness d_s, deposited on a semi-infinite backing material b. Then

$$\frac{N_{2\pi}}{N_0} = \frac{1}{2}\left(1 - \frac{d_a}{R_a} - \frac{d_s}{2R_s}\right) + f_s, \tag{3.28}$$

where R_a and R_s are the alpha-particle ranges in absorber and source, respectively, and f_s is the scattered fraction. For "weightless" sources and no absorbing material, f_s reduces to f_b, which is the backscattering due to the backing material. These results have been discussed in

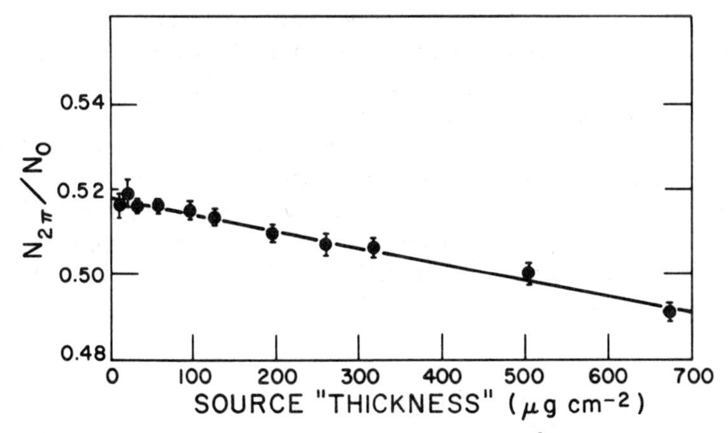

Fig. 34. $N_{2\pi}/N_0$ as a function of surface density in $\mu g\ cm^{-2}$ for UF_4 sources. Error bars are representative of one standard deviation. (From Gold *et al.*, 1968)

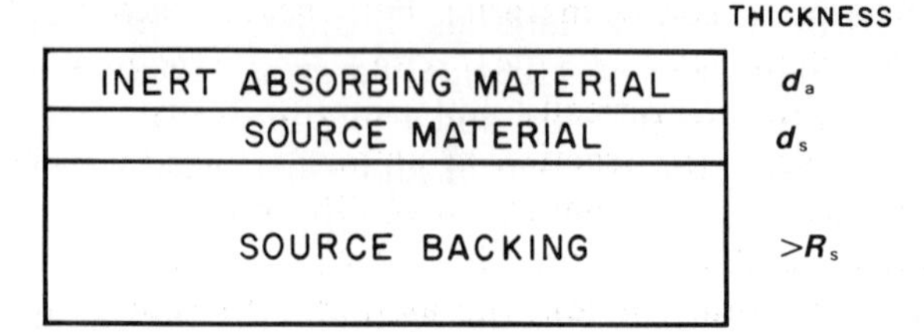

Fig. 35. Schematic illustration of a covered, deposited, alpha-particle source.

Lucas and Hutchinson (1976) and Hutchinson *et al.* (1976a). Figure 36 shows their experimental values and the theoretical curve for $N_{2\pi}/N_0$ *vs.* d_s/R_s for sources of $^{235}UO_2$ deposited on polished platinum.

3.7.6 *The Robinson Counter*

Robinson, H. (1960) has developed a counter, Figure 37, with an accurately defined geometrical efficiency close to 0.8π steradians, for alpha particles emitted in a direction normal to the plane of the source. Multiply scattered alpha particles, essentially all of which are emitted at grazing angles to the source, are not detected. The difficult problem of estimating the backscattering contribution, encountered in 2π counting, is thus avoided. The detector is usually of scintillating plastic, CaF(Eu), ZnS, or CsI(Tl) (Hutchinson *et al.,* 1968), and a baffle at its center is so designed that variations in source positioning of as much as 1 cm result in negligible changes in the overall detection efficiency. Consequently, the counter can be used to measure relatively large low-activity sources with accuracies of approximately 0.5 percent. In this geometry less than 0.2 percent of the counts are lost due to source self-absorption perpendicular to the source, if its thickness is less than 0.25 R_s. Such counters, during operation, are usually evacuated or filled with hydrogen at atmospheric pressure.

The detection efficiency, $\varepsilon_{0.8\pi}$, of the 0.8π counter was deduced from measurements on thin sources with the 2π and 0.8π counters for different values of atomic number, Z, of the backing material (Hutchinson *et al.*, 1968). The 2π count rate, $N_{2\pi}$, divided by the 0.8π count rate, $N_{0.8\pi}$, was plotted as a function of Z, and extrapolated to Z=0 where, theoretically, $\varepsilon_{0.8\pi} = N_{0.8\pi}/2N_{2\pi}$.

3.7.7 *Low-Solid-Angle Counters*

As mentioned in Section 1.5.2, the defined low-solid-angle counter, although often just used for relative measurements, has proved to be

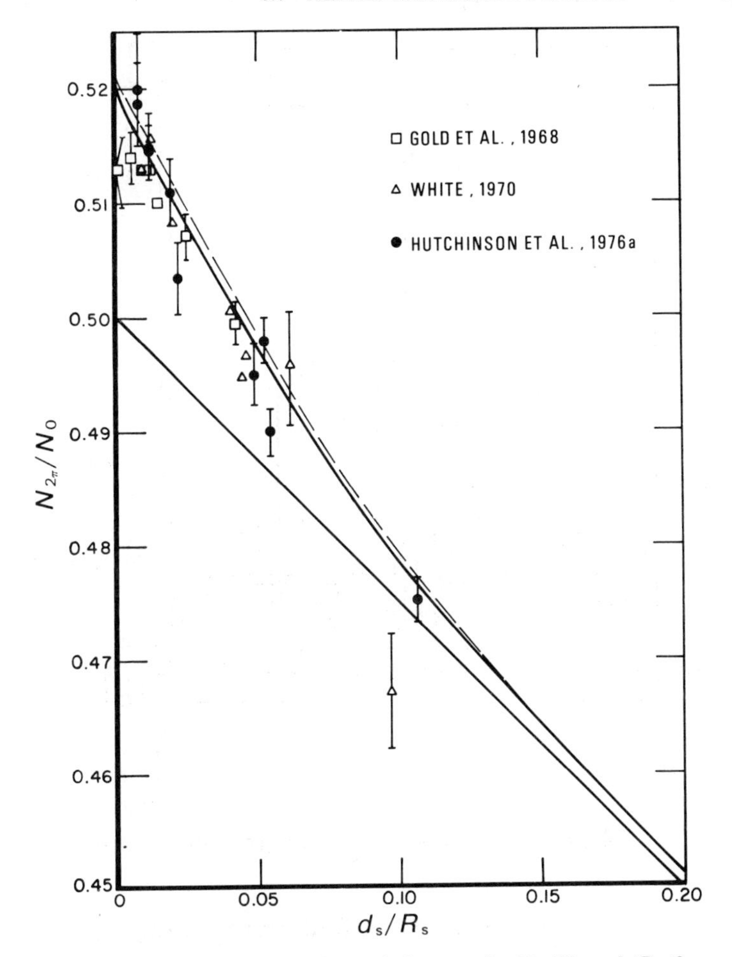

Fig. 36. Experimental values and theoretical curves for $N_{2\pi}/N_0$ *vs* d_s/R_s, the source thickness expressed as a fraction of the range. The diagonal straight line corresponds to no scattering. The lower heavy curved line was calculated for $^{235}UO_2$ and the upper dashed curved line was calculated for a source of $^{238}UO_2$, both on polished platinum. Error bars represent plus and minus one standard error.

powerful as a tool for the measurement of alpha-particle-emitting radionuclides. Such a counter consists essentially of a detector, behind a diaphragm of known area, and a source mount, designed so that the source may be positioned at known distances from the diaphragm. The silicon surface-barrier-detector system, described in Section 4.5.3, is an example of such a system for *relative* measurements in which the unknown area of the detector replaces the diaphragm. Examples of low-solid-angle counters have been described by Zumwalt (1950), Jaffey (1954), and others.

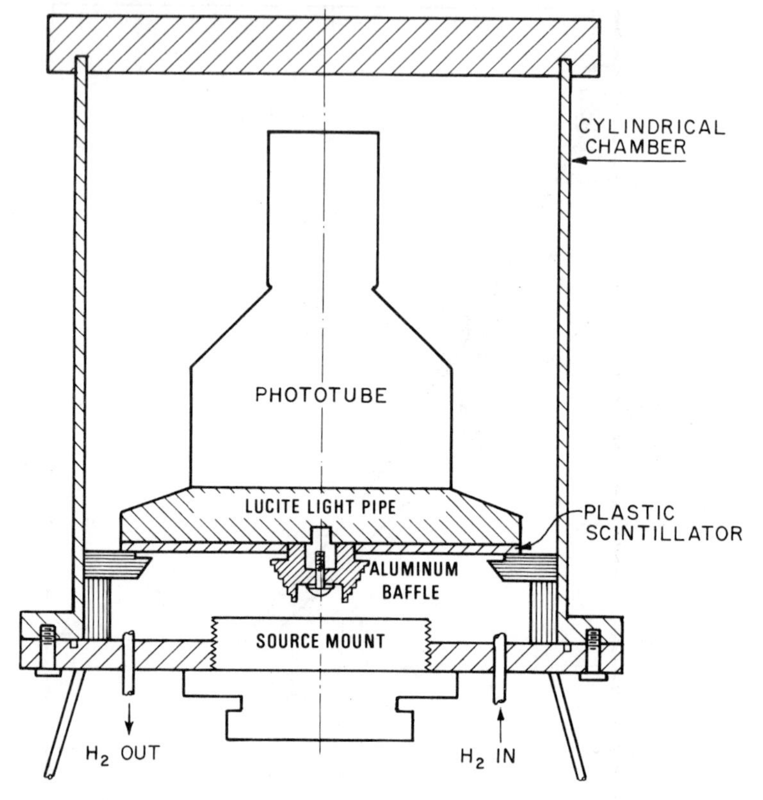

Fig. 37. Schematic diagram of the NBS Robinson counter.

3.8 Liquid-Scintillation Counting

3.8.1 *Detection Sensitivity*

The physical basis for the method of liquid-scintillation counting is discussed in Section 2.4.5. While liquid-scintillation counting is useful in making relative measurements between samples of the same beta-particle emitters of all energies, and the direct measurement of alpha-particle emitters, it does not compare to gas counting when it comes to the direct assay of beta-particle emitters of low energy, as the average energy that must be deposited in the scintillator to result in the ejection of one electron from the photocathode of the phototube is of the order of 2 keV, compared with the approximately 35 eV required to produce one ion pair in a gas.

This lack of sensitivity to low-energy radiations does not present a problem in the measurement of monoenergetic alpha particles, for which the detection efficiency is generally 100 percent, if losses due to all "wall effects" can be neglected. An early example of this method is the work of Basson and Steyn (1954) in standardizing ^{210}Po in solution.

A major disadvantage of alpha-particle liquid-scintillation counting has been that of poor energy resolution. With most commercial liquid-scintillation counters it is difficult to resolve the peaks of the alpha particles of interest from contributions due to impurities or beta-particle-emitting daughters. McDowell (1975), however, has reported recently on a system that has much improved resolution, and he (*loc. cit.*) and McBeth *et al.* (1971) have used the pulse shape to discriminate against unwanted beta-particle pulses.

Liquid-scintillation counting is frequently used for the direct assay of high-energy beta emitters such as ^{32}P and ^{90}Y, but many low-energy beta particles escape detection. The amplitudes of output pulses from the phototube will therefore give a spectrum, from which an integral discrimination curve may be obtained that, at low energies, will fall below that expected from the true beta-particle spectrum. For high-energy beta emitters, the proportion of such low-energy beta particles escaping detection may be sufficiently small that a simple linear extrapolation of the plot of integral-count-rate against discriminator-voltage, to zero discriminator volts, will give a value of the activity to well within the accuracy required. To obtain a better fit for lower-energy beta emitters, such as ^{3}H, Flynn *et al.* (1971) have adopted an exponential decrease in count rate with decreasing energy in extrapolating to zero energy.

3.8.2 *The Non-Detection Probability*

For radionuclides of low and intermediate average beta-particle energy, it is necessary to consider the distribution of pulses from the phototube arising from decay events in the scintillator in the range of 0 to 10 keV. Because there is a cut-off or threshold to the sensitivity of the system in the region of 1 to 2 keV (corresponding to the emission of only one electron from the photocathode), it is necessary to calculate the non-detection probability in this region in order to arrive at a more accurate value of the radioactivity concentration of the beta emitter in the scintillator.

In efforts to extend the liquid-scintillation-counting techniques to the assay of lower-energy radionuclides such as ^{3}H and ^{55}Fe, some workers (Gibson and Marshall, 1972; Houtermans, 1973) have calculated the non-detection probability defined as the fraction of decays that produce zero photoelectrons at the photocathode of the phototube. The statistical calculations used in arriving at this probability in the case of ^3H are given in these references.

The first step in these methods is to determine a distribution of output pulses from the phototube due to single photoelectrons (so-called "monos") being emitted from the photocathode. This distribution can be determined experimentally by exposing the phototube to a low-intensity source of light from an incandescent lamp or light-emitting diode. The output distribution corresponding to two or more photoelectrons being simultaneously emitted from the photocathode can be constructed from the "mono" distribution, as shown by Houtermans (1973).

In the liquid-scintillation counting of low-energy events, errors may arise due to afterpulsing caused by delayed light emission or by effects in the phototube (Williams and Smith, 1973).

3.8.3 *Coincidence-Counting Techniques*

Liquid-scintillation detectors have also been used for the x-ray, alpha-particle, and electron-detection channels for 4π coincidence-counting systems (Section 3.2 and Steyn, 1973). Several participants in an international intercomparison of ^{241}Am, organized by the Bureau International des Poids et Mesures (BIPM) in 1963 (Rytz, 1964), successfully used liquid-scintillation counting for both direct alpha and alpha-gamma coincidence measurements. A good example of more recent work is the $4\pi(LS)\beta$-γ measurement of radioactive concentrations of ^{134}Cs solutions by Smith and Stuart (1975).

Efficiency-tracing techniques (Section 3.2.11) employing liquid-scintillation detectors have also been reported; see, for example, the work of Jones and NcNair (1970) in which ^{14}C and ^{35}S are standardized using ^{60}Co and ^{95}Nb, respectively, as tracers.

An extensive list of references concerning the use of liquid-scintillation detectors for high-precision activity measurements is given in Vaninbroukx and Stanef (1973).

3.9 Calorimetry

The calorimetric method has been described earlier in this report as a "near-direct" method in that, in order to determine radioactivity from the rate of energy emission from a radionuclide, it is necessary to know its average energy per disintegration. In the case of measurements of radium and its daughters, calorimetric measurements are, however, relative.

The case of radium is probably unique in that radium standards are not radioactivity standards but are *defined* mass standards, usually in milligrams of ^{226}Ra content as of a certain date, and are analogous to the *defined* standard of mass, the kilogram. All other radium standards, in varying degrees of known equilibrium with their daughters, are measured relative to the international radium standards by a comparison of some measurable quantity such as gamma-ray emission rates, or rates of energy emission.

Many calorimeters utilize two similar receptacles (twin-cup calorimeters) in which various radioactive sources can be placed, the heat generated being balanced by electric heating coils within, or wound around, the receptacles. In some cases, the rates of energy emission from the radioactive sources are balanced by passing an appropriate electric current in such a direction as to cool Peltier junctions attached to the cups of a twin-cup calorimeter. The Bunsen ice calorimeter can be used to measure the rate of energy emission, of the order of 1 watt, from more active sources. The low specific heat of materials at low temperatures ($<$4.2K) permits measurement of small amounts of radioactive materials in calorimeters designed for operation in this temperature range by the Centre d'Etudes Nucleaires (LMRI, 1975).

Calorimetric, or microcalorimetric, methods have been used in the measurement of radioactive effects since the experiments of Curie and Laborde in 1903. It is of interest to note that the historic calorimetric measurement of the rate of energy emission from radium E by C. D. Ellis and W. A. Wooster led to the postulation of the neutrino by W. Pauli in 1931. A quite comprehensive survey of the measurement of radioactivity by calorimetric methods, with general references covering the years 1903 through 1958, was prepared for the Encyclopaedic Dictionary of Physics (Mann, 1962). Very comprehensive reviews of work subsequent to 1949 have been written by Gunn (1964, 1970), and recent calorimetric applications were described at the Herceg Novi International Summer School (Ramthun, 1973; Mann, 1973).

Microcalorimetry has been used at the National Bureau of Standards both for the intercomparison of national radium standards and for the standardization of artificially produced radionuclides and ^{210}Po. A precision of measurement of about 0.1 to 0.2 percent is normally attainable. The twin-cup compensated microcalorimeters that were used to make these measurements (Mann, 1954a, 1954b) employ the Peltier-cooling effect to balance the energy emission of the radioactive source.

The National Bureau of Standards radiation balance, as this twin microcalorimeter has been called, has been used to prepare standards of ^{3}H (Mann *et al.*, 1964), ^{35}S (Seliger *et al.*, 1958), and ^{63}Ni, together with a half-life measurement using mass spectrometry (Barnes *et al.*, 1971), and to calibrate a ^{210}Po source in connection with the measurement of the branching ratio in the decay of ^{210}Po (Hayward *et al.*, 1955). A knowledge of this ratio, namely 1.21×10^{-5} gamma quantum per alpha particle, enables a strong ^{210}Po source to be assayed simply by measurement of the number of 0.804-MeV gamma rays from the daughter nucleus, ^{206}Pb.

NBS solution standards prepared from the calorimetrically assayed ^{63}Ni were also measured by the method of efficiency tracing (see Section 3.2.7), by the National Research Council of Canada (Baerg and Bowes, 1971), and Atomic Energy of Canada Limited (Merritt and Taylor, 1971), and by the Australian Atomic Energy Commission Research Establishment (Lowenthal *et al.*, 1973) with complete agreement between two laboratories, and a total range of 2.7 percent in the measurements (Lowenthal *et al.*, 1973).

The NBS radiation balance was also used to calibrate a source of radium chloride in the preparation of radium solution standards in the ranges 10^{-6} g of ^{226}Ra per 5 ml, and 10^{-11} g of ^{226}Ra per 100 ml, of carrier solution (Mann *et al.*, 1959). In this calibration, reproducibility within the range of 0.004 mg in 6.106 mg in the mass of ^{226}Ra was obtained. The name "radiation balance" is based on the fact that a constant source of power in one cup, either radioactive or electrical, can be balanced against one or more similar sources in the other cup of the twin-cup microcalorimeter. An apt example of this aspect of the potential of the NBS radiation balance was its use in calibrating, relative to the NBS 38-mg Hönigschmid radium standard, the 100-mg radium source used by Attix and Ritz (1957) in their measurement of the gamma-ray constant of radium. The Hönigschmid standard was balanced first against a 65-mg radium source which, in turn, was balanced against the 100-mg source.

With the advent of nuclear power, calorimetry is also used to assay plutonium derived from spent nuclear fuel. Calorimeters of the type developed by Eichelberger *et al.* (1954), from the original design of Rutherford and Robinson (1904), have been widely used to assay such special nuclear materials, and recommended procedures for making such measurements have been published by the American National Standards Institute (ANSI, 1975).

Ditmars (1976), using the NBS Bunsen ice calorimeter, has calibrated approximately 0.23-watt and 1.4-watt sealed sources of special nuclear material, comprised chiefly of ^{238}Pu, but also containing ^{239}Pu, ^{240}Pu, ^{241}Pu, ^{242}Pu, and ^{241}Am. To assay the individual nuclides, the abundances of americium and each of the plutonium isotopes present in such special nuclear materials must be determined by either mass spectrometry or gamma-ray spectrometry (ANSI, 1975). A number of half-life and specific-power measurements utilizing the calorimetric method are also referenced in ANSI (1975).

3.10 Mass Spectrometry

The measurement of radioactivity by mass spectrometry exploits the radioactive decay law, $dN/dt = -\lambda N$. If λ is known and N is measured, then the activity, dN/dt, may be determined. Conversely, the half-life of a long-lived radionuclide can be determined, if dN/dt and N are measured. Examples are the measurements of the half-lives of ^{14}C and ^{63}Ni, respectively 5730 years and 96 years, which are too long to be determined by direct observation. In the case of ^{14}C, the activity was determined by internal gas counting (Section 3.6), and the abundance of ^{14}C relative to ^{12}C and ^{13}C in the quantitative carbon-dioxide filling in the gas counters was determined by mass spectrometry (Watt *et al.*, 1961; Mann *et al.*, 1961; Olsson *et al.*, 1962; Hughes and Mann, 1964). The activity of the ^{63}Ni, in the form of a small 1.3-g cylinder of nickel containing about 1.3 Ci of ^{63}Ni, was determined by microcalorimetry (Section 3.9) The rate of energy emission was about 133 μW, and the activity was calculated using a computed average energy of 17.23 keV per disintegration. The nickel cylinder was then dissolved and the abundance of ^{63}Ni relative to the stable isotopes of nickel was determined mass-spectrometrically (Barnes *et al.*, 1971). From these measurements, both radioactivity standards of ^{63}Ni and a value for its half-life were obtained.

Mass spectrometry has been used to measure the half lives of ^{3}H by measuring the rate of growth of ^{3}He (Jones, 1955); of ^{81}Kr, in conjunction with an isotope separator and the measurement of activity (Eastwood *et al.*, 1964); of ^{129}I together with activity measurements (Katcoff *et al.*, 1951); of ^{134}Cs and ^{137}Cs by measuring the number of ^{134}Cs and ^{137}Cs atoms as a function of time over a period of 2 and 11 years, respectively, without activity measurements (Dietz and Pachucki, 1973); and of ^{239}Pu by several laboratories (see papers in Lucas and Mann, 1978).

"Burn-up standards" of ^{137}Cs have been prepared at the National Bureau of Standards by use of mass spectrometry, by the method of mass-spectrometric isotope dilution using ^{133}Cs. These standards are calibrated for concentration, in terms of ^{137}Cs atoms per gram of solution, for the measurement of burn-up of nuclear fuels. Results are quoted with an estimated overall uncertainty of ± 0.5 percent, and a standard error at the 99 percent confidence level of 0.1 percent.

3.11 Standardization of Gamma-Ray Emitters by Dosimetry Measurements

3.11.1 *Measurements by Ionization Chambers and Sources Calibrated in Terms of Exposure*

A useful relationship exists between the activity of a radionuclide source and the exposure rate at some stated distance. In order to derive this relationship, we note that the exposure rate (due to a gamma-ray fluence), at a given point in space, is the rate at which gamma-ray energy is transformed into ionization in free air, per unit mass of air.[4]

Consider a source with activity A nuclear transitions per unit time. Let P_i be the fraction of the nuclear transitions which give rise to photons of energy E_i. Then the gamma-ray energy fluence at distance l from the source, due to photons of energy E_i, is $A\ P_i\ E_i/4\pi l^2$. The rate at which the energy fluence is transformed into kinetic energy of moving electrons, per unit volume of air, is obtained by multiplying the energy fluence by the linear energy-absorption coefficient in air, $\mu_{en,i}$. Kinetic energy per unit volume is converted into kinetic energy per unit mass of air by dividing by ρ, the density of air. Finally we convert to the rate of creation of charge per unit mass of air by dividing by $\overline{W}/e$, where $\overline{W}$ is the mean energy expended in air to form one ion pair and e is the electronic charge. Thus, the exposure rate at the specified point, due to photons of energy E_i, is

$$\dot{X}_i = \frac{A\ P_i E_i\ \mu_{en,i}}{4\ \pi\ \rho\ l^2\ (\overline{W}/e)}\ . \tag{3.29}$$

It is assumed that (1) the distance l is sufficiently large so that the exposure rate is independent of source size, and (2) there is neither attenuation nor buildup of the photon fluence in the distance l.

In general, radionuclides emit photons over a range of energies, and a summation over these energies must be made to obtain the exposure rate due to the source. Some of the photons may, however, be of such a low energy that they are absorbed in the source itself or in the intervening air, so that the exposure rate of interest may involve only photons with energies above some minimum value, represented by δ.

[4] The approximate definition of exposure given here is adequate for the heuristic derivation of the desired relationship between activity and exposure rate. A rigorous and complete definition of exposure, and also of the other dosimetric quantities used in Section 3.11, is given in ICRU (1971).

If, then, this summation is carried out in Eq. 3.29 and the result solved for A, we get

$$A = l^2\dot{X}_\delta/\Gamma_\delta, \tag{3.30}$$

where

$$\Gamma_\delta = \frac{1}{4\pi(\overline{W}/e)} \sum_i P_i E_i(\mu_{en,i}/\rho). \tag{3.31}$$

The summation is carried out over those values of i for which $E_i > \delta$. Equation 3.30 is the desired relationship between activity and exposure rate. The *exposure-rate constant* Γ_δ (ICRU, 1971) is the exposure rate at unit distance from a source of unit activity due to photons of energy greater than δ, subject to the two conditions stated following Eq. 3.29. The exposure-rate constant is characteristic of the radionuclide and independent of the material properties of a particular source. Having specified δ, Γ_δ can be calculated from the decay parameters of the radionuclide. In choosing that value of δ most useful for a particular application, the attenuation properties of the source, the intervening medium, and the detector or object irradiated should be taken into account; photons of too low an energy to make a significant contribution to the measurement or irradiation should be excluded by choosing an appropriate value of δ.

For $\delta = 0$, the exposure-rate constant Γ_0 includes all the photons emitted by the radionuclide, including gamma rays and x rays. Except for the inclusion of x rays, Γ_0 is the same as a parameter which has been variously known as the specific gamma-ray emission, the specific gamma-ray constant, and the gamma-ray dose-rate constant, is generally represented by Γ, and has been tabulated by a number of authors (NCRP, 1961, Appendix A; Hine, 1956, Appendix; Johns and Cunningham, 1969, Table A-8; Shalek and Stovall, 1969).

If the decay scheme of the radionuclide is known (i.e., for known values of P_i and E_i) and once the value of δ has been specified, Γ_δ can be calculated from Eq. 3.31. The calculation can be performed using the mass energy-absorption coefficient $\mu_{en,i}/\rho$ (ICRU, 1964; Hubbell, 1969), in which case Eq. 3.31 takes the form

$$\Gamma_\delta = \mathrm{k}_1 \sum_i P_i E_i\,(\mu_{en,i}/\rho), \tag{3.32}$$

where the numerical value of k_1 depends on the choice of units. Alternatively, the calculation can be performed using the linear energy-absorption coefficient, shown in Figure 38 as a function of energy, in which case Eq. 3.31 takes the form

$$\Gamma_\delta = \mathrm{k}_2 \sum_i P_i E_i\,\mu_{en,i}\,, \tag{3.33}$$

where $k_2 = k_1/\rho_{STP}$, and ρ_{STP} is the density of air at standard temperature and pressure (STP). At one standard atmosphere and 0 °C, $\rho_{STP} = 1.293\ mg/cm^3$. The presently recommended value of $\overline{W}$ is 33.7 eV per ion pair formed in air (ICRU, 1964), which gives for the average energy expended per unit charge $\overline{W}/e = 33.7$ J/C. If SI units are used (see Section 1.3)

$$k_1 = 1/4\pi(\overline{W}/e) = 2.36 \times 10^{-3}\ C/J. \qquad (3.34)$$

In this case, the energy per nuclear transition P_iE_i has units of joules per becquerel second[5] (J/Bq s), $\mu_{en,i}/\rho$ has units of meters squared per kilogram (m^2/kg), and the exposure-rate constant Γ_δ has dimensions of exposure times length squared per nuclear transition, corresponding to units of coulombs per kilogram times meters squared per becquerel second ((C/kg) m^2/Bq s). When using Eq. 3.30 with SI units, length l is in meters (m), exposure rate, $\dot{X}$, is in coulombs per kilogram per second (C/kg s), and the activity, A, is obtained in becquerels (Bq). Equations 3.29 and Eq. 3.31 are correct for SI units as written; if other units are used, an additional numerical factor may be necessary, as indicated in Eq. 3.32 and Eq. 3.33.

It is frequently convenient to use SI units for all quantities except the photon energy E_i, for which the unit mega-electron-volts (MeV) is used. For this case $k_1 = 3.78 \times 10^{-16}$ C/MeV.

A frequently used combination of units is activity, A, in millicuries (mCi), exposure rate, $\dot{X}$, in roentgens per hour (R/h), distance, l, in centimeters (cm), and photon energy, E_i, in mega-electron-volts (MeV). Then the product P_iE_i has units of mega-electron-volts per becquerel second (MeV/Bq s), $\mu_{en,i}/\rho$ has units of centimeters squared per gram (cm^2/g), and $\mu_{en,i}$ has units of reciprocal centimeters (cm^{-1}). Using these units, the numerical constants in Eq. 3.32 and Eq. 3.33 become

$$k_1 = 195.2\ R\ g\ Bq\ s/mCi\ h\ MeV$$

and (3.35)

$$k_2 = 1.510 \times 10^5\ R\ cm^3\ Bq\ s/mCi\ h\ MeV.$$

Using these units, the exposure-rate constant Γ_δ has units of roentgens centimeters squared per millicurie hour (R cm^2/mCi h).

It is seen in Figure 38 that the linear energy-absorption coefficient $\mu_{en,i}$ is approximately constant for photon energies from about 0.06 to 2 MeV. As a result, in this photon energy range, the exposure-rate constant is approximated to within about ± 15 percent by

[5] Since the becquerel is one nuclear transition per second, one becquerel second is one nuclear transition, and 1 J/Bq s = 1 J per nuclear transition.

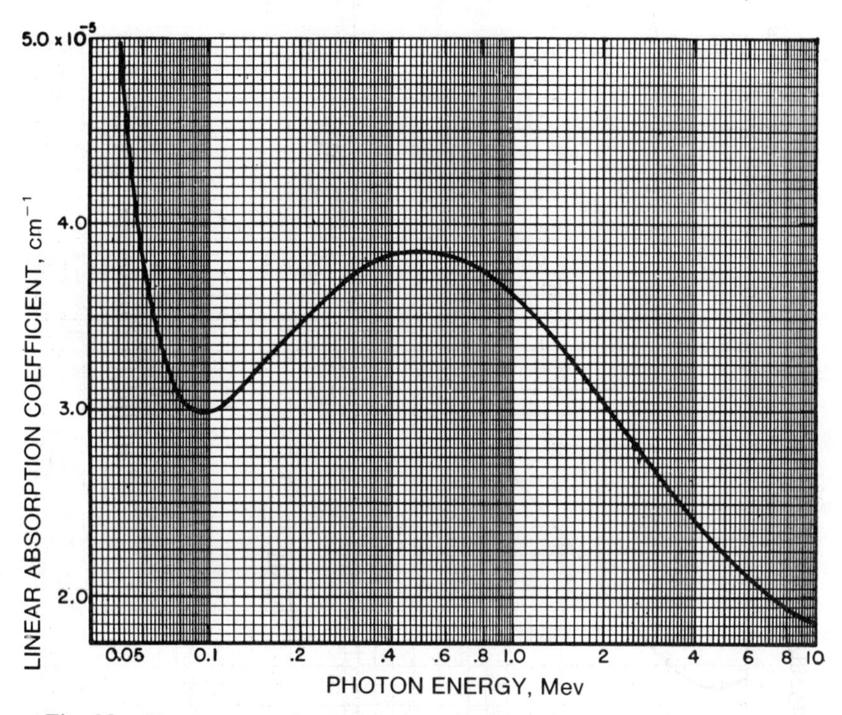

Fig. 38. Linear energy-absorption coefficient for photons in air at standard temperature and pressure. (From data in ICRU, 1961)

$$\Gamma_\delta \approx 5.2 \sum_i P_i E_i \,, \tag{3.36}$$

where E_i is in mega-electron volts (MeV), and Γ_δ is in roentgens centimeters squared per millicurie hour (R cm^2/mCi h).

The exposure-rate constant Γ_δ can be calculated from Eqs. 3.32, 3.33, or 3.36, using the appropriate numerical value of k_1 or k_2. Alternatively, the exposure-rate constant Γ_i per photon of energy E_i can be read directly from Figure 39, and then Γ_δ obtained from the equation

$$\Gamma_\delta = \sum_i P_i \Gamma_i \,, \tag{3.37}$$

where, as usual, the summation is taken over those photon energies for which $E_i > \delta$. The units of Γ_δ are the same as the units of Γ_i.

If a calibrated ionization chamber is available to measure the exposure rate $\dot{X}_\delta$ at a known distance l from a radionuclide source for which the exposure-rate constant Γ_δ is known or can be calculated, the activity can be calculated from Eq. 3.30. Suitable ionization chambers can be calibrated in terms of exposure rate at the National Bureau of Standards (NBS), or at other calibration laboratories which possess secondary exposure standards which have been calibrated by NBS.

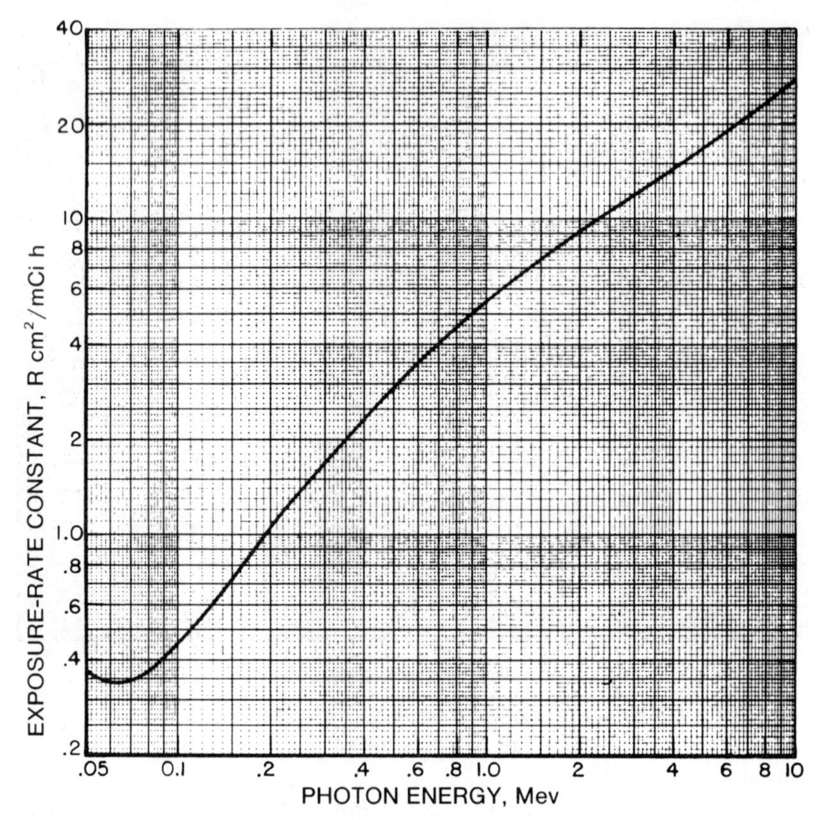

Fig. 39. Exposure-rate constant Γ for a point source emitting one photon of energy E per nuclear transition, as a function of E. The values of Γ were calculated by K. W. Geiger for NCRP Report No. 28 (NCRP, 1961) from the data given in Table 8.1 of NBS Handbook 78 (ICRU, 1961), using a value $W = 34.0$ eV per ion pair. If the presently recommended value of 33.7 eV per ion pair (ICRU, 1964) had been used, the value of Γ shown in the graph would be 1 percent larger. Values of Γ_δ are obtained by summing the individual Γ values using Eq. 3.37. The ordinate must be multiplied by 1.94×10^{-19} to convert to the SI units C m^2 kg^{-1}, or by 0.194 to convert to aC m^2 kg^{-1}.

Exposure calibration factors are supplied for certain photon energies (e.g., those of the gamma-ray beams of ^{60}Co and ^{137}Cs), or for filtered x-ray beams generated by potentials less than 250 kV. In general, ionization-chambers are to some extent energy dependent, and it may be necessary to interpolate between known calibration factors in order to obtain a calibration factor appropriate for use with the calculated value of Γ_δ. In addition, it may be necessary to correct for attenuation of the gamma-ray beam in the source, attenuation and buildup in the intervening air, and scatter from the environment. It should be noted that attenuation and scatter in the ionization chamber wall is included in the calibration factor supplied by the calibration laboratory; no correction for these effects is necessary unless the chamber is used under different circumstances from those in which it was calibrated.

An alternative to the use of a calibrated chamber is the use of gamma-ray sources or photon beams calibrated in terms of exposure rate. Encapsulated brachytherapy sources of ^{137}Cs and ^{60}Co are calibrated by NBS in terms of exposure rate at 1 meter in air; encapsulated and filtered sources of radium are calibrated by NBS in terms of mass of radium, and values of Γ_δ for radium sources are given in the tabulations of Γ cited after Eq. 3.31. X-ray and gamma-ray beams used in medical practice are commonly calibrated in terms of exposure rate, using instruments the calibration of which is directly traceable to NBS exposure standards. Thus, a variety of photon beams is available for calibrating an ionization chamber in terms of exposure rate, the disadvantage of this method being that the calibration factor so obtained is one or more steps further removed from NBS standards than would be the case with a chamber calibrated at a calibration laboratory. If this method is used, it is highly desirable to use several sources or beams, in order to assess the energy dependence of the chamber being calibrated and thus choose a calibration factor suitable to the photon spectrum of the activity being measured. It may be necessary to correct for attenuation and scatter in the source, the intervening air, and the environment, but not for that in the walls of the ionization chamber.

Still another alternative is the use of a gamma-ray source calibrated in terms of activity. If such a source is available, with a known exposure-rate constant, the activity of another gamma-ray source can be obtained by comparing the two using an ionization chamber. The comparison gives the unknown activity in terms of the known activity by use of Eq. 3.30, and involves the square of the ratio of the distances, the inverse of the ratio of the exposure-rate constants, and the ratio of the measured currents. The last ratio is equal to the ratio of the exposure rates, after correcting for differences in ionization chamber response due to differences between the photon spectra. It may also be necessary to correct for self absorption in one or both of the sources. Radium standards are often used, since accurately standardized sources are widely available and are of long half-life. Table 8 summarizes the corrections needed in determining the activity of a few radionuclides by comparison with a radium standard. The corrections for absorption in a typical small ampoule and in the wall of the ionization chamber are seen to be small except at the lowest photon energies.

In carrying out ionization-chamber measurements, care should be taken that the wall thickness is sufficient to exclude all electrons arising outside the chamber walls. The operation of ionization chambers is also discussed in Sections 2.1 and 4.4.

TABLE 8—*Typical correction factors for ionization-chamber gamma-ray measurements*

Radionuclide	Absorption in typical ampoule	Chamber wall absorption relative to radium	Overall correction factor
^{170}Tm	1.07	1.10	1.18
^{99m}Tc	1.05	1.07	1.12
^{131}I	1.025	1.025	1.05
^{137}Cs, ^{82}Br, ^{60}Co	1.02	1.00	1.02
^{24}Na	1.015	1.00	1.015

3.11.2 *Measurement by Ionization Chamber of Known Volume*

The use of an ionization chamber of known volume for the determination of activity was first described in detail by Gray (1949). The rate at which energy is absorbed within the air of a cavity chamber can be calculated from the ionization produced, assuming a value for the mean energy expended in air per ion pair. If the cavity is small enough to produce no appreciable disturbance of the electron fluence generated by the gamma rays in the neighborhood of the cavity, the rate of energy absorption per unit mass in the chamber wall is related to that in the air by the ratio of the mean stopping power of the wall relative to that of air, provided the chamber walls are thick enough to exclude all electrons arising outside the walls. The rate of energy absorption in the chamber wall can be related to the source activity in a manner closely analogous to the relationship derived in Section 3.11.1 between activity and exposure rate, except that the present case involves the mass energy-absorption coefficients of the chamber wall material instead of air and does not involve the mean energy expended in air per ion pair. Equating the two independent calculations of the rate of energy absorption in the chamber wall allows the source activity to be expressed in terms of the measured ionization current. The relevant equations have been given by Gray (1949), and in considerable detail in an investigation of the relationship between source activity, source power, and exposure rate at unit distance, by Petree *et al.* (1975) and Roux (1974).

A determination of activity by this method involves not only a knowledge of the decay scheme of the radionuclide, but also the accurate determination of the source-to-chamber distance, chamber volume, air density, and ionization current. Allowances must be made for absorption of the gamma rays in the source, chamber wall, and intervening air, and for the relative stopping power of wall to air. The measurement of effective distance presents no great difficulties; the inverse square law holds until the distance is only a few times the chamber dimensions (Sinclair *et al.*, 1954; Kondo and Randolph, 1960). Dependence on the distance can be eliminated entirely by using a spherical extrapolation chamber surrounding the source (Gross *et al.*,

1957b). Absorption within the source can be allowed for by the methods of Evans and Evans (1948). Absorption in the chamber walls can be estimated from observation of the effect of increasing the wall thickness. There is an uncertainty, however, leading to an error that is usually less than 1 percent, in deciding the wall thickness to which the observations should be extrapolated, since the secondary electrons arise at various depths within the walls (Attix and Ritz, 1957; Loftus and Weaver, 1974). The correction for relative stopping power can be made quite small by choosing a wall material whose atomic number is close to air; carbon in the form of graphite is often used for this purpose.

The use of an ionization chamber of known volume to determine the activity of a gamma-ray source is considerably more demanding than the use of a calibrated chamber, and is likely to be the method of choice only when special circumstances make traceability to a dosimetry standard undesirable.

3.12 Loss-of-Charge Method

Measurements of the charge carried away from a source of radium have been made by several investigators (Wien, 1903; Rutherford, 1905; Makower, 1909; and Moseley, 1912). Subsequently, however, this method was applied to the standardization of beta-emitters (Clarke, 1950; Gross and Failla, 1950; Keene, 1950). Using a spherical geometry, Clarke (1950) achieved agreement with National Bureau of Standards measurements of standard samples of ^{32}P to within 5 percent, but the method has not since been in general use. Gross and Failla (1950) used a parallel-plate geometry where the source was a very thin deposit of the emitter in question on one plate of a parallel-plate collecting system, the second plate being sufficiently close to balance the secondary electron emission and backscattering effects. Low-energy secondary electrons were bent back by means of a weak magnetic field, and the measurement was made in a vacuum to eliminate air ionization. Agreement with the National Bureau of Standards measurements of standard samples of ^{32}P to within ± 2 percent was obtained by this method by Gross and Failla over a number of years (Seliger and Schwebel, 1954).

3.13 Calculation of Activity of Artificially Produced Radionuclides From Irradiation Data

The disintegration rate of radionuclides produced by bombardment with neutrons or charged particles may be calculated if the cross section for interaction with the target nuclide (stable) and the flux of bombarding particles are known (Hughes, 1953; Halliday, 1955). Neutron reactions only will be considered here, since their treatment is typical of particle bombardments.

Thermal, or "slow" neutrons in a reactor are most often used for such activation. The neutron flux is usually determined by use of a monitor containing a known amount of a nuclide whose thermal-neutron-activation cross section is well established. Cobalt (37.2 barns; 37.2×10^{-24} cm^2), gold (98.8 barns), and manganese (13.3 barns) are convenient standards. A tabulation of cross sections is available (Mughabghab and Barber, 1973).

For calculation of the neutron flux density by use of the monitor, one may use the equation

$$f = \frac{A_m}{N_m \sigma_m e^{-\lambda_m t_d}(1 - e^{-\lambda_m t_b})}, \qquad (3.38)$$

where

f = flux density, in cm^{-2} s^{-1},
A_m = disintegration rate of monitor, in s^{-1},
N_m = number of atoms of stable monitor nuclide,
σ_m = thermal-neutron-activation cross section of monitor nuclide, in cm^2,
λ_m = decay constant of monitor activity, in s^{-1},
t_d = time between end of bombardment and measurement, in s, and
t_b = bombardment time, in s.

The disintegration rate of the desired radionuclide produced by bombardment of an appropriate target may be calculated from the relation

$$A = N\sigma f e^{-\lambda t_d}(1 - e^{-\lambda t_b}), \qquad (3.39)$$

where N and σ now refer to the target and A and λ to the activity produced. The introduction of the sample must not affect the flux distribution appreciably.

In general, the use of bombarded samples as standards is not as accurate as most of the methods previously discussed, chiefly because of uncertainty in knowledge of the energy distribution of neutrons in reactors. The usually well-known thermal-neutron cross section therefore has to be modified to include activation by epithermal or fast neutrons. Further, it is necessary to measure the disintegration rate of the monitor; such standards might therefore be considered to be derived standards.

4. Indirect Or Comparative Measurements of Activity in Radioactive Decay

4.1 General

The simplest way to prepare a calibrated relative (or derived, or indirect, or "secondary") standard of radioactivity in terms of a directly measured (or "absolute", or "primary") standard is to measure, by means of a substitution method under reproducible conditions, the relative activities using any instrument that records either individual radiations or else measures their ionizing effects. With instruments of adequate precision, such methods can give relative values having a standard deviation of from 0.03 to 1.0 percent, which is often less than the uncertainty with which the activity of the direct standard is known. It is also to be understood that within the meaning of "reproducible conditions" is to be included the geometry of the equipment and the disposition and quantity of any material absorbing or scattering the radiations. In these respects, the characteristics of the source itself are no less important than those of the measuring instrument.

The recording device used in such a substitution method can be a gold-leaf or quartz-fiber electroscope; a beta-ionization chamber or gamma-ionization chamber used in conjunction with a Lindemann-Ryerson, or vibrating-reed electrometer, or any other low-current-measuring device; or any kind of alpha-, beta-, gamma-, or x-ray detector.

4.2 Radionuclide Identification and Assay by Energy Spectrometry

4.2.1 *General*

As discussed in Section 1, the decay of radioactive nuclei results in the emission of charged particles (alpha and beta particles, and electrons), gamma rays, and x rays due to interactions with bound atomic electrons. It should be noted that the beta-decay process results in the emission of an energy continuum. The alpha particles and photons emitted exhibit discrete energies corresponding to the amount of energy given off as the nucleus or atom de-excites from one allowed energy state to another. Thus, if we observe the energy spectrum of alpha particles or gamma rays emitted in the decay of a given radionuclide, a pattern of energies and intensities will be seen that is unique. Just as the discrete structure of optical spectra resulting from electronic transitions in atoms forms the basis for elemental analysis, charged-particle- and photon-energy spectrometry may be used to analyse radioactive material for radionuclidic content.

Pulse ionization chambers and gas proportional counters (see section 2.2) provide a means of measuring the energy of charged particles. Because the range of both alpha particles and beta particles is finite in a gas medium, these counters offer a means of measuring the energy of individual particles by pulse-amplitude spectrometry. Since the particle ionizes the gas in the counting chamber, the collection of charge resulting from the total energy absorption of an incoming particle will provide a measure of the kinetic energy of that particle. The electrical pulse appearing momentarily at the anode of the counter chamber will have an amplitude proportional to the energy of the particle detected. As described in Sections 2.4 and 2.5, both scintillation and semiconductor detectors exhibit the same energy-measurement properties. All these devices are extensively used for the measurement of both charged-particle- and photon-energy spectra.

4.2.2 *Interaction of Photons with Matter*

A knowledge of the basic processes by which a photon interacts with matter is essential to an understanding of the response of a gamma-ray spectrometer and the techniques of pulse-amplitude (or pulse-height) analysis. Although many processes are involved in the chain of events that produces an electrical pulse at the output of the detector, the major features of a differential pulse-amplitude spectrum resulting from the detection of gamma rays may be interpreted in terms of the basic interactions that occur within the detector. As indicated in Section 1.4, the three primary processes by which photons interact with matter are (1) the photoelectric effect, (2) Compton scattering, and (3) the production of positron-electron pairs. These processes are represented diagrammatically in Figure 40. Photoelectric absorption predominates in the low-energy region with its probability decreasing rapidly with increasing energy. Figure 41 is a plot of the probability

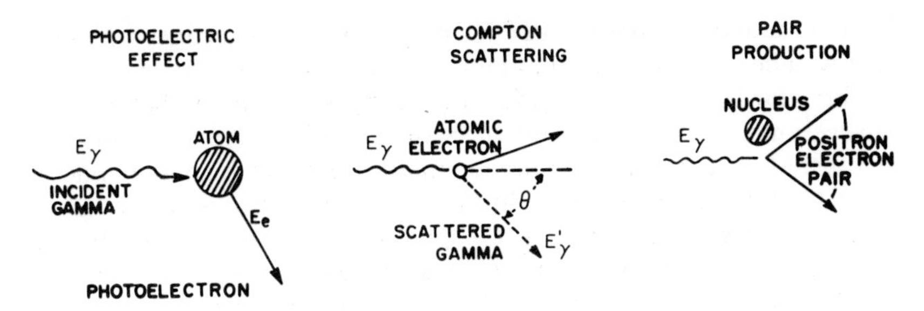

Fig. 40. Illustration of major gamma-ray interaction processes.

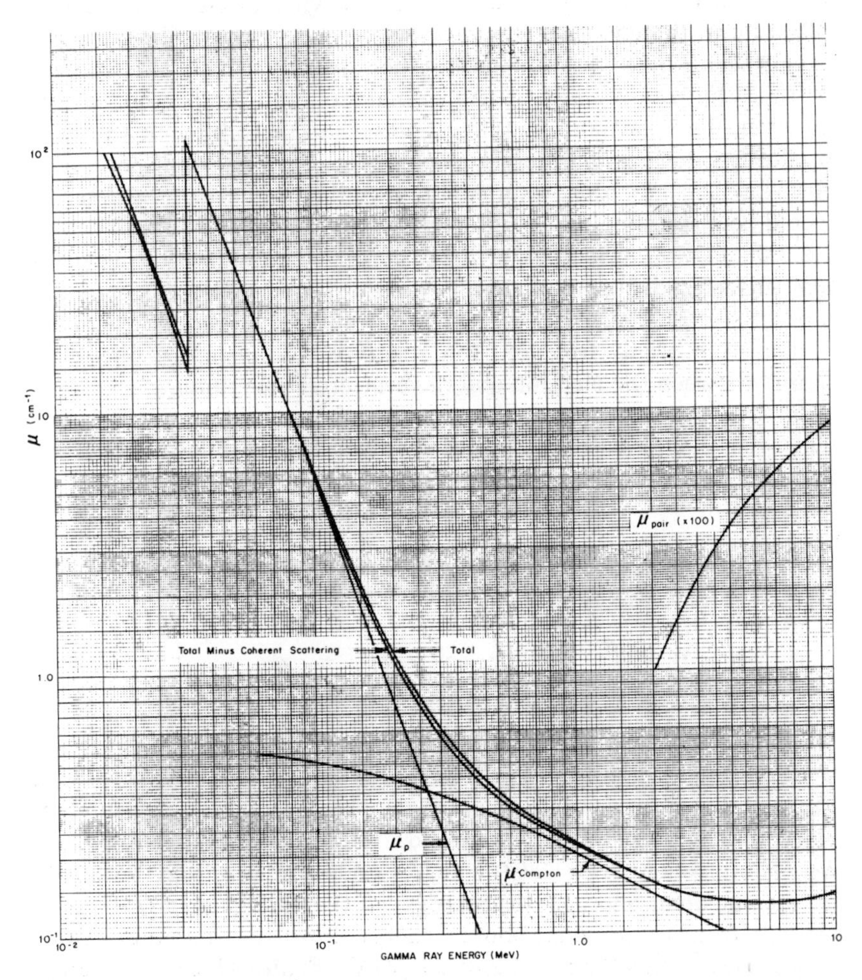

Fig. 41. Plot of linear attenuation coefficient for NaI as a function of gamma-ray energy showing contributions from photoelectric effect, Compton scattering, and pair production.

for interaction of photons in NaI as a function of primary photon energy. The probability for interaction by the Compton process is seen to vary slowly up to several MeV in energy, while the probability for interaction by the pair process increases rapidly above the threshold at 1.022 MeV to become the predominant process at higher energies.

4.2.2.1 *Photoelectric Effect.* In the photoelectric process, absorption occurs by interaction between the photon and an inner-shell bound electron in the absorbing material. The energy of this ejected electron will be equal to the difference between the energy of the incident photon and the binding energy of the shell from which the electron was ejected. As a result, the atom is left with an electron vacancy, resulting in the emission of x rays or Auger electrons. This series of events occurs in less than 10^{-12}s and the x rays are usually also stopped, so that the total energy of the primary photon is absorbed in the detector. Thus, a monoenergetic photon interacting by this process produces electrons totalling in energy that of the incident photon. If this were the only process for energy loss, the response of the detector would be quite simple.

4.2.2.2 *Compton-Scattering Process.* In Compton scattering, photons suffer partial losses of energy when scattered by electrons. In this process scattering generally occurs with outer-shell atomic electrons that are essentially free, and the energy of the incoming photon is shared between the electron and the scattered quantum.

In the Compton process the energies of the scattered photon and electron, E_γ' and E_e, respectively, are given by the following relationships:

$$E_\gamma' = \frac{E_\gamma}{1 + \alpha(1 - \cos\theta)} , \tag{4.1}$$

$$\text{and} \quad E_e = E_\gamma - \frac{E_\gamma}{1 + \alpha(1 - \cos\theta)} , \tag{4.2}$$

where E_γ is the incident gamma-ray energy, α is the ratio E_γ/mc^2, and θ is the angle between the direction of the primary and scattered photons. From these relationships it may be deduced that a Compton-electron-energy spectrum will extend from zero energy ($\theta = 0°$) up to a maximum energy ($\theta = 180°$) which is somewhat less than the energy of the incident photon. The energy of the scattered photon then extends from the original photon energy down to a minimum value that is always less than $mc^2/2$ (0.256 MeV). Figure 42 shows a Compton-electron-energy distribution obtained by integrating the differential scattering cross section over all angles from a primary photon energy of 0.511 MeV. A monoenergetic source of gamma radiation will produce such an energy distribution of electrons as a result of interaction by the Compton process.

4.2.2.3 *Pair Production.* If the incident photon has an energy in excess of the rest mass of a positron-electron pair (1.022 MeV), then pair production is possible. In this process, which occurs in the presence of the Coulomb field of a nucleus, the gamma ray disappears and a positron-electron pair is created; their kinetic energy will be equal to the total energy of the gamma ray minus the rest energy of the two particles ($2mc^2$). When the positron comes to rest in the field of an electron, annihilation of the two particles occurs with the emission of two photons equal in energy to the rest mass of the particles (0.511 MeV). Interaction by the pair process in a detector will therefore deposit energy at least equal to the kinetic energy of the positron-electron pair. As will be discussed below, the possibility also exists of detecting one or both of the annihilation quanta by either the photoelectric or Compton process. These alternatives result in a complex electron-energy distribution for the pair process. A series of events can result in an energy deposited in the detector of any energy between E_γ−1.022 MeV and E_γ, the energy of the primary photon.

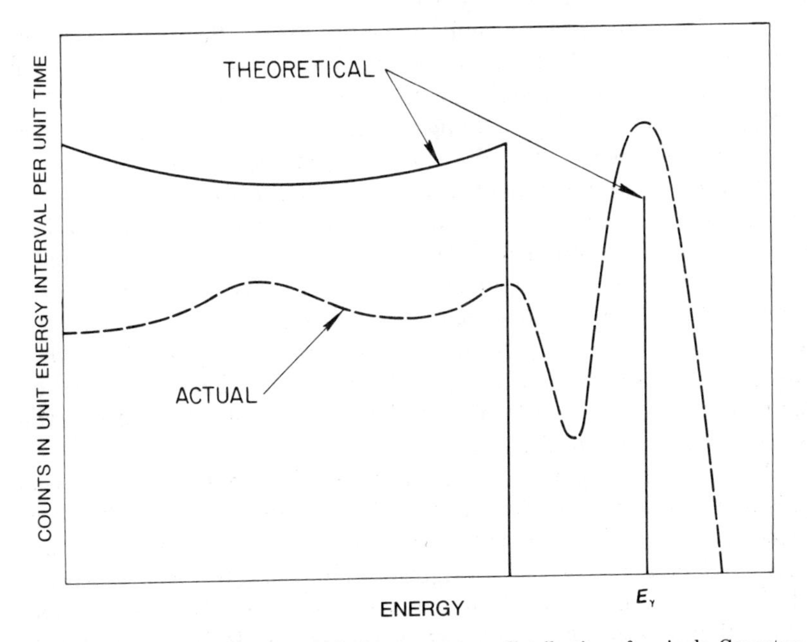

Fig. 42. Theoretical and actual electron-energy distributions for single Compton and photoelectric interactions in a 3″ × 3″ NaI(Tl) detector using a 0.511-MeV photon source. (From Heath, 1964)

The total probability for detection of a gamma ray, expressed as the total linear attenuation coefficient (less the coherent contribution), will then be given by:

$$\mu = \mu_{\text{phot}} + \mu_{\text{Compt}} + \mu_{\text{pair}} \tag{4.3}$$

The total linear attenuation coefficient for NaI(Tl) and the contribution from the photoelectric, Compton, and pair processes is shown in Figure 41.

4.2.3 *The Pulse-Height Spectrum*

To provide a basis for describing a pulse-height distribution observed at the output of the scintillation or semiconductor detector in terms of the basic processes described above, let us briefly review the succession of events that produce it. Assume that a monoenergetic source of gamma radiation of approximately 0.5 MeV is incident upon a NaI(Tl)-crystal-phototube combination or semiconductor detector. Inspection of the linear attenuation coefficient for NaI(Tl) shown in Figure 41 indicates that these gamma rays will interact with the NaI(Tl) detector by both Compton scattering and the photoelectric effect in the ratio of about 6:1. If a sufficient number of gamma rays is detected to give a reasonable statistical sample, an electron energy distribution similar to that portrayed by the solid line in Figure 42 will be produced within the volume of the detector. We see a monoenergetic line of photoelectrons and a continuous distribution of Compton electrons from zero energy up to a sharp cut-off at an energy somewhat below the "photoline". Shown for comparison is the pulse-height distribution from a 3″ × 3″ NaI(Tl) detector obtained with a differential pulse-height analyzer. Although the pulse-amplitude spectrum is similar in character, we observe that the energy distribution has been smeared by what appears to be a Gaussian function. This is due to fluctuations in the light output from the phosphor and to the statistical nature of the processes occurring in the electron multiplier.

The peak resulting from total energy loss (photopeak) is the distinguishing characteristic of all spectra. The location of this peak and its intensity are used to determine the energy and intensity of gamma rays producing a given pulse-height distribution. The energy response of NaI(Tl) is approximately linear, hence the peak location is about proportional to the photon energy. Non-linearity effects in NaI(Tl) are discussed in Heath (1964). The width of this peak is a measure of the energy resolution of the detector—a subject that will be treated in more detail in a later section.

In addition to the amplitude smearing of the original electron spectrum, it is apparent that a much larger fraction of the total number of events has appeared in the "photopeak" than would be predicted from the ratio of the photoelectric and Compton cross sections. In a detector of this size, the probability that a Compton-scattered gamma ray will escape the volume of the detector is somewhat reduced and many multiple events (e.g., Compton scattering followed by a photo event) can occur that result in total energy loss. Since this chain of events occurs well within the decay time of the phosphor, the energy loss from all events will sum to produce a pulse in the full-energy peak. As will be shown later, the relative probability for total energy loss in the detector due to the occurrence of multiple events will increase as the dimensions of the detector are increased.

The complicated pulse-height spectrum produced from monoenergetic photons incident upon a detector presents the basic problem in the interpretation and analysis of data obtained with a gamma-ray spectrometer. Let us now examine variations in the shape of the pulse-height spectrum as the incident gamma-ray energy is varied. The lower portion of Figure 43 shows the energy positions of four gamma rays of 0.06-, 0.32-, 0.83-, and 1.92-MeV incident energy, all of equal intensity. Above this are the pulse-height spectra for each of these gamma rays obtained with a 3″ × 3″ NaI(Tl) detector. At 0.06 MeV, gamma rays interact with the detector almost entirely by the photoelectric process and the detector response is essentially a single peak, nearly Gaussian in shape. The peak of lesser intensity on the low-energy side of the full-energy peak is attributed to the escape of iodine x rays from the surface of the detector following a photoelectric interaction. This effect will be discussed in more detail in a later section. For the gamma-ray energy equal to 0.32 MeV, the pulse-height distribution becomes more complicated. In addition to the photopeak, we see the continuous distribution of pulses resulting from the detection of Compton electrons. For the gamma ray of energy 0.83 MeV, the fraction of pulses in the photopeak is reduced as the Compton cross section becomes still more dominant. At 1.92 MeV, even more features appear in the pulse-height distribution. In addition to the Compton-electron distribution and the full-energy peak, several satellite peaks are superimposed upon the Compton distribution as a result of interactions involving the pair-production process. In this process, all energy in excess of the 1.022-MeV threshold is converted into kinetic energy of the positron-electron pair, which will normally be deposited in the crystal. If the energies of

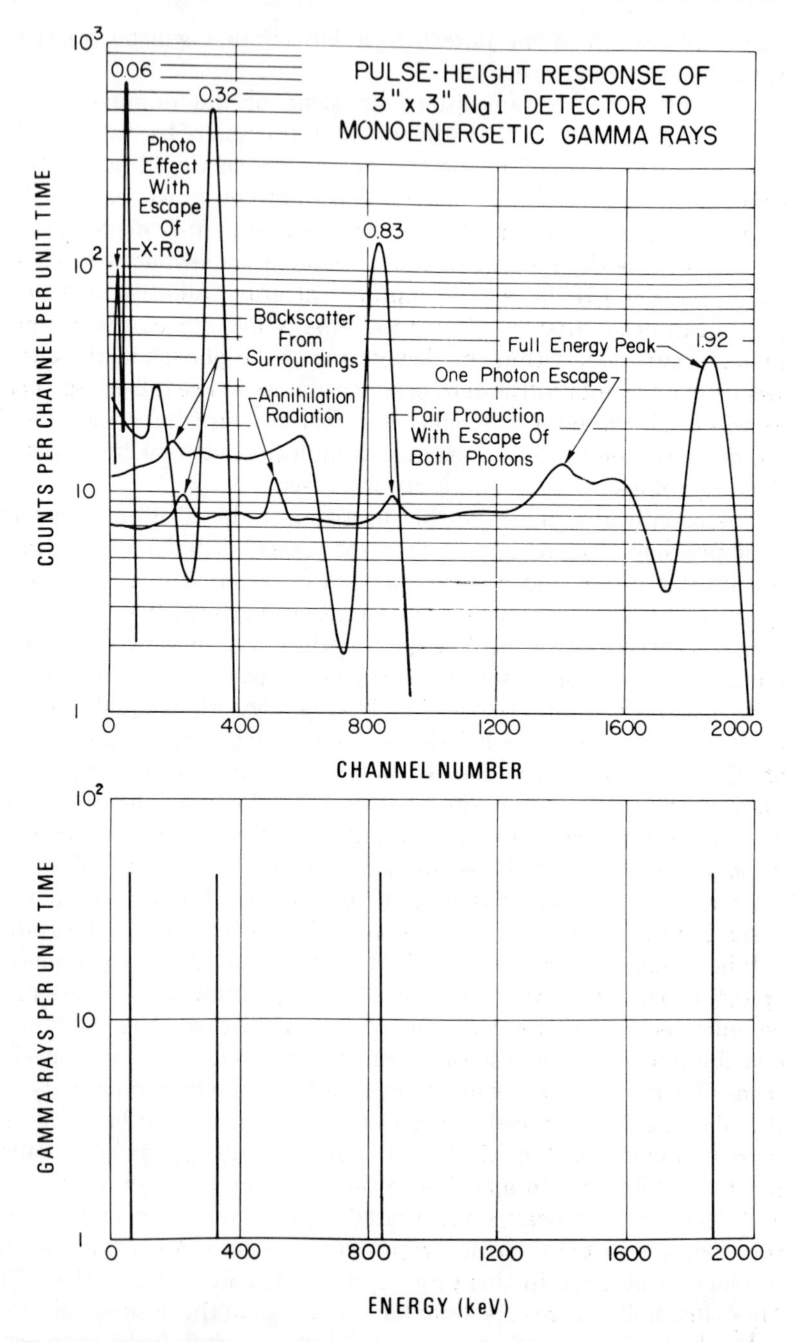

Fig. 43. Pulse-height distributions obtained with a 3″ × 3″ NaI(Tl) detector from monoenergetic gamma-ray sources of 0.06, 0.32, 0.83, and 1.92 MeV normalized to equal emission rates. Pertinent features of these spectra are noted, including the decrease in detector efficiency with increasing energy. (After Heath, 1964)

both annihilation quanta are totally absorbed in the crystal, then the energy deposited by all processes will produce a pulse in the full-energy peak. Subsequent annihilation of the positron will create two 0.511-MeV photons, one or both of which may escape from the crystal without detection. These events correspond, respectively, to the "single-escape" and "double-escape" peaks at 0.511 and 1.022 MeV below the full-energy peak (i.e., at 1.41 and 0.90 MeV). The total result of interactions involving the pair-production process will then be a distribution of pulses ranging from the full-energy peak to 1.022 MeV less than the photopeak, including the three prominent peaks just described. The backscatter peaks and the annihilation-radiation peak at 0.511 MeV arise from interactions with material surrounding the detector. For comparison, Figure 44 shows a pulse-height spectrum of gamma rays emitted in the decay of ^{88}Y, obtained with a Ge(Li) spectrometer. Apart from the improved energy resolution, the general character of the detector response is seen to be similar to that obtained with NaI(Tl) detectors.

4.2.4 *Pulse-Height Spectrometers*

As outlined in the previous section, the proportional response of solid scintillators and semiconductor detectors affords a means of determining the energy of photons interacting with the material of the detector by measuring the energy deposited in the detector in single events. The detection of a single photon results in the production of a small amount of electrical charge that is collected at the output of the detector as a "pulse" of current whose amplitude is proportional to the energy deposited in the detector and hence to the energy of the detected photon. A modern pulse-height analysis system employed for particle- or photon-energy spectrometry consists of the following functional elements as shown in Figure 45: (1) scintillation or semiconductor detector; (2) linear pulse amplifier; and (3) a multichannel pulse-height analyzer with suitable display and readout devices. Although a detailed description of nuclear electronics is beyond the scope of this discussion, a brief description of the design considerations that are important to the operation of these systems is presented. The reader is referred to references Fairstein and Hahn, 1965; Chase, 1961; Nicholson, 1974; and Heath, 1964, for a more complete treatment of this subject.

4.2.4.1 *Detector.* Pulse-height analysis systems for photon or particle-energy spectrometry generally employ NaI(Tl) scintillation or Ge and Si semiconductor detectors. Special-purpose high-voltage power

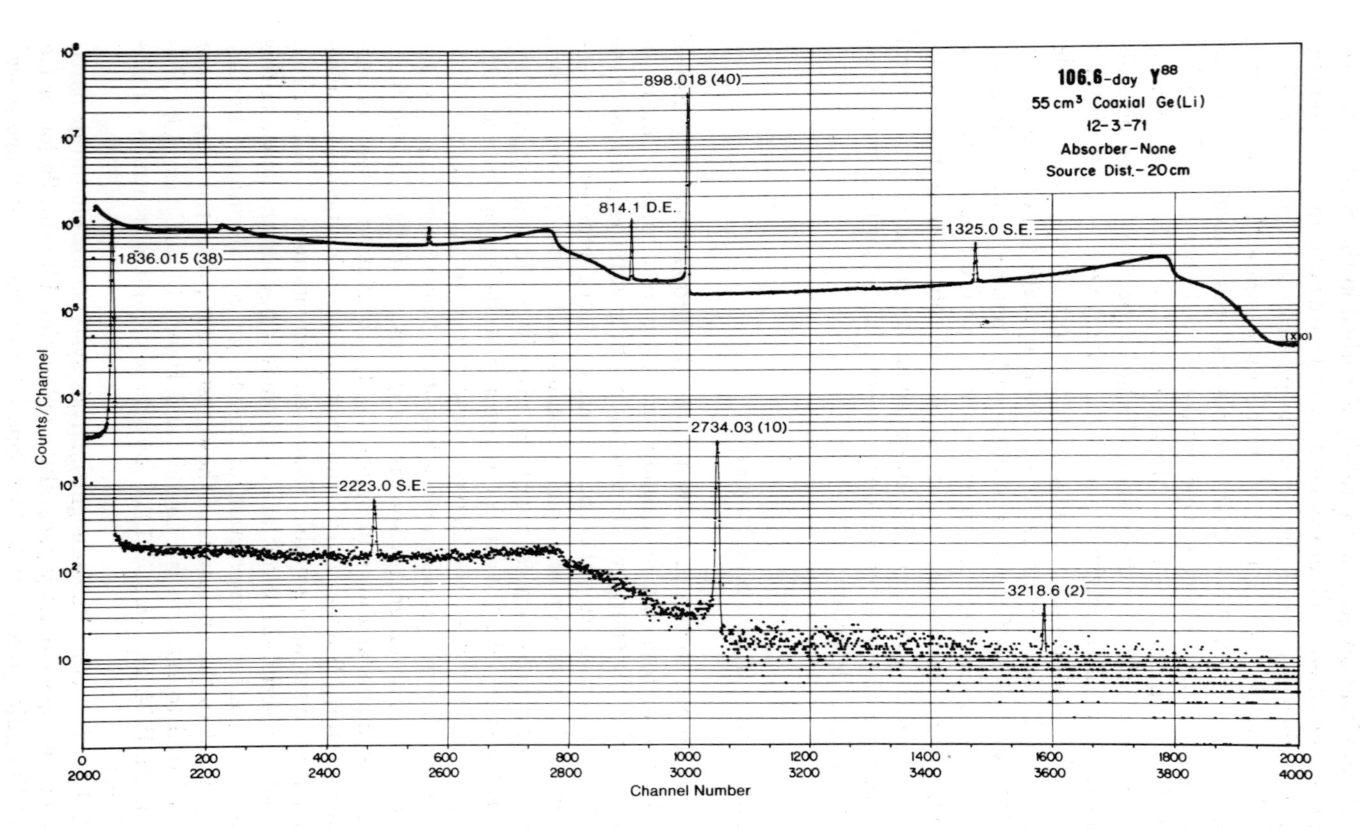

Fig. 44. Gamma-ray spectrum of radiation emitted in the decay of Y_{88} observed with a Ge(Li) detector system. The letters S.E. and D.E. refer to single and double escape peaks, respectively, and the values in parentheses are the uncertainties in the last digits. (From Heath, 1974)

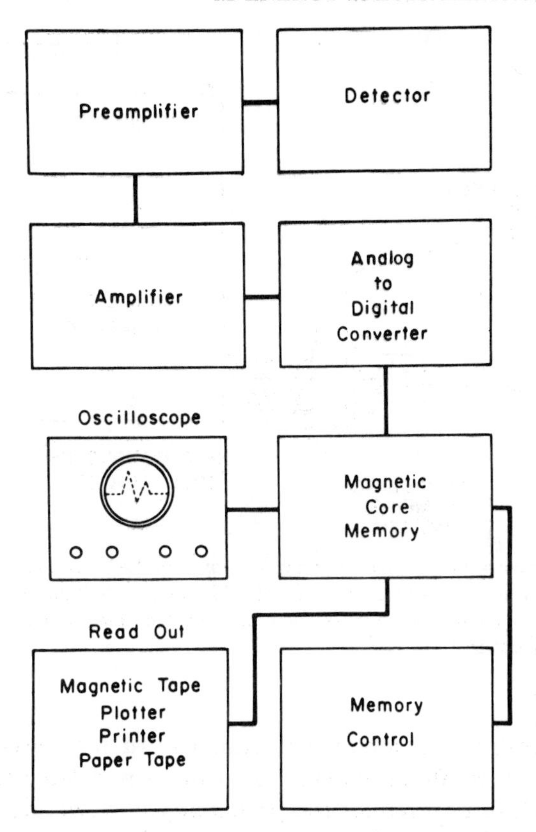

Fig. 45. Block diagram of a conventional "hard-wired" gamma-ray spectrometer system used for pulse-amplitude analysis of the output from scintillation and semiconductor radiation detectors.

supplies are required for both types of detector. In the case of the scintillation detector, a very stable high-voltage source is needed to provide accelerating voltage for the electron-multiplier structure of the photo-tube. Semiconductor devices require a low-noise source of several thousand volts to insure collection of charge created in the detector. The operation of semiconductor detectors at liquid-nitrogen temperatures requires the use of a vacuum-insulated (Dewar) container. The semiconductor device itself is contained in an evacuated chamber, mounted on the end of a long metal rod, or "cold finger", designed to conduct heat from the detector to the liquid nitrogen. A typical detector arrangement for a Ge(Li) detector is shown in Figure 46. As indicated, the vacuum in the detector housing is usually main-

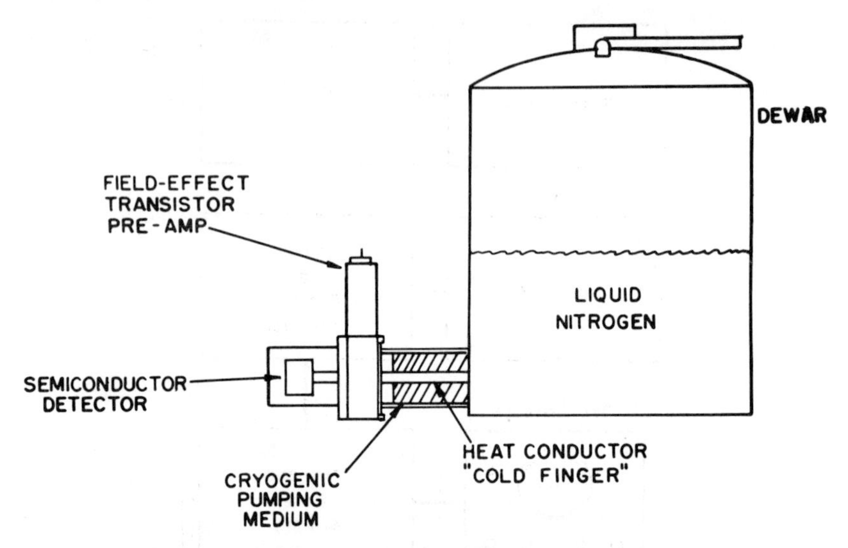

Fig. 46. Illustration of typical method employed in the operation of semiconductor radiation detectors. To maintain integrity of lithium-ion-drifted devices and to achieve optimum electronic-noise levels these detectors must be maintained and operated at cryogenic temperatures.

tained by absorption pumping techniques. In the measurement of radioactivity, it is often necessary to operate these detectors in a large shielded enclosure to reduce the probability of detecting background radiation. A typical shield installation for a laboratory spectrometer is shown in Figure 47.

4.2.4.2 *Pulse Amplifier.* A linear pulse amplifier is used to amplify the small electrical pulses created in the detectors to a level sufficient for measurement of pulse amplitude. In the scintillation detector, the phototube not only converts the light emitted by the scintillator to an electrical impulse, but serves as a very fast, low-noise amplifier. Phototubes that are used with NaI(Tl) scintillation detectors have current gains ranging approximately from 10^6 to 10^8. The very minute amount of energy associated with the detection of single photons in radiation detectors requires that special care be employed to amplify these signals with no addition of electrical noise from the amplifier circuits. The principal effect of the addition of noise to the signal will be a deterioration in the energy resolution of the system. In the case of semiconductor detectors, the level of signal produced at the output of the detector is so minute that very sophisticated techniques must be employed in the design of low-noise-amplifier circuits to realize the energy-resolution capabilities of these devices.

A basic understanding of the importance of the characteristics of amplifiers used in these systems is important to an appreciation of the pulse-amplitude measurement technique. To achieve optimum signal-to-noise ratios, the first stages of the amplifier system are frequently mounted inside the detector cryostat (as shown in Figure 46) with field-effect transistors (FETs) operated at temperatures in the range of 100 to 130K to reduce the thermal-noise level in the transistors. The development of this technology, pioneered for nuclear-radiation detectors, has found other applications in the detection and measurement of very minute electrical signals.

Following the first stages of amplification in the "preamplifier", the pulses are further amplified and shaped by integration and differentiation circuitry. Figure 48 shows examples of pulses from a typical detector at various stages in the amplification process. Part (a) of this figure shows pulses at the output of the preamplifier. Typically, pulses at this point in the system are characterized by a fast-rising leading edge (<0.1 μs) and a slow decay (typically 100 μs). Parts (b) and (c) show typical output pulses from a linear pulse amplifier, after they have been shaped to produce symmetrical pulses that return to zero in

Fig. 47. Photograph of typical shield arrangement used in the operation of a laboratory Ge(Li) gamma-ray spectrometer.

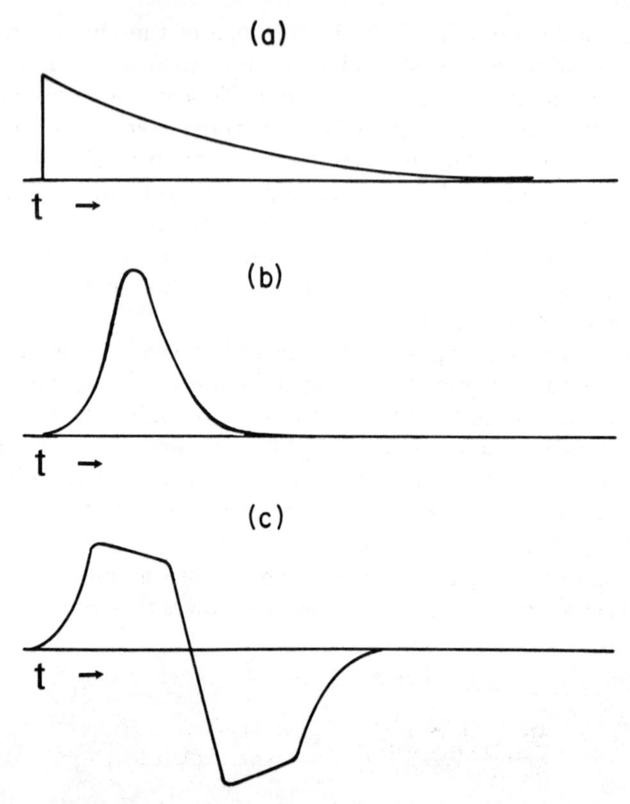

Fig. 48. Pulse wave forms observed in detector-amplifier systems: (a) preamplifier output pulse, (b) unipolar output pulse from shaping amplifier using RC circuit elements, (c) bipolar or "double-differentiated" shaping amplifier output.

a finite time. Part (b) is the pulse shape normally associated with low-noise amplifiers used with semiconductor detectors. Since it rises and returns to zero and ideally does not cross the zero level, it is termed a unipolar or "single-differentiated" pulse shape. Part (c) illustrates a bipolar or "double-differentiated" pulse shape that is frequently employed with NaI(Tl) scintillation detectors. There are many experimental variables which influence the choice of pulse-shaping networks and specialized pulse-conditioning circuitry employed in pulse-counting systems. Of these, the most important are noise performance (related to energy resolution) and pulse width (related to time resolution of counting systems). The choice of circuit parameters that determine the pulse width can be varied to satisfy the measurement requirements. Typically, NaI(Tl) spectrometers use bipolar shaping to produce pulse widths of 2 to 3 μs. In the use of semiconductor detectors, the shaping "time constants" will generally vary from 2 to 10 μs. This results in pulses whose width at the base will vary from 5 to 25 μs, with consequent increased counting losses, compared with NaI(Tl). The general subject of counting losses is discussed in Section 2.7. Specific counting-loss problems in the use of semiconductor detectors are discussed in Cohen (1974).

4.2.4.3 *Multichannel Pulse-Height Analyzer.* The multichannel pulse-height analyzer (PHA), the heart of the pulse-height-analysis system, provides a means of quantitatively measuring the maximum amplitude reached during an electrical impulse and recording this as a digital number. This process is termed "analog-to-digital conversion", and the circuitry developed for such measurements is termed an "analog-to-digital converter" (ADC). The approach that is used is to divide the voltage range to be measured into n equal intervals or bins, so that a 0- to 10-V range of possible pulse sizes can be divided into 10, 100, or 1000 equal intervals. In this manner, if the peak amplitude reached for a single pulse falls within one of these intervals, or channels, its amplitude is assigned the number of that interval within the 0- to 10-V amplitude range.

The first instruments of this type employed a series of amplitude discriminators that would indicate whether or not the amplitude of a pulse fell between two voltage levels. For example, a 0- to 10-V range might be divided into 10 equal divisions of 1 V. A pulse having an amplitude between 5 and 6 V would then exceed the voltage levels of the first 5 discriminators but not the sixth discriminator. A coincidence circuit was then employed to produce an output only from the discriminator at the 5-V level. Thus, the term "channel" signifies that a pulse has a measured maximum amplitude that resides somewhere between the boundaries of voltage increment associated with channel number 5. As the art of pulse-amplitude analysis advanced and detector energy resolution improved, there was a need for improved accuracy in amplitude measurement and a requirement for dividing the voltage range into finer increments or more "channels". Gamma-ray spectrometers using NaI(Tl) scintillation detectors generally divide the amplifier output into 256 or 512 channels or voltage increments. High-resolution Ge(Li) spectrometers frequently employ an amplitude scale with 4096 or even 8192 channels.

The modern pulse-height analyzer most frequently used in nuclear spectroscopy utilizes a technique for analog-to-digital conversion that is based upon an amplitude-to-time concept first developed by Wilkinson (1950). In this type of ADC, the pulse input to the circuit is first "stretched" or held at its maximum amplitude as shown in Figure 49c. A capacitor is then charged to a voltage proportional to this peak

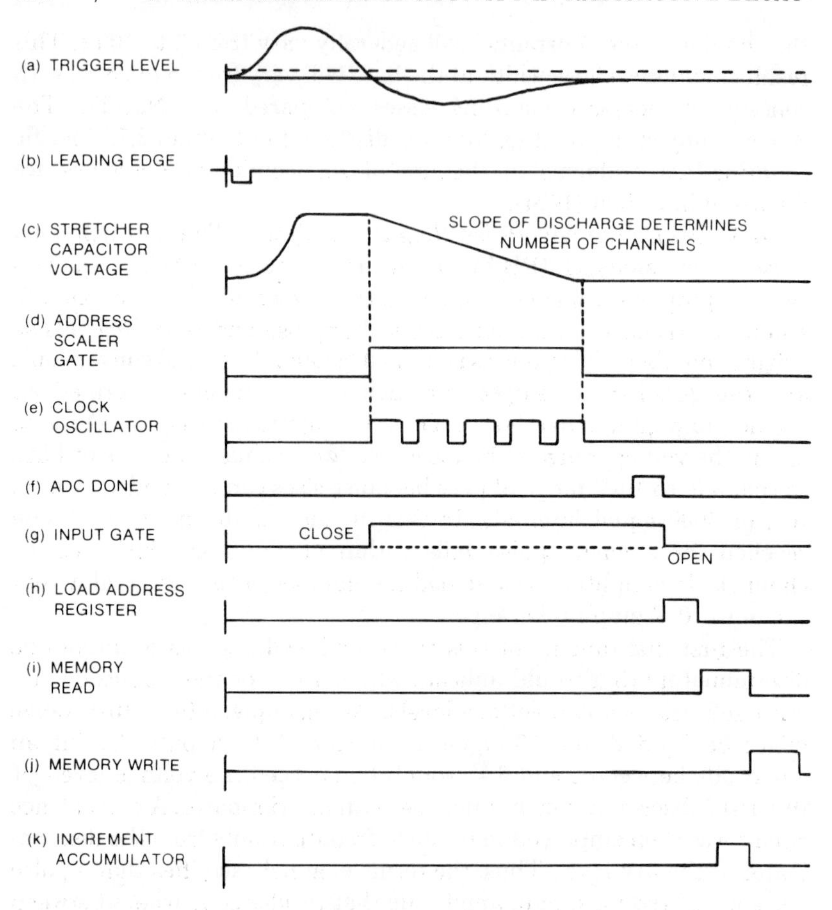

Fig. 49. Timing diagram of time-conversion type analog-to-digital converter (ADC) indicating sequence of events in the processing of a pulse to produce a digital number indicating relative amplitude of the maximum voltage.

amplitude and allowed to discharge at a constant rate, and so produces a linear sloping ramp as shown in the figure. During the voltage discharge, pulses from a constant frequency oscillator are counted in a scaler for the time required for the capacitor to discharge to zero, as shown in Figure 49c, d, and e. In this manner a digital number is accumulated in the scaler whose value is proportional to the amplitude of the stretched pulse.

The remaining operations of Figure 49 indicate logic signals controlling the accumulation of data in the analyzer memory. As previously described, a pulse-amplitude spectrum is obtained by measuring the amplitude of a large number of pulses and accumulating a histogram or pulse-amplitude distribution that is a tally of all pulse amplitudes. Figure 50 is a simplified block diagram of a multichannel pulse-height analyzer. As shown, it consists of three functional segments: (1) an ADC to provide a means of measuring pulse amplitude; (2) a magnetic-core or semiconductor memory consisting of a number of registers (one for each amplitude channel) to tally the number of pulses having an amplitude within a given voltage increment; and (3) an input/output section that provides for oscilloscope display of data and output of digital information in printed form or on other data storage media such as paper tape or magnetic tape. A photograph of

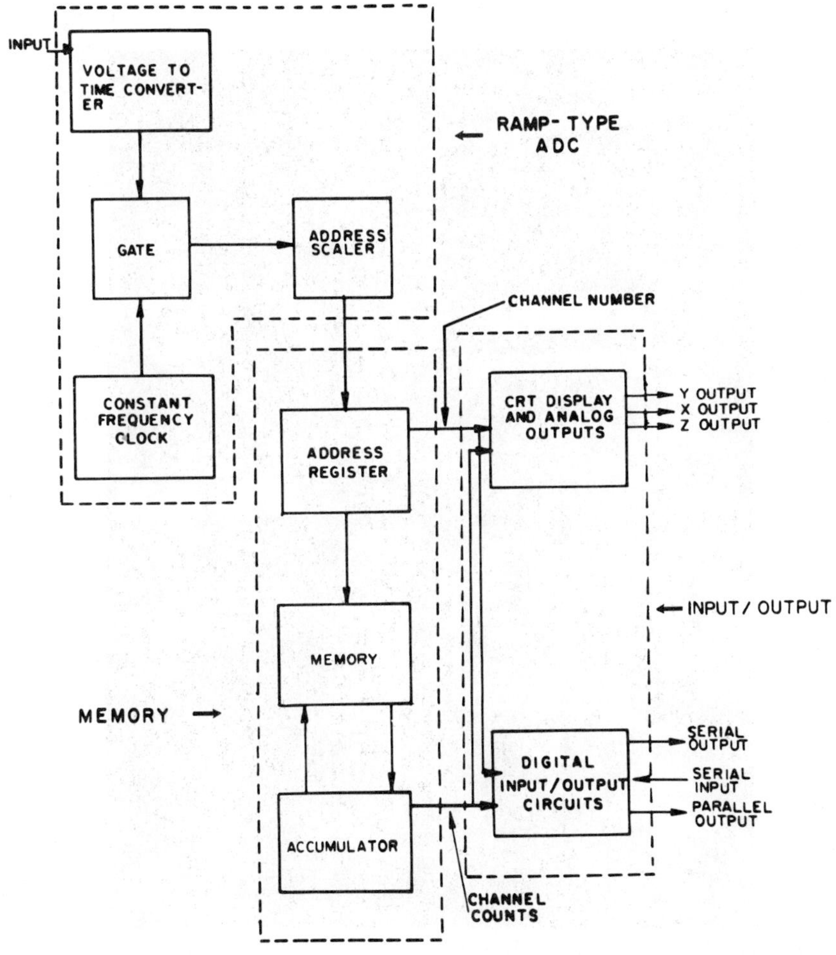

Fig. 50. Simplified block diagram of a multichannel pulse-height analyzer used in photon spectrometry.

a laboratory pulse-height analyzer is shown in Figure 51.

To assist in understanding the operation of a PHA, reference is made to the timing diagram shown in Figure 49, which illustrates the sequence of events that occurs when a pulse appears at the input of the analyzer. At the end of the capacitor discharge the gated scaler contains a number that represents the amplitude of the pulse, as shown in Figure 49c, d, and e. This number corresponds to one of the "channels" in the memory. The memory "bin" of this channel is identified and incremented by "1" to complete the measurement of an individual pulse. In this manner, as events occur in the detector, the amplitude of each pulse is measured and recorded in the PHA memory.

Fig. 51. Photograph showing typical commercial laboratory pulse-height analyzer system.

The measurement of many events will then result in the collection of a differential-amplitude spectrum.

4.2.4.4 *Instrumental Dead Time.* An examination of the timing diagram of Figure 49 indicates that the input to the analog-to-digital converter, Figure 49g, is closed during the processing of a pulse, so that any pulses arriving at the input of the ADC during this "dead time" will not be processed. If we are interested in the quantitative measurement of radiation striking the detector, this "counting loss" must be accounted for. The amount of "dead time" required in this type of ADC will be dependent upon the amplitude of the pulse which is being measured in terms of clock cycles. Typical ADCs use an oscillator frequency of 100 MHz, so that the time required to measure a pulse amplitude will vary from 0 to 40 μs for a 4000-channel amplitude scale. To this we must add the fixed time for memory addition and other "housekeeping" operations, which typically require 5 to 10 μs. In view of the difficulties involved in calculating an effective "dead time" or fraction of clock time during which incoming pulses cannot be measured, specialized circuitry is incorporated into PHAs to correct experimental data for this instrumental dead time. This correction for counting losses is accomplished with a crystal-controlled clock oscillator whose output pulses are stored in a register during a measurement. During the processing of pulses by the ADC the clock oscillator is turned off. Thus, in the absence of pulses at the input of the ADC, this register accumulates "real clock time". During a measurement, the action of the gate will only allow the clock to store pulses during the amount of time that the ADC is available for processing of data. In this manner the clock register accumulates elapsed "live time" to provide a running correction for loss of time due to ADC processing. In present pulse-height analyzers, these "live timers" provide automatic correction of ADC counting losses with an accuracy better than 1 percent over the normal operating range of input counting rate. Corrections for combined dead-time and pulse-pile-up losses have been considered in detail by Cohen (1974). For other methods of determining losses, see Sections 2.7.2.3 and 2.7.2.4.8.

4.2.5 *Experimental Effects in Photon Spectrometry*

In the analysis of x- and gamma-ray spectra obtained with scintillation and semiconductor systems, there are a number of experimental effects that must be considered in order to obtain correct photon energies and intensities. Important considerations include: radiation scattered into the detector from surrounding material, distortion of the pulse spectrum at high counting rates due to random coincidence effects, and limitations in the electronic system. A brief description of these problems is presented to provide insight into these experimental problems.

4.2.5.1 *Effect of Environment on Detector Response.* If we could isolate the source and detector from all surrounding material, the shape and magnitude of the observed pulse-height spectrum would be dependent only upon the energy of the gamma ray, the physical properties of the source and detector, and the geometrical arrangement. In practice, the shape of the observed pulse-height distribution is influenced by many factors. Since the response of the detector to its experimental environment must be understood in any attempt to analyze data, let us examine the many experimental variables that can influence the response of the detector. Figure 52 shows the experimental pulse-height spectrum resulting from the detection of a monoenergetic gamma-ray source (0.478 MeV) by a 3″ × 3″ NaI(Tl) detector. This spectrum was taken under experimental conditions that have been optimized to reduce extraneous effects to a minimum. The spectrum has been measured with and without a 670 mg/cm^2 polystyrene absorber that is normally interposed between detector and source to prevent the entry of beta particles into the detector. Extraneous contributions are shown in the spectrum that do not represent the response of the detector to gamma radiation directly incident on the detector.

The small peak appearing on the Compton distribution results from gamma rays Compton-scattered from the surrounding materials in the detector shield, air, source, and source mount. The calculated energies of these scattered gamma rays is equal to or less than 0.25 MeV. The magnitude and shape of the contribution from scattered radiation is dependent upon the geometrical arrangement. Figure 53 shows the pulse-height spectrum for a 0.835-MeV gamma ray measured in the three different shield configurations illustrated in Figure 54. Inspection of the scatter spectrum shows that the intensity and shape of the distribution of pulses attributable to scattering is inversely proportional to the size of the shield and to the Z of the absorbing material. The reduction in scattering from lead as opposed to iron is due to the relatively high photoelectric-absorption cross section of the high-Z material. It is of interest to inspect the shape of the scattered spectrum from the small iron shield. Note the presence of two distinct peaks in this distribution. These correspond to 180° single scattering and to processes involving two successive large-angle Compton events before striking the detector. An important deduction from these measurements is that appreciable scattering from surrounding material considerably degrades the quality of the spectrum. The presence of a large

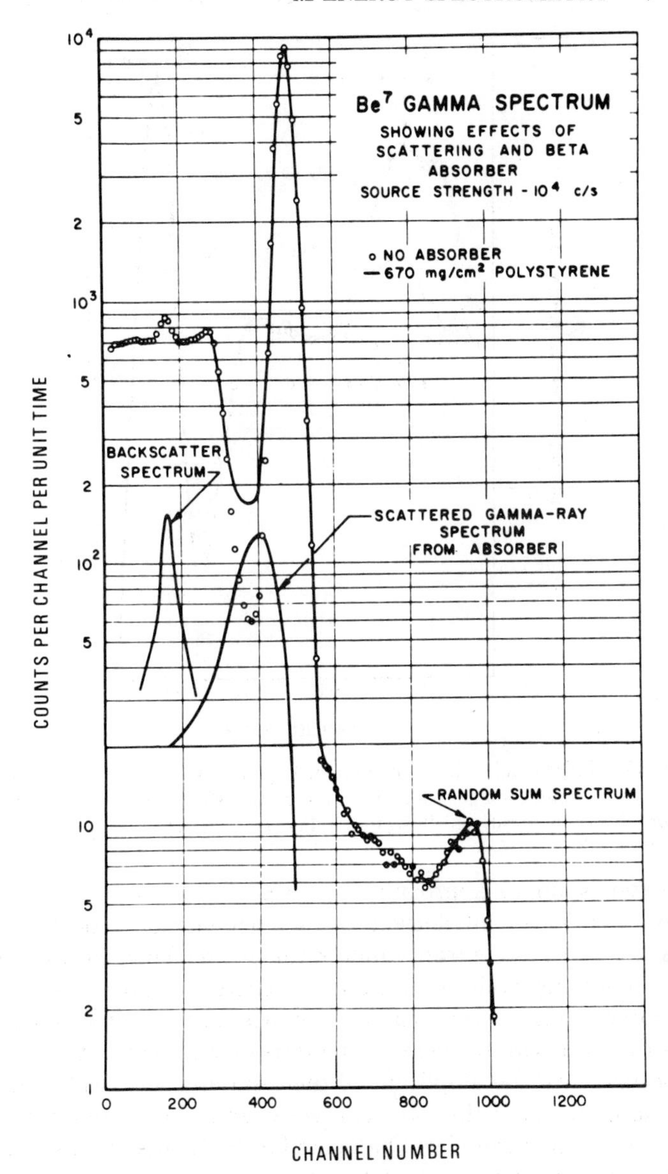

Fig. 52. Experimental response of NaI(Tl) detector to a monoenergetic gamma-ray source showing contribution to observed pulse-height spectrum from scattering and other experimental effects. (From Heath, 1964)

contribution to the pulse-height spectrum from scattering will reduce the "uniqueness" of the energy response.

4.2.5.2 *Summation Effects.* If we again examine Figure 52 in the region above the photopeak, we see a continuous distribution of pulses

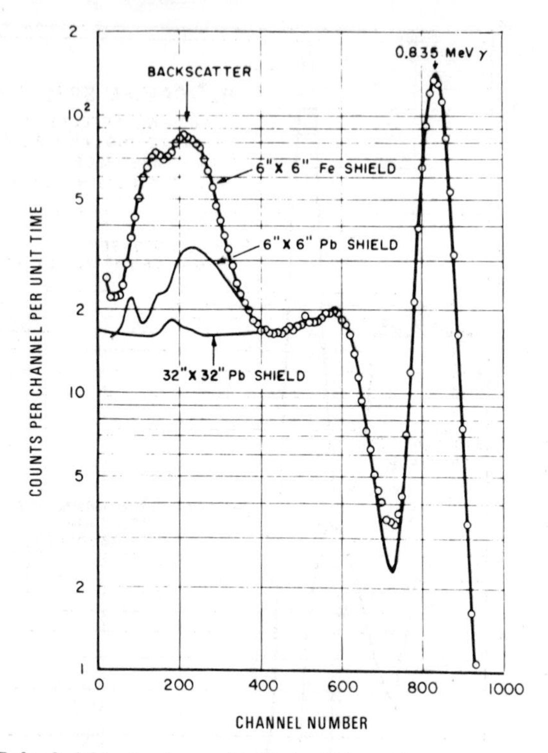

Fig. 53. Pulse-height spectrum obtained with a 3″ × 3″ NaI(Tl) scintillation detector from a monoenergetic gamma-ray source, ^{54}Mn, measured in three different detector-shield configurations, illustrated in Fig. 54. (After Heath, 1964)

that extends up to approximately twice the amplitude of the full-energy peak. This contribution to the monoenergetic energy response of the detector results from random time-coincidence between pulses. Since radiations emitted from a radioactive source are randomly distributed in time, there is a finite chance that two events may occur in the detector within the resolving time of the electronic circuitry. If this occurs, then the amplitude of the resultant pulse will be the sum of the amplitudes of the two separate events. The count rate in this "random sum spectrum" will be related to the source intensity and the system resolving time by the following simplified relationship:

$$N_{\mathrm{rss}} = 2\tau N^2, \tag{4.4}$$

where N is the input pulse rate, τ is the resolving time of the electronic system for pulse pairs, and N_{rss} is the total sum-pulse rate in the sum spectrum. Figure 55 illustrates the superposition of two pulses that occur within the sampling time of the pulse-height analyzer. Two

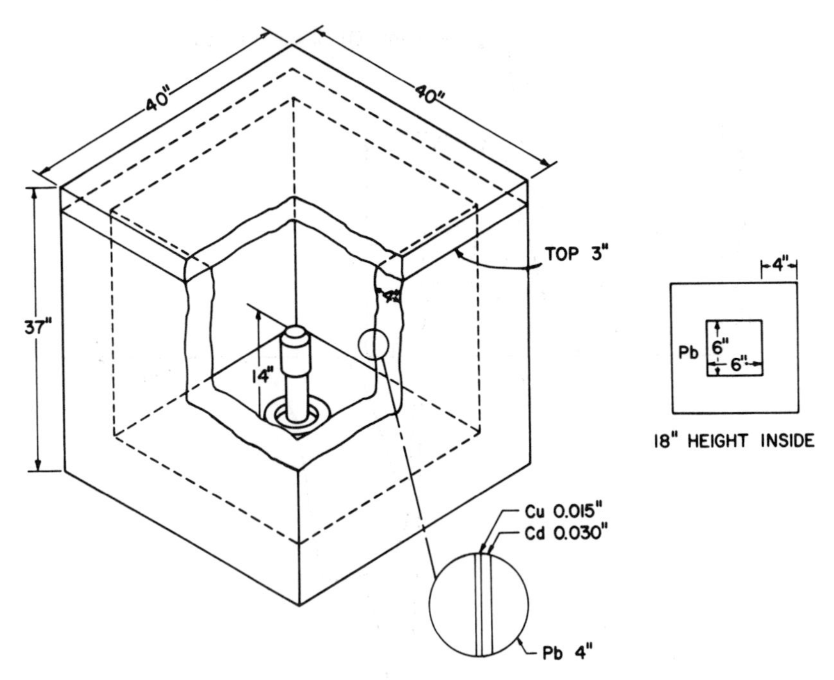

Fig. 54. Detector-shield configurations used to illustrate the effects of scattered radiation shown in Fig. 53. The large lead shield encloses a 32-inch-sided cubic volume. The dimensions shown in the smaller shield represent those of the two 6″ × 6″ lead and iron shields. The inside surfaces of the large lead shield are graded with cadmium and copper as shown in the insert. (After Heath, 1964)

pulses, with amplitudes V_1 and V_2 are shown partially superimposed in time. The sum of these two pulses is shown by the dashed line. Clearly, the shape of the resultant pulse will vary with the time interval between the two pulses and their respective amplitudes. The complex pulse that results can differ markedly from the normal pulse shape presented to the pulse-height analyzer. Depending upon the method used in the analog-to-digital converter to determine pulse amplitude, the recorded amplitude for such pulses can vary from V_1 to the highest amplitude reached in the region of overlap. The result is the observed "smeared" pulse-height distribution. Pulses in this sum spectrum represent the detection of photons directly incident upon the detector and so must be considered to be part of the energy response of the detector.

This random sum spectrum should not be confused with the so-called "coincidence sum spectrum", which is illustrated in Figure 56. This figure shows the spectrum of radiations emitted by ^{94}Nb. In the

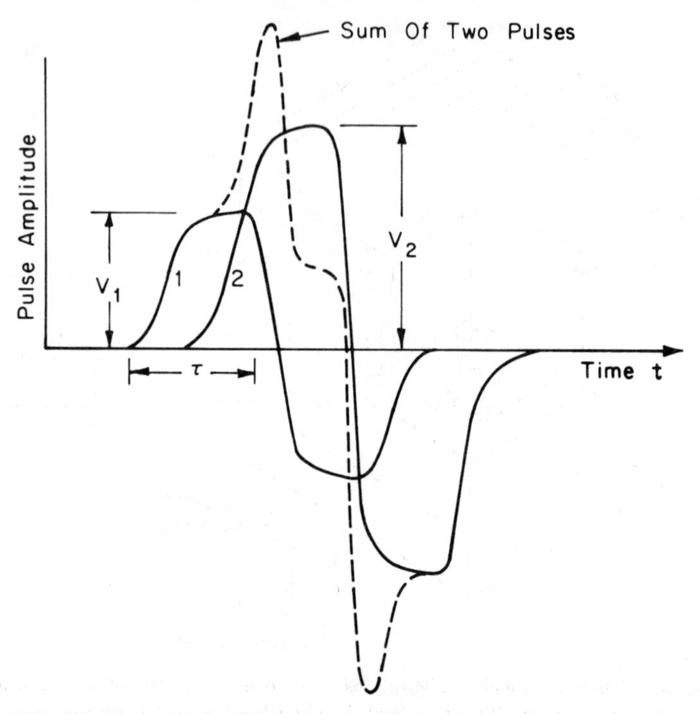

Fig. 55. Illustration of the effect of superposition of two pulses within the sampling time of the ADC. Variations in overlap will result in smearing of the random-coincidence sum spectrum. (From Heath, 1964)

decay of a nucleus of this radionuclide, two gamma rays are emitted simultaneously. In addition to the response of the detector to the two gamma rays individually, we also see a distribution of pulses that results from the simultaneous detection of both coincident gamma rays. The resultant sum spectrum is really a convolution of the two response functions for the individual gamma rays. The most prominent feature of this spectrum is the so-called sum peak that results from the coincident detection of pulses from the photopeaks of the two coincident gamma rays. The probability for the detection of pulses in this sum spectrum, N_{css}, is given by the following expression:

$$N_{css} = N_0 \epsilon_1 \epsilon_2 \overline{W}(\theta), \tag{4.5}$$

where N_0 is the disintegration rate of the source, ϵ_1 and ϵ_2 are the total detector efficiencies for the detection of gamma rays 1 and 2, respectively, and $\overline{W}(\theta)$ is an averaged factor to account for the angular distribution of the coincident gamma rays. Corrections for peak losses

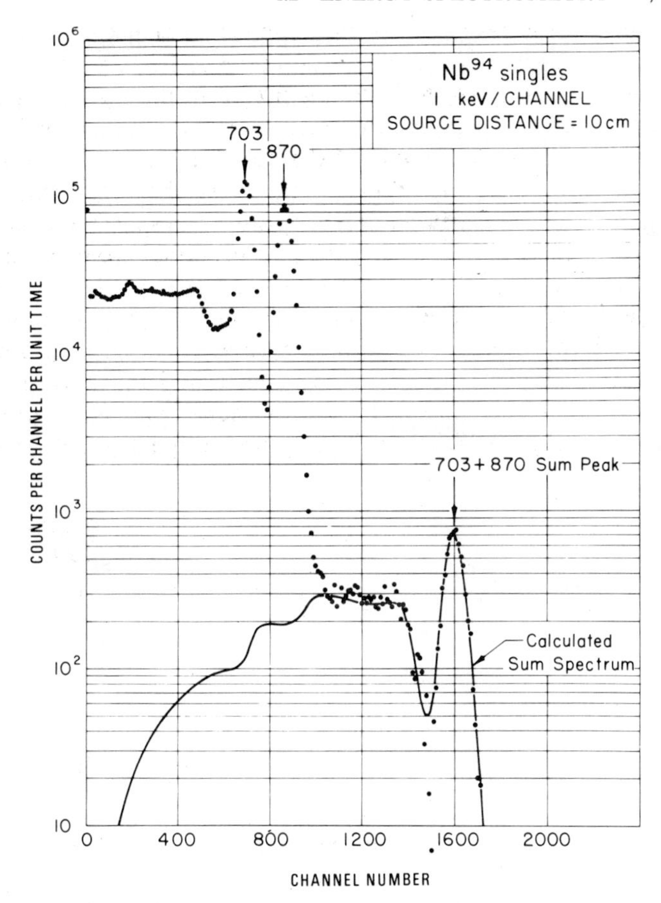

Fig. 56. Pulse-height spectrum of radiation emitted in the decay of ^{94}Nb showing the "true" coincidence spectrum (calculated sum spectrum) due to cascade emission of gamma rays in the decay of a single nucleus. The source is located 10 cm from a 3″ × 3″ NaI(Tl) detector. (After Heath, 1964)

and gains due to coincident summing in semiconductor detectors are given by McCallum and Coote (1975).

To summarize, we have seen that the shape of the pulse-height spectrum resulting from the detection of monoenergetic gamma rays may be influenced by many parameters. These include source preparation, source-detector geometry, size and shape of the detector, input count rate to the spectrometer, and the environment in which the measurement is performed. In addition, other radiations emitted by the source and the time relationships between emitted radiations may produce extraneous effects that must be considered in any analysis of the data. For these reasons it is most important that serious thought be given to the design of the experimental arrangement.

4.2.5.3 *Problems at High Counting Rates.* In the use of high-resolution spectrometers, experimental conditions are frequently far from ideal for optimum performance of the electronic systems employed. Perhaps the most difficult problems are those that arise when high counting rates are encountered. Without the use of special precautions, operation at high counting rates will result in serious shifts of zero and system gain and degradation of resolution. Most of these observed effects are the result of fluctuations in the zero reference baseline at the input of the analog-to-digital converter, produced by random fluctuations. Some insight into these effects and the nature of the specialized circuitry employed to correct these problems is provided by reference to Figure 57. To achieve optimum signal-to-noise ratio in

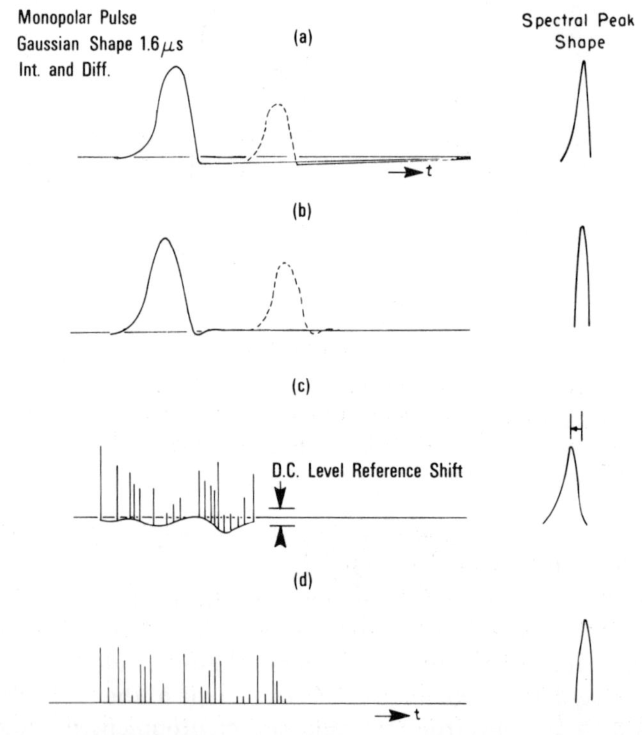

Fig. 57. The improved response of spectrometer systems with two circuit modifications. Parts (a) and (b) illustrate the response of a system at low count rates, with and without pole-zero cancellation, while (c) and (d) illustrate high-count rate performance of a system with and without d.c. restoration, in addition to pole-zero cancellation. (After Heath, 1969)

high-resolution systems, a monopolar pulse shape with equal RC integration and differentiation time constants is generally employed (Figure 57). The output pulse shape from such a filter network will exhibit a more-or-less Gaussian shape with equal rise and fall times. Following the pulse there will be a negative undershoot with a long recovery time constant (generally several hundred microseconds). At low count rates, the pulse will have returned to the original baseline and will have no influence on succeeding events. If, however, at high count rates, another pulse (indicated by the dotted line) occurs, the negative tail of the preceding pulse will result in the ADC recording a reduced amplitude. At reasonable rates (a few thousand pulses per second) the net effect of this will be an asymmetry of spectral peak shapes as indicated at the far right of Figure 57a.

A circuit technique, first proposed by Nowlin and Blankenship (1965), quite effectively reduces the undershoot from a monopolar pulse. The net effect of this technique, called pole-zero cancellation, is shown in Figure 57b. Here we see that the undershoot returns more quickly to the original baseline, so that the next pulse can follow more closely without reduction in amplitude. It should be pointed out that pole-zero cancellation must be accomplished on the complete preamplifier-amplifier system to be effective. The use of such networks in low-noise-amplifier systems is now quite common and, if adjusted properly in combination with a given detector, will provide considerable reduction in spectrum degradation from undershoot at moderate counting rates.

At high counting rates (in excess of 5000 s^{-1}), we still have problems from fluctuations in the baseline as shown in Figure 57c. In this figure, we have reduced the time scale to show the long-term character of the baseline shift. As a result of residual charge on coupling capacitors within the a.c.-coupled amplifier system, the baseline will vary in a random manner to produce a net negative shift in the zero reference and a general broadening of peaks in the spectrum, toward lower energy, as indicated on the right side of the figure. This effect can be reduced by the use of d.c.-baseline restoring circuitry. Typical circuits have been described by Chase and Poulo (1967). The action of these circuits is to remove the random fluctuations in the zero reference level. It should be stated that the input time constant on the restorer must be optimized for best performance at high rates. The most recent d.c. restorers have switch selection of this time constant to permit selection of the best operating conditions for a given experiment, although noise fluctuations will cause a small loss in resolution at low pulse rates for any setting. The result of restoring action, properly optimized, is illustrated in Figure 57d. Here we see that the zero reference baseline has been maintained at a constant level and the degradation of the peak resolution, seen without the restorer, has been largely removed.

4.2.6 *Quantitative Analysis of Gamma-Ray Spectra*

4.2.6.1 *Detector Efficiency.* The total detection efficiency ϵ_t, the fraction of gamma rays of energy E_γ, emitted from the source that interacts with the detector, can be calculated for known values of the absorption cross section μ_{en} for NaI(Tl) under a well-defined source-detector geometry. The equation derived for the case of a point source of radiation located on the extended axis of a right circular cylindrical detector is shown in Figure 58. In this expression, t_0 is the thickness of

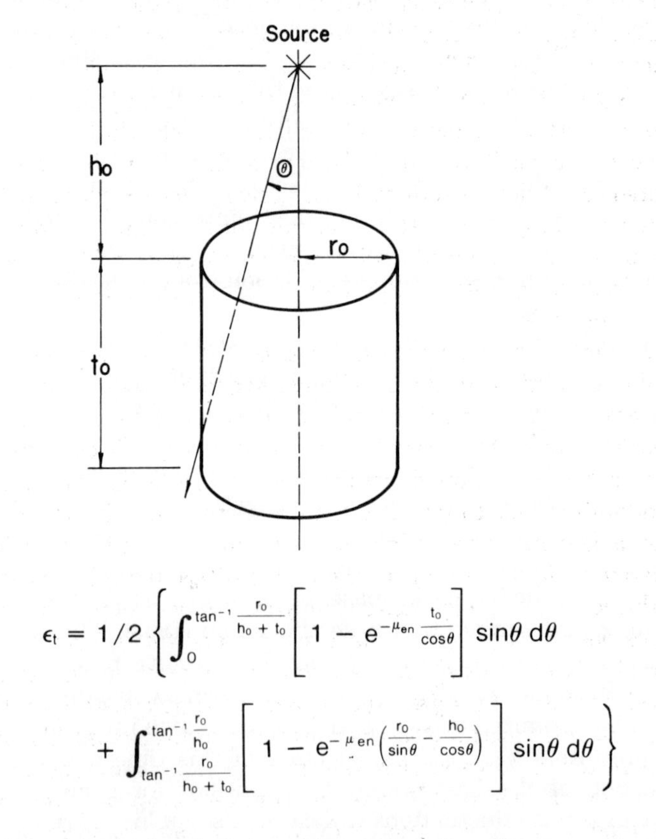

$$\epsilon_t = 1/2 \left\{ \int_0^{\tan^{-1}\frac{r_0}{h_0+t_0}} \left[1 - e^{-\mu_{en}\frac{t_0}{\cos\theta}} \right] \sin\theta \, d\theta + \int_{\tan^{-1}\frac{r_0}{h_0+t_0}}^{\tan^{-1}\frac{r_0}{h_0}} \left[1 - e^{-\mu_{en}\left(\frac{r_0}{\sin\theta} - \frac{h_0}{\cos\theta}\right)} \right] \sin\theta \, d\theta \right\}$$

Fig. 58. Expression for the calculation of the total detection efficiency, ϵ_t, of a right-circular cylindrical detector for a point source of radiation. (From Heath, 1964)

the detector, r_0 is the radius of the detector, and h_0 is the perpendicular distance between the source and the top surface of the detector. Extensive calculations have also been made for point, disk, and line sources located on the central axis of the detector (Heath, 1964).

In a solid material such as NaI(Tl), which has a density of 3.67 g/cm^3, the sensitive volume is very clearly defined. Since the amount of material that a secondary electron, produced in the solid phosphor, must traverse to leave a measurable amount of energy is negligible, edge effects are insignificant. Any error to be expected will be due to uncertainty in the value of μ_{en} used in the calculation.

4.2.6.2 *Photopeak, or Full-Energy-Peak, Efficiency.* Consider a point source of monoenergetic radiation and a detector of specified size that is isolated from all material that might scatter gamma rays into the detector that would not originally have been emitted into the cone of solid angle subtended by the detector. The number of events in the observed pulse-height spectrum will then be related to the emission rate from the source by the equation given in Figure 58. If, however, there is a contribution of scattered radiation to the detector response, it is evident that this relationship will hold only if this contribution is negligible. In view of the complex nature of the detector response, it is more convenient and precise to work in terms of the photopeak efficiency. This efficiency, ϵ_p, is defined as the probability that a gamma ray of energy E_γ, emitted from the source, will appear in the photopeak of the observed pulse-height spectrum. The efficiency and the photopeak area provide a measure of emission rate that is independent of scattering geometry.

4.2.6.3 *Peak-to-Total Ratio.* It would be difficult to calculate ϵ_p directly because of the large number of multiple processes that occur in a scintillation detector. For this reason it is convenient to use the following expression:

$$\epsilon_p = \epsilon_t P \tag{4.6}$$

where P is defined as the fraction of the total number of events in the pulse-height spectrum that appears in the photopeak (the peak-to-total ratio). The peak-to-total ratio has been determined experimentally by careful measurement of selected sources under experimental conditions that reduce scattered radiation to negligible levels. As an example, a plot of experimental values of the peak-to-total ratio for a 3″ × 3″ NaI(Tl) detector as a function of energy is shown in Figure 59.

4.2.6.4 *Emission-Rate and Energy Determination.* The emission rate of a single gamma ray will be given by the following relationship

$$N_0 = \frac{N_p}{\epsilon_t P A}, \tag{4.7}$$

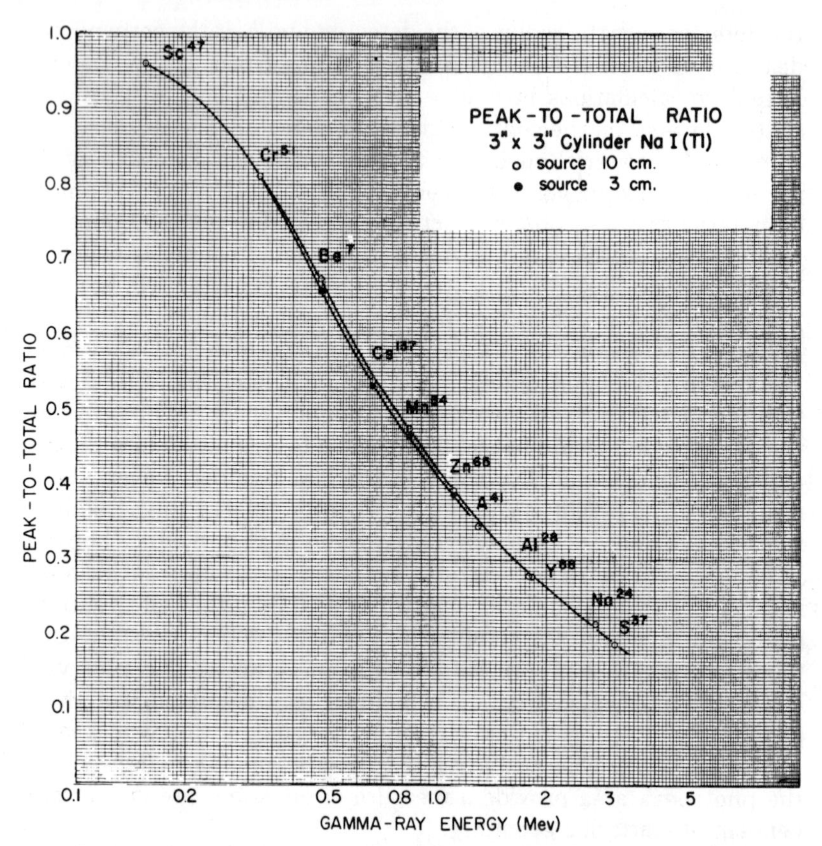

Fig. 59. Plot of experimental values of P, the peak-to-total ratio for a 3″ × 3″ cylindrical NaI(Tl) scintillation detector for source-detector distances of 3 and 10 cm. (From Heath, 1964)

where N_0 is the number of gamma rays emitted in unit time from the source, N_p (as shown in Figure 60) is the area under the photopeak, ϵ_t is the total detection efficiency for the source-detector geometry used, P is the appropriate value for the peak-to-total ratio, and A is a correction factor for absorption of the gamma radiation by any beta-particle absorber used in the measurement. This absorption correction is usually determined experimentally in order to account properly for measurement geometry. Values for ϵ_t, P, and A have been determined with considerable precision for a 3″ × 3″ cylindrical NaI(Tl) scintillation detector and are tabulated in Heath (1964).

As outlined above, a monoenergetic source of photons will produce a unique pulse-height distribution in a scintillation- or semiconductor-spectrometer system from which the energy and intensity of the photon source can be deduced. For most radionuclides the gamma-ray spectrum observed is generally more complex than that for a monoenergetic source. The observed detector response represents the summation of the response of the detector to individual gamma rays. To illustrate the methods required to interpret complex spectra, the response of a NaI(Tl) spectrometer to a source emitting several gamma-ray energies is shown in Figure 61. The radiation source was a composite of three radioisotopes each emitting a single gamma ray. As shown, the three sources were measured individually and in combination. If reference spectra for each radionuclide are available, then the composite spectrum can be decomposed by successively subtracting each component spectrum, normalizing in turn to the highest-energy peak. This "stripping" process can be used to analyze complex

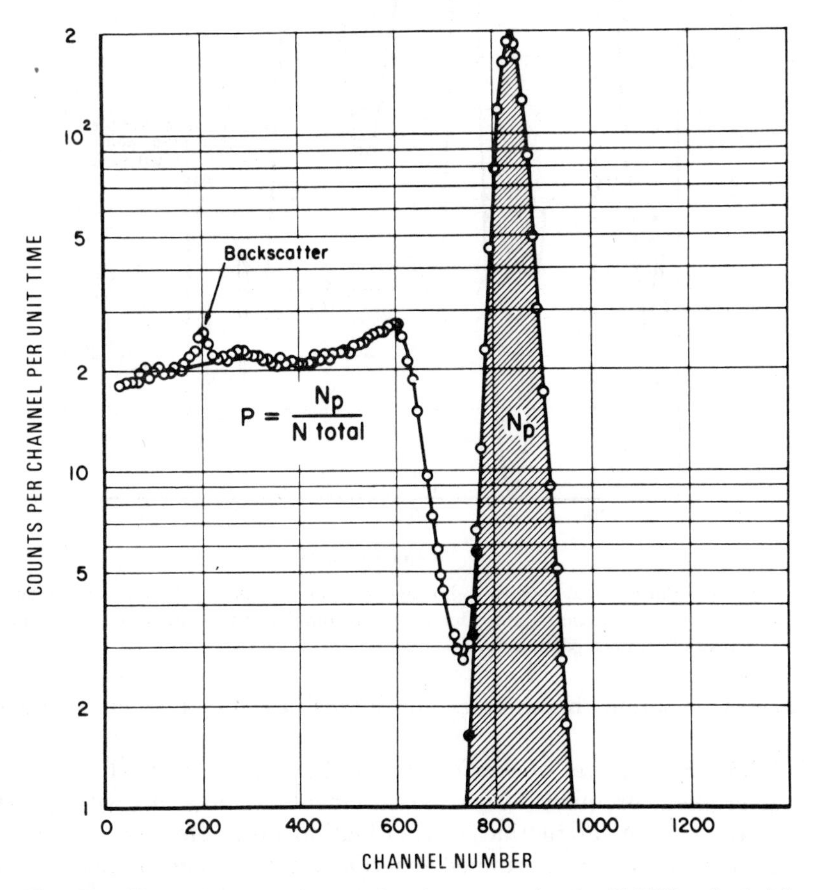

Fig. 60. Illustration of peak-to-total ratio concept for the NaI(Tl) pulse-height spectrum of ^{54}Mn. (From Heath, 1964)

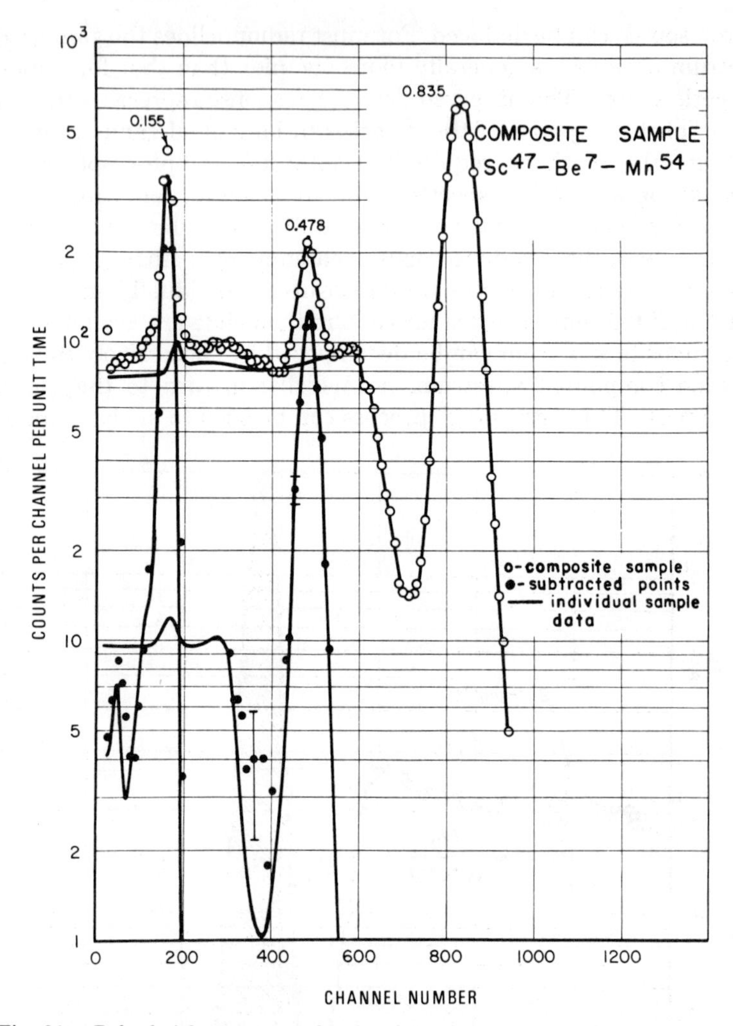

Fig. 61. Pulse-height spectrum showing detector response to three monoenergetic gamma-ray sources measured individually and in combination to illustrate approach to the analysis of complex gamma-ray spectra.

spectra to obtain the energies and intensities of component gamma rays.

4.2.6.5 *Efficiencies of Semiconductor Detectors.* Although the general problem or quantitative measurement of radiation with these devices is identical to that of scintillation detectors, the detection efficiency is usually determined experimentally by using a selected set of gamma-ray sources with precisely known gamma-ray emission rates (McNelles and Campbell, 1973; Debertin *et al.*, 1976; Hirshfeld *et al.*,

1976). The gamma-ray spectrum of each source is measured at a specific source-to-detector distance. Unlike the large-volume scintillator, where the sensitive volume of the detector can be determined with high precision, the semiconductor device has a poorly-defined sensitive volume. Significant edge effects and scattering components in the response function make it difficult to define the complete response of such a detector to monoenergetic radiation with sufficient precision. For this reason it is customary to determine the photopeak efficiency as a function of energy and measurement geometry by using standard reference sources. A typical photopeak, or full-energy peak, efficiency curve, as a function of photon energy, for a large coaxial Ge(Li) detector is shown in Figure 62. This plot presents the photopeak efficiency for this detector for point sources of radiation measured at a distance of 10 cm from the detector face. To permit calibration of Si(Li) and Ge(Li) detectors, calibrated gamma-ray sources are available from the National Bureau of Standards and several commercial suppliers. A set of standards that has been widely used to calibrate these detectors is shown in Table 9.

Source-detector geometry is an important consideration in the calibration of detector systems and must be reproducible with good precision. Rate-dependent factors, especially dead-time and summa-

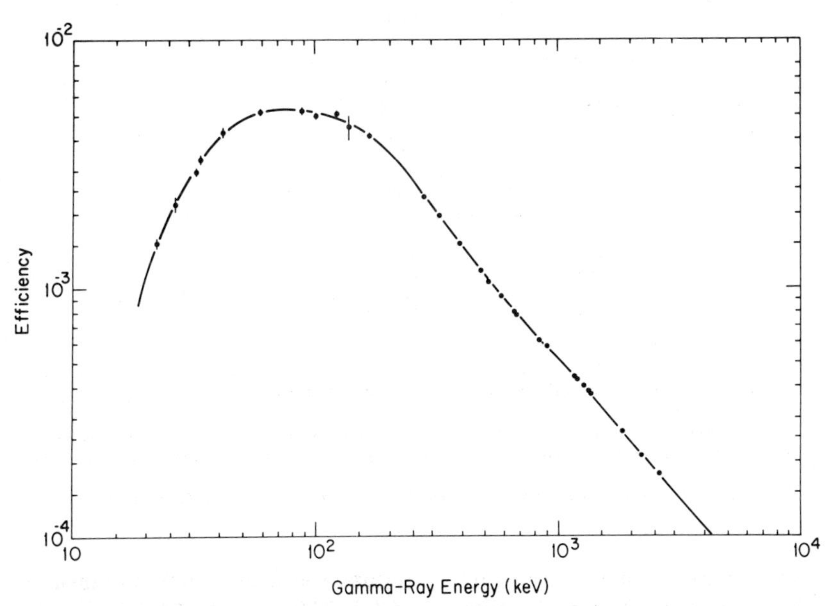

Fig. 62. Full-energy-peak efficiency measured, as a function of photon energy, for a 50-cm^3 coaxial Ge(Li) detector.

TABLE 9—*Photon point-source standard reference materials (SRM's) available from the National Bureau of Standards*[a]

Gamma-Ray Sources[b]

Parent radionuclide	Half life	Photon energy	Photon probability per decay[c]
		keV	
^{241}Am	433 ± 2 y	26.35	0.0258 ± 0.0022
		59.54	0.363 ± 0.004
^{109}Cd[d]	453 ± 2 d	88.04	—
^{57}Co[d]	270.9 ± 0.6 d	122.06	0.8559 ± 0.0019
^{139}Ce[d]	137.65 ± 0.05 d	165.85	0.8006 ± 0.0013
^{203}Hg[d]	46.59 ± 0.05 d	279.19	0.815 ± 0.008
^{51}Cr	27.704 ± 0.002 d	320.08	0.0980 ± 0.0010
^{113}Sn[d]	114.9 ± 0.1 d	391.69	0.6490 ± 0.0020
^{85}Sr[d]	64.85 ± 0.03 d	513.99	0.980 ± 0.010
^{207}Bi	38 ± 3 y	569.67	0.978 ± 0.005
		1063.62	0.74 ± 0.03
		1770.22	0.073 ± 0.004
^{137}Cs[d]	30.0 ± 0.2 y	661.65	—
^{94}Nb	(2.03 ± 0.16) × 10^4 y	702.63	1
		871.10	1
^{54}Mn	312.5 ± 0.5 d	834.83	0.999760 ± 0.000002
^{88}Y[d]	107 ± 1 d	898.02	0.934 ± 0.007
		1836.04	0.9935 ± 0.0003
^{65}Zn	244.1 ± 0.2 d	1115.52	0.5075 ± 0.0010
^{60}Co[d]	5.271 ± 0.001 y	1173.21	0.99900 ± 0.00020
		1332.46	1
^{22}Na	2.602 ± 0.002 y	1274.54	0.99940 ± 0.00020

X-Ray Sources

Parent radionuclide	Half life	Photon energy	Photon probability per decay
^{55}Fe	2.7 ± 0.1 y	5.89 (K_{α_2}), 5.90 (K_{α_1}), 6.49 (K_β)	See appendix A.3
^{85}Sr	64.85 ± 0.03 d	13.34 (K_{α_2}), 13.40 (K_{α_1}), 15 (K_β)	See appendix A.3
^{109}Cd	453 ± 2 d	21.99 (K_{α_2}), 22.16 (K_{α_1}), 24.9 (K_β)	See appendix A.3
^{125}I	60.14 ± 0.11 d	27.20 (K_{α_2}), 27.47 (K_{α_1}), 31 (K_β), 35.46 (γ)	

Gamma-Ray Sources of Radionuclides with Complex Decay Schemes

Parent radionuclide	Half life	Photon energy	Photon probability per decay
^{75}Se	120 ± 1 d	See Appendix A.3	See Appendix A.3
^{110m}Ag	250.8 ± 0.3 d	See Appendix A.3	See Appendix A.3
^{152}Eu	13.6 ± 0.2 y	See Appendix A.3	See Appendix A.3
^{228}Th[e]	1.91313 ±0.00088 y	—	—

[a] Detailed decay schemes for most of these radionuclides are given in Appendix A.3.
[b] In order of increasing gamma-ray energy.
[c] Photon probabilities per decay are not given for those SRM's which are certified in terms of photon emission rate.
[d] A mixed radionuclide source containing these nine radionuclides is distributed on an annual basis (Cavallo *et al*, 1973).
[e] ^{228}Th is primarily used as a standard because of the 2.6125-MeV gamma ray from the ^{208}Tl daughter.

tion effects, are also important in accurate efficiency determinations. These are described in Sections 4.2.4.4, 4.2.5.2, and 4.2.5.3. Section 2.7 treats rate-dependent corrections in general and Section 2.7.2.4.8 deals specifically with their application to gamma-ray spectrometry.

4.2.7 *Methods of Spectral Analysis—NaI(Tl) Detectors*

Typical pulse-height spectra obtained from gamma-ray spectrometers may result from the detection of a discrete set of monoenergetic gamma rays, or a gamma-ray continuum such as bremsstrahlung, or both in combination. The method of analysis to be applied in a given case will depend upon which type of spectrum is being measured.

4.2.7.1 *Graphical Analysis by Subtraction.* Graphical analysis has been the basic method employed for the analysis of pulse-height spectra by nuclear spectroscopists for some time. A plot of the pulse-height distribution is inspected and the component with the highest energy identified by its photopeak. The pulse-height distribution corresponding to the experimental response of the spectrometer to this component is then subtracted from the composite spectrum as indicated in Figure 61. The resulting difference is then inspected and the process repeated, in succession of decreasing energy, for each remaining component identified in the spectrum. This method is rather subjective and the accuracy of the result is highly dependent upon the number of gamma rays present in the spectrum, their relative intensities, the photopeak-to-total ratios, and the skill of the analyst. Estimation of error is highly dependent upon many experimental factors, such as gain shifts, different scattering conditions, and pulse pile-up.

4.2.7.2 *Linear Least-Squares Methods.* For the analysis of spectra that exhibit well-defined photopeaks the so-called least-squares method of analysis offers a useful alternative to the graphical technique described above. This method can be programmed for computer use and has proved to be very satisfactory for the analysis of many classes of data (Heath, 1966). It assumes that the pulse-height spectrum to be analyzed consists of a sum, from 1 through c, of the gamma rays of known energy from the radionuclides j, each represented as a pulse-height spectrum of i channels, a through n. It is further assumed that *standard spectra*, representing the response of the detector to these radionuclides, are available for comparison. The total count rate in channel i, C_i, can be represented by the composite of these standard spectra, where the count rate in channel i from the jth component will be S_{ij}. If the standard spectra are all normalized to the same disintegration rate, then it is necessary to include a normalizing factor, α_j, which is the ratio of the intensity of component j in the composite to the normalized standards. We may then write

$$C_i = \sum_{j=1}^{c} \alpha_j S_{ij} + \delta_i, \qquad (4.8)$$

where δ_i is the residual, or departure of the synthesized value from the observed rate in the ith channel.

If we assume that the only error is the random fluctuation in the channel count, the principle of least squares may be applied. Then the sum of the weighted squares of the residuals, δ_i, can be minimized over all channels, i.e.,

$$\text{minimize} \sum_{i=a}^{n} \left(C_i - \sum_{j=1}^{c} \alpha_j S_{ij}\right)^2 w_i, \tag{4.9}$$

where w_i is a weighting factor.

We then have a set of linear simultaneous equations that may be solved to obtain values for the α_j's. This is most conveniently done by a matrix technique that can be readily programmed for machine solution.

A number of computer programs using the simple linear least-squares technique have been applied to this problem (Salmon, 1961; Schonfeld *et al.,* 1966; Trombka and Schmadebeck, 1968). More sophisticated non-linear regression techniques have been developed that yield higher precision. An example is a program called SHIFTY (Helmer *et al.,* 1967b; see also Heath *et al.,* 1967) that allows the gain, as well as the intensity, of a whole spectrum, or of the portions of the spectrum for each radionuclidic component, to be varied during the fitting process.

This program computes intensities by a linear least-squares program. In addition, the best gain scale for each component is determined by a "direct-search" technique that operates as follows. Suppose a composite is being fitted with two components and both intensities and both gains are variable. The intensities of the two spectral components are computed using an initial value for the energy scale. Then, the sum of the weighted squares of the residuals is computed. With initial computed intensity, the first component is gain-shifted by a small increment, and a new value of residuals is computed. If the new value for the residual is smaller than the previous value, the new gain is retained; if not, a small decrement is tried. Next, the second component is increased or decreased in gain by small amounts. After the best gains among this set have been determined in this manner, a new set of intensities is calculated. This procedure continues until the shifts cease to have significant effect on the comparison between the original data and the sum of assumed components.

Figure 63 shows an example of an application of this program. A five-component spectrum containing ten gamma rays was analyzed with this program by using a library of experimental spectra as response functions for the individual radionuclidic components. For this purpose, the reference spectra must be first normalized to a standard gain or energy scale. To demonstrate the method, the five components were first measured individually, and were then combined

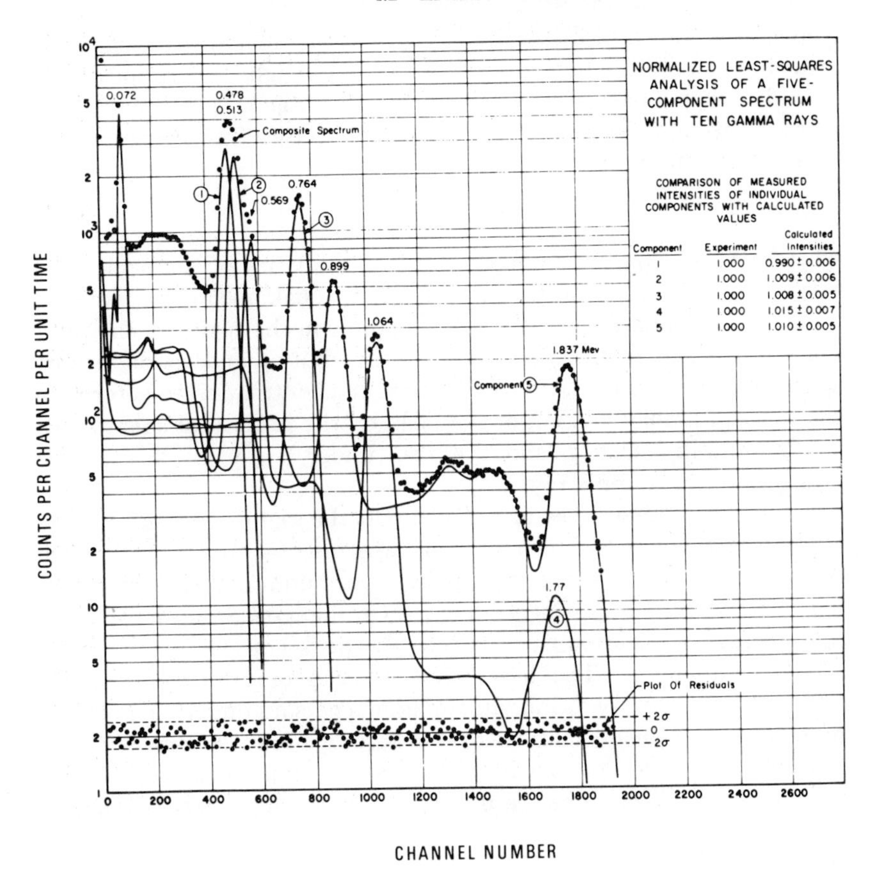

Fig. 63. Example of least-squares analysis of a five-component spectrum by using computer-data-analysis techniques.

to produce a composite spectrum. As indicated by the residuals in the figure, the least-squares analysis shows little systematic error.

4.2.8 *Methods of Spectral Analysis—Ge(Li) Detectors*

As previously indicated, spectra obtained with Ge(Li) detectors are different from those obtained with scintillation detectors in several ways. First, the energy resolution is greatly enhanced so that individual photopeaks are usually resolved. For this reason an important simplification can be made in the interpretation of pulse-height spectra obtained with these detectors. The essential information, gamma-ray energy and intensity, is represented by the position and magnitude of individual photopeaks. A section of a complex Ge(Li) gamma-ray spectrum is shown in Figure 64, in which individual x rays or gamma rays appear as peaks having finite width and height superimposed on a slowly varying background. The general approach to the analysis of spectra is to represent the peaks in a spectrum by a Gaussian function, $Y(x)$, having the form:

$$Y(x) = y_0 e^{-(x-x_0)^2/b_0} \tag{4.10}$$

where x is the channel number, x_0 is the midpoint of the peak, b_0 is related to the width of the peak, and y_0 is the peak maximum. By using a linear least-squares procedure, data points in the region of the peak may be fitted to a combined Gaussian distribution and a straight line or other functional form (for the background continuum) to determine the most probable location of the peak on the horizontal axis to give the gamma-ray energy, and the area under the peak to give the intensity of the gamma ray.

In Figure 64 two cases are shown. On the right-hand side of the figure a single peak is shown, with all experimental quantities defined. A Gaussian function and a straight-line continuum (background) have been fitted to the experimental points. On the left-hand side we see a more complex situation. In this case the peak consists of two unresolved components. The best representation of this region of the spectrum would be two Gaussian functions plus an underlying continuum. To accomplish this fitting procedure in a least-squares sense, it is necessary to resort to what is termed a "non-linear regression" technique. This technique is quite sophisticated and generally requires

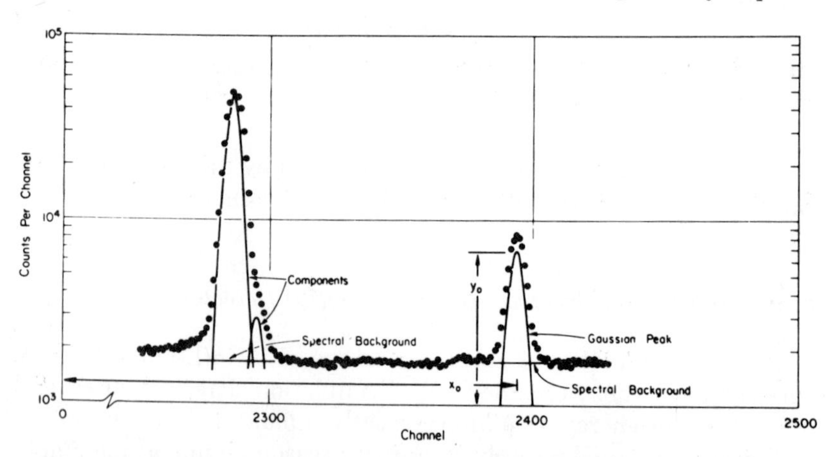

Fig. 64. Segment of a Ge(Li) gamma-ray spectrum. The first peak illustrates an unresolved pair. The second peak indicates the essential parameters used in the analysis of a spectrum to obtain peak location, width, and area.

the use of rather powerful computer programs. It has been described in detail by Helmer *et al.* (1967a).

The fitting of a photon peak with a Gaussian distribution enables one to locate the peak channel of the multichannel spectrum for the photon of interest. It is, however, necessary to extract from the raw data the emission rate and energy of each photon, and then to proceed, in many cases, to an identification of the radionuclide in the source giving rise to the observed spectrum. To obtain this information from the raw data it is necessary to perform the following operations, which are described in more detail in the next sections:

(1) locate the median positions of the peaks in the pulse-height distribution by fitting a Gaussian distribution as described above;
(2) establish a precise energy *vs.* channel-number scale by suitable calibration techniques;
(3) determine the energy of the photons observed in the pulse-height spectrum from the information obtained in the first two steps;
(4) determine the area (intensity) of the peaks in question after subtraction of the underlying continuum;
(5) compare the calculated energies of the peaks with a list of energies of all the characteristic x rays and gamma rays to identify the radionuclide sources of the photons detected in the spectrum;
(6) determine the concentration of each radionuclide present in the sample from the peak areas by suitable calibration techniques.

4.2.8.1 *Energy vs. Pulse-Height Scale.* The essential problem involved in the measurement of gamma-ray energy with Ge(Li) spectrometers is to establish a precise energy *vs.* pulse-height scale. This requires that the departures from linearity of the pulse-height analysis system be measured with a high degree of precision, to typically 0.01 channel, to permit interpolation. For this purpose it is customary to use radioactive sources emitting photons of precisely known energies and to identify the peak channel number corresponding to each photon energy. A typical calibration curve is shown in Figure 65, in which the departure from linearity over a range from about channel 200 to channel 4000 is not greater than ±0.3 channel.

Having calibrated the entire spectrometer system for non-linearity of response to energy deposited in the detector, the actual energy calibration of the system for the particular gain chosen is carried out by exposing the detector to photon-emitting reference sources with accurately known photon energies. The spectrometer system can now

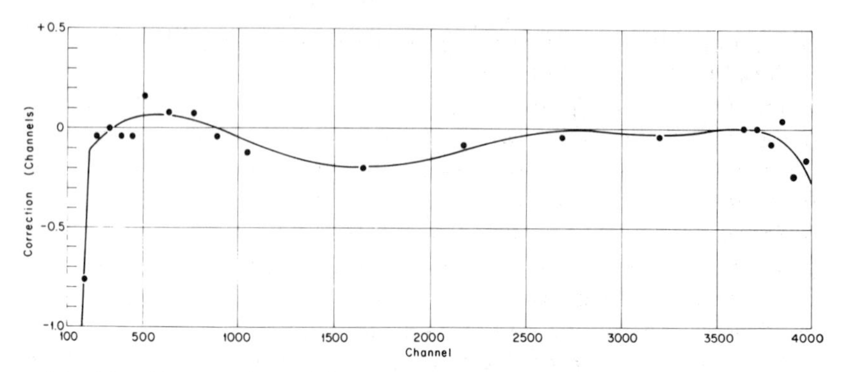

Fig. 65. Measured integral linearity of a 4096-channel gamma-ray-spectrometer system. (From Heath, 1969)

be used to measure photon energies and intensities over a wide energy range.

If it is required to make a more precise energy measurement of an unknown spectral line, an interpolation technique, illustrated in Figure 66, is employed. In this technique, the unknown line is recorded *simultaneously* with several lines of slightly greater and slightly less, and accurately known, energy. The energy of the unknown can then be determined as indicated in Figure 66, and the calibration curve of Figure 65, for the appropriate region of interest, can then be used to obtain a more precise interpolation for the energy of the unknown.

A list of reference calibration sources for which the energies have been measured with high precision is given in Table 10.

A block diagram of typical programs used for analysis of Ge(Li) pulse-height spectra is presented in Figure 67. After collection of a spectrum, the following procedure is used: (1) an algorithm is used to search the spectrum with a Gaussian function to determine the location of all significant peaks; (2) the contribution to each peak from the underlying background continuum is determined; (3) a least-squares fit of the data points, after background subtraction, in the vicinity of each peak is performed to determine the centroid and area of each peak; (4) instrumental linearity and energy calibration information is utilized to establish an energy *vs.* channel-number scale and the energy of each peak is determined from the peak position; and finally (5) the energies determined for each peak are compared with a reference list of known energies for all x rays and gamma rays emitted by each radionuclide to provide identification. A reference list of gamma-ray energies emitted by each radionuclide is provided in order of ascending energy for use with the "isotopic identification search routine". Detailed descriptions of computer programs and analysis techniques used

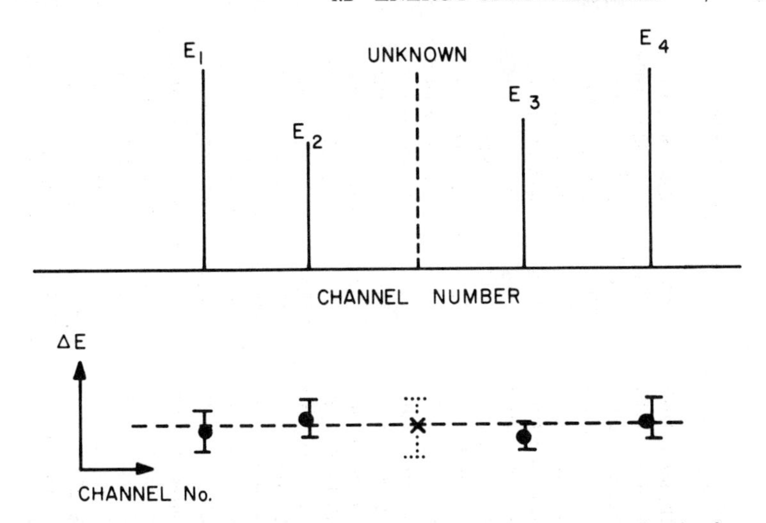

Fig. 66. Illustration of the measurement of a gamma-ray energy by simultaneous comparison with reference energy standards. ΔE is the deviation of the reference energy from the fitted function. (From Heath, 1969)

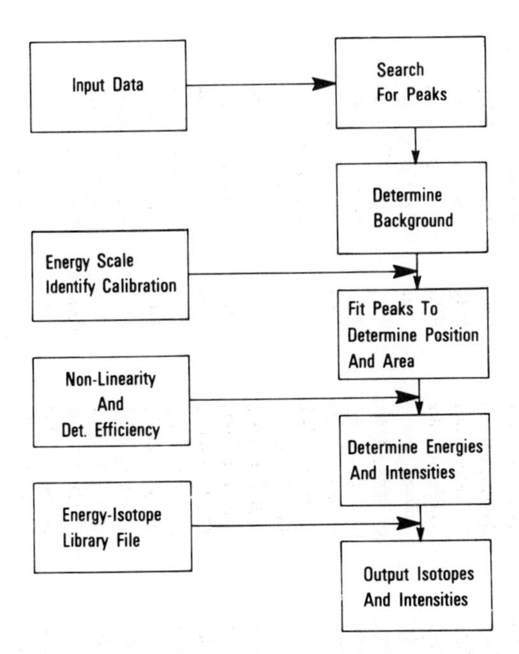

Fig. 67. Block diagram of data-analysis computer programs used in the analysis of Ge(Li) detector gamma-ray spectra for quantitative radionuclidic analysis of radioactive samples.

TABLE 10—*Gamma rays used as reference standards for energy calibration*[a]

Parent nuclide	Gamma-ray energy	Parent nuclide	Gamma-ray energy	Parent nuclide	Gamma-ray energy	Parent nuclide	Gamma-ray energy
	keV		keV		keV		keV
^{183}Ta	52.596 ± 0.001	^{182}TA	264.072 ± 0.006	^{94}Nb	702.627 ± 0.019	^{56}Co	1238.256 ± 0.027
^{241}Am	59.537 ± 0.001	^{75}Se	264.651 ± 0.008	^{110m}Ag	706.669 ± 0.020	^{182}Ta	1257.390 ± 0.028
^{75}Se	66.055 ± 0.009	^{133}Ba	276.397 ± 0.012	^{124}Sb	713.761 ± 0.018	^{160}Tb	1271.850 ± 0.026
^{182}Ta	67.750 ± 0.001	^{203}Hg	279.188 ± 0.006	^{185}Os	717.424 ± 0.018	^{182}Ta	1273.703 ± 0.028
^{153}Gd	69.676 ± 0.002	^{203}Pb	279.188 ± 0.006	^{124}Sb	722.767 ± 0.018	^{22}Na	1274.511 ± 0.028
^{133}Ba	80.998 ± 0.008	^{75}Se	279.528 ± 0.008	^{95}Zr-^{95}Nb	724.184 ± 0.018	^{182}Ta	1289.126 ± 0.029
^{183}Ta	82.919 ± 0.001	^{183}Ta	291.724 ± 0.006	^{110m}Ag	744.254 ± 0.020	^{59}Fe	1291.564 ± 0.028
^{170}Tm	84.254 ± 0.003	^{192}Ir	295.949 ± 0.006	^{95}Zr-^{95}Nb	756.715 ± 0.019	^{82}Br	1317.440 ± 0.027
^{182}Ta	84.680 ± 0.002	^{160}Tb	298.572 ± 0.006	^{110m}Ag	763.928 ± 0.019	^{132}Cs	1317.890 ± 0.028
^{183}Ta	84.712 ± 0.002	^{133}Ba	302.851 ± 0.015	^{95}Zr-^{95}Nb	765.786 ± 0.019	^{124}Sb	1325.478 ± 0.029
^{160}Tb	86.786 ± 0.002	^{75}Se	303.913 ± 0.007	^{82}Br	776.502 ± 0.016	^{60}Co	1332.464 ± 0.028
^{109}Cd	88.037 ± 0.005	^{192}Ir	308.445 ± 0.007	^{95m}Tc	786.184 ± 0.017	^{56}Co	1360.175 ± 0.029
^{160}Tb	93.917 ± 0.006	^{192}Ir	316.497 ± 0.007	^{58}Co	810.757 ± 0.021	^{124}Sb	1368.130 ± 0.029
^{75}Se	96.733 ± 0.002	^{51}Cr	320.078 ± 0.008	^{110m}Ag	818.018 ± 0.022	^{24}Na	1368.599 ± 0.029
^{153}Gd	97.432 ± 0.003	^{183}Ta	353.999 ± 0.004	^{95m}Tc	820.608 ± 0.019	^{182}Ta	1373.807 ± 0.030
^{183}Ta	99.080 ± 0.002	^{133}Ba	356.005 ± 0.017	^{82}Br	827.809 ± 0.019	^{110m}Ag	1384.267 ± 0.029
^{182}Ta	100.105 ± 0.001	^{183}Ta	365.615 ± 0.007	^{54}Mn	834.827 ± 0.021	^{182}Ta	1387.376 ± 0.030
^{153}Gd	103.180 ± 0.002	^{133}Ba	383.851 ± 0.020	^{95m}Tc	835.132 ± 0.018	^{82}Br	1474.853 ± 0.030
^{183}Ta	107.932 ± 0.002	^{113}Sn	391.688 ± 0.010	^{56}Co	846.751 ± 0.019	^{110m}Ag	1475.757 ± 0.034
^{177}Lu	112.954 ± 0.003	^{75}Se	400.646 ± 0.009	^{58}Co	863.935 ± 0.018	^{144}Ce	1489.124 ± 0.032
^{182}Ta	113.673 ± 0.002	^{203}Pb	401.315 ± 0.013	^{94}Nb	871.099 ± 0.018	^{110m}Ag	1505.006 ± 0.032
^{182}Ta	116.418 ± 0.002	^{183}Ta	406.589 ± 0.006	^{185}Os	874.814 ± 0.019	^{110m}Ag	1562.264 ± 0.033
^{75}Se	121.115 ± 0.003	^{198}Au	411.794 ± 0.008	^{160}Tb	879.364 ± 0.018	^{82}Br	1650.296 ± 0.034
^{57}Co	122.063 ± 0.004	^{110m}Ag	446.791 ± 0.010	^{185}Os	880.272 ± 0.019	^{58}Co	1674.679 ± 0.036
^{185}Os	125.358 ± 0.004	^{192}Ir	468.062 ± 0.010	^{84}Rb	881.595 ± 0.018	^{124}Sb	1690.942 ± 0.036

^{75}Se	136.000 ± 0.005	^{7}Be	477.593 ± 0.012	^{192}Ir	884.523 ± 0.018	^{207}Bi	1770.188 ± 0.037
^{57}Co	136.473 ± 0.004	^{192}Ir	484.570 ± 0.011	^{110m}Ag	884.667 ± 0.018	^{56}Co	1771.303 ± 0.037
^{59}Fe	142.648 ± 0.004	^{85}Sr	513.996 ± 0.016	^{46}Sc	889.258 ± 0.018	^{125}Sn	1806.652 ± 0.038
^{183}Ta	144.127 ± 0.002	^{82}Br	554.334 ± 0.012	^{88}Y	898.021 ± 0.019	^{56}Co	1810.672 ± 0.039
^{141}Ce	145.440 ± 0.003	^{207}Bi	569.689 ± 0.013	^{110m}Ag	937.483 ± 0.020	^{88}Y	1836.014 ± 0.037
^{182}Ta	152.434 ± 0.002	^{95m}Tc	582.068 ± 0.013	^{160}Tb	962.295 ± 0.020	^{125}Sn	1889.844 ± 0.039
^{182}Ta	156.387 ± 0.002	^{228}Th	583.174 ± 0.013	^{160}Tb	966.151 ± 0.020	^{84}Rb	1897.727 ± 0.038
^{199}Au	158.370 ± 0.003	^{192}Ir	588.572 ± 0.012	^{124}Sb	968.188 ± 0.022	^{56}Co	1963.669 ± 0.040
^{185}Os	162.854 ± 0.008	^{185}Os	592.066 ± 0.014	^{84}Rb	1016.143 ± 0.021	^{132}Cs	1985.581 ± 0.041
^{139}Ce	165.853 ± 0.007	^{124}Sb	602.715 ± 0.013	^{56}Co	1037.815 ± 0.022	^{125}Sn	2002.089 ± 0.043
^{182}Ta	179.393 ± 0.003	^{192}Ir	604.401 ± 0.012	^{95m}Tc	1039.247 ± 0.022	^{56}Co	2015.133 ± 0.040
^{59}Fe	192.344 ± 0.006	^{192}Ir	612.450 ± 0.013	^{82}Br	1043.973 ± 0.022	^{56}Co	2034.706 ± 0.041
^{183}Ta	192.646 ± 0.005	^{82}Br	619.088 ± 0.013	^{124}Sb	1045.106 ± 0.022	^{124}Sb	2090.889 ± 0.044
^{160}Tb	197.030 ± 0.004	^{110m}Ag	620.342 ± 0.016	^{207}Bi	1063.635 ± 0.024	^{56}Co	2113.049 ± 0.045
^{182}Ta	198.356 ± 0.004	^{124}Sb	645.835 ± 0.017	^{198}Au	1087.633 ± 0.024	^{144}Ce	2185.608 ± 0.046
^{75}Se	198.596 ± 0.007	^{185}Os	646.111 ± 0.017	^{59}Fe	1099.224 ± 0.025	^{125}Sn	2200.965 ± 0.047
^{95m}Tc	204.117 ± 0.005	^{110m}Ag	657.744 ± 0.017	^{65}Zn	1115.518 ± 0.025	^{56}Co	2212.862 ± 0.049
^{199}Au	208.196 ± 0.005	^{137}Cs	661.638 ± 0.019	^{46}Sc	1120.516 ± 0.025	^{125}Sn	2275.710 ± 0.048
^{177}Lu	208.362 ± 0.010	^{132}Cs	667.698 ± 0.017	^{182}Ta	1121.272 ± 0.026	^{56}Co	2598.400 ± 0.053
^{160}Tb	215.641 ± 0.004	^{198}Au	675.871 ± 0.018	^{60}Co	1173.208 ± 0.025	^{228}Th	2614.471 ± 0.054
^{82}Br	221.476 ± 0.005	^{110m}Ag	677.601 ± 0.018	^{56}Co	1175.067 ± 0.026	^{24}Na	2753.965 ± 0.056
^{182}Ta	222.110 ± 0.003	^{203}Pb	680.495 ± 0.017	^{160}Tb	1177.934 ± 0.024	^{56}Co	3201.878 ± 0.064
^{182}Ta	229.322 ± 0.006	^{110m}Ag	686.998 ± 0.019	^{182}Ta	1189.022 ± 0.027	^{56}Co	3253.341 ± 0.065
^{185}Os	234.158 ± 0.010	^{144}Ce	696.492 ± 0.019	^{182}Ta	1221.376 ± 0.027	^{56}Co	3272.912 ± 0.065
^{95m}Tc	253.066 ± 0.006	^{82}Br	698.358 ± 0.016	^{182}Ta	1230.989 ± 0.028	^{56}Co	3451.064 ± 0.069

[a] From Heath (1974).

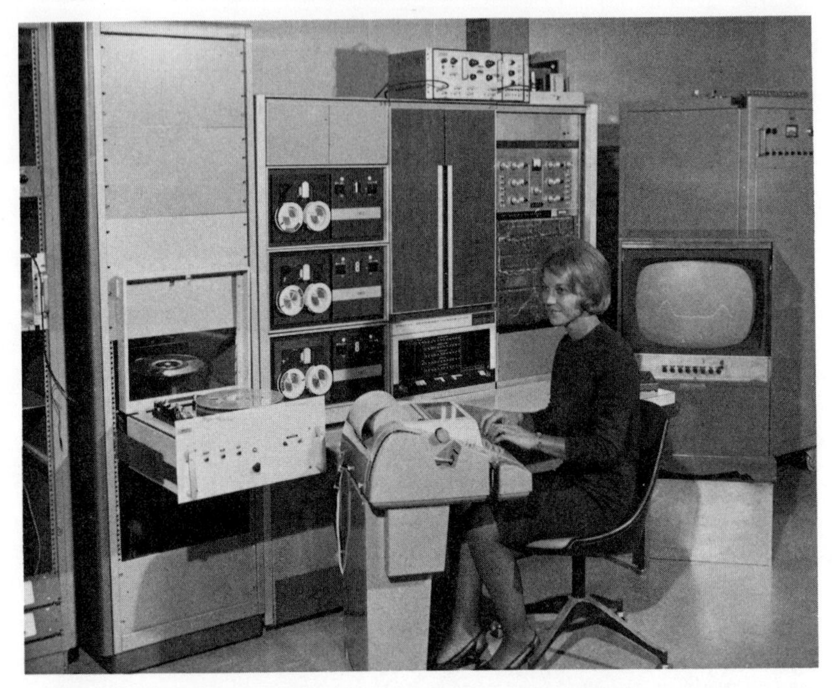

Fig. 68. Photograph showing a modern gamma-ray-spectrometer system employing a small digital computer for real-time data acquisition and analysis. (From Heath, 1969)

for routine analysis of complex gamma-ray spectra for radionuclide assay are given by Helmer and Putnam (1972), Cline *et al.* (1973), Helmer *et al.* (1975), and Meyer *et al.* (1976).

4.2.8.2 *Computer-Based Data Systems.* The problems encountered in the analysis of pulse-height data from x-ray and gamma-ray spectrometers require the application of rather sophisticated techniques to obtain the desired results. The advent of small, inexpensive digital processors has resulted in the development of automated techniques for the analysis and acquisition of pulse-height spectra. On-line computer data systems have been successfully employed for the analysis of pulse-height spectra in nuclear physics applications for some time (Heath, 1969; Gunnink and Niday, 1976). Of particular importance has been the development of graphic techniques employing the use of large-display oscilloscopes, function keyboards, and other peripheral devices to permit operator interaction with the data to assist in the analysis procedure. As an example, Figure 68 is a photograph of a small computer data system that utilizes a small digital processor. This system incorporates magnetic tape, a large-area CRT display oscilloscope, function keyboard, and control panel. The major difference between such a system and a conventional "hard-wired" pulse-height

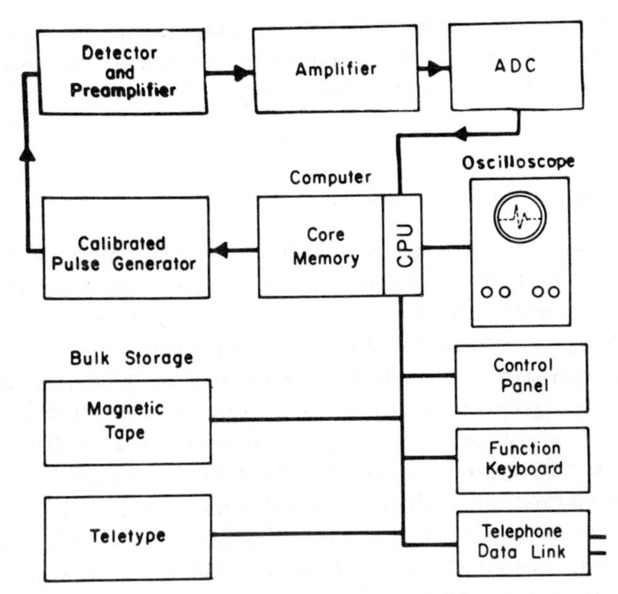

Fig. 69. Typical computer-coupled pulse-height spectrometer.

analyzer lies in the ability of the processor to modify information provided by the digital-to-analog converter under control of a programmed sequence of operations that may assume any form the user desires. It is this versatility that makes a digital processor desirable for data acquisition. Figure 69 shows a block diagram of a typical processor data acquisition and analysis system suitable for gamma-ray spectral analysis. In a conventional "hard-wired" pulse-height analyzer, analysis of a pulse from the detector produces a number corresponding to a particular address in the analyzer memory. This address is uniquely defined by the memory control circuitry and the procedure is then to initiate a process that adds the number "1" to the contents of that memory location. If this same number representing the amplitude of a given pulse is entered into the accumulator of a digital processor, this information can be modified at will. For example, it might be stored in a number of different locations for different purposes, modified to change the energy scale or to correct for non-linearities or instabilities in the electronic system, and so on.

4.3 Liquid-Scintillation Counting

Indirect or comparative measurement of radionuclides can be advantageously carried out by liquid-scintillation counting. Although the activity of essentially all radionuclides can be measured by liquid scintillation, this technique is particularly useful for the assay of low-energy beta-emitting radionuclides such as ^{3}H and ^{14}C.

4.3.1 *Composition of Liquid Scintillator*

As discussed in Section 2.4.5, liquid scintillation is based on the interaction of ionizing radiation with a multicomponent mixture consisting of a solvent and one or two scintillators, the second normally the wavelength shifter. The solubilization of the sample containing the radionuclide requires the addition of solubilizing, chelating, and other agents. Although *p*-xylene is the most efficient solvent, toluene is more frequently used because it is more economical. In addition, if the instrument is refrigerated, the melting point of +12°C of *p*-xylene compares unfavorably with the melting point of toluene of −95°C.

The most commonly used primary scintillator is 2,5-diphenyloxazole (PPO). This scintillator is efficient, sufficiently soluble in common solvents, available in sufficient purity, and reasonably economical. Terphenyl is somewhat less efficient than PPO, but it is substantially more economical. Terphenyl cannot be used with dioxane-naphthalene systems because the terphenyl fluorescence band overlaps the absorption spectrum of naphthalene. There are many other scintillators with properties suitable for specific applications (Birks, 1964; Kowalski *et al.,* 1967).

One of the earliest secondary scintillators was 1,4-*bis*-(2-(5-phenyloxazolyl))-benzene (POPOP) (see Section 2.4). This scintillator is still extensively used. The dimethyl derivative of POPOP is somewhat more soluble than POPOP; however, its spectral-matching response is poor for bialkali phototubes. The most efficient secondary scintillator is presently *p*-*bis*-(*o*-methylstyryl)-benzene (*bis*-MSB), which has acceptable solubilities in common solvents.

Primary scintillators are used in concentrations ranging from 4 to 10 g/l of solvent depending upon the degree of quenching. The concentration of the secondary scintillator is 10 to 20 percent, by mass, of that of the primary scintillator.

4.3.2 *Measurement of Aqueous Samples*

Since toluene and other aromatic solvents are not miscible with water, numerous attempts have been made to develop alternative mixtures in order to incorporate aqueous samples, including tritiated water, into the scintillation liquid. The oldest generally applicable technique used dioxane as the primary solvent. This solvent is miscible with water in any ratio and, if 100 to 120 g naphthalene/l is added to it, it is almost as efficient as aromatic solvents. Because of the addition of naphthalene there is, however, a practical limitation, of approximately 25 percent by volume, to the amount of water that can be added to the dioxane-naphthalene mixture. In addition, the solubility of inorganic compounds in this mixture is limited. The addition of dioxane, various alcohols, and certain other solvents to toluene makes it possible to dissolve water in toluene up to a certain limit.

These ternary mixtures are no longer used, since more convenient and more stable emulsion systems have been developed. The emulsion system consists of an aromatic solvent containing appropriate scintillators mixed with a detergent (Patterson and Greene, 1965). A large number of detergents have been tested (Lieberman and Moghissi, 1970) and the results indicate that alkyl phenyl polyethylene glycol ethers are the most desirable. (Many of these are known by the Rohm and Haas trade name of Triton.) Usually the alkyl group contains eight or nine carbon atoms, although mixtures of higher and lower homologs have been used. These materials are available under various brand names from a number of commercial suppliers as are, also, premixed scintillators. The great advantage of these premixed scintillators is their ability to incorporate up to 50 percent of water, by volume, while maintaining reasonable efficiencies, e.g., 20 percent for tritium in counting samples containing 40 percent of water. Also, such water-detergent-solvent emulsions have the ability to incorporate considerable quantities of various salts (Moghissi, 1970).

Solubilization or combustion of tissue samples is described in Sher *et al.* (1971) and in a recent NCRP publication (NCRP, 1976a), and will not be further described here.

4.3.3 *Measurement of* ^{14}C

^{14}C, which is widely used in the life sciences, can be conveniently assayed by liquid-scintillation counting. Essentially all compounds containing carbon can be labeled with ^{14}C and a large number of labeled compounds are commercially available. In many cases the compound to be measured is converted to CO_2 which is subsequently introduced into a scintillation liquid. The solubility of CO_2 gas in toluene is relatively small. However, up to 50 cm^3 of CO_2 (at STP) corresponding to 27 mg of carbon can be introduced into 20 ml of scintillation liquid provided the mixture is deaerated (Horrocks, 1968). A more common technique for introduction of CO_2 into a scintillation liquid is the absorption of CO_2 by a base that is subsequently dissolved in the proper mixture. The choice of bases is often arbitrary. Accordingly, a large number of bases has been used. An efficient method developed by Woeller (1961) uses phenethylamine as the absorbent for CO_2. Woeller's mixture consisted of 46 percent toluene, 27 percent ethanol, and 27 percent phenethylamine containing appropriate scintillators. This mixture is capable of absorbing over 500 mg of carbon corresponding to approximately 930 cm^3 of CO_2 in about 22 ml of the liquid scintillator. The absorption of CO_2 in this mixture is, however, slow. Ethanolamine is also used as an absorbent for CO_2 (Jeffay and Alvarez, 1961), as is also methoxyethylamine. The latter is particularly effective because both its carbamate and carbonate are soluble in toluene.

A more convenient method was developed by Moghissi *et al.* (1971). This method uses NaOH and is capable of absorbing up to 100 cm^3 of CO_2. A detergent mixture consisting of 380 g of Triton QS44 per liter of Triton N101 is used. The ratio of detergent to solvent, containing appropriate scintillators, is approximately 1:2 by volume. Sodium carbonate is introduced as an aqueous solution into the mixture. As the quenching properties of NaOH and Na_2CO_3 in this mixture are similar, the quantity of added CO_2 does not influence the quenching of the mixture. Numerous other bases have been proposed and used for absorption of CO_2.

4.3.4 *Measurement of Other Radionuclides*

The activities of many other radionuclides such as ^{32}P, ^{35}S, ^{55}Fe, ^{63}Ni, and ^{125}I can be advantageously measured by liquid-scintillation counting. The techniques employed are essentially the same as those described in Sections 4.3.2 and 4.3.3. In the assay of ^{35}S, SO_2 can be used in much the same way as CO_2. There are, however, certain radionuclides that are of particular interest and, or, require specific preparation. In the measurement of the Auger electrons resulting from the decay of, say, ^{55}Fe, care must be taken to minimize quenching effects.

Alpha-particle emitters are often assayed by liquid-scintillation counting. Lindenbaum and Smyth (1971), for example, measured ^{239}Pu and ^{241}Am in animal tissues using an emulsion scintillator. Other measurements of actinides in environmental samples are described by McDowell (1971) and in Section 6.1.2. In the case of plutonium, the composition of the liquid scintillator must be carefully controlled to prevent polymerization and deposition, or precipitation, of the plutonium.

4.3.5 *Background*

Consideration of background in liquid scintillation is more significant than in many other systems, since certain reactions may occur in the scintillation liquid that produce light at levels exceeding that resulting from radioactive decay of the sample.

Since most liquid-scintillation counters operate in the coincidence mode, background originating from thermionic emission from the photocathode is low and generally of the order of one count per minute, or less. Somewhat more significant is the contribution of electronic noise based on interactions between two phototubes that are facing each other. More details on this phenomenon, often referred to as cross-talk, as well as a discussion of other contributions to the background due to the phototubes, are given in Birks (1964).

The contribution of cosmic radiation to background is largely eliminated by shielding and in a modern liquid-scintillation counter does not exceed a few counts per minute.

There are several other phenomena that may contribute to the background. Excitation of the glass vial, vial cap, and scintillation liquid by sunlight or fluorescent light may lead to delayed light emission. The decay half-time of this phosphorescence can exceed hours and thus a high irreproducible background may be observed.

Chemiluminescence results from interactions of various compounds present in the sample. Bases are particularly known to produce chemiluminescence. This process is, however, not limited to bases; many other compounds produce light when added to the scintillation liquid (Hercules, 1970; Kalbhen, 1970).

As chemiluminescence decay is temperature dependent, preheating of the sample results in a deexcitation, whereas reduction of the temperature prolongs the decay half life (Moghissi, 1970). In either case, a lower count rate is observed, obviously for different reasons.

Both phosphorescence and chemiluminescence are monophotonic phenomena and thus are greatly suppressed in the coincidence-counting mode. However, because the light production in both processes can be large, random coincidences do occur if the sample has been subjected to prolonged or intense sunlight, or whenever chemiluminescencing chemicals are present.

4.3.6 *Selection of Counting Vials*

Proper selection of counting vials contributes significantly to optimization of the counting system. Methods for the determination of counting efficiency are usually based on the presumption of vials having identical properties. Garfinkel *et al.* (1965) have shown that differences in properties of commercially available vials can contribute to inaccuracy of measurement. These inaccuracies can be magnified if an external standardization technique is used, as such techniques are affected by differences in the physical properties of the vial.

Suggestions for improvements in vials and their properties are numerous and the reader is referred to Horrocks (1974) and Peng (1975) for reviews of the subject. In practice, only vials made from ordinary glass, polyethylene, nylon, and sometimes quartz, are used. Because of their ^{40}K content, glass vials show a higher background than plastic vials. However, permeation of various solvents through plastics is significant and can cause swelling or distortion of the vials. Table 11 contains data on the loss of various solvents through scintillation vials. Counting efficiencies are similar for all vials; however, thin-wall plastic vials tend to give slightly higher counting efficiencies than thick-wall plastic vials, which, in turn, give slightly higher efficiencies than glass vials. (See also Horrocks, 1974 and Painter, 1974).

4.3.7 *Determination of Counting Efficiency*

Due to the unpredictable nature of many materials, the counting efficiency, ϵ, must be determined for essentially every sample. If ϵ has been determined previously for an identical sample, this requirement is reduced to the routine maintenance of the instrument, i.e., corrections for instrument drift and changes in the background. However, by far the greatest number of samples require a determination of ϵ due to the differences in their composition and degree of quenching.

TABLE 11—*Loss of compounds through liquid-scintillation vials*[a]

	Volume	*p*-Xylene	Toluene	Dioxane	H_2O
	ml	mg/d	mg/d	mg/d	mg/d
Polyethylene ("Teflon" coated)	25	130	190	4	~0.5
Polyethylene	25	120	190	3	~0.1
Polyethylene (thick wall)	20	80	100	1	~0.3
Nylon	25	5	10	15	30

[a] Adapted from Lieberman and Moghissi (1970).

4.3.7.1 *Internal-Standard Method.* Although the number of proposed methods for quench correction is large, only two have gained acceptance. The older and more accurate method is the internal-standard method and its many variations. It consists of a comparison between the sample count rate, C, and that of the sample plus an added activity, S, of the same radionuclide. The efficiency, ϵ, is then calculated by the following equation:

$$\epsilon = \frac{(C + s) - C}{S} = \frac{s}{S}, \tag{4.11}$$

where s is the observed increase in count rate due to the activity, S, of the added standard. For the accurate determination of C, s should be equal to, or greater than C. The original technique consisted of counting a sample, adding an appropriate, but small, quantity, or "spike", of the standard, and recounting it. The added standard should therefore have a high radioactive concentration to avoid any drastic change in the characteristics of the original counting sample. This technique has been improved by Garfinkel *et al.* (1965) for preparation of a tritiated toluene standard with an accuracy comparable to the tritiated water standard of the National Bureau of Standards. The generalized method (Moghissi and Carter, 1968) consists of the preparation of two samples that are identical in their chemical composition, except for the replacement of some of the stable isotope by an appropriate radioisotope.

The practical application of the method for the assay of tritium in aqueous media would consist of the preparation of the liquid scintillator, dividing it in half, and adding a certain quantity of water to one half and an equal quantity of tritiated water to the other half. The tritiated water should have a known specific activity. Subsequently, equal aliquots of the unknown sample are added to the two prepared scintillation vials, and counted. There are many variations to this technique. Standardized tritiated water and inactive water can be added to two equal aliquots of the counting sample.

If a high degree of accuracy is desired, variations in vials must be taken into account (Garfinkel *et al.*, 1965). For the majority of measurements where precision of 2 to 3 percent is adequate, these variations are tolerable if vials from the same production batch are used.

4.3.7.2 *External-Standard Methods.* The second method, which is often applied to the determination of the counting efficiency, ϵ, is considerably more convenient. It is called the external-standard channels-ratio (ESCR) method, and has evolved from two techniques, namely the channels-ratio method (Baille, 1960) and the external-standard method (Higashimura *et al.*, 1962).

The first step in all of these methods is to prepare a set of vials of "quenched" working standards of known activity, in which the volume fraction of the sample is varied from, say, 1 to 20 percent, while the total volume of sample plus scintillator solution is held constant.

The channels-ratio method is based on splitting the spectrum observed for a given sample into upper (u) and lower (l) parts by setting the discriminators at levels which will remain fixed. The effect of quenching on this sample would be to shift the spectral energy response toward lower energy so that the observed count rate u, in window u, will decrease. The l/u-count-rate ratios and counting efficiency for the whole spectrum are then determined for each of the differently quenched working standards. These efficiencies are then plotted against the count-rate ratios to obtain a quench calibration curve, from which the efficiency of an unknown sample can be obtained by interpolation. The disadvantage of this method is that high-activity samples are required to obtain accurate values for the channels ratio in a reasonable counting time.

The external-standard method is based on the measurement of that component of the count rate arising from Compton electrons generated in the counting sample by an external gamma-ray source. The count rates due to the external source are measured for the working standards, for which counting efficiencies have been previously determined (Section 4.3.7.1). As in the last method, a quench calibration curve is prepared in which the "automatic-external-standard (AES)" count rate is plotted against ϵ for the working standards. The counting efficiency for an unknown sample is then obtained by reference to the quench calibration curve after a measurement of its AES count rate.

Although the channels-ratio and external-standard techniques lack general applicability, their combination (ESCR) is a useful and convenient technique. In practice, three wide-energy channels are used; one covers the spectrum of the beta-particle-emitting radionuclide. The other two are chosen above the sample spectrum and are used to determine the channels ratio for the sample. (After each sample is counted, the external gamma-ray source is, in most commercial systems, automatically positioned near the sample for a one-minute count.) Quench calibration curves are prepared as before for the series of working standards and these are used to obtain the counting efficiency for the unknown sample.

There are several problems associated with the ESCR method. A change in the chemical composition of the samples requires a new calibration curve derived from new working standards. Vial thickness, total volume, and several other factors affect the accuracy of the results (Horrocks, 1975). Although the ESCR method has many inherent sources of error, its convenience of operation has made it a popular technique.

4.3.8 *Simultaneous Measurement of the Activities of Radionuclide Mixtures by Liquid Scintillation*

4.3.8.1 *General.* Although energy resolution is poor for liquid scintillators (Seliger, 1960), simultaneous measurement of the activities of several radionuclides by this technique is still possible. Measurements based on differences of half-lives of two radionuclides such as ^{125}I and ^{131}I will not be elaborated upon here, as they follow general rules of radioactivity and thus are common to all radiation-detection systems. This method has been recently discussed by Horrocks (1974).

4.3.8.2 *Mixtures of Alpha-Particle-Emitting Radionuclides.* Horrocks (1966) has designed an alpha-particle spectroscopy system using a single phototube system. Approximately 200 μl of liquid scintillator are used to dissolve a mixture of ^{233}U and ^{238}Pu. Due to the reasonably high energy of the alpha particles emitted by these radionuclides and because of the small volume of the liquid scintillator, tolerable background count rates are obtained under the peaks. If the activities of alpha- and beta-emitting radionuclides are to be measured, it should be kept in mind that the light output of an alpha particle is only approximately one-tenth of that of a beta particle with identical energy (see Section 2.4.5.1).

4.3.8.3 *Dual-Label Assays of Beta-Particle-Emitting Radionuclides*

4.3.8.3.1 *General.* A widely used application of liquid-scintillation counting in the life sciences is the assay of samples containing two different beta-emitting radionuclides. In the majority of cases the samples are labeled with ^{3}H and ^{14}C, although several other pairs have been assayed, e.g., ^{3}H and ^{35}S, ^{3}H and ^{32}P, ^{14}C and ^{36}Cl. For any given pair, the extent of the overlap of the beta-particle spectra determines the accuracy of the measurement. If the $E_{\beta\max}$ for the two differ by less than a factor of two, accurate measurement is extremely difficult.

In recent measurements, multichannel analyzers have been used with the liquid-scintillation counter, thereby allowing comparison of the total beta spectrum for the sample with the spectra for the individual components. It is then a relatively easy matter to assign counting windows to the higher- and lower-energy radionuclides (Randolph, 1975). Computers have also been used to fit the composite spectrum with library spectra of the individual constituents (Oller and Plato, 1972).

In most dual-label assays made with commercial counters, however, the two counting windows must be selected by the user. Because of the importance of this type of measurement, instrument manufacturers have devoted much effort to developing automated methods for selecting the proper channels so as to reduce spectral overlap while still maintaining reasonable detection efficiencies for each radionuclide. In the simplest instance, the total beta-ray spectrum is divided into two channels by the use of only three discriminators. For ^{3}H and ^{14}C, the lower discriminator is set just above the instrument "noise" and the upper discriminator is set just beyond the ^{14}C endpoint. The spectrum is then split into channels 1 and 2 by setting the third discriminator near the ^{3}H endpoint. This latter discriminator is then adjusted to allow a preselected overlap of the ^{3}H and ^{14}C spectra.

4.3.8.3.2 *The "spectrum-exclusion" method.* The spectrum-exclusion method consists of raising the intermediate discriminator beyond the ^{3}H endpoint such that only ^{14}C events will be recorded in channel 2. This can be expressed in a general mathematical form as follows. Let $N_1 = h_1 + c_1$ and $N_2 = h_2 + c_2$. where N_1 and N_2 are total count rates in channels 1 and 2 respectively, and h_1 and h_2, c_1 and c_2 are count rates due to ^{3}H and ^{14}C in the indicated channels.

It then follows that the ^{3}H count rate in the lower channel and the ^{14}C count rate in the upper channel are given by the equations (Peng, 1975):

$$h_1 = \frac{N_1\left(\frac{c_2}{c_1}\right) - N_2}{\frac{c_2}{c_1} - \frac{h_2}{h_1}}, \tag{4.12}$$

and

$$c_2 = \frac{\frac{c_2}{c_1}\left[N_2 - N_1\left(\frac{h_2}{h_1}\right)\right]}{\frac{c_2}{c_1} - \frac{h_2}{h_1}}. \tag{4.13}$$

In the spectrum-exclusion method, $h_2 = 0$, and these equations are simplified to:

$$h_1 = N_1 - N_2\left(\frac{c_1}{c_2}\right), \tag{4.14}$$

and

$$c_2 = N_2. \tag{4.15}$$

Using this method, however, results in a reduction in the detection efficiency for ^{14}C by about a factor of 2 for an unquenched sample. In practice then, it is normal to accept some ^{3}H contribution in the upper channel.

4.3.8.3.3 *Method for selecting the optimum channels.* Such methods are reviewed in Kobayashi and Maudsley (1970). One method can be described using the ^{3}H and ^{14}C illustration used in the previous section. Two counting samples are required: an unquenched ^{3}H sample and an unquenched ^{14}C sample, both of known activity.

The lower and upper discriminators are set as described in 4.3.8.3.1, and ^{3}H and ^{14}C counting efficiencies are determined in both channels at different settings of the intermediate-level discriminator. The ^{3}H efficiency in the ^{3}H channel is then plotted against the ^{14}C efficiency in the ^{3}H channel as shown in Figure 70. The point at which the curve begins to deviate significantly from a straight line through the origin, corresponds to the point beyond which the ^{14}C counting efficiency in the lower channel increases more rapidly than the ^{3}H efficiency.

4.3.8.3.4 *Determination of sample activity.* Quench-correction curves must be prepared in which the ^{14}C efficiency in channel 2 ($\epsilon_2(^{14}C)$) and the ^{3}H efficiency in channel 1 ($\epsilon_1(^{3}H)$) are each plotted against the external-standard channels-ratio (ESCR). These curves are prepared in essentially the same manner as described for single-label samples in Section 4.3.7.2, except that two separate sets of vials, one containing ^{3}H and the other ^{14}C, are required.

When the ESCR, which is a measure of the degree of quenching, is determined for an unknown sample, the ^{3}H and ^{14}C counting efficiencies for that sample are taken from the quench-correction curves, and the net count rates due to ^{3}H in the lower channel (h_1) and ^{14}C in the upper channel (c_2) are computed from Eqs. 4.12 and 4.13. The sample activities are then given by:

$$N_0(^{3}H) = h_1/\epsilon_1(^{3}H), \tag{4.16}$$

and

$$N_0(^{14}C) = c_2/\epsilon_2(^{14}C). \tag{4.17}$$

It is very difficult to estimate the statistical errors due to spectral overlap in dual-label assays. If the spectrum-exclusion method is used

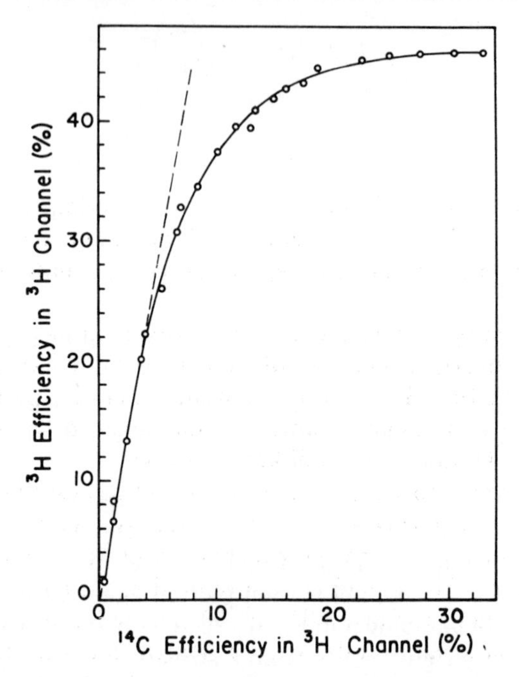

Fig. 70. Plot of ^{3}H efficiency *vs* ^{14}C efficiency in the ^{3}H channel at different settings of the intermediate-level discriminator. (After Kobayashi and Maudsley, 1970)

for ^{3}H and ^{14}C, the ^{14}C can be assayed with no uncertainty due to the ^{3}H. However, there is still uncertainty in the ^{3}H channel. If near optimum conditions are set as described in Section 4.3.8.3.3, and the activity ratio ^{3}H/^{14}C is varied from 100 to 0.01, errors resulting from spectral overlap in the assay of a given component can vary from less than 5 percent to more than 100 percent, respectively.

4.4 Ionization Chambers

4.4.1 *General*

The physics of gas ionization chambers has been discussed in Section 2.1. The phenomenon of the production of ions in gases under the influence of radiation has been used since the earliest experiments in radioactivity to quantify radiation. The gold-leaf electroscope led to the Lauritsen electroscope and Carmichael "Chalk River" electroscopes, the pocket dosimeter, the free-air chamber, the cavity chamber, the re-entrant (or 4π) ionization chambers, and the internal ionization chambers for the comparative assay of radioactive gases. Solid or liquid sources may be external to the chamber or placed within it. In principle, all gas ionization chambers are instruments in which an electric field is applied, by means of two insulated electrodes, across a confined volume of a gas. The electrons and positive and negative ions collected by these two electrodes constitute an ionization current that is a measure of the radiation incident upon the gas.

These instruments are widely used for the detection and assay of radionuclides and in applications to health physics, as, for example, monitors of neutrons (Manley *et al.*, 1946) and other radiations, their design and gas filling being dependent upon the type and intensity of the radiation to be measured.

Ionization chambers, once calibrated, can be used to measure or calibrate further samples of the same radionuclide (and some for which direct standards are not available, see page 185) to give derived or secondary standards at any later date, or to check the consistency of subsequent direct measurements. All relative standardization procedures inherently require calibration of an instrument in terms of a calibration standard, and succeeding measurements of samples need to be made under conditions identical to those of calibration. Certain precautions should be observed if accuracy is to be achieved.

The geometrical factors involved (for example, the distance from source to detector and the dimensions of the sample) should be maintained constant. The arrangement of scattering material in the vicinity of the source and detector must be reproducible (for example, shielding used to protect the operator or equipment from stray radiation should not be moved since a resulting change in the scattered radiation from the shielding will change both the calibration and the background). Where completely fixed conditions cannot be achieved (for example, with unavoidable variation in sample size) corrections should be made from data obtained by investigation of changes in the pertinent parameters. The contribution of the background should always be carefully considered, and if large or variable or both, frequent measurements may be necessary. The material to be assayed should be examined for impurities; the effect of a radionuclidic impurity may be inconsequential or large, depending on the radiations of the principal and impurity radionuclides, their half lives, the amount of impurity present, and the instrument. Energy-selective systems will have to be employed to identify and quantify impurities present, and the overall ionization-chamber response corrected for the contributions, if any, due to the contaminants.

The stability of an ionization-chamber system is usually verified by measuring its response to a reference source of some long-lived radioactive material, such as ^{226}Ra in equilibrium with its daughters, whenever calibrations, or measurements, are made. In practice, one may take the activity of the sample to be proportional either to the direct response of the ionization-chamber system, or to the ratio of its response to the sample to that produced by a reference source. In the latter case, small variations in the sensitivity of the system may be of less consequence insofar as they affect the response to both the sample and reference source nearly equally. In the first method, the appropriate calibration factor, K, for a particular radionuclide is the ratio of its activity to the ionization current, or observed response R, for a source of that radionuclide in a reproducible configuration. In the second method, the relative calibration factor, K_R, is equal to AR_R/R, where R_R/R is the ratio of the measured responses, i.e., the ratio of the ionization current due to the reference source in a given geometry to the ionization current due to the particular radionuclide in a similar and reproducible geometry, and A is the activity of the radionuclide.

It should be noted that the K factor (A/R) is inversely proportional to "efficiency" that has been used for other detectors in previous sections of this report. In the case of a system that counts individual events, the count-rate response is divided by the efficiency, ϵ, to give total emission rate. The K factor has, however, classical origins and multiplies the response to give activity. Therefore, to obtain the equivalent of an efficiency curve as a function of energy, the reciprocal of the K factor should be plotted against energy. The relative factor, K_R, has the dimensions of activity and is, in fact, the "equivalent activity" referred to in Section 4.4.7.

The number of ion pairs formed per unit of path length is a function of the density of the gas in an ionization chamber. In unsealed chambers, readings may have to be corrected to standard temperature and pressure if the range of the primary radiation exceeds the dimensions of the chamber. For beta-ray and low-energy photon emitters, a simple correction to STP may result in over-correction. To effect a correction in these two cases, one can make measurements on specific radionuclides (relative to a reference source) on days with different air densities and modify the STP correction by the empirical factor calculated from those measurements. For the most accurate work, it is desirable to have a chamber in which the gas pressure can be varied to permit investigation of the dependence of response on gas density. This correction is not necessary with sealed ionization chambers containing, for example, argon at 20 atmospheres.

Ionization chambers, employing any of a number of suitable current-measuring systems and using the precautions mentioned above, can be used to maintain standards or to make relative measurements of radioactivity with precisions as good as 0.02 percent, which is usually less than the uncertainty associated with the activity of the standard used to calibrate the instrument.

4.4.2 *Beta-Particle Ionization Chambers*

Beta-particle ionization chambers are simple in construction and in function, but in their use one might encounter difficulties, especially with low-energy beta-particle emitters, due to variations in source preparation or configuration of source containers. However, ionization-chamber measurements of evaporated sources are less sensitive to source self absorption than are Geiger-Müller or proportional counter measurements. Each high-energy beta particle may contribute thousands of ion pairs to the ionization current, but only one count in the counter. On the other hand, the absorption of one beta particle in the source represents the negligible loss of perhaps only one or two ions from the ion current but still represents a whole count that is lost from the counter. The relative effect on the current is, therefore, much less than it is on the count rate. One should prepare sources and perform the measurements in exactly the same manner as for the calibration of the instrument. Beta-particle ionization chambers can be used for the calibration of solid sources or of solutions.

In preparing solid sources, the following precautions should be observed:

(1) Use highly polished, flat, pitless disks of a metal that will not react chemically with the solutions to be deposited. Solutions may be deposited on thin polyvinylchloride-polyvinylacetate copolymer (VYNS) or collodion films, allowed to dry, then affixed to disks of any metal.

(2) Maintain the same solids content and pH for the solution of a particular radionuclide to be deposited as used in the calibration.

(3) Add a seeding or wetting agent to the liquid deposit (see also Section 5.2.1).

(4) When sources are prepared from some solutions by deposition and evaporation, there may be loss due to volatilization. If this is likely, source preparation by precipitation followed by evaporation may be necessary (see Section 5.2.1).

(5) Dry sources slowly (e.g., at ambient temperature, in a gentle flow of warm air, or with a heat lamp, and take care to avoid spattering).

In preparing liquid sources:

(1) Use containers of uniform and reproducible dimensions and composition.
(2) Use the same volume.
(3) Work carefully and rapidly to minimize evaporation.
(4) Maintain the same solution composition for the sample as for the calibration sources.

In making measurements (after checking the purity of the material):

(1) Measure sources by comparison with some long-lived radionuclide. ^{226}Ra (in equilibrium with its daughters), contained in platinum-iridium cells, encapsulated in stainless steel disks of the same diameter as source disks or dishes, can be used as reference sources. Inasmuch as a range of activities can be measured in the chamber, a set of reference sources should be intercompared, maintained, and corrected for radioactive decay.
(2) Determine corrections for source self absorption from measurements of solid sources prepared with different masses of the radioactive solution, maintaining the same source diameter by adding drops of conductivity water to the radioactive solution on the source mount, before drying. Extrapolate to zero mass.
(3) Work in a room with controlled temperature and humidity.
(4) Use care in positioning of source.

An early design, but a chamber still functioning reliably, is a $2\pi\beta$-ionization chamber shown diagrammatically in Figure 71. At the base of the spherical body is a cylindrical shoulder that slides over the source-mounting ring. The source-mounting ring is recessed to hold, reproducibly, disks on which aliquots of an active solution can be evaporated. The length of the 0.3-cm diameter brass collecting electrode has been adjusted empirically so that a point source can be displaced slightly from the center of the source mount without affecting the observed ionization current.

Widely used as a secondary instrument for both beta-particle and gamma-ray emitting radionuclides is the type 1383A ionization chamber[6] designed at the National Physical Laboratory (NPL), England (see Figure 72). It is a combination re-entrant cylindrical chamber for gamma-ray measurements and parallel-plate chamber for beta-particle measurements. Designed to be manufactured to give constant characteristics, each chamber is provided with the calibration factors for a number of radionuclides (Dale, 1961; Dale *et al.*, 1961; Woods, 1970; Woods and Lucas, 1975; Woods *et al.*, 1975). Results obtainable with one chamber are applicable to others of the same type if uniform

[6] Available from Elliot Bros. (London) Ltd., Century Works, Covington Road, London SE13 7LN.

procedures for source preparation and measurements are practiced. Both solid and liquid sources that fit into the circular depression on the slide may be used for beta-particle measurements. (See Lowenthal, 1969, for a summary review of beta ionization chambers.)

4.4.3 *Gamma-Ray Ionization Chambers*

Important relevant characteristics of these instruments are sensitivity, long-term stability, and simplicity of operation. Also, source preparation is not elaborate. These chambers can be used to maintain results of direct standardizations and to check the consistency between

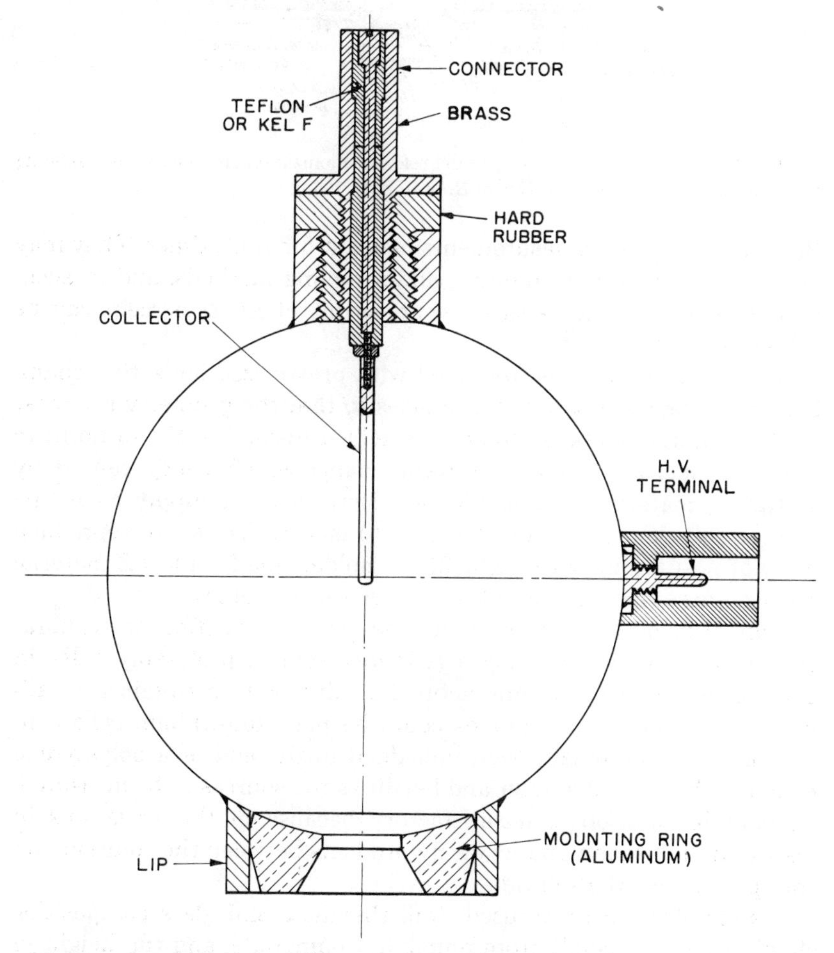

Fig. 71. Schematic drawing showing the construction details of the NBS $2\pi\beta$-ionization chamber (Seliger and Schwebel, 1954)

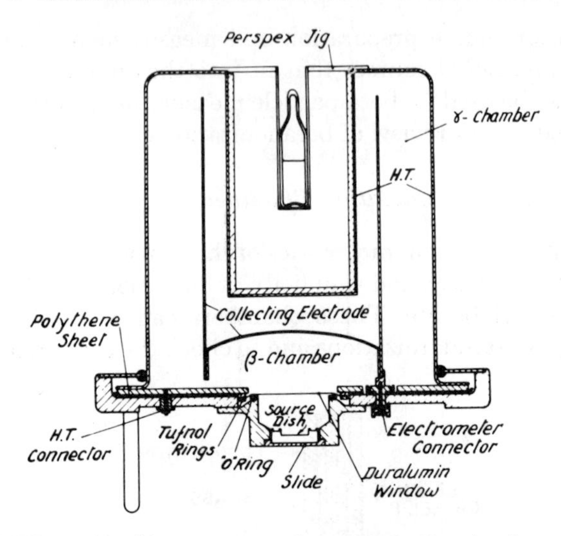

Fig. 72. NPL combined beta-gamma reference ionization chamber with plastic jig for support of small ampoules. (From Robinson, B. 1960)

the results of direct measurements made at different times. They may also be used in tests of dilution and sampling methods, and to accumulate half-life data. With care, precisions of ±0.02 percent can be achieved.

Best results have been obtained with pressurized ionization chambers with reentrant source receptacles so that the geometry is almost 4π. Ionization chambers should be shielded (using Pb, Hg, or both) to reduce background and to prevent change in efficiency caused by scattering material in their vicinity. Batteries can supply collecting voltage sufficient for saturation (200 to 600 V). To insure reproducibility of positioning, rigid tight-fitting holders made of low-Z material are used for ampoules, samples, and reference sources.

Standards of particular radionuclides are used to effect the calibration of a chamber, by using a reference source, preferably ^{226}Ra in equilibrium, under the same defined conditions to maintain the calibration. The reference sources could be platinum-iridium cells containing the radium salt, each imbedded in the end of a polystyrene cylinder, the same diameter and height as the sources to be measured. Radium in solution is unsatisfactory because of the variations in distribution of radon and its decay products between the solution and the space above the solution.

If ampoules are to be used, wall thickness and glass composition should be reproducible from ampoule to ampoule, and the height of solution should be constant in order to avoid variation in gamma-ray attenuation. Alternatively, corrections can be empirically determined for different solution heights, the function being only slightly dependent on the energy of the radiation. Again, the solutions to be assayed should have as nearly the same composition as the standards used to calibrate the chamber.

It may be necessary to apply corrections for chamber saturation-current losses to measurements of high activity sources; the loss, if any, can be determined by measuring the sources at increasing collecting voltage (Weiss, 1973). Another method for detecting saturation-current losses (usually caused by ion recombination) is to prepare, by dilution, and to measure a series of sources, of equal volumes of solution in identical ampoules, of the same long-lived gamma-ray-emitting nuclide, with different radioactivity concentrations giving responses in the range of ionization current to be used. On plotting response against radioactivity concentration any departure from linearity at the higher levels of activity will indicate the occurrence of saturation-current losses through general recombination. (For the definitions of *initial* and *general* recombination, see Boag, 1966.)

It is possible to estimate the calibration factors for solution samples of some photon-emitting radionuclides for which standards are not available because a direct measurement is difficult, provided that the photon probability per decay for each photon is known and the number of photons with energies below about 200 keV is small. First, it is necessary to determine the efficiency of the chamber as a function of gamma-ray energy, E, by measuring the response of the chamber to equal volumes of calibrated solution standards of monoenergetic photon-emitting radionuclides over the appropriate range of energy. The K factor, $K_{\mathrm{R},\gamma}$, relative to that of a long-lived reference source, now refers only to photon-emission rates and is given by

$$K_{\mathrm{R},\gamma} = P_\gamma A \frac{R_\mathrm{R}}{R}, \qquad (4.18)$$

whence

$$A = \frac{K_{\mathrm{R},\gamma} R}{P_\gamma R_\mathrm{R}}, \qquad (4.19)$$

where R is the reading for the calibrated source emitting a photon, of energy E, with a probability of emission per decay P_γ, A is the activity of the calibrated source, and R_R is the reading for the reference source, normally of ^{226}Ra. The readings R and R_R could be, typically, the time rate of change of voltage on a standard condenser.

If one now plots and fits the best curve to $1/K_R$ as a function of E, it is possible to calculate the effective K factor for radionuclides emitting n photons of different energies and of known probabilities per decay, by simply taking $1/K_{R,\gamma}$ for each of these photon energies from the curve and multiplying it by the appropriate probability per decay. The activity of the radionuclide is then given by

$$A = \frac{R}{R_R} \sum_{i}^{n} \left(\frac{K_{R,\gamma}}{P_\gamma} \right)_i , \qquad (4.20)$$

where the sum is carried out for all i photons. The efficiency-energy curve is often linear but a linear extrapolation below 200 keV may be incorrect. Equation 4.20 can be compared with Eq. 6.5 in Section 6.4.3.2.3 on dose calibrators. Further information has also been given by Weiss (1973).

Figure 73 is a schematic drawing of a 15-liter unsealed "4π" or "re-entrant-well", atmospheric-pressure, chamber based on that described by Shonka and Stephenson (1949), but with slight modification in chamber design. The cost in construction is small compared to that of a pressure chamber. A thin brass cylinder is used as the collection electrode in place of the usual wire cage. The response to 5 ml of a solution (containing a gamma-ray-emitting nuclide) in a standard 5-ml ampoule is approximately 99½ percent of that of a 1-ml sample with

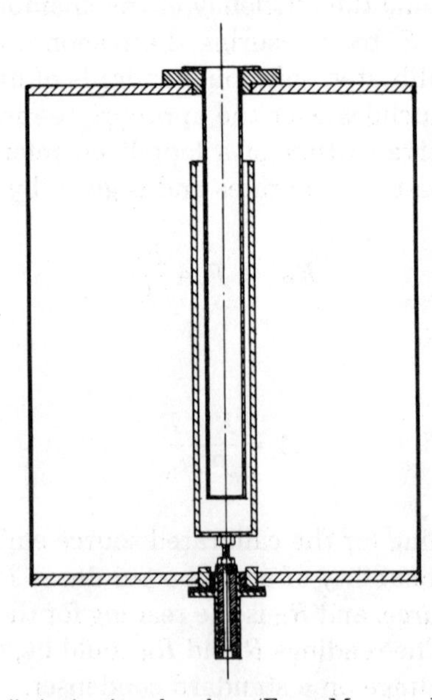

Fig. 73. A "4π"γ ionization chamber. (From Mann and Seliger, 1958)

the same total activity in a standard 5-ml ampoule. The chamber, in use, is surrounded by a cylindrical lead shield 1.5-inches thick.

Figure 74 shows the important design considerations of the Stephenson "high-precision", high-pressure ionization chamber such as the locations and construction of the insulators, guard ring, and guard-

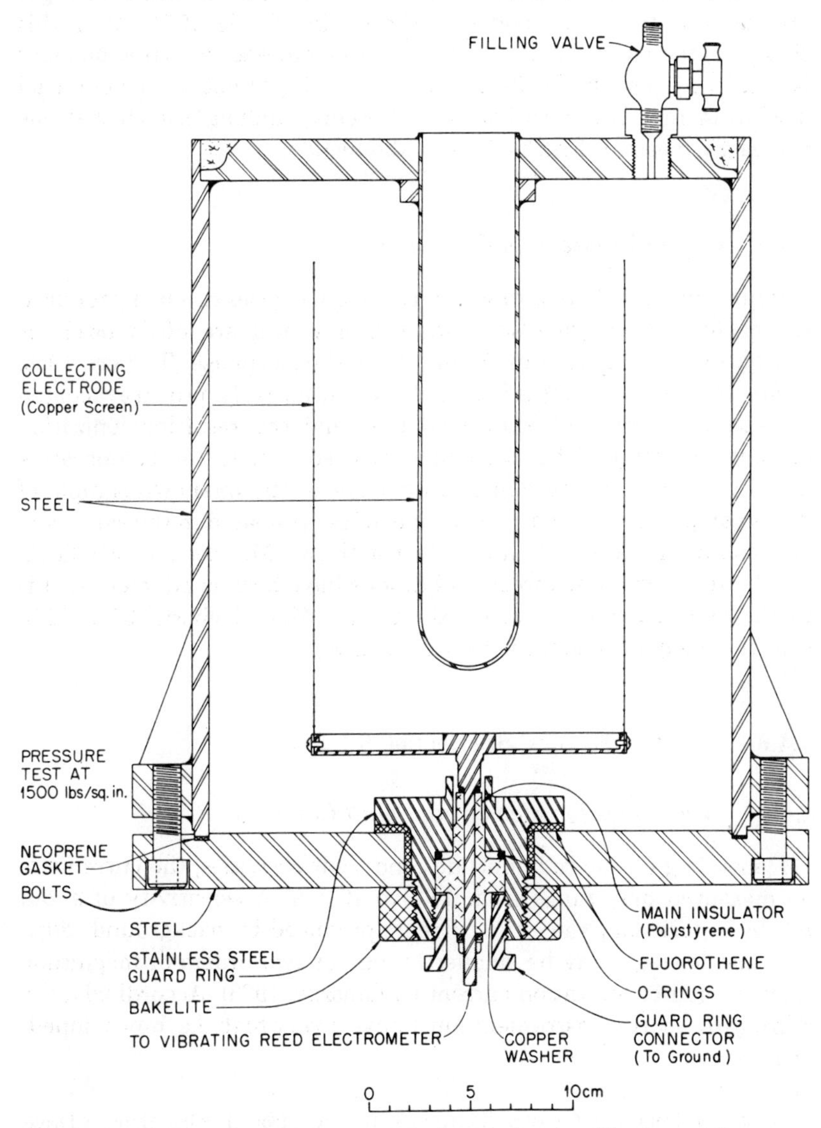

Fig. 74. High-precision, high-pressure ionization chamber (Shonka and Stevenson, 1949). (From O'Kelley, 1962)

ring connector (Shonka and Stephenson, 1949). The chamber is filled with 40 atmospheres of dry argon. The high sensitivity of such a chamber necessitates enclosure in a 4-inch thick lead housing to reduce environmental background effects.

Figure 75 illustrates the design features and internal construction of the T.P.A. Mk.II. ionization chamber[7] which can be filled with gas pressures up to 20 atmospheres (Sharpe and Wade, 1951). With this chamber, the variation in sensitivity for gamma-ray-emitting nuclides is within ±1 percent for 9-cm and within 0.5 percent for 6.5-cm axial displacements. The sensitivity to ^{32}P bremsstrahlung is such that one can easily measure 1 mCi of that radionuclide.

4.4.4 *Internal Ionization Chambers*

Ionization chambers into which radioactive gases can be introduced are useful for assaying such gas samples, and are often used for comparing the activities of ^{3}H and ^{14}C in the gas phase (Tolbert, 1956; 1963). The gaseous sample is admitted quantitatively into an evacuated ionization chamber of known volume, and the resulting ionization current measured. This technique is used chiefly for comparative measurements, i.e., the ionization current of the unknown sample of gas is compared with that of a sample of gas prepared in the same way and containing a known amount of activity (Marlow and Medlock, 1960). However, some direct calibrations have been conducted by this method (Wilzbach *et al.*, 1954; Rieck *et al.*, 1956; Tolbert, 1956, 1958). (For the assay of radon, see Section 6.1.2.3.)

4.4.5 *Dose Calibrators.* See Section 6.4.3.2.3

4.4.6 *Techniques for Measuring Small Currents*

Depending on the radionuclides and their activities, the currents to be measured may range from 10^{-7} to 10^{-13}A. A sensitivity of about 10^{-15}A is necessary since the currents produced by background radiation and leakage may be smaller by one or two orders of magnitude than the source ionization current (Zsdánszky, 1973). Accordingly, the current-measuring instrument must have a very high d.c. input impedance.

[7] Available from 20th-Century Electronics, Ltd. (Centronic), King Henry's Drive, New Addington, Croydon CR9 0BG, England.

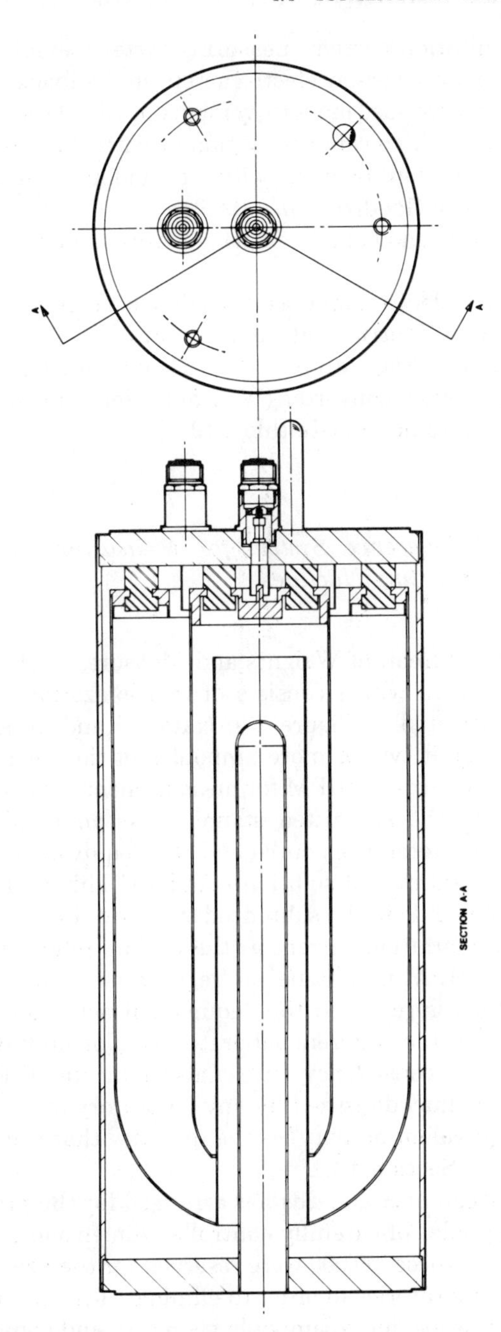

Fig. 75. Sectional view of a one-inch-diameter IG 11 well-type pressure ionization chamber. (From Centronic, 20th Century Electronics Ltd., King Henry's Drive, New Addington, Croydon CR9 0BG, England)

Many precise ionization-current measuring systems employ vibrating-reed or Lindemann-Ryerson electrometers in feedback-type current integrators, that are just modern versions of the classical Townsend induction balance. Typical of these and other techniques are:

(1) Townsend induction balance with continuous compensation (Townsend, 1903; Goodier *et al.*, 1965).
(2) Automatic Townsend balance with stepwise compensation (Garfinkel, 1959).
(3) Vibrating-reed electrometer as a feedback current integrator with preset time (Merritt and Taylor, 1967a).
(4) Time-averaged voltage drop across a high resistance and a voltage to frequency converter (Walz and Weiss, 1970).
(5) Solid-state electrometers (Keithley, 1977).

4.4.7 *International Reference System for Measuring Activity of Gamma-Ray-Emitting Nuclides*

The International Bureau of Weights and Measures (BIPM) maintains a reference system which consists of two ionization chambers (similar to that illustrated in Figure 75). National and international laboratories may submit two or more ampoules of their gamma-ray-emitting solution standards to BIPM for measurement. The ionization current produced by the submitted standard is compared to that produced by a radium reference source and the "equivalent activity" is calculated for the submitted solution standard. This is the stated activity of the radionuclide in the submitted standard that would have produced the same ionization current as the radium reference source at a specified time. BIPM then issues a "registration table" for each radionuclide in which is recorded the "equivalent activities" for the samples submitted by the different laboratories. Not only does this maintain international consistency, but it also indicates if the standards of a given radionuclide issued by any laboratory are consistent with time. The "equivalent activity" is the quantity that is called the relative factor, K_R, in Section 4.4.1.

The National Bureau of Standards has arranged for the production of 127,000 glass ampoules of carefully controlled dimensions, from the same melt of glass. Of these, 69,000 were reserved for the use of BIPM and other national laboratories in order to eliminate systematic differences in intercomparisons due to ampoule geometry and composition.

The International Atomic Energy Agency (IAEA) of the United Nations maintains a similar service for its member states.

4.5 Other Counting Methods

4.5.1 *Čerenkov Counting*

Some recent reviews describe various aspects of Čerenkov counting (Ross, 1970; Francois, 1973). This method is applicable, in aqueous solutions, only for beta particles having energies greater than 0.26 MeV (see Section 2.6). In practice, the sample is placed in a liquid-scintillation-counting vial, plastic or glass, 8 to 15 ml of water is added, and the sample is counted in a manner similar to liquid-scintillation counting. An advantage of the method lies in the relative ease of source preparation.

Due to the mechanism of light production in Čerenkov counting, chemical quenching, which is an important factor in liquid-scintillation counting, is non-existent in Čerenkov counting. Color quenching is a problem common to both techniques, and similar methods are used to correct for quenching.

Table 12 shows optimum counting efficiencies for various radionuclides measured in a plastic vial. Glass vials exhibit a somewhat lower efficiency, probably because of problems associated with the light transmission of glass in the ultraviolet region.

TABLE 12—*Optimum Čerenkov-counting efficiencies for various beta-particle-emitting nuclides*[a]

Radionuclide	$\bar{E}$	E_{max}	Detection efficiency
	MeV	MeV	%
^{204}Tl	0.240	0.764 (97.7%)	3
^{143}Pr	0.314	0.932	6.2
^{198}Au	0.312	0.961 (98.7%)	6.7
		0.285 (1.3%)	
^{24}Na	0.554	1.390	32.5
^{40}K	0.561	1.312	38.4
^{89}Sr	0.583	1.488	28.3
^{91}Y	0.607	1.545	35.5
^{32}P	0.695	1.710	35.1
^{90}Y	0.935	2.284	50.6
^{76}As	1.064	1.75 (16%)	
		2.41 (31%)	59
		2.97 (50%)	
^{42}K	1.430	1.996 (17.5%)	70.3
		3.521 (82.1%)	

[a] After Francois (1973). Measurements were performed at 4°C on solutions having various volumes (8 to 14 ml, 4 samples at each volume). The solutions consisted of the radionuclide and carrier in water, or, if a carrier was unavailable, in a 1-percent NaCl solution. The beta-particle "branching" is 100 percent unless stated parenthetically in the third column.

The fact that Čerenkov radiation is emitted anisotropically means, in practice, that the usual liquid-scintillation-counting system, with its two diametrically-opposed phototubes in coincidence, will have a *lower* counting efficiency than when operated in the non-coincidence mode, or with a smaller angle between the phototubes. Some wavelength shifters, such as 4-methyl-umbelliferone, increase the coincidence-counting efficiency to greater than that for non-coincidence through a reduction of the anisotropy of the re-emitted light (Ross, 1971). However, improper use of these compounds can result in color quenching that increases with time.

Although the detection efficiency of Čerenkov-radiation counting is less than that of liquid-scintillation counting, the ease of operation and, particularly, the ability to recover liquid samples unaltered make Čerenkov-radiation counting a useful technique.

The preparation of samples for Čerenkov counting is described in Section 5.4.1.5, and low-level-counting applications are described in Section 6.1.3.3.

4.5.2 *Solution Counting*

Several radionuclides have been successfully counted in solution by using an inert, low-vapor-pressure solvent such as formamide, dimethyl formamide, or ethylene glycol (Schwebel *et al.*, 1951). The sample is dissolved in the inert solvent and an aliquot is pipetted into a cupped planchette. The activity is then measured in a $2\pi\beta$ windowless gas-flow counter, and compared with the activity of a solution prepared in the same manner, but containing a known amount of the radionuclide (for sample preparation, see Section 5.4.2). Formamide containing up to 10 percent by volume of water can be introduced into a proportional counter without impairment of its electrical or counting characteristics.

This "formamide" counting procedure has been found useful for the relative standardization of ^{14}C as glucose and benzoic acid, ^{32}P as sodium phosphate, ^{60}Co as cobalt chloride, ^{90}Sr as strontium chloride, ^{131}I as potassium iodide, ^{204}Tl as thallic nitrate, ^{22}Na and ^{24}Na as sodium chloride, and ^{35}S as sulfuric acid.

For any given counter and planchette it is possible also to obtain relative efficiencies for the above radionuclides. These efficiencies may then be used to assay any of these radionuclides in terms of, say, a ^{14}C source.

Proportional or Geiger-Müller counters of the dipping or jacketed type may also be used to assay aqueous solutions.

4.5.3 *Surface-Barrier Detectors for Alpha-Particle Counting*

Silicon surface-barrier detectors (see also Section 6.1.2.4) are "solid-state" devices that, in radioactivity applications, are used for the detection and measurement of alpha particles. As with ionization chambers and Ge(Li) detectors, electrons released in the "sensitive region" as the result of incident radiation are swept to the anode by an applied electric field. This charge is then deposited on a capacitor, that is shunted by a high resistance, and the resulting voltage pulse is amplified. The alpha-particle spectrum, which is usually recorded with a pulse-height analyzer, gives, typically, an energy resolution, indicated by full width of the peak at half maximum (FWHM), from a "weightless" source, of 12 to 20 keV. The counts per channel in the low-energy tail of the alpha-particle spectrum for such a system is typically of the order of 10^{-4} to 10^{-3} of the counts per channel in the peak. The low response to beta particles and gamma rays, combined with the good energy resolution and high peak-to-tail ratio, makes these detectors powerful instruments for impurity analysis and measurement.

This type of detector is also extremely useful for the relative measurement of the activities of alpha-particle-emitting sources. Because alpha particles emitted by radioactive sources are totally absorbed in the detector, the system can be calibrated by using radioactivity standards that have alpha-particle energies that are not necessarily the same as those of the unknown. For example, in a recent experiment at the National Bureau of Standards, a system calibrated with an ^{241}Am standard was used to calibrate ^{148}Gd alpha-particle point-source standards. The alpha-particle energies for the two radionuclides are approximately 5.4 and 3.2 MeV, respectively.

Gibson *et al.* (1965) have listed the advantages and disadvantages of silicon surface-barrier detectors in nuclear physics applications. An updated list drawn in part from their article, is given below:

Advantages

a) Excellent energy resolution;
b) Linear response over a wide range of particle types and energies, even though the response to β particles and γ rays is low;
c) Insensitivity of pulse height to counting rate;
d) Fast pulse rise time;
e) Essentially windowless operation;
f) Selectivity for alpha particles with respect to other radiations;
g) Small size, making them easy to use;
h) Insensitivity to magnetic fields;
i) Ability to operate down to 0°C.

Disadvantages

a) Light sensitivity;
b) Relative fragility—care must be taken not to touch the gold surface layer of the detector, although very recently a rugged Al coated surface-barrier detector has been developed that allows cleaning of the surface;
c) Detector sizes are limited to a maximum of a few square inches;
d) Must be operated in a vacuum of 50×10^{-3}mm of Hg (6.7 Pa) or less because surface-current leakage may impair the detector response;
e) Variable thickness of the sensitive region, depending on bias voltage.

A word should be said in this section about diffused-junction-diode detectors whose principle of operation is the same as for surface-barrier detectors. The method of obtaining the depletion region is different, however, and has been discussed in Chapter 4 of a monograph by J. M. Taylor (1963). In spite of the fact that the energy resolution is worse than that for surface-barrier detectors, diffused-junction detectors are produced commercially because of two factors: namely, they are more rugged—the surface can be wiped—and, generally, the cost is less.

4.5.4 *Assay of Radioactive X-Ray Sources*

Measurements of the x-ray-emission rates of radioactive sources are sometimes needed in nuclear medicine as well as in investigations pertaining to the fundamental physics of x-ray emissions. Fink (1973) has listed forty-five radionuclides that are suitable as radioactive x-ray standards emitting x rays (and gamma rays) in the energy range from 4.8 to approximately 150 keV. Chew *et al.* (1973) have described a method using a multiwire proportional counter to calibrate a number of such radionuclides. Hansen *et al.* (1973) have described the calibration procedure for determining the photopeak efficiency of a Si(Li) detector as a function of energy in this region by fitting points, obtained with radioactivity standards, to a theoretical curve. Bambynek (1967) has discussed the calibration of a defined-solid-angle, low-geometry counter for the measurement of low-energy-photon, as well as alpha-particle, sources. Such systems are being used in the production of radioactivity standards in the Laboratoire de Metrologie des Rayonnements Ionisants (LMRI), Saclay, France, and the NBS (Hutchinson *et al.*, 1976b). Emission rates of low-energy (<10-keV) x rays have been determined using variable-pressure 4π proportional counters (Section 3.5.6.3).

The assay of low-energy x-ray sources requires that particular attention be paid to absorption by even small amounts of material in the path of the radiation, by scattering, and by x-ray fluorescence from materials in the region of the source and detector. Absorption, in 2.5 cm of air at atmospheric pressure, of 5.9-keV x rays emitted in the decay of ^{55}Fe is of the order of 6.8 percent. Backscattering of 22-keV x rays emitted in the decay of ^{109}Cd as a function of the atomic number of the source backing material is given in Hutchinson *et al.* (1976b). With decreasing photon energy, attenuation is caused predominantly by photoelectric interactions so that, for example, with 5.9-keV x rays there is less than 0.1-percent backscattering from steel backings. Examples of fluorescence encountered in recent experiments are gold L-x-ray lines observed in a spectrum from a Si(Li) detector which has a gold covering layer (Keith and Loomis, 1976), zinc x rays emitted by sources deposited on galvanized iron, platinum L x rays produced in the edges of a platinum aperture, and so on.

5. Techniques for the Preparation of Standard Sources for Radioactivity Measurements

5.1 Liquid Sources

5.1.1 *Chemically Stable Radioactive Solutions*

Most radioactivity solution standards consist of trace quantities of radionuclides of known radioactivity concentrations, but only approximately known concentrations of carriers, in flame-sealed glass ampoules. A carrier is "a substance in an appreciable amount which, when associated with a trace of another substance, will carry the trace with it through a chemical or physical process", especially a precipitation process. If the added substance is a different element from the trace, the carrier is called a non-isotopic carrier (ANSI, 1976). The carrier concentrations and solution media for standards are chosen to satisfy two criteria: (1) that the solution standards remain stable throughout their useful lives and (2) that low-absorption sources can be prepared from them—therefore, in diluting, the diluent should, usually, have the same carrier concentration and composition as the standard. The minimum amount of carrier that can safely be used is found from experience. As a general guide, about 100 μg of the element per gram of about 0.1-M acid solution is adequate, but concentrations above and below this value may be used successfully for specific radionuclides. Among solutions of metals, a few exceptions to this general guide are noteworthy: niobium, tantalum, zirconium, and hafnium solutions are prepared as fluoride or oxalate complexes, and tin solutions must be in $\geq$1M HCl to avoid hydrolysis. Table 13 gives chemical data for various radionuclide solution standards and may serve as a guide for the preparation and dilution of chemically stable radioactive solutions. A few very high carrier concentrations listed in the table result from the low specific activities that were commercially available for some radionuclides (e.g., ^{26}Al), and not from added carrier.

The chemistry of stable and radioactive nuclides is the same except at high levels of activity where the composition of the solvent can be substantially changed by the radiation. (As a rule of thumb, it takes an absorbed dose of about 10^8 rad to effect such changes.) If possible, inactive material added to serve as carrier should be in the same chemical form as the radionuclide, to permit them to exchange and thus follow the same chemistry. Exchange readily takes place in solutions for those elements that are normally found in only one valence state. However, it can be very slow among different valence states of multivalent elements and is best achieved through an oxidation-reduction cycle. For elements that do not have stable isotopes, those with similar chemical behavior can be used as carriers, e.g., barium for radium or lanthanum for promethium.

Elements such as krypton are usually found to be radioisotopically contaminated with man-made radionuclides. Salts of lanthanum, samarium, and lutetium contain naturally occurring ^{138}La, ^{147}Sm, and ^{176}Lu respectively. Salts of lanthanum and cerium have been found to be contaminated with ^{227}Ac as well as ^{232}Th. Distilled water, nitric, perchloric, and hydrofluoric acids have been found to contain "smidgen[8] quantities of uranium" (DeVoe, 1961). Consequently, analytical chemical reagents should be checked for radioactive contamination, and uncontaminated salts of elements with similar chemical behavior used, if necessary.

In preparing solutions and dilutions, the chemical and physical cleanliness of all vessels used with both the radioactive and inactive solutions is paramount. If chromate or sulfate is left on glassware from cleaning solutions or is present in the water, precipitation or radiocolloidal behavior may occur with elements such as strontium and radium. Water purified by distillation combined with ion exchange should be used. This two-step procedure removes particulate matter and most dissolved substances, both ionized and non-ionized.

Some processes occur in stored solutions that result in changes in concentrations. Several of these changes may not be noted if the concentration of the solution is 10^{-3} mole per liter or more, but others produce significant alteration (Korenman, 1966). The following will be more significant at low concentrations: hydrolysis resulting in gradual

[8] "Smidgen" in this reference equals 10^{-9}g (1 ng).

TABLE 13—*Source preparation techniques and chemical data on some standards solutions*

Radionuclide	Decay modes	β_{max} for principal β	Ionic form	Carrier solution	Approximate inactive to active ion ratio	Some methods of preparation of minimum self-absorption sources
		MeV				
^{3}H	β^-	0.0186	H^+	H_2O	10^{10}	
^{14}C	β^-	0.1561	CO_3^{2-}	2.12 g/l Na_2CO_3 in 0.001 M NaOH	10^2	
^{22}Na	β^+ ϵ γ	0.5455	Na^+	0.10 g/l NaCl in 1 M HCl	10^6	Ludox[b] added, dried in air, placed under glass exposed to infrared lamp (~50 °C)
^{24}Na	β^- $e_2^{\pm}$ [a] γ	1.3902	Na^+	0.10 g/l NaCl in 1 M HCl	10^6	Ludox added, dried in air, placed under glass exposed to infrared lamp (~50 °C)
^{26}Al	β^+ ϵ γ	1.1740	Al^{3+}	60 g/l $AlCl_3$ in 1 M HCl (see section 5.1.1)	10^5	
^{32}P	β^-	1.7110	PO_4^{3-}	0.098 g/l H_3PO_4 in H_2O	10^6	Ludox added, dried in air, placed under glass exposed to infrared lamp (~50 °C)
^{35}S	β^-	0.1674	SO_4^{2-}	0.1 M HCl	10	Ludox added, dried in air, then placed in oven at 70 °C for several days.
^{36}Cl	β^- ϵ	0.7093	Cl^-	0.2 g/l NaCl in H_2O	20	Ludox added, dried in air, placed under glass exposed to infrared lamp (~50 °C)
^{37}Ar	ϵ γ			Stable argon	10^9	
^{42}K	β^- γ	3.521	K^+	0.10 g/l KCl in 1 M HCl	10^7	Ludox added, dried in air.

[a] $e_2^{\pm}$ symbolizes an internal electron-positron pair associated with the 2.754-MeV transition in ^{24}Mg.

[b] Ludox: A 1:10^4 dilution of Ludox-SM 15 (15-percent suspension of silica) is most often used. Ludox is available from E. I. Dupont, Industrial Engineering Dept., Wilmington, Delaware.

TABLE 13—*Continued*

Radionuclide	Decay modes	β_{max} for principal β	Ionic form	Carrier solution	Approximate inactive to active ion ratio	Some methods of preparation of minimum self-absorption sources
^{45}Ca	β^-	0.259	Ca^{2+}	0.04 g/l $CaCl_2$ in 1 M HCl	10^5	One drop ludox + 1 drop 0.001M H_3PO_4 added, dried in air, redissolved with H_2O and dried in NH_3 atmosphere.
^{46}Sc	β^- ce γ	0.357	Sc^{3+}	0.10 g/l $ScCl_3 \cdot H_2O$ or $YCl_3 \cdot H_2O$ in 0.1 M HCl	10^4	Ludox added, dried in air, redissolved with H_2O and dried in NH_3 atmosphere.
^{51}Cr	ϵ ce γ		Cr^{3+}	0.10 g/l $CrCl_3 \cdot 6H_2O$ in 0.1 M HCl	10^4	Ludox added, dried in air, redissolved with H_2O and dried in NH_3 atmosphere.
^{54}Mn	ϵ ce γ		Mn^{2+}	0.01 g/l $MnCl_2 \cdot 4H_2O$ in 1 M HCl	10^5	Ludox added, dried in air, placed under glass exposed to infrared lamp (~50 °C) Also, autoelectroplated (non-quantitative).
^{55}Fe	ϵ		Fe^{3+}	0.01 g/l $FeCl_3 \cdot 6H_2O$ in 1 M HCl	10^2	Ludox added, dried in air, placed under-glass exposed to infrared lamp (~50 °C). Also, autoelectroplated (non-quantitative).
^{56}Mn	β^- ce γ	2.850	Mn^{2+}	0.01 g/l $MnSO_4$ in H_2O	10^5	Ludox added, dried under infrared lamp (~50 °C).
^{57}Co	ϵ ce γ		Co^{2+}	0.05 g/l $CoCl_2 \cdot 6H_2O$ in 1 M HCl	10^4	Ludox added, dried in air, redissolved with water, dried in air, placed under glass exposed to infrared lamp (~50 °C) for several hours.
^{59}Fe	β^- ce γ	0.466	Fe^{3+}	0.10 g/l $FeCl_3 \cdot 6H_2O$ in 1 M HCl	10^5	Ludox added, dried in air, placed under glass exposed to infrared lamp (~50 °C) for several hours.

TABLE 13—*Continued*

Radionuclide	Decay modes	β_{max} for principal β	Ionic form	Carrier solution	Approximate inactive to active ion ratio	Some methods of preparation of minimum self-absorption sources
^{60}Co	β^- ce γ	0.3179	Co^{2+}	0.05 g/l $CoCl_2 \cdot 6H_2O$ in 1 M HCl	10^4	Ludox added, dried in air, redissolved with water, dried in air, placed under glass exposed to infrared lamp (~50 °C) for several hours.
^{63}Ni	β^-	0.06587	Ni^{2+}	0.08 g/l $NiCl_2$ in 1 M HCl	10^2	Ludox added, dried in stream of warm air, redissolved with water, placed under glass exposed to infrared lamp (~50 °C) for several days.
^{65}Zn	ϵ β^+ ce γ	 0.330	Zn^{2+}	0.01 g/l $ZnCl_2$ in 1 M HCl	10^3	Ludox added, dried in air, redissolved with water, dried in air, placed under glass exposed to infrared lamp (~50 °C) for several days.
^{75}Se	ϵ ce γ		SeO_3^{2-}	0.33 g/l H_2SeO_3 in 1 M HCl	10^5	Ludox added, dried in air, placed under glass exposed to infrared lamp (~50 °C) for several hours.
^{85}Kr	β^- γ	0.687		Stable krypton	10^3	
^{85}Sr	ϵ ce γ		Sr^{2+}	0.09 g/l $SrCl_2 \cdot H_2O$ in 1 M HCl	10^6	Ludox added, dried in air, placed under glass exposed to infrared lamp (~50 °C) for several hours.
^{88}Y	ϵ ce β^+ γ	 0.755	Y^{3+}	0.10 g/l $YCl_3 \cdot H_2O$ in 1 M HCl	10^3	Ludox added, dried in air, placed under glass exposed to infrared lamp (~50 °C) for several hours.
^{89}Sr	β^- γ	1.492	Sr^{2+}	0.09 g/l $SrCl_2 \cdot H_2O$ in 1 M HCl	10^6	Ludox added, dried in air, placed under glass exposed to infrared lamp (~50 °C) for several hours.

TABLE 13—*Continued*

Radionuclide	Decay modes	β_{max} for principal β	Ionic form	Carrier solution	Approximate inactive to active ion ratio	Some methods of preparation of minimum self-absorption sources
^{90}Sr-^{90}Y	β^- β^- ce	0.546 2.284	Sr^{2+} Y^{3+}	0.027 g/l $SrCl_2 \cdot 6H_2O$ and 0.022 g/l $YCl_3 \cdot 6H_2O$ in 1 M HCl	10^4 10^7	Ludox added, dried in air, placed under glass exposed to infrared lamp (~50 °C) for several days.
^{95}Nb	β^- ce γ	0.1597	Nb^{5+}	0.2 g/l Nb $(HC_2O_4)_5$ in 3 percent $H_2C_2O_4$	10^5	Ludox added, dried in air, placed under glass exposed to infrared lamp (~50 °C) for several hours.
^{99m}Tc	ce γ		TcO_4^-	3 M HCl		Ludox added and dried in atmosphere of H_2S.
^{109}Cd-^{109m}Ag	ϵ ce γ		Cd^{2+} Ag^+	1 g/l $CdCl_2 \cdot 2\frac{1}{2}H_2O$ in 2 M HCl	10^4	Ludox added, dried in air, placed under glass exposed to infrared lamp (~50 °C) for several days.
^{110}Ag-^{110m}Ag	β^- ϵ ce γ	2.893	Ag^+	1 g/l $AgNO_3$ in ~0.5 M HNO_3	10^4	1. Ludox added, dried in air. 2. Dried in NH_3 atmosphere.
^{111}In	ϵ ce γ		$InCl_2^+$	3 M HCl	—	Dilute solution with 0.3 M HCl, deposit solution, add a drop of H_2O and one of Ludox, dry in H_2S atmosphere
^{113}Sn-^{113m}In	ϵ ce γ		Sn^{4+} In^{3+}	0.74 g/l $SnCl_4$ in 2.5 M HCl	10^5	1. H_2O_2 and water added, dried in air. 2. Ludox added, dried in air.
^{123}I	ϵ ce γ		I^-	Same as for ^{125}I and ^{131}I	10^4	Same as for ^{125}I and ^{131}I
^{125}I	ϵ ce γ		I^-	0.02 g/l Na_2SO_3 + 0.02 g/l LiOH + 0.05 g/l KI in H_2O	10^4	^{125}I solution is deposited into a drop of $AgNO_3$ solution and dried in air.

TABLE 13—*Continued*

Radionuclide	Decay modes	β_{max} for principal β	Ionic form	Carrier solution	Approximate inactive to active ion ratio	Some methods of preparation of minimum self-absorption sources
^{129}I	β^- ce γ	0.152	I^-	NaI, $NaHSO_3$, NaOH; pH 11.5 (~2 mg Na per gram of ^{129}I solution)	1	Solution absorbed into anionic exchange paper (in the chloride form).
^{131}I	β^- ce γ	0.6063	I^-	0.02 g/l Na_2SO_3 + 0.02 g/l LiOH + 0.05 g/l KI in H_2O	10^6	$AgNO_3$ solution added immediately to deposit of ^{131}I solution, and dried in air. Alternatively $AgNo_3$ can be deposited first.
^{131m}Xe	ce γ			Stable xenon	10^7	
^{133}Xe	β^- ce γ	0.346		Stable xenon	10^3	Freeze to point source.
^{137}Cs-^{137m}Ba	β^- ce γ	0.514	Cs^+ Ba^{2+}	0.028 g/l CsCl in 2.7 M HCl	10^2	Ludox added, dried in air, then placed under glass exposed to infrared lamp (~50 °C) for several days. Vacuum evaporated sources (non-quantitative).
^{139}Ce	ϵ ce γ		Ce^{3+}	0.01 g/l $CeCl_3$ in 1 M HCl	10^3	
^{141}Ce	β^- ce γ	0.435	Ce^{3+}	0.1 g/l $Ce(NO_3)_3$ in 1 M HNO_3	10^6	Ludox added, dried in air, then place under glass exposed to infrared lamp (~50 °C) for several hours.
^{144}Ce-^{144}Pr	β^- ce γ	0.3155 2.996	Ce^{3+} Pr^{3+}	0.01 g/l $CeCl_3$ in 1 M HCl	10^3	

TABLE 13—*Continued*

Radionuclide	Decay modes	β_{max} for principal β	Ionic form	Carrier solution	Approximate inactive to active ion ratio	Some methods of preparation of minimum self-absorption sources
^{147}Pm	β^-	0.2246	Pm^{3+}	1 M HCl	—	Ludox added, dried in air, placed in desiccator (NaOH), then desiccator placed under infrared lamp for several days. Sources redissolved with water, dried in air; replaced in desiccator under infrared lamp for at least one week.
^{182}Ta	β^- ce γ	0.522	TaF_7^{2-}	Very dilute HF	10^5	Dried in air, redissolved with water, dried, redissolved again with water and dried in NH_3 atmosphere.
^{197}Hg	ϵ ce γ		Hg^{2+}	0.4 g/l $Hg(NO_3)_2 \cdot H_2O$ in 0.1 M HNO_3	10^4	Ludox added, dried in atmosphere of H_2S.
^{198}Au	β^- ce γ	0.9612	$Au(CN_4)^-$	0.1 g/l $KAu(CN)_4$ + 0.001 g/l KCN	10^7	Ludox added, dried in air.
^{203}Hg	β^- ce γ	0.212	Hg^{2+}	0.4 g/l $Hg(NO_3)_2 \cdot H_2O$ in 0.1 M HNO_3	10^4	Ludox added, dried in atmosphere of H_2S.
^{201}Tl	ϵ ce γ		Tl^{3+}	0.01 g/l $TlNO_3$ in 1 M HNO_3	10^4	Ludox added, dried in air.
^{204}Tl	ϵ β^-	0.7634	Tl^{3+}	0.017 g/l $Tl(NO_3)_3$ in 1 M HNO_3	10^4	Ludox added, dried in air, then placed in oven at 50°C for one week.
^{207}Bi	ϵ ce γ		Bi^{3+}	4 M HNO_3		
^{241}Am	α ce γ		Am^{3+}	0.001 g/l $EuCl_3$ in 0.1 M HCl		Ludox added, dried in air.

precipitation, the sorption of hydrolysis products, or polymerization; oxidation of reducing agents by dissolved oxygen; change in pH if carbon dioxide is or was present; decomposition of some compounds when exposed to light; absorption of activity by microorganisms; solubility of glass. These effects may be diminished or eliminated by the following techniques: 1) *acidification of the solution*, which curtails colloid formation and polymerization and thereby minimizes adsorption losses; 2) *increased carrier concentration*, which in many cases reduces, or may eliminate, detectable adsorption; 3) *complex-ion formation*, which insures stable solutions of some elements such as gold, tantalum, thorium; 4) *sterilization of solutions* (by heating, if solutions are not adversely affected by heating), after flame-sealing in glass, which inhibits the growth of microorganisms; 5) *steam treatment of clean pyrex or quartz vessels*, which lowers the solubility of the glass (Korenman, 1966). A further useful reference is the "Users' Guides for Radioactivity Standards", a compilation of brief guides to chemical and counting problems for standards of many elements (NAS-NRC, 1974).

5.1.2 *Quantitative Sampling of a Solution*

To make use of solution radioactivity standards, it is necessary to take quantitative samples. This involves leaving the standard in the laboratory for a sufficient length of time to attain thermal equilibrium. Then the solution is mixed by gently shaking and inverting the ampoule to avoid error caused by condensation on the upper walls of the vessel. It is important to allow the liquid to drain from the neck after the last inversion, then to open the ampoule and transfer quickly most of the solution to another vessel, which is closed immediately. Such simple precautions will avoid the most common sources of error in sampling a solution, i.e., inadequate mixing and losses due to evaporation. If the sample is to be taken volumetrically, the vessel to which the solution is transferred is usually a small volumetric flask whereas, for gravimetric sampling, the solution is transferred to a pycnometer if the dispensing is to be done forthwith.

Whether volumetric- or gravimetric-sampling methods should be employed depends chiefly upon the certified data for the standard. If the standard to be used is certified in terms of activity, emission rate or number of atoms per unit weight of solution, samples should be taken gravimetrically unless the density of the solution has been stated or measured.

Volumetric sampling has the advantages of simplicity and low-cost

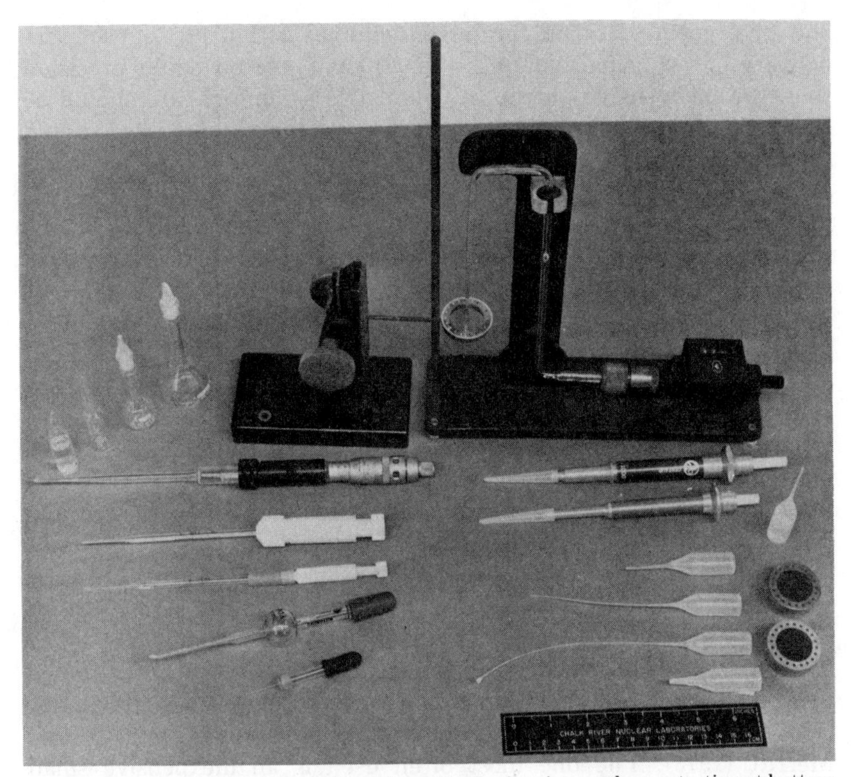

Fig. 76. Source Preparation Paraphernalia. In the picture above, starting at bottom left and proceeding clockwise, are a disposable capillary-tube pipet, a Grunbaum pipet, 2 Kirk pipets, a micropipet with precision-machined plunger, 2 ampoules, 2 volumetric flasks, a microburet with precision-machined plunger, 2 disposable polyethylene pipets, polyethylene pycnometers in various stages of preparation, and at bottom right, 2 source mounts of metallized thin plastic film.

equipment, but is less accurate than gravimetric sampling. Volumetric techniques have been reviewed in the literature (Moran, 1967; Merritt, 1973). For dispensing small samples for source preparation, micropipets and microburets are used and several types are shown in Figure 76. The least expensive, costing only a few cents, are self-filling capillary tubes from which the contained solution is ejected using a rubber bulb or syringe. These give a precision of from one to a few percent (Emanuel, 1973), are used only once, and discarded. Other types, such as the Grunbaum and Kirk micropipets, are normally cleaned and reused. For more accurate work they may be calibrated (Marinenko and Taylor, 1966; Moran, 1967). The same is true for micropipets or microburets in which a precision-machined plunger displaces a volume that is read on a micrometer dial. These devices are rather expensive and give good precision for larger volumes (≈100 μl), but for the delivery of very small volumes (≈1 μl) they give no better precision than the inexpensive pipets. In volumetric sampling, care should be taken to have all apparatus and solutions at thermal equilibrium and attention should be given to the directions for use of the particular equipment selected, as some pipets are calibrated "to deliver" a specified volume and others "to contain" it. In the latter case they must be rinsed, usually with the dilution medium of the solution being sampled.

Most radioactivity standards are certified as to activity per unit weight of solution and therefore gravimetric-sampling methods have assumed increased importance. Furthermore, gravimetric methods can give higher accuracy, e.g., <0.1 percent (Campion *et. al.,* 1964; Merritt, 1973). The essential equipment is a high-precision balance of 20-g capacity. Merritt (1973) has emphasized that care should be taken to place the balance on a sturdy heavy table (to minimize vibration) in a location free from drafts, preferably in a temperature-regulated and humidity-controlled room; and that routine tests of balance performance and calibration procedures are necessary as an assurance against systematic error.

Mass dispensing involves difference weighings from small bottles or pycnometers. Different designs of pycnometers have been discussed by Mann and Seliger (1958), Campion *et al.* (1964), Merritt and Taylor (1967), Bowes and Baerg (1970), Lowenthal and Page (1970), and Merritt (1973). The one most often used is an inexpensive small polyethylene ampoule[9] with the neck drawn out to form a capillary. Such a pycnometer is shown in Figure 76; at the lower right is the ampoule as received, and just above it is shown in the following three stages: with capillary (i) drawn out and sealed until required; (ii) cut to a suitable length for filling; and (iii) cut to a shorter length for weighing. Error caused by evaporation of liquid from such a pycnometer has been shown to be negligible (Merritt and Taylor, 1967; Bowes and Baerg, 1970). The radioactive solution (>1 g) is introduced into the pycnometer, the capillary tube is cut to a convenient length, and it is placed in the balance room for about ½ hour to attain thermal equilibrium. The balance operator should be seated in front of the balance for the last 10 to 15 minutes. A simple stepwise procedure with advisable precautions has been outlined by Merritt (1973), in which the pycnometer is weighed before and after a sample of solution is dispensed; the difference between the two weighings gives the sample size. By dispensing the solution sample to a weighed dilution flask, adding carrier solution, and finally weighing the flask plus contents, a dilution can be prepared (Merritt, 1973).

[9] Supplied by Canus Equipment Limited, 340 Gladstone Avenue, Ottawa, Canada.

5.2 Solid Sources

5.2.1 *Quantitative Source Preparation*

Many counting methods, such as alpha-, beta-, x-, and gamma-ray or coincidence counting, require that sources be in the solid phase, normally as thin deposits covering an area of approximately 1 cm^2. The simplest way to prepare them is to deliver the solution sample to a source mount and then evaporate to dryness by placing either under an infrared lamp, or in a covered box through which warm air flows (Bowes and Baerg, 1970).

Thin electrically-conducting source mounts are necessary for $4\pi\alpha$ and $4\pi\beta$, as well as 4π coincidence-counting methods. The preparation and effectiveness of metal-coated plastic films stretched on metal support rings have been reported by Roy and Turcotte (1967). Those described are both collodion and VYNS (a polyvinylchloride-polyvinylacetate copolymer) of surface density $\approx 10\ \mu g\ cm^{-2}$, coated with gold or gold (82 percent)-palladium (18 percent) alloy of surface density $\approx 20\ \mu g\ cm^{-2}$. The preparation of other thin films has been reviewed by Yaffe (1962). As a precautionary measure, many laboratories cover the dried active deposit with a similar metal-coated plastic film to form a stronger "sandwich" source and to inhibit the loss of small particles during manipulations.

Simple evaporation of the radioactive solution is not always an adequate source preparation technique, because the deposit tends to be in the form of large crystals, i.e., it tends to be non-uniform and self-absorption can be a serious problem (Merritt *et al.*, 1959). Furthermore, for some elements the deposit produced by simple evaporation is unstable because the resulting chemical compound is volatile or hygroscopic. In such cases, the precipitation method is effective in producing a stable source of improved uniformity. Normally a fine deposit of the radionuclide is precipitated *in situ* on the source mount by delivering the active-solution sample into a slight excess of precipitating agent (van der Eijk *et al.*, 1973). However, chemical problems encountered in specific applications may require additional steps. For example, to precipitate the radioelement in question as an insoluble hydroxide from an acid solution, it is necessary first to evaporate the radioactive solution on the source mount, then to dissolve the residue by adding a drop or two of water to the source mount, and finally to expose it to ammonia vapor during its final drying period. Some examples can be seen in Table 13. Helpful guides in judging the volatility of a precipitate are thermogravimetric data (Duval, 1963).

A wetting or seeding agent can improve source uniformity. Among the wetting agents employed are insulin, Catanac SN[10], Tween[11], and Teepol[12]. An area ≈1 cm^2 of the source mount is covered with a dilute solution of the wetting agent and then as much as possible of it is removed. The radioactive solution sample is delivered to the hydrophilic area, and then evaporated to dryness. Seeding agents used are Ludox-SM[13] and Teflon dispersion No. 30[14]. The active-solution sample is delivered into a drop of a 1:10^4 dilution of either of these sols containing ≈0.2 μg solids that serve as nucleation sites where many small crystals are produced upon drying.

Still more uniform sources can be prepared by other techniques that tend to be time-consuming. Freeze-drying combined with the use of a wetting agent has given good results (Keith and Watt, 1966). New approaches employ special source mounts onto which thin layers of a seeding agent (SiO) (Bellemare *et al.*, 1972; Lachance and Roy, 1972), or a mixture of ion-exchange resins (Lowenthal *et al.*, 1973; Lowenthal and Wyllie, 1973a, 1973b) have been added.

Although they are not ordinarily relevant to radioactivity standards, there are other source-preparation methods that give good source uniformity but less than 100-percent yields. They have been used, however, in the assay of radioactive solutions if a yield determination can be made by relative gamma-ray counting against a quantitatively prepared source. Among these methods are the formation of monomolecular layers (Dobrilovič and Simovič, 1973), vacuum evaporation, electrodeposition with and without an external source of e.m.f., electrospraying, chemical transport reactions, and ion implantation, all of which are discussed in a review paper by van der Eijk *et al.* (1973).

An alpha-particle source should be thin, uniform, and adherent to its substrate in order to minimize or eliminate scattering in the source, degradation of the spectrum, and change in the emission rate, respectively.

Alpha-particle emission gives rise to recoil atoms or even small conglomerations of atoms, and recoil contamination of detectors can occur, especially with those radionuclides with short-lived alpha-particle-emitting daughters. (Beta-particle emission yields too little recoil energy to cause noticeable contamination under usual measurement conditions.) Sources of high activity should be covered with metallized mylar, collodion, or VYNS films to reduce the chances of loss of

[10] Supplied by American Cyanamid Co., Bound Brook, N.J.
[11] Supplied by Atlas Powder Co., Wilmington, Delaware.
[12] Supplied by Shell Chemical Co., Houston, Texas.
[13] Supplied by E. I. DuPont de Nemours and Co., (Inc.), Wilmington, Delaware.
[14] Supplied by Interchemical Corp., Lodi, N.J.

material, to protect the surface from abrasion, and to avoid corrosion by the atmosphere (but see Section 5.2.2 concerning radiation damage to plastics). If sources must be used uncovered, such as for high-resolution alpha-particle spectrometry where spectrum degradation is a concern, the use of weak sources is advised to mitigate against possible problems due to radioactive contamination.

There are a number of source-preparation techniques widely employed, among which are electrodeposition, molecular plating, electrophoresis, electrospraying, vacuum evaporation, sputtering, and simple evaporation of a stable solution (see Greene *et al.*, 1972, for lists of references to the various techniques). Each method has at least one advantage: with electrodeposition and molecular plating, recovery is almost quantitative; with electrophoresis, mg cm^{-2} amounts can be adherently deposited; by electrospraying, thin layers with large areas can be made; vacuum evaporation and sputtering give good adherence; and simple evaporation of a stable solution is quantitative. The evaporation-of-solution technique should be used with wetting or seeding agents such as those mentioned in Section 5.2.1, and in the prescribed manner. A variety of backing materials can be used such as VYNS and collodion of surface density 10 $\mu g\ cm^{-2}$, coated with gold or gold (82 percent)-palladium (18 percent) alloy of surface density 10 $\mu g\ cm^{-2}$, metallized mylar, gold, silver, platinum, and stainless steel. If the backing material is clean and smooth, and, in the case of metals, polished, the source material will adhere to it better (Greene *et al.*, 1972).

5.2.2 *Handling and Storage of Thin Sources*

Thin sources are fragile and must be manipulated carefully and stored properly. In humid climates source deterioration caused by absorption of water vapor occurs and is particularly noticeable if the active deposit is a deliquescent salt. As indicated in Table 13, it can be helpful to store the sources at elevated temperatures until they have been counted.

On exposure to warm humid air for a few days, the thin metallic coating on VYNS or collodion films deteriorates, and adversely affects the electrical conductivity of the source mount. This occurs less in the case of Au-Pd-alloy coatings than for those of pure Au (Lowenthal and Smith, 1964). The problem is insidious in that the degradation usually is not visible, although in extreme cases it can be seen that the metallic coating has coalesced. The electrical conductivity of a source can be restored by the addition of another layer of metallized plastic film. The problem can be avoided by storing the sources in a dry atmosphere such as that in a desiccator.

Radiation damage to thin plastic source mounts is especially severe for alpha emitters and essentially prevents the long-term storage of such high-activity sources (>1 μCi). Although weak alpha-active sources on thin films have survived for several years, such source mounts become weaker with time. For this reason and because the health hazard associated with inhaled alpha activity is generally high, extra precaution should be taken in handling such sources. Garfinkel and Hutchinson (1970) have described sources of alpha emitters for use as gamma-ray standards in which the active material is enclosed between thin gold foils that in turn are sandwiched between layers of plastic. This technique prevents radiation damage to the plastic, but, unfortunately, can be used only for cases where thicker source mounts are tolerable.

5.3 Gaseous Sources

5.3.1 *Preparation of Gases for Liquid-Scintillation Counting*

Introduction of radioactive gases into the scintillation liquid for determination of activities is often convenient and simple. The solubility of gases is usually sufficient to permit their radioassay by liquid scintillation.

There are two distinct techniques for the radioassay of gases by liquid scintillation. If accurate results are desired, the entire gas sample must be dissolved in the liquid. However, it is sometimes convenient to fill the vial only partially with the liquid and to correct for losses resulting from the radioactive decay in the gas phase. In both cases, temperature control is important, particularly for the second method, because the solubility of gases is temperature dependent.

An important requirement for the liquid-scintillation counting of gases is the deaeration of the liquid prior to introduction of the gas. According to Shuping *et al.* (1969), the scintillation liquid should be refluxed for about 30 minutes and drawn up into a syringe that has a Luer fitting and appropriate stopcocks. The counting vial, provided with a Luer fitting and removable Luer stopcock, is evacuated and filled with the gas at a pressure that corresponds to slightly less than the limit of solubility of the gas in the solvent. Subsequently, the vial is connected to the syringe and the liquid is slowly forced into the vial. The vial is filled to the top with the scintillation liquid, the stopcock is removed, and a Luer cap is placed on the vial. Care should be exercised to avoid changes in temperature. Environmental monitoring for ^{85}Kr is routinely carried out by using this technique.

The gaseous products of oxidation and combustion of many types of samples can be trapped in suitable media such as water for ^{3}H, ethanolamine for ^{14}C, peroxide for ^{35}S, and then incorporated in scintillator solutions. Such methods are described in Jeffay and Alvarez (1961), Eastham *et al.* (1962), and Bransome (1970).

5.3.2 *Preparation of Gaseous Radioactivity Standards and Samples for Internal-Gas-Counter and Internal-Ionization-Chamber Systems*

Gaseous radioactivity standards are normally contained in glass ampoules, the sample being accessible by means of a glass break-seal or stopcock. The calibration is in terms of activity per mole of active gas and isotopic carrier, or total activity per sample. The latter is derived from the former simply by containing a sample of the former gas, at a known pressure and temperature, in a vessel of known volume.

Other radioactivity standards calibrated by gas counting or by internal ionization chamber (Section 4.4.4) may actually be distributed as solutions. Examples are ^{3}H as water, ^{14}C as Na_2CO_3 in aqueous solution (calibrated as ^{3}H and $^{14}CO_2$ by gas counting), and ^{14}C as benzoic-acid-7-^{14}C in toluene (calibrated as $^{14}CO_2$ in an internal ionization chamber). In the preparation of these standards, the water was reduced to hydrogen by passing the water vapor over a bed of hot zinc (Mann *et al.*, 1964), and the sodium carbonate was converted to $^{14}CO_2$ by the action of perchloric acid (Mann *et al.*, 1961). The benzoic acid in toluene was converted to $^{14}CO_2$ and water by combustion with a slight excess of oxygen in a Paar bomb (Marlow and Medlock, 1960). Samples of the various gases were then trapped cryogenically in glass bulbs of known volume, at known pressures and temperatures, for assay in internal gas counters (Section 3.6).

In the case of the NBS noble-gas radioactivity standards, namely ^{37}Ar, ^{85}Kr, ^{131m}Xe, and ^{133}Xe, samples of these gases are prepared for gas counting by quantitative mixing with appropriate quantities of natural argon, krypton, and xenon, respectively. ^{131m}Xe and ^{133}Xe, which are obtained from an irradiation of uranium in a nuclear reactor, are separated in the NBS isotope separator. The separated 60-keV ion beams are allowed to impinge on aluminum foil in which they are trapped (Figure 77). The aluminum strips containing the two isotopes are cut out and each aluminum strip is melted in an evacuated vycor glass tube, thereby releasing very pure samples of each radioactive xenon isotope.

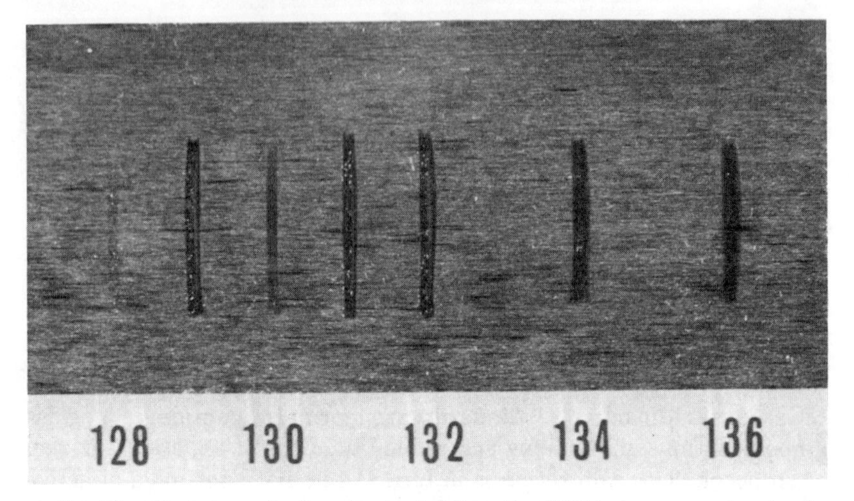

Fig. 77. A photograph of a collector foil from the NBS isotope separator showing the implanted stable isotopes of Xe. The mass regions are visually enhanced by means of a thin coating of vacuum grease.

In the preparation of noble-gas radioactivity standards for distribution, the radioactive gas sample is mixed with inactive isotopic carrier in a gas-handling system such as that shown in Figure 78. The mixing is performed by repeated freezing and evaporation by using either liquid nitrogen or liquid helium, the former when the vapor pressure of the radioactive gas sample at 77 K is sufficiently low for adequate freezing. Up to two hundred ampoules may be filled at one time, together with a special volume-calibrated flask (Figure 79), which is then attached to the gas-handling system of the NBS internal gas-counting system (Figure 31). If, however, the activity in the individual standard ampoules of approximately 10 ml is sufficient, the gas in them may be totally transferred, cryogenically, to the gas-counting system for measurement of the activity.

After transfer to the internal gas-counting system, the quantity of the gas sample is calculated from pressure and temperature measurements, by using the known volumes of the gas-handling system and van der Waals' equation. It is then mixed with a known quantity of a suitable counting gas, usually methane for proportional counting, together with some 2 to 5 percent by volume of the inactive noble gas in question (or natural hydrogen for ^{3}H, or natural CO_2 for ^{14}C). In the process of mixing it is, however, important first to flush the previously evacuated counters with some of the counting gas mixed with inactive isotopic carrier in order to mitigate against the possible subsequent adsorption of active gas onto the walls of the counters.

Fig. 78. The gas handling rack used at NBS for mixing gaseous radionuclides with inactive carrier gas, which is usually isotopic, and for filling the ampoules shown.

5.4 Preparation of Liquid Sources for Specific Methods of Measurement

5.4.1 *Liquid-Scintillation Counting*

5.4.1.1 *Solution and Emulsion Scintillators.* The principles of liquid-scintillation counting were dealt with in Section 2.4.5 and some

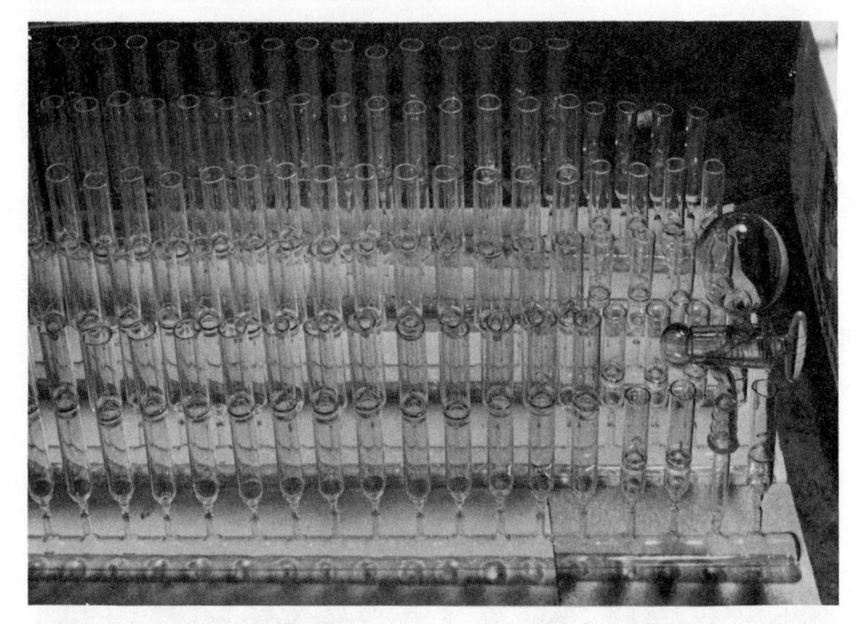

Fig. 79. Partial view of ampoule racks containing approximately 200 ampoules. The calibrated volume flask is seen on the right above the vacuum stopcock.

applications of solution and emulsion scintillators were reviewed in Section 4.3. In most instances, counting samples are prepared by pipetting the scintillator formulation into the vial and then adding (either gravimetrically or volumetrically) a known amount of an aqueous solution of the sample. Procedures for accurately dispensing the aqueous sample have been reviewed recently by Merritt (1973). Techniques for preparing stable solution scintillators are given in Vaninbroukx and Spernol (1965), while precautions to be observed when preparing emulsions are given in Lieberman and Moghissi (1970), Handler and Romberger (1972), and Benson (1976).

5.4.1.2 *Suspended Scintillators.* Scintillators such as anthracene, calcium fluoride, and plastics in different shapes may be added to aqueous radioactive solutions to form two-phase counting systems. When it is essential to concentrate small quantities of inorganic ions from large volumes of aqueous solution prior to measurement, ion-exchange resins, both of the scintillation and conventional type, can be used. Ion-exchange resins charged with radioactive ions can be measured by using suspension techniques (see Section 5.4.1.3). The counting efficiency with suspended-scintillator systems is more dependent on the scintillator-volume to sample-volume relationship than on the salts in, or the pH of, the solution (Bransome, 1970). One distinct advantage of this method (excluding the use of conventional ion-exchange resins) is the easy recoverability of the sample (Heimbuch, 1961).

5.4.1.3 *Gel Scintillators.* This method is applicable to insoluble or not conveniently solubilized material, to solutions with high solids content, and to materials that are strong quenchers in solution (Helf *et al.*, 1960). With the addition of a thixotropic agent to the scintillator liquid, finely divided solid samples can be suspended, a procedure that eliminates particle or sample settling. The most frequently used gelling agent is Cab-O-Sil[15] (Ott *et al.*, 1959). Benakis (1971) describes his experiences with another thixotropic agent, Poly-Gel-B[16], which offers the advantages of long-term stability of suspensions even at room temperature, ease in handling, and reduction of adsorption on surfaces of glass counting vials.

5.4.1.4 *Counting with Solid-Source Supports.* Substances that are precipitated, filtered, or absorbed on supports such as cellulose and nitrocellulose membranes, or lens, filter, and anion-exchange papers can be assayed by liquid-scintillation counting. The counting efficiency will depend on a number of variables, such as the position of the solid support in the counting vials, and the chemical and physical form of the sample material. Internal standards should be used for quench compensation; for the counting of low-energy beta-ray emitters, the procedure should be "rigorously standardized" for the specific kind of sample and support used (Bransome, 1970).

5.4.2 *Čerenkov Counting*

For materials containing beta-ray emitters with energies greater than 0.26 MeV, sample preparation for Čerenkov counting is simple and the source can be recovered. Solid samples may be prepared by dissolving the material in a suitable solvent. Detection efficiency will be dependent on the beta-ray energy of the radionuclide, the characteristics of the phototube, the geometrical and optical arrangement of the counting vials, the refractive index, and density of the solvent (Parker and Elrick, 1970). With the addition of wavelength shifters, coincidence-counting efficiency can be improved, and the method used for lower-energy beta particles. However, sample preparation becomes more complicated. Stability of the solution is dependent on the particular wavelength shifter used, its concentration, and pH. Ross (1971) recommends 4-methylumbelliferone as the "overall best" wavelength shifter for Čerenkov counting because solutions with it appear to be

[15] Supplied by G. L. Cabot, Inc., Boston, Massachusetts.

[16] French patent No. 1,590,762.

stable with changes in pH. Work by Ross (1970) gives evidence that Čerenkov-counting techniques can be extended to ^{14}C with the use of solvents of high refractive index (see also Section 4.5.1).

5.4.3 *Formamide Counting*

Solution samples for counting in a $2\pi\beta$ windowless gas-flow counter are prepared by dissolving the radioactive material in an inert, low-vapor-pressure organic solvent such as formamide (Schwebel *et al.*, 1951, 1954). If the radioactive material is not readily soluble in the organic solvent, it may be first dissolved in a liquid that is miscible with the organic solvent. Suitable dilutions are from 10 to 100 times, depending on the activity of the radioactive solution. Up to 10-percent water content in the formamide can be tolerated without change in the counting characteristics or efficiency of the internal gas counter. A sample of the counting solution, generally 1 ml, is pipetted into a machined counting cell of specific dimensions and counted in a $2\pi\beta$ windowless gas-flow counter. The counting solution should contain 1 percent of an appropriate carrier to prevent adsorption losses and contamination of the counting cell. If the dimensions of the counting cells used are always the same and the same volume of solution is always counted, the depth of solution is always the same and any effects due to gamma rays will remain constant.

6. The Assay of Radioactivity and the Identification of Radionuclides in Environmental, Medical, and Industrial Laboratories

6.1 Environmental and Low-Level Radioactivity Measurements

6.1.1 *General Considerations*

6.1.1.1 *Introduction.* Radioactivity measurements on environmental samples are usually made to assure that levels are kept below current maximum permissible concentrations. Radioactive tracer studies have also been used to measure transport phenomena. The large number of radionuclides found in the environment can be divided into three categories: namely, (i) "natural" radioactivity from uranium, thorium, and other primordial radionuclides; (ii) radionuclides produced by cosmic-ray interactions in the atmosphere; and (iii) the man-made contaminants from nuclear-weapons fallout and effluent from nuclear facilities. Some typical concentrations of long-lived radionuclides in ground-level air and surface soil are given in Tables 14 and 15, respectively. Table 16 lists some of the radionuclides found in effluents from nuclear-power facilities.

The instruments used in low-level radioactivity measurements are basically the same as those described in previous chapters. The special problem in low-level counting is that the ratios of counting rates of sample to background are often so low that significant modifications to the measurement system are needed to improve sensitivity. Some extensive general references in this field are Watt and Ramsden (1964), ICRU (1972), Harley (1972), and NCRP (1976b).

Normally there are four steps in the measurement of environmental radioactivity:

(i) Selecting the measurement system;
(ii) Collecting a suitable sample;
(iii) Separating and, or, concentrating the desired species by radiochemical and, or, physical means;
(iv) Assaying the radioactivity in the source with a detector that has been calibrated under similar counting conditions with appropriate radioactivity standards.

Although this section is primarily concerned with (iv), specific measurements systems for alpha-particle, beta-particle, and gamma-ray emitters, some general remarks concerning (i) are in order.

6.1.1.2 *Selection of a Low-Level-Measurement System.* The choice of a measurement system for a particular application is based on a consideration of such factors as: cost, amount of sample available, type and intensity of radiation to be measured by the detector, time available for counting, background, and efficiency. Many expressions have been developed that are useful in comparing different detectors or different measurements systems. For example, Moghissi (1970) derived an expression often used for comparing liquid-scintillation-counting procedures. He defined a Y value as the minimum radioactivity concentration detectable for a one-minute counting time as

$$Y(\text{pCi/g}) = \frac{\sqrt{B}}{2.22\, M\epsilon}, \tag{6.1}$$

where

B = background count rate (cpm),
ϵ = counting efficiency (cpm/dpm),
M = sample mass (g),
and the constant 2.22 is the number of disintegrations per minute per picocurie.

The three variables on the right hand side of Equation 6.1 are usually interrelated (background and efficiency depend on the sample size) and this must be considered in optimizing the sensitivity of the measurement process.

It can be seen that Equation 6.1 is of the general form

$$Y(\text{pCi/g}) = \frac{\text{Constant} \times \text{Detection Limit}}{\text{Sample Size} \times \text{Efficiency}}, \tag{6.2}$$

where Moghissi used $\sqrt{B}$ as the detection limit. In the more general case, detection limits and efficiency terms will depend on the system under consideration. Some comments concerning detection limits, sample size, and background that may aid in evaluating a measurement procedure are given in the following paragraphs.

6.1.1.3 *Detection Limits.* The problem of assigning an *a priori* detection limit to a measurement involves the theory of Hypothesis Testing (Section 7). Basically one must decide whether a given response is statistically different (at prescribed confidence limits) from the background response. This problem has been examined by Altshuler and Pasternack (1963), Currie (1968), Pasternack and Harley (1971), and Donn and Wolke (1977). A derivation of detection limits along the lines developed by these papers is given in Section 7. It is sufficient to say here that the detection limit is directly proportional to the standard deviation of the background counting rate σ_B. Thus, it is not only necessary that the background be reduced as much as possible, but also that it be free from fluctuations that would increase σ_B. Sources of background counts are discussed in Section 6.1.1.5.

6.1.1.4 *Sample Size.* The sample size will usually be limited by such practical aspects as cost, shipping, and storage space. In the laboratory, other limitations on sample size may be imposed by the analytical procedure or physical dimensions of the counter and shielding. For each counter, there will also be an optimum sample size that will depend on considerations such as the average distance of penetration of the radiation within the source, and on geometry.

6.1.1.5 *Background.* The background may be defined as the count rate observed when measuring a "blank" that simulates as closely as possible in chemical composition and physical form the source being measured. Care should be taken that the structural materials and reagents used in the preparation of the blank and the counting sample contain minimal activity. Statistical considerations involving background and minimum detectable activity are treated in Sections 7.1.2 to 7.1.4.

Examples of sources of background count rate are natural and fallout radioactivity, nearby nuclear facilities, cosmic radiation, instrument noise, and radioactivity in the detector materials. Some general comments concerning each of these are given in the following paragraphs.

a. *Natural and Fallout Radioactivity*

The common radionuclides found in the environment are listed in Tables 14 to 16. These species can contribute to the background in several ways depending on the nature of the detector. Their presence, for example, in soil, air, concrete, and other environmental matrices, gives rise to a continuous (and variable) background of radiation. Shielding alone will not necessarily eliminate some of this contribution because, for example, gaseous radionuclides such as ^{220}Rn and ^{222}Rn may be present inside the shielding, and their daughter products may be deposited upon it and the counting equipment.

b. *Nearby Nuclear Facilities*

The presence of a nearby nuclear facility may pose serious problems for environmental measurements. In addition to the off-gases listed in Table 16, reactors release some short-lived gaseous radionuclides that may contribute to the background count rate in the local area.

Neutron-generating facilities in the vicinity may also have an adverse effect on the counter background due to radiation arising from neutron capture. In semi-conductor detectors, for instance, Bunting

TABLE 14—*Typical radioactivity concentrations of long-lived radionuclides in ground-level air*

Radionuclide	Gas[a]	Particles[b]	Monthly deposition[c]
	pCi/m^3	pCi/m^3	pCi/m^2
Naturally occurring			
^{222}Rn	100	—	—
^{220}Rn	1	—	—
^{210}Po	—	0.003	—
^{210}Pb	—	0.01	—
^{7}Be	—	0.1	3,000
Man-Made			
^{239}Pu	—	3×10^{-5}	—
^{238}Pu	—	1×10^{-5}	—
^{144}Ce	—	0.01	200
^{137}Cs	—	0.002	50
^{90}Sr	—	0.001	30
^{85}Kr	17	—	—
^{3}H	1	—	20,000

[a] Mean ^{222}Rn, ^{220}Rn and ^{85}Kr values from NCRP (1976b).

[b] Data for New York City in 1972 (HASL, 1973) except ^{210}Pb values for Cincinnati in April—May 1961 (Gold *et al.,* 1964). Units are pCi/m^3 of filtered air.

[c] Data for Cincinnati in 1973, extracted from monthly reports of the Radiochemistry and Nuclear Engineering Laboratory of the Environmental Protection Agency, Cincinnati, Ohio. The value of ^{3}H is based on an average concentration of 0.2 nCi/l in drinking water and an average monthly rainfall of 10 cm (EPA, 1973).

TABLE 15—*Radioactivity concentrations of radionuclides in soil*[a]

Radionuclide	Concentration
	pCi/g
Naturally occuring[b]	
^{40}K	10
^{87}Rb	4
^{226}Ra	0.8
^{232}Th	0.6
^{238}U	0.7
Fallout[c]	
^{90}Sr	~1
^{137}Cs	~1

[a] From Chapter 2 of NCRP (1976b).

[b] Values for the natural radionuclides are reported as world average concentrations (NCRP, 1976b), although variations as large as a factor of 4 between different locations are quite common.

[c] Values shown for fallout radionuclides are typical for surface soil (that is, the sample is taken to a depth of about 1 cm). Depth distributions for fallout ^{137}Cs in soil are given by Ragaini (1976).

and Kraushaar (1974) observed ^{73m}Ge, ^{75m}Ge, and ^{28}Al, the last being produced by irradiation of the detector housing. The problem is much more serious, however, for NaI(Tl) detectors (ICRU, 1972) where ^{24m}Na and ^{128}I are produced.

In addition to the effects described above, a neutron flux produces a number of high-energy electrons and photons in the surroundings that may subsequently interact within a detector system.

c. *Cosmic Rays*

The cosmic-ray flux in the upper atmosphere consists of a "galactic" component and a "solar" component. The galactic radiation (NCRP, 1975) is composed of about 87 percent protons, 11 percent alpha particles, 1 percent heavier nuclei, and 1 percent electrons, many of which have energies in excess of 10^3 MeV. The sea level cosmic-ray flux consists of both muonic and nucleonic components. The effects of cosmic rays on detector backgrounds have been examined by May and Marinelli (1960) and Tanaka *et al.* (1965). Extensive reviews are available in ICRU (1972), NCRP (1975), and Adams and Lowder (1964). The counts observed in detectors are principally due to the secondary particles generated by the interaction of these high-energy particles with matter.

TABLE 16—*Radionuclides in effluents from typical nuclear facilities* ($T_{1/2} > 1$ *d*)

Nuclear power stations[a]		
in gases:		
^{3}H	^{85}Kr	^{133}Xe
^{14}C	^{131}I	^{133m}Xe
in airborne particles:		
^{51}Cr	^{60}Co	^{134}Cs
^{54}Mn	^{89}Sr	^{137}Cs-^{137m}Ba
^{55}Fe	^{90}Sr-^{90}Y	^{140}Ba-^{140}La
^{59}Fe	^{99}Mo-^{99m}Tc	^{141}Ce
^{58}Co	^{131}I	
in liquids:		
^{3}H	^{58}Co	^{124}Sb
^{14}C	^{60}Co	^{131}I
^{32}P	^{63}Ni	^{133}Xe
^{51}Cr	^{65}Zn	^{133m}Xe
^{54}Mn	^{89}Sr	^{134}Cs
^{55}Fe	^{90}Sr-^{90}Y	^{137}Cs-^{137m}Ba
^{59}Fe	^{95}Zr-^{95}Nb	^{140}Ba-^{140}La
^{57}Co	^{99}Mo-^{99m}Tc	^{141}Ce
	^{110m}Ag	^{144}Ce-^{144}Pr
Nuclear fuel reprocessing plants[b]		
in gases:		
^{3}H	^{85}Kr	^{129}I
in liquids:		
^{3}H	^{134}Cs	^{238}U
^{14}C	^{137}Cs-^{137m}Ba	^{238}Pu
^{22}Na	^{144}Ce-^{144}Pr	^{239}Pu
^{54}Mn	^{147}Pm	^{240}Pu
^{60}Co	^{155}Eu	^{241}Pu
^{90}Sr-^{90}Y	^{233}U	^{241}Am
^{95}Zr-^{95}Nb	^{234}U	^{242}Cm
^{125}Sb	^{235}U	^{244}Cm

[a] From Kahn *et al.* (1970, 1971).

[b] From Magno *et al.* (1970); Cochran *et al.* (1970); Holm and Persson (1976); Cuddihy *et al.* (1976).

d. *Instrument Noise*

The contribution to background from instrument noise is very dependent on the type of detector and energy region of interest. For instance, in scintillation detectors, spurious counts are observed at low pulse height due to noise in the electron-multiplier phototubes. Thermal noise of this nature, once a significant problem in liquid-scintillation counters, has been reduced by improved tube design (see Section 4.3.5).

Variations in the high-voltage supply may also pose a problem. All counters should be isolated from electro-mechanical equipment such as pumps, compressors, motors, and Tesla coils by an efficient filter in the main power supply in order to prevent spurious counts. A single efficient electrical ground may also be required to prevent 60-Hz noise induced between interconnected instruments at differing ground potentials. Other examples of instrument noise will be discussed in relation to specific measurements systems.

e. *Radioactivity in Detector Materials*

The contamination of the detector materials with radioactivity is a serious problem and one that is difficult to correct (Rodriguez-Pasqués, 1971). A common problem is that of ^{40}K in glass phototubes. The contribution from this source is greatly reduced by using quartz rather than glass windows. Measurements of radioactivity in materials used in low-level-counting systems have been reported by DeVoe (1961), Weller (1964), and Rodriguez-Pasqués *et al.* (1972); and in Section 2 of ICRU (1963).

6.1.1.6 *Interference Due to Other Radioactive Constituents.* As stated earlier, environmental samples are often mixtures of several radionuclides. In such cases, the sensitivity with which one can measure a particular constituent depends on the degree of interference from the other radionuclides in the sample. If only one species is of interest, it may be possible to effect a partial chemical separation during sampling. For instance, one might extract material of interest from several hundred liters of water using a cation-exchange column. In the laboratory, radionuclides may be separated from the bulk matrix and each other by radiochemical methods (Harley, 1972; Johns, 1975; NCRP, 1976b).

It is often possible to assay several radionuclides in the same sample by using spectrometric techniques. However, interferences may still be encountered, such as a Compton continuum caused by more energetic photons from other components in the sample. These interferences may often be minimized by instrumental techniques. Examples such as coincidence, anticoincidence, and pulse-shape discrimination will be discussed in relation to specific measurement systems in the following three subsections.

6.1.2 *Low-Level Measurement of Alpha-Particle-Emitting Nuclides*

6.1.2.1 *Introduction.* The very short ranges of alpha particles dictate that the amount of matter between the radioactive source being measured and the alpha-particle detector be minimized. Different methods for achieving this end include close physical proximity of source to detector, placing them in a vacuum chamber, and mounting the source inside the detector. The detector and source holder should be constructed of materials having very low alpha-particle activity (see also references in Section 6.1.1.5). Even materials having negligible intrinsic alpha-particle activity can indirectly contribute to the observed alpha-particle background if other background radiations cause nuclear reactions in these materials that produce alpha particles, e.g., (n,α) reactions in nylon sample planchets (Rodriguez-Pasqués *et al.*, 1972). Materials that are clean initially can become contaminated in time by exposure to atmospheres containing alpha-particle-emitting daughters of ^{222}Rn and ^{220}Rn that "plate out" on surfaces.

The short range of alpha particles in liquid and solid samples, less than 0.01 cm, usually requires the removal by physical and, or, chemical means of as much as possible of the inactive material originally present. For gross alpha-particle measurements, maximum sensitivity is attained when the prepared sample is at least as thick as the range of the alpha particles, i.e., so-called "infinitely thick" (Watt and Ramsden, 1964). For alpha-particle spectrometry, the source should be in a very thin layer to provide maximum resolution. If absorption corrections are to be less than 1 percent, the source thickness should be less than 1 percent of the average range of the alpha particles when counting in a 2π configuration (Lucas and Hutchinson, 1976). This criterion means that a sample of low atomic weight should have a surface density of approximately 100 $\mu g/cm^2$ or less. (Surface density is obtained by multiplying the density of the sample by its thickness.) A number of methods for preparation of alpha-particle-emitting sources has been described in NCRP (1961), by Yaffe (1962) and in references therein, and by Watt and Ramsden (1964). For a discussion of alpha-particle-emission rates from solid sources and range-energy relationships, see Evans (1955) and ICRU (1972).

Specific methods for measuring low-level alpha-particle radiation are discussed, according to type of detector, in Sections 6.1.2.2 through 6.1.2.5. The principles of some of these detectors are concisely given by Johnson *et al.*, (1963).

6.1.2.2 *Scintillation Counting.* One of the earliest methods for detecting alpha particles used a ZnS screen and a low-power microscope to view the tiny flashes of light. Discs of polyester film coated with silver-activated ZnS and phototubes with appropriate electronic circuitry are now used. Developed by Hallden and Harley (1960), such detectors have a geometrical efficiency of approximately 50 percent. They are not suitable for spectrometric analysis. The background counting rates over the entire alpha-particle-energy range are quite low, and of the order of one event per hour for ZnS (Ag) on a thin plastic disc of 24-mm diameter. Figure 80a shows the configuration used for measuring thick sources. Thin sources can be placed between the planchet and the phosphor in the configuration shown in Figure 80b, which is also used for background measurements.

"Infinitely thick" powdered sources containing isotopes of radium and their daughters can be assayed by scintillation counting, provided the sample is sealed to prevent the loss of radon. The fraction of the total activity arising from the decay of ^{226}Ra may be found by following the ingrowth of ^{222}Rn. Biological and geological samples have been assayed in this way (Turner *et al.*, 1958; Cherry, 1963).

The contributions of two of the natural radioactivity series to the total alpha-particle activity in thin sources can be measured by the so-called "fast-pairs" technique. ^{215}Po in the actinium series and ^{216}Po in the thorium series have half lives of 1.78×10^{-3} s and 0.15 s, respectively. As each of these is formed and decays, pairs of nearly coincident alpha particles are emitted which can interact with the detector, giving rise to output pulses closely related in time. These fast pairs of pulses are presented to two scalers, one that records all pulses, and the other that is insensitive to any other pulse for a preset time, typically 0.3 to 0.4 s, after the arrival of the first pulse. The difference between the count rates of the two scalers gives the number of pairs in unit time. A variation of the method, that gives the proportion of pairs directly, is for the second scaler to be sensitive to the second member of the

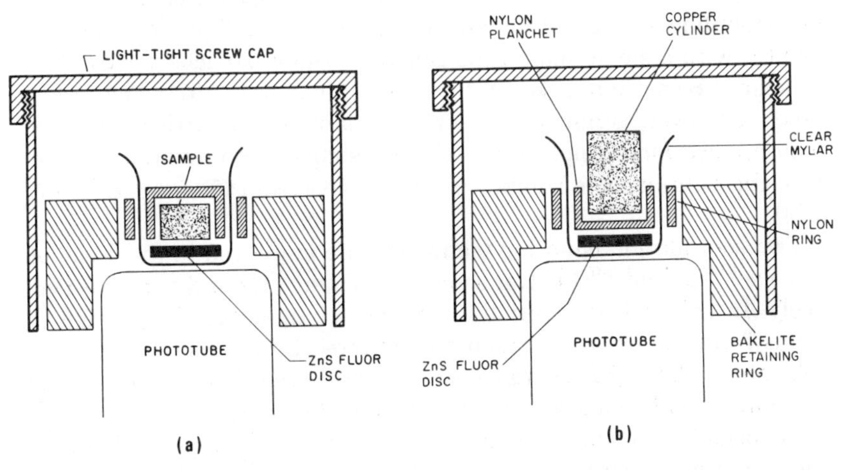

Fig. 80. Schematic diagram of the zinc-sulfide scintillator with phototube for the detection of alpha particles. The configuration in (a) is for count-rate measurements of thick samples, while that in (b) is for thin-sample and background measurements. (From Rodriguez-Pasqués *et al.*, 1972)

pair for a preset time after the arrival of the first. Except for very thin sources, however, a correction must be calculated to account for the differences between the ranges of the pairs of alpha particles in the sample, hence the technique is only an approximate one (Cherry, 1963, and references therein).

Other inorganic scintillators that are useful for the detection of alpha particles are CsI(Tl) and CaF_2(Eu). CsI(Tl) is only very slightly, but reversibly, hygroscopic so that no encapsulation is required. Exposure to air at high relative humidity should be avoided. The resolution for 5.29-MeV alpha particles with small crystals, 5-mm diameter by 0.3-mm thick, is approximately 200 keV full width at half maximum (FWHM) (Watt and Ramsden, 1964), but is poorer for larger crystals. A recently developed scintillator, CaF_2(Eu), is not hygroscopic, and can be cleaned with water. It is less sensitive to background gamma radiation than CsI(Tl) because it is composed of lower atomic-weight elements.

Conventional liquid-scintillation counting systems are suitable for the assay of alpha-particle-emitting nuclides in some samples, but the background count rates may be so high as to preclude their application, in reasonable counting times, to very low-level sources. A low-background liquid-scintillation counter with a single phototube has been developed by McDowell (1975). High-detection efficiencies, that can approach 100 percent, are an advantage of liquid-scintillation counting. ^{222}Rn and its decay products have been assayed by liquid-scintillation counting (Horrocks and Studier, 1964; Rolle, 1970; Darrall *et al.*, 1973). The coupling of a multi-channel analyzer to a liquid-scintillation counter with a single phototube can yield alpha-particle spectra with an energy resolution of several percent (McDowell, 1971; Figure 81), which, however, is not satisfactory for many alpha-particle spectroscopy applications. Application of pulse-shape discrimination can reduce the beta-particle component of the spectrum (Thorngate *et al.*, 1974).

Two scintillation-counting techniques are employed in the assay of gaseous ^{222}Rn and ^{220}Rn after separation from ^{226}Ra and ^{224}Ra, respectively. In the first, the initially positively-charged recoiling daughter atoms from the decay of radon are collected by means of an electric field, on to a ZnS (Ag)-coated electrode within the chamber containing the sample and filling gases (Bryant and Michaelis, 1952). The electrode, usually a screen, may be viewed from outside the chamber by a phototube as shown in Figure 82 (Mann and Seliger, 1958; Roberts and Davies, 1966), or removed and placed in an alpha-particle detector after a collection period (Kuchta *et al.*, 1976). In the second technique, the gas is held in a smaller chamber and the photons, produced when

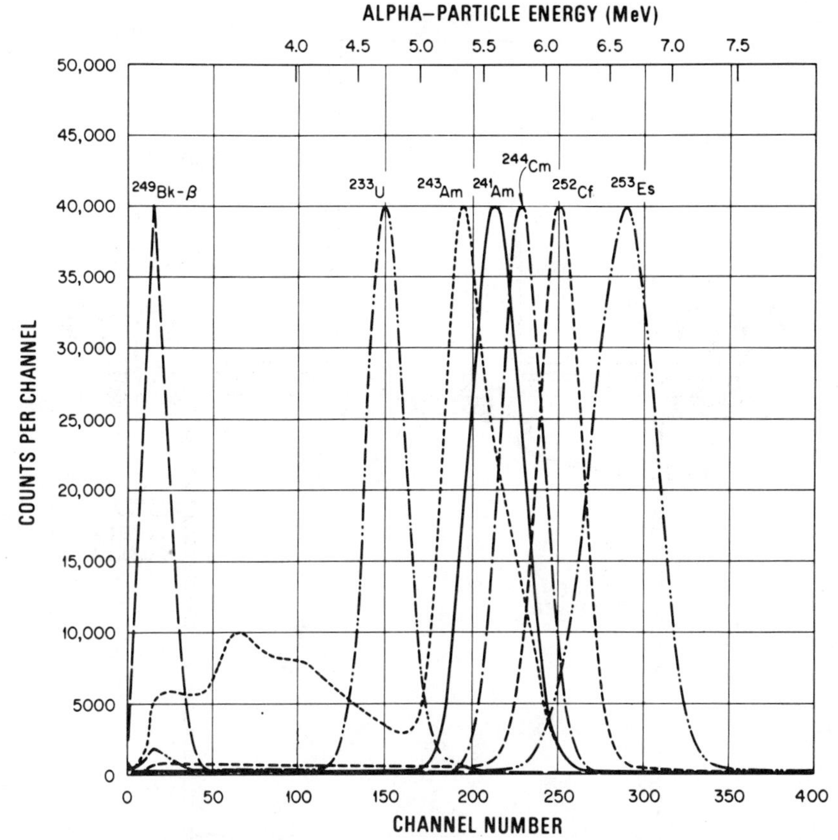

Fig. 81. Liquid-scintillation spectra of selected actinides using a single-phototube detector, with the peak-channel counts of each major peak preset to 40,000. (From McDowell, 1971)

the alpha-particles hit the ZnS(Ag)-coated walls, are detected by one (Lucas, 1957) or two (Kraner *et al.*, 1964) phototubes.

6.1.2.3 *Gas-Filled Detectors.* The assay of ^{226}Ra can be accomplished by measuring its separated gaseous daughter ^{222}Rn in an ionization chamber. Ionization chambers, both direct-current and pulse, and the associated gas-handling and purification systems, were developed for the assay of radon in solution, breath and air samples, and of radium in solutions of rock and ore samples, by Evans (1935, 1937), Curtiss and Davis (1943), and Harding *et al.* (1958). A radon-analysis system that is similar in principle has been described by Harley (1972).

A gas-filled proportional detector (see, e.g., McDaniel *et al.*, 1956) can be used to detect alpha particles from thick or thin solid sources. The source may be inside the detector, or outside next to a thin window. Some systems have the capability for the simultaneous re-

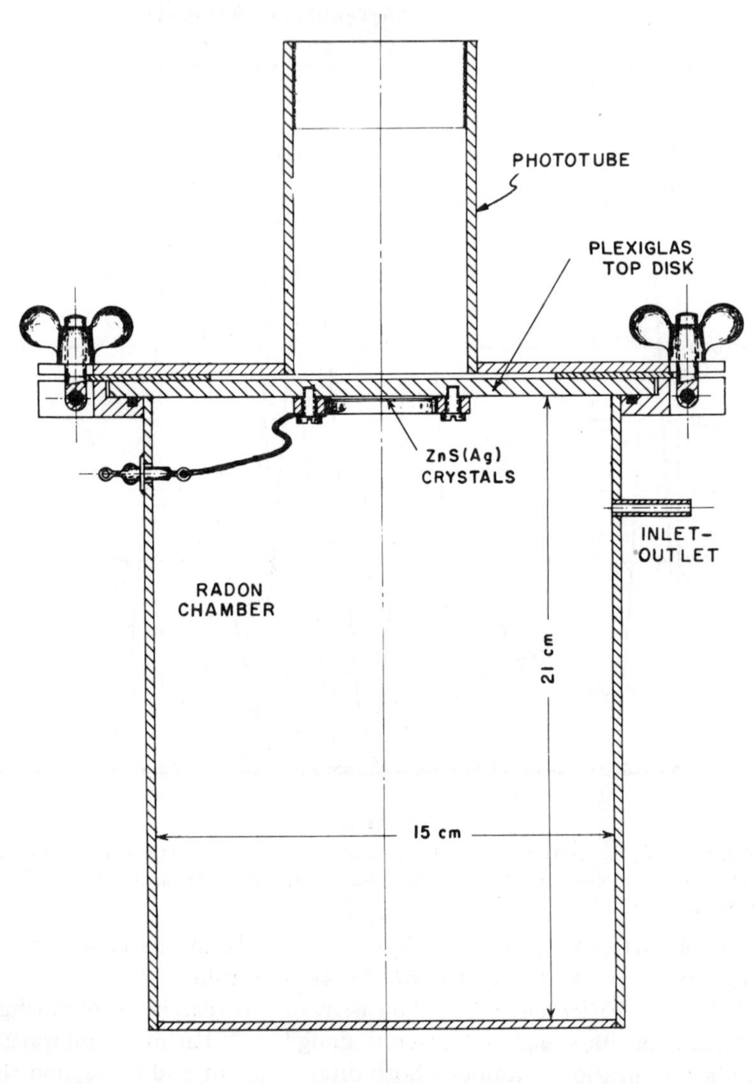

Fig. 82. Cross-sectional drawing of scintillation counter for radon analysis constructed at the National Bureau of Standards and based on the method of Bryant and Michaelis (1952). (From Mann and Seliger, 1958)

cording of alpha and beta particles by using appropriate pulse-height discrimination.

6.1.2.4 *Alpha-Particle Spectrometry.* The two major types of alpha-particle spectrometers are the pulse-ionization chamber and the semiconductor detector. Pulse-ionization chambers of many designs have been built, but nearly all can be classified as either parallel-plate or cylindrical, and each class can be divided into gridded or gridless varieties. Cylindrical chambers have the advantage in low-level measurements of providing a large surface area for the sample without building up such large capacitance between electrodes as to degrade resolution. A Frisch-type grid that shields the collecting electrode is often employed to improve resolution by reducing the pulse-widening effect of the positive ions (see, e.g., Bunemann *et al.*, 1949). Cherry (1963) and Watt and Ramsden (1964) elaborate on the merits and construction principles of pulse-ionization chambers. For good examples of spectrometers with very large area (15,000 cm^2) and high resolution (down to 40-keV FWHM), see Hill (1961) and Osborne and Hill (1964), respectively.

Silicon surface-barrier semiconductor detectors, that were previously fragile and costly, are now probably the most widely used in alpha-particle spectrometry. A 1-cm^2 active-area detector typically has a resolution of 15- to 20-keV FWHM for 5.5-MeV alpha particles emitted from a source 1.5 detector-diameters away. Placing the source closer to the detector increases the detection efficiency at the expense of degraded resolution. Compared to most pulse-ionization chambers, the larger semiconductor detectors, with an active area of about 20 cm^2, are still very small. This fact may necessitate more extensive sample preparation prior to counting than for pulse-ionization detection. Electrodeposition and evaporation are frequently used as the final step to concentrate the activity on to a small area (see Section 5.2; Yaffe, 1962; Watt and Ramsden, 1964). Contamination of silicon detectors by recoil atoms or by volatilization of ^{210}Po can raise otherwise very low background-count rates to unacceptable levels. These problems and solutions for them have been investigated by Sill and Olson (1970).

Mass spectrometry is another analytical method that has been applied to low-level quantitative identification of alpha-particle-emitting nuclides (NCRP, 1976b). By this method Diamond *et al.* (1960) assayed transuranic elements in thermonuclear-bomb debris, Krey and Hardy (1970) determined the isotopic composition of plutonium in contaminated soil, and Fields *et al.* (1972) assayed for several actinides in lunar materials.

6.1.2.5 *Nuclear-Track Techniques.* Special photographic emulsions, commonly referred to as nuclear emulsions, can detect alpha particles and fission fragments. These emulsions can detect very low activities provided that the exposure periods are sufficiently long, typically days or longer. The method is a form of autoradiography. A major drawback is the time-consuming counting of tracks in the developed emulsion.

Elements assayed include thorium (Picciotto and Wilgain, 1954) and plutonium (Geiger *et al.*, 1961). A broad survey of the technique was provided by Yagoda (1949).

Alpha particles also leave tracks upon passing through various plastics, glasses, and micas. Although too small to be observed directly, the tracks can be enlarged by chemical etching to the point where they can be seen with a microscope. Three-dimensional recording of tracks is possible with thicker pieces of plastic. Counting of the tracks can sometimes be accomplished with a flying-spot microscope (Young and Roberts, 1951) or a spark counter (Fleischer *et al.*, 1972). This technique has applications, for example, in health physics (Becker, 1969), and the assay of uranium in ores, water, biological materials, and fossils (Fleischer *et al.*, 1972, and references therein).

6.1.3 *Low-Level Measurement of Beta-Particle-Emitting Nuclides*

6.1.3.1 *Introduction.* The necessity for counting beta particles lies in the fact that a number of radionuclides either emit no gamma or x rays, or emit them at so low a probability per decay as to make assay by photon detection impossible. Such radionuclides that are frequently assayed (and their maximum beta-particle energies in MeV) include ^{3}H (0.0186), ^{14}C (0.156), ^{32}P (1.71), ^{35}S (0.167), ^{45}Ca (0.259), ^{85}Kr (0.69), ^{89}Sr (1.49), ^{90}Sr (0.546), and ^{90}Y (2.28). Of these, only tritium emits beta particles of such low energy as to require special detection techniques.

Assay by beta-particle counting may be the method of choice even when low-level radionuclides emit both beta particles and gamma rays. When only one or, possibly, two such radionuclides are present in a sample, assay by beta-particle counting may be preferable to assay by gamma-ray counting, because of the higher counting efficiencies of most beta-particle detectors. Gamma-ray spectrometry may, nonetheless, also be of use to identify radionuclides that may be present.

Because of the continuous nature of a beta-particle-energy spectrum, spectrometric identification and quantification of a sample containing several beta-particle emitters are generally difficult. Thus, radiochemical separations are often necessary in the assay of such samples. An exception to this is where radionuclides that emit conversion electrons are present in the source (Section 6.1.3.5).

For the majority of detectors, physical shielding is required primarily for the reduction of background due to gamma rays and cosmic rays. Considerable further reduction is frequently possible through the use of anticoincidence shielding. The comments made in Section 6.1.1.5e with regard to radioactivity in materials used in counter construction are also valid here.

Due to the relatively short range of beta particles in solids and liquids, it is generally desirable to prepare thin sources. Techniques for this are described in NCRP (1961), by Yaffe (1962), by Watt and Ramsden (1964), and in Section 5.2.

Absorption and scattering of beta particles by the sample matrix in thick sources must be determined for quantitative measurements, the corrections being greater for solid sources than for gaseous ones or liquid sources dispersed in a scintillator solution. Frequently, the detection efficiency as a function of sample surface density must be determined (see, e.g., Nervik and Stevenson, 1952), unless an "infinitely thick" sample is utilized.

Specific methods for measuring low-level beta-particle radiation are discussed, according to type of detector, in Sections 6.1.3.2 through 6.1.3.5.

6.1.3.2 *Liquid- and Solid-Scintillation Counting.* Liquid-scintillation-counting systems are widely used for the assay of low levels of beta-particle-emitting nuclides. High-sensitivity phototubes with bi-alkali (Na-Cs-Sb) photocathodes and quartz faces are desirable. Lower backgrounds can be obtained with sample vials made of quartz, low-potassium glass, or plastic, although the last type exhibits some permeability to organic compounds. Coupling the vial optically to the phototube with fluid or a plastic light pipe will increase the light-collection efficiency, but the former technique is not practical for systems having an automatic sample changer. Degassing the sample and then sealing the vial will reduce quenching by oxygen. A technique for the quantitative transfer of volatile radioactive solutions has been developed by Garfinkel *et al.* (1965).

Extensive work has gone into the study of chemical and color quenching and the development of better scintillator mixtures, or "cocktails" (Rapkin, 1967; Parmentier and Ten Haaf, 1969; Bransome, 1970; Horrocks and Peng, 1971; Wang and Willis, 1975). Some of this effort is germane to low-level applications, even though not specifically directed toward it. (See also Section 4.3.)

All of the beta-particle emitters listed in Section 6.1.3.1 can be assayed by the method of liquid-scintillation counting. This method is frequently employed for tritium analyses (Moghissi and Carter, 1973), although considerable concentration and processing of some kinds of samples may be required to convert them to a suitable form that is soluble or miscible in, usually, water or benzene (Tamers *et al.*, 1962; Cameron, 1967). A system that can quantify very small amounts of tritium in samples as large as 100 ml has been developed by Noakes *et al.* (1973). ^{14}C can be assayed as $^{14}CO_2$ contained in an organic base, or $^{14}CO_2$ can be converted to $BaCO_3$ and counted as a suspension (Cluley, 1962; Rapkin, 1967; Section 4.3.3). The ^{3}H or ^{14}C sample can also be dispersed or dissolved directly in the scintillator solution (Fox, 1968). Inorganic compounds containing ^{35}S, ^{45}Ca, ^{89}Sr, ^{90}Sr-^{90}Y, and ^{90}Y can also be prepared and suspended in the "cocktail". For some details and references, see ICRU (1972) and Section 4.3.4. Chapters 6 and 7 in Bransome (1970) give conditions for counting some additional beta-particle emitters and advice on counting various pairs of beta-particle emitters. Assays of mixtures of three pure beta-emitting nuclides are sometimes possible with the aid of signal processing, a multi-channel analyzer, and, or, computer spectrum analysis (Oller and Plato, 1972; Piltingsrud and Stencel, 1972).

Solid organic scintillators, usually plastics but also organic crystals such as anthracene and trans-stilbene, are used both for measuring gross beta-particle count rates and for beta-particle spectrometry. For gross counting, a thin plastic phosphor, typically with a surface density of tens of mg/cm^2, is placed between the sample and phototube as in Figure 80. With a simple mercury shield the background count rate averaged about $0.01\ s^{-1}$ (Harley *et al.*, 1962). When the phototube and sample were fitted into the well of a reentrant-anode Geiger-Müller counter, operated in anticoincidence, the background was reduced by nearly a factor of two (HASL, 1969).

For beta-particle spectrometry, a 1- to 2-cm-thick scintillator for complete absorption of the beta-particle energies is essential. The use of plastic scintillators with shapes other than discs can reduce the scattering of beta particles out of the detector and the distortion of spectral shapes, and increase the detection efficiency (see, e.g., Gardner and Meinke, 1958). The probability of gamma rays interacting with these specially-shaped scintillators is greater because of their larger size. Hence, these scintillators have higher backgrounds than simple disc-shaped scintillators. An elegant solution to the increased background problem, shown in Figure 83, was achieved by constructing a small Geiger-Müller counter into the front side of the scintillator (Tanaka, 1961). To obtain beta-ray spectra, of medium to high energy, scintillator pulses from only those events which are coincident in the two detectors are recorded by a pulse-height analyzer. With the addition of a logarithmic amplifier to the system, Tanaka, E., *et al.* (1967) were able to interpret, without complicated calculations, beta-particle spectra from radionuclides with beta-particle end-point energies as low as approximately 150 keV, as for example, ^{35}S and ^{14}C.

6.1.3.3 *Čerenkov-Radiation Counting.* Čerenkov radiation is pro-

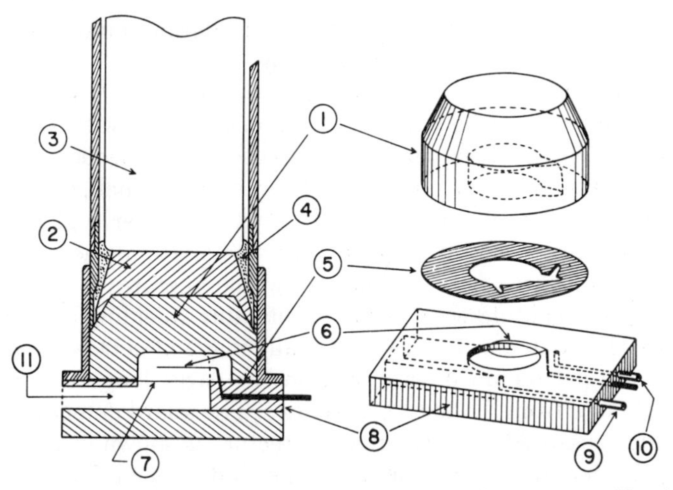

Fig. 83. Cross-section and exploded view showing the detector assembly of a beta-particle scintillation spectrometer using a coincidence method with a Geiger-Müller counter: (1) plastic scintillator; (2) lucite light pipe; (3) phototube (EMI 6097B); (4) MgO powder light reflector; (5) vinyl packing; (6) anode wire of the G-M counter; (7) window of the G-M counter (gold-plated Mylar film 1.3 mg/cm^2 surface density); (8) Lucite plate; (9) counting gas inlet tube; (10) counting gas outlet tube; (11) sample inlet. (From Tanaka, 1961)

duced by a charged particle traveling through a medium at a velocity greater than that of light in the medium. The photons thus produced are usually detected by the phototubes in conventional liquid-scintillation-counting systems. Not all beta-particle emitters can be assayed by this technique because an electron must initially have sufficient energy to be relativistic in the medium. The minimum or threshold energy for electrons in water is 0.26 MeV, and photon yields are given in Table 7. However, most other solvents used in Čerenkov-radiation counting have a higher index of refraction and thus have lower threshold energies (Figure 17). Much of the radiation emitted is in the ultraviolet region and is absorbed by the glass window of the phototube and the glass vials. This drawback can be partly overcome by the use of wavelength shifters that can increase the detection efficiency several fold (Parker and Elrick, 1970).

The reader should consult Sections 2.6, 4.5.1, and 5.4.1.5 for general information on Čerenkov radiation counting. Although the method is usually no more sensitive than alternative ones, an advantage lies in the relative ease of preparation for many samples. An example of the uses of the method is the measurement, by Yamada (1962), of Čerenkov radiation arising from ^{90}Y beta particles, to assay for ^{90}Sr in radioactive waste water. Randolph (1975) also used the method of consecutive

Čerenkov-radiation counting and liquid-scintillation counting of a sample to assay for ^{89}Sr and ^{90}Sr, respectively.

6.1.3.4 *Gas-Multiplication Detectors.* Counters with gas-multiplication detectors may be divided into those for use with solid beta-particle-emitting sources, and those for use with gaseous sources. In the former category, the usual design incorporates a small, thin-window detector that operates in the proportional or Geiger region, with a continuous flow of counting gas at just above atmospheric pressure through the detector, a guard detector which operates in anticoincidence with the main detector, and considerable shielding around both detectors. The detector bodies are usually made of petroleum-derived plastics, the petroleum having been in the earth long enough to be essentially free from ^3H and ^{14}C activities. The guard detector may be physically separated from the main detector, as shown in Figure 84a, or may be incorporated into the same body, as in Figure 84b. If separate, the guard may be constructed of other materials, such as glass or metal, because it is not necessary for the guard to be made of very low background-activity material.

The source is placed close to the window outside the detector, whose net background is generally less than 0.01 s^{-1} with a 2.5-cm diameter window, and less than 0.02 s^{-1} with a 5-cm diameter window. Windows are usually metalized polyethylene terephthalate (Mylar) films with surface densities of 0.1 to 1 mg/cm^2. Except in the case of tritium, the window does not seriously reduce the counting efficiency of the detector. Hence, windowless detectors, with the increased possibility of detector contamination, are less often used. Widely used detector-filling gases are, for proportional counters, 90 percent argon plus 10 percent methane or 100 percent methane, and, for Geiger-Müller counters, 99.05 percent helium plus 0.95 percent isobutane or 98.7 percent helium plus 1.3 percent n-butane.

Counting systems for solid sources that use this type of detector combined with automatic sample changing and data readout have been extensively developed commercially. Special laboratory-built systems are superior for certain applications, such as those described in Watt and Ramsden (1964).

Perhaps the greatest problem associated with the assay of thick, solid beta-particle-emitting sources with gas-multiplication detectors is to insure that the detection efficiency is known with an accuracy sufficient for the purpose of the measurement. In addition to the absorption and scattering in the source mentioned in Section 6.1.3.1, the effects of backscattering from the source mount, source-to-detector distance, and thickness of the source must be evaluated, and the

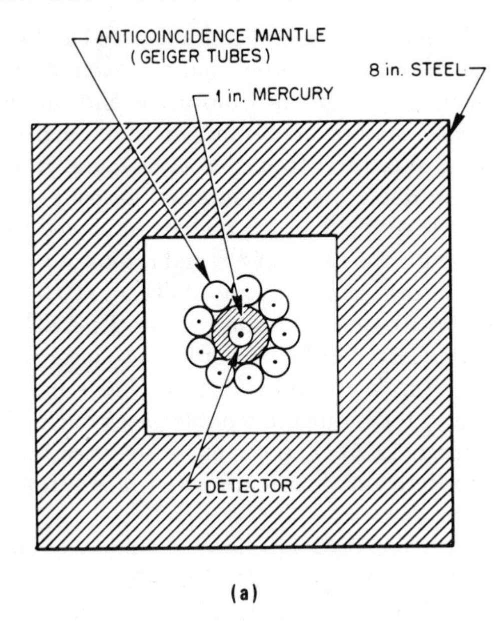

(a)

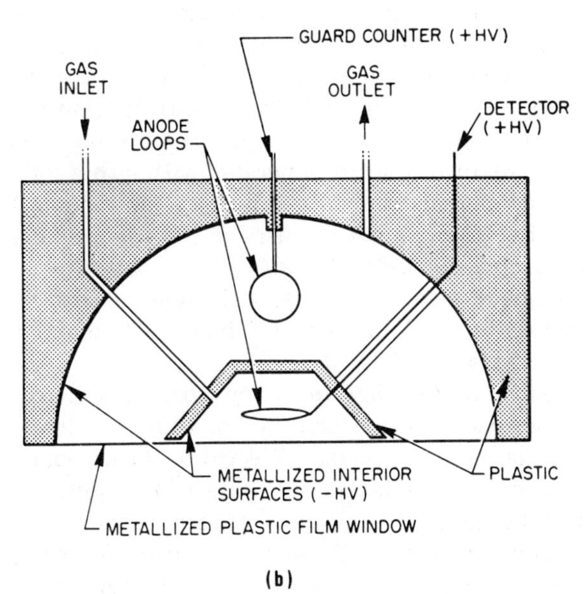

(b)

Fig. 84. Low-background gas-multiplication detectors for beta particles from solid sources: a. external guard detector; b. guard detector in same body. (From Johnson *et al.*, 1963)

system calibrated under known and reproducible conditions. Furthermore, these experimental conditions must be maintained as nearly constant as possible for both the calibrating and unknown sources. Discussions of these factors are found in such references as Snell (1962), Price (1964), Watt and Ramsden (1964), and Robinson (1967).

Gas-multiplication detectors with internal gaseous sources, i.e., internal gas counters (Section 3.6), are widely employed in the assay of ^{3}H and ^{14}C, and also for the assay of ^{35}S and radioisotopes of the noble gases. Low-level internal gas counting of ^{3}H and ^{14}C has been thoroughly reviewed (Watt and Ramsden, 1964; USAEC, 1965; IAEA, 1967b; Moghissi and Carter, 1973). References to internal gas counting for ^{35}S may be found in Watt and Ramsden (1964) and ICRU Report 22 (ICRU, 1972), and for radioactive noble gases in Currie (1972) and Stanley and Moghissi (1975). The gas to be analyzed may be the only gas present in the detector, but more often it is mixed with a nonradioactive counting gas.

These detectors are usually long, closed cylinders of metal or quartz with an axially centered 0.0025- to 0.0051-cm-diameter anode wire, and their volumes range from only a few milliliters to several liters. Anticoincidence shielding often employs separate detectors (see, e.g., Watt and Ramsden, 1964). Another design has the guard detector externally concentric to the detector in a single unit in which the two detectors may be separated only electrically by a multiwire grid or perforated, metalized plastic film (Drever *et al.*, 1957; Houtermans and Oeschger, 1958; Section 3.6), or physically separated with a thin foil. In the last case, all the sample gas can be confined to the inner detector, and counting gas, free from activity, is used in the guard detector (Charalambus and Goebel, 1963; Schell, 1970).

The technique of pulse-shape discrimination has also been applied to reduce background (Culhane and Fabian, 1972; Sudar *et al.*, 1973; Currie and Lindstrom, 1975).

6.1.3.5 *Semiconductor Detectors.* Semiconductor detectors can be used advantageously for the analyses of thin samples containing a radionuclide that emits conversion electrons. Background count rates under the sharp conversion-electron peaks will be quite low compared to the total background. In one instance (Folsom *et al.*, 1965), radiochemically separated ^{137}Cs was assayed by measuring the 624-keV conversion-electron activity of its ^{137m}Ba daughter. The background in that energy region of their silicon detector was less than 4 counts per hour. Even though nearly all the conversion electrons reaching a semiconductor detector will be counted, the fraction of decays by this route for most radionuclides is so low as to preclude widespread use of this detection method.

6.1.4 *Low-Level Measurement of X-Ray- and Gamma-Ray-Emitting Nuclides*

6.1.4.1 *Introduction.* One of the most valuable techniques for low-level radioactivity measurements is gamma-ray spectrometry. The various systems, consisting of scintillation and, or, semiconductor detectors coupled to multichannel analyzers, provide for rapid simultaneous measurement of many radionuclides in the same sample. The fundamental characteristics of scintillation and semiconductor detectors are described in Sections 2.4 and 2.5, respectively, and those of gas proportional counters used in the assay of x-ray emitters in Section 2.2. Such characteristics as energy resolution and intrinsic detector efficiency are important in choosing the detector for a particular measurement (see Section 4.2). Israel *et al.* (1971) have compared the characteristics for several detectors in the energy range from 10 to 300 keV.

The choice of measurement system for gamma-ray emitters is also based on considerations such as sample size available, the level of activity in the sample, and time available for measurement (see also Section 6.1.1). Some of the more commonly encountered types of samples and suggested measurements systems are discussed below:

(i) *Large Sample of Low Radioactivity Concentration*—Example: Fallout ^{54}Mn in Seawater. In an example such as this where great sensitivity is required, the radioelement of interest may be concentrated by radiochemical techniques, and the resultant sample counted by using a high-efficiency NaI(Tl) detector (Folsom *et al.*, 1963).

(ii) *X-Ray-Emitting Radionuclides in the Environment*—Example: Fallout ^{55}Fe in Biological Samples. Due to the short range of ^{55}Fe x rays, it is usually necessary to separate iron from the sample matrix prior to counting. Thus, in the assay of ^{55}Fe in serum, an electrodeposited source may be counted in an argon-methane proportional counter (Palmer and Beasley, 1967).

(iii) *Analysis of Extraterrestrial Materials*—Example: Meteorites. When a nondestructive assay of a number of radionuclides in a small sample is required, a Ge(Li) semiconductor is clearly the best choice due to the superior resolution. (See, for example, Perkins *et al.*, 1970.)

(iv) *Effluent Monitoring at Nuclear Facilities*—Example: Mixed Fission and Corrosion Products. Ge(Li) detectors have largely replaced NaI(Tl) in this application because of superior resolution. Time requirements usually prohibit chemical separation of individual elements in the case of routine samples.

(v) *Measurement of Gamma Rays in the Region of 60 to 150 keV*—Example: ^{155}Eu in soil, ^{133}Xe in air. The efficiency of a typical coaxial

Ge(Li) detector is difficult to determine in this energy region. Si(Li) detectors or planar Ge or Ge(Li) detectors may be most advantageous for this type of measurement.

6.1.4.2 *Inorganic Scintillation Detectors.* As mentioned previously, NaI(Tl) is the most common scintillator used in low-level assay. The following discussion of counting efficiency, standardization, data processing, and background reduction techniques, while assuming the use of NaI(Tl) detectors with multichannel analyzers, is generally applicable to other types of scintillation counters as well.

6.1.4.2.1 *Counting Efficiency.* The overall counting efficiency may be increased by increasing either the sample size or the size of the detector. Heath (1964) has tabulated efficiencies for point sources at different distances from 7.6 x 7.6-cm to 12.7 x 12.7-cm cylindrical NaI(Tl) detectors. The effects of detector size have also been evaluated by May and Marinelli (1960).

For small samples, the efficiency is greatly increased by the use of well crystals (Hine, 1967; Lloyd *et al.*, 1967) that give almost 4π geometry. For larger samples the source may be distributed around the detector in a Marinelli beaker (May and Marinelli, 1960). With such a geometry as this, where photopeak efficiencies are very high, the probability of summing for coincident events will also be high. This does not present a problem if one is measuring the integral count rate for a sample containing a single radionuclide for which the efficiency is known. Summing corrections must be made, however, if count rate measurements are made on a single photopeak for which the corresponding gamma ray is emitted in coincidence with other radiations.

It is difficult to predict the efficiency for the large variety of source detector arrangements used. Thus, the most common procedure is to calibrate each system by using radioactivity standards in the same geometry as the sample. (Such standards are listed in Table 9.) The calibration is usually accomplished by incorporating several radionuclides, such as ^{57}Co, ^{54}Mn, ^{137}Cs, ^{60}Co, and ^{22}Na, into a matrix similar to the sample (e.g., soil, water, or biological matrix). The results are then used to prepare photopeak-efficiency curves, such as the curve shown for point sources in Figure 85, using a NaI(Tl) detector of practically 4π geometry (Hutchinson *et al.*, 1973b). The efficiency appropriate to the energy, or energies, of the unknown is taken from the curve. In practice, however, quantitative interpretation of NaI(Tl) scintillation-spectrometer data, for a source emitting several gamma rays, is difficult because of poor resolution (see, for example, Figure 86). The various methods used for peak-area analysis and spectrum "stripping tech-

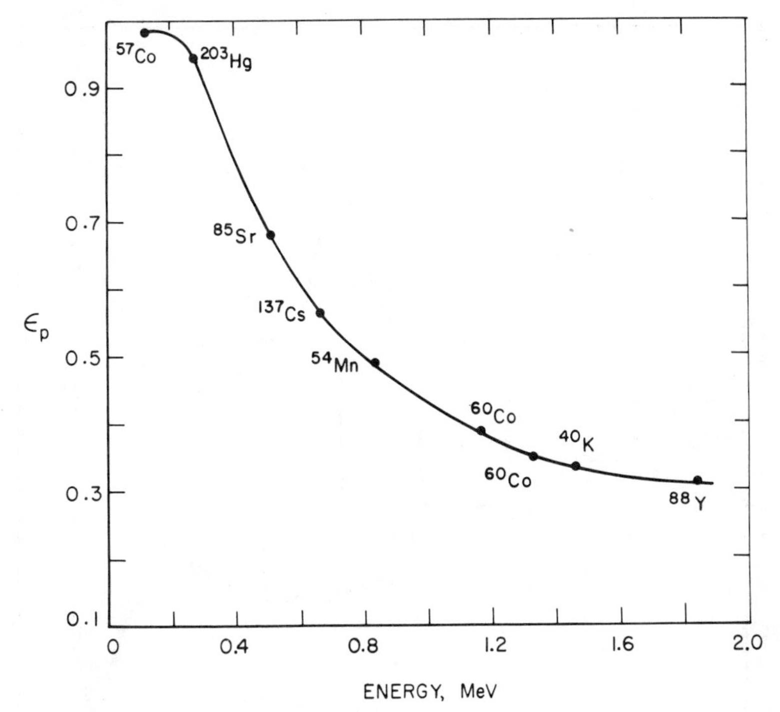

Fig. 85. Photopeak-efficiency curve obtained using point-source standards in the NBS $4\pi\gamma$ 8-inch-crystal system. (From Hutchinson *et al.*, 1973b)

niques" are beyond the scope of this section. General characteristics of such methods are reviewed in Section 4.2.7 and in ICRU (1972).

6.1.4.2.2 *Background.* The major sources of background counts were discussed in Section 6.1.1. Major techniques for reducing the contributions from these sources to the background for a scintillation counter are siting, air filtration, and physical and anticoincidence shielding.

(i) *Siting.* Siting of any low-level-measurements system is an important consideration. Counting rooms should be located below ground since the covering materials (earth, water, and concrete, for example) will provide some shielding from cosmic radiation. Wogman *et al.* (1969), for instance, describe the background of a counting room located at the base of the Grand Coulee Dam. As the authors point out, however, the covering material alone does not eliminate the need for shielding from the muon and neutron components of the cosmic radiation [see (iii) below]. When a counting room is under construction, building materials and nearby soil should be checked to insure that

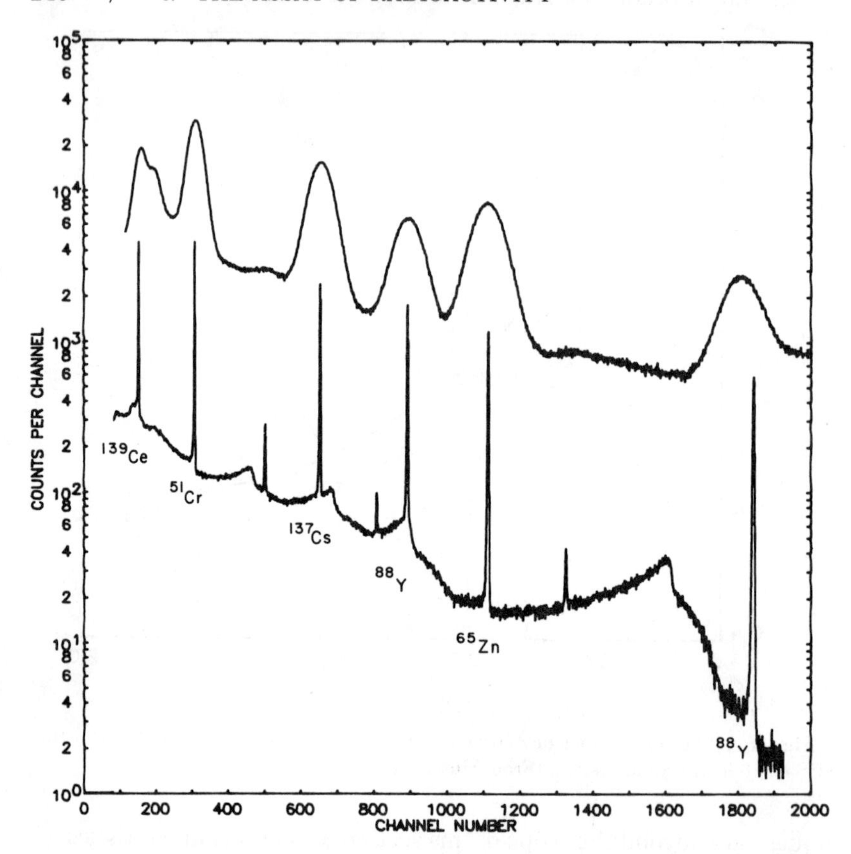

Fig. 86. Gamma-ray spectra of a 5-ml mixed-radionuclide-solution source taken with the source within a 5-inch NaI(Tl) well crystal (upper curve) and at the face of a 60-cm^3 Ge(Li) detector (lower curve). The counting time in each case was 2000 s. (From measurements made at the National Bureau of Standards)

they do not contain excessive amounts of natural or fallout radioactivity. Siting of the room away from reactors, accelerators, and x-ray equipment is, of course, desirable.

(ii) *Ventilation.* The counting room should be equipped with air-filtration systems to remove particulates that may contain radioactivity. To avoid radon- and thoron-daughter deposition within the shielding for detectors of the NASA Lunar Sample Laboratory, Lucas (1964, 1965) designed a counting room in which fresh air was introduced through a low-temperature charcoal-filtration system to trap the radon.

(iii) *Shielding.* For large scintillation detectors, massive shielding is required to reduce the background contribution from environmental gamma-ray emitters. Shielding requirements are less cumbersome for small NaI(Tl) wafers or CaI_2(Eu) scintillators since they have low efficiency for the more penetrating gamma rays. The most common shielding materials are lead, steel, mercury, and water. The materials used should be checked for contamination (Section 6.1.1.6). This is one advantage of using water; although it is low-Z, large quantities of pure material are easily prepared.

In the design of a shield, some thought must be given to the scattering and attenuation characteristics of the materials used. Steel, for example, is a common shielding material which serves to scatter and degrade incoming photons but has a low photoelectric cross section for photons above 100 keV. Thus, in the absence of other shielding, some degraded photons, usually in the energy region 100 to 300 keV, will interact with the detector. Also backscatter peaks may be detected in this energy region (Heath, 1957) due to source gamma rays scattered through 180°.

Another consideration in the design of a compact, efficient shield is a process known as "grading" (ICRU, 1972) in which materials of successively lower Z are used to absorb the x rays from outer layers. For instance, if lead shielding is used, then a cadmium liner is used to absorb the Pb x rays having energies between 73 and 88 keV. The Cd x rays of approximately 22 keV are then absorbed by a thin liner of copper or steel.

The design of a typical shield is described in ICRU, 1972. The outer shield is fabricated from carefully selected low-activity steel or lead. Inside this shield, one may use a neutron shield of borated paraffin wax, in which the hydrogen serves to moderate neutrons while the boron has a high capture-cross section for those at thermal energies. Within the wax shield, one may use an anticoincidence shield, as described in (iv) below, and finally, closest to the detector, a shield of selected lead, steel, or mercury.

If the outer component of the innermost shield is steel, it should be lined with about 3 mm of lead to remove the degraded photons. The lead itself must then be lined with about 1 mm each of cadmium and copper. If lead alone is used for the shield, the cadmium and copper linings are of course still required.

(iv) *Background Reduction by Instrument Techniques.* A reduction in both the background contribution from external sources and of the Compton background due to higher-energy gamma rays scattered out of the detector can be obtained by closely surrounding the main detector with an anticoincidence shield such as a plastic annulus containing *p*-terphenyl, or by a NaI(Tl) annulus (Perkins *et al.*, 1960).

(v) *Applications.* Examples of NaI(Tl) counting systems that incorporate most of the background-reduction features discussed here have been reported by Perkins (1965), Nielsen and Perkins (1967), Tanaka, S., *et al.* (1967), Wogman *et al.* (1969), Hutchinson *et al.* (1972, 1973b), and Eldridge *et al.* (1973). Figure 87 gives a schematic of the anti-Compton, anticoincidence-shielded NaI(Tl) system at the NBS (described in Section 3.3).

One of the most sophisticated systems reported to date is a "multi-dimensional" γ-γ coincidence counter by Wogman *et al.* (1969) that uses two NaI(Tl) detectors. The method was subsequently extended to β-γ-γ coincidence counting by Wogman and Brodzinski (1973) by employing two NaI(Tl) detectors and two gas proportional counters. In addition, each half of the main detector is surrounded by an anticoincidence shield of a plastic phosphor. Brauer *et al.* (1970)

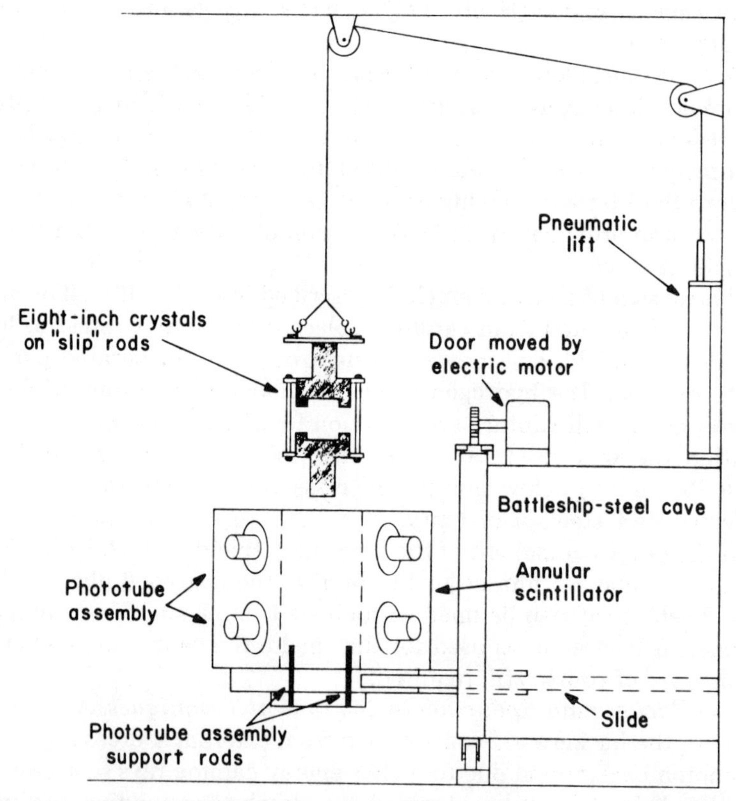

Fig. 87. Two 8″ × 8″ NaI(Tl) well detectors surrounded by a 30″-o.d. by 20″-long plastic scintillator anti-coincidence anti-Compton shield. The detectors enclose a cavity 5″ in diameter and 1.5″ deep. (From Hutchinson *et al.*, 1973b)

reported similar β-γ coincidence systems in which plastic phosphors were employed as the beta-particle detectors.

6.1.4.3 *Semiconductor Detectors.* The most powerful measurement technique used in low-level assay at present is gamma-ray spectrometry using semiconductor detectors. Most of the current systems utilize Ge(Li) detectors (Section 2.5) that have counting efficiencies of the order of 1 to 25 percent relative to 7.6 x 7.6-cm NaI(Tl) scintillation detectors for 1.3-MeV gamma rays. In the future, however, the drifted semiconductors may be replaced by high purity materials such as Ge, CdTe, or HgI_2. Comparisons of gamma-ray response by different types of detectors have been made by Israel *et al.* (1971) and Malm *et al.* (1972). Future prospects for the use of pure Ge crystals are discussed by Hansen and Haller (1973). A review of detection systems currently used for low-level gamma-ray measurements has been made by Wogman (1974). Additional references may be found in Nielsen (1972).

6.1.4.3.1 *Energy Calibration.* Calibrations of analyzer channel number in terms of energy for Ge(Li) detector systems were discussed in Section 4.2.4. Such calibrations may be accomplished by counting sources that emit several gamma rays whose transition energies are well known. Unknown radionuclides may be identified by comparing observed gamma-ray energies with tabulations from Oak Ridge National Laboratory (see Appendix A). Compilations of gamma-ray spectra such as Adams and Dams (1970), Cline (1971), and Heath (1974) are also useful in the identification of unknowns.

6.1.4.3.2 *Counting Efficiency.* The principles of calibration for counting efficiency of low-level Ge(Li) gamma-ray spectrometers are the same as for systems using NaI(Tl) detectors (Section 6.1.4.2). Consideration must be given to the selection of the optimum counting geometry and to the preparation of energy-photopeak efficiency curves.

In general, the efficiency of the detection system may be increased by using a larger detector or optimizing the sample-detector geometry. The size of the detector is limited by present state-of-the-art techniques in semiconductor fabrication. Near optimum counting geometries have been obtained in two ways. Latner and Sanderson (1972) have used Marinelli-type beakers in which the source surrounds the detector, while Brauer and Kaye (1974) place the source between two detectors.

The photopeak efficiencies for various source-detector configurations may be predicted using Monte Carlo calculations (e.g., Coley and Frigerio, 1975), but more often the system is calibrated directly by using one or more radioactivity standards (Legrand, 1973). A single standard, such us ^{166m}Ho (Lavi, 1973), that emits many gamma rays of different energies may be used, or a number of standards such as ^{22}Na, ^{54}Mn, ^{57}Co, ^{60}Co, and ^{137}Cs may be individually incorporated into "blank" sample matrices (Laichter *et al.*, 1973). Systems may be calibrated for liquid and point sources with mixed radionuclide standards from the National Bureau of Standards (Cavallo *et al.*, 1973). A gamma-ray spectrum of one of these standards is shown in Figure 88.

6.1.4.3.3 *Background.* As with other types of low-level assay the central problem in Ge(Li)-gamma-ray spectrometry is to achieve the optimum signal-to-noise ratio in the energy region of interest. General references on the sensitivity of such systems are Cooper (1970) and Armantrout *et al.* (1972). One can discern three interrelated topics for discussion: (i) detection limits; (ii) spectrum processing; and (iii) experimental methods used to enhance the signal-to-noise ratio.

The general problem of detection limits was discussed in Section 6.1.1.3. Pasternack and Harley (1971) examined the problem of detecting gamma-ray photopeaks in multicomponent gamma-ray spectra, and Wood and Palms (1974) have examined gamma-ray spectrum stripping techniques used for such spectra. Commercial multichannel analyzers with minicomputers are now available with programs for analyzing the data, listing the radionuclides found, their activities, and gamma-ray peak energies and intensities. Users of such systems should

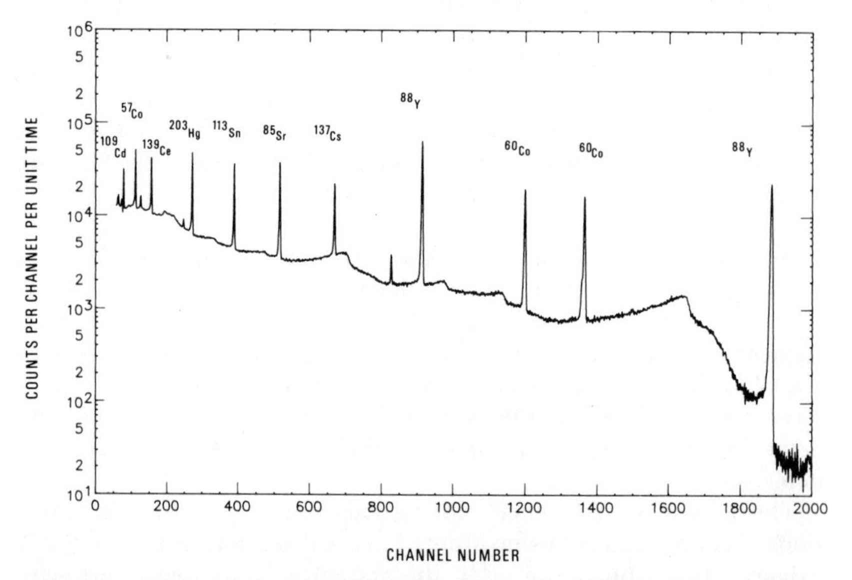

Fig. 88. Gamma-ray spectrum of an NBS mixed-radionuclide gamma-ray emission-rate standard. This 50-ml solution source was counted at 5 cm from a 60-cm^3 Ge(Li) detector. (From Coursey, 1976)

be familiar with the detection criteria and peak-area-evaluation methods used in these computer programs. Guidance on standard methods for such analyses may be provided by the ANSI Subcommittee on Gamma-Ray Spectrometry (ANSI, 1977). (See, also, Section 4.2.6.)

As discussed in Section 6.1.4.2, background reduction for gamma-ray spectrometry is generally accomplished by a combination of massive shielding and instrumental techniques. The more sophisticated systems employ one or two main detectors of Ge(Li) surrounded by one or more plastic or NaI(Tl) anticoincidence shields (Phelps and Hamby, 1972). Less widely used are summing Compton spectrometers (SCS) (e.g., Ivanov and Shipilov, 1974) that consist of a single germanium ingot with two different lithium-drifted regions. A common characteristic of all of these systems is that one can utilize the decay scheme of the radionuclide of interest to reduce interfering background from other sources. A typical system is shown in Figures 89 and 90. When, in the spectral analysis of monoenergetic gamma-ray-emitting radionuclides, the NaI(Tl) shield and Ge(Li) detector are operated in the anticoincidence mode, contributions from both Compton events and radionuclides emitting coincident gamma rays will be dramatically reduced. This reduction is demonstrated in Figure 91, which shows the gamma-ray spectra of a mixture of ^{137}Cs and ^{60}Co in the normal and anticoincidence modes.

6.1.4.4 *X-Ray Detection Systems.* Radionuclides of environmental interest that are detected by x-ray or low-energy gamma-ray measurements include ^{37}Ar, ^{55}Fe, ^{129}I, and ^{125}I. The assay of low-energy-photon emitters in bulk environmental samples, without concentration, is usually complicated by attenuation of the photons within the sample.

The detectors used for x-ray measurements include Si(Li) and pure Ge, proportional counters, gas proportional scintillation counters, and liquid-scintillation counters. The operation of these instruments is covered elsewhere in this report, such as, for example, in Sections 2.2, 2.4.5, 2.4.6, 4.2, and 4.3.

6.1.4.4.1 *Semiconductor Detectors.* The use of semiconductor detectors in the x-ray region is discussed by Israel *et al.* (1971) and Swierkowski *et al.* (1974), and the calibration of such systems is reviewed by Fink (1973) and Hansen *et al.* (1973). Background-reduction techniques need not be as elaborate as for gamma-ray spectrometers because detection efficiencies are low for background gamma rays with energies greater than 0.2 MeV. Cosmic rays may, however, produce photons in surrounding material that contribute background counts in the x-ray region.

6.1.4.4.2 *Gas Proportional Counters.* Gas proportional counters are often used for the assay of x-ray emitters. Palmer and Beasley (1967)

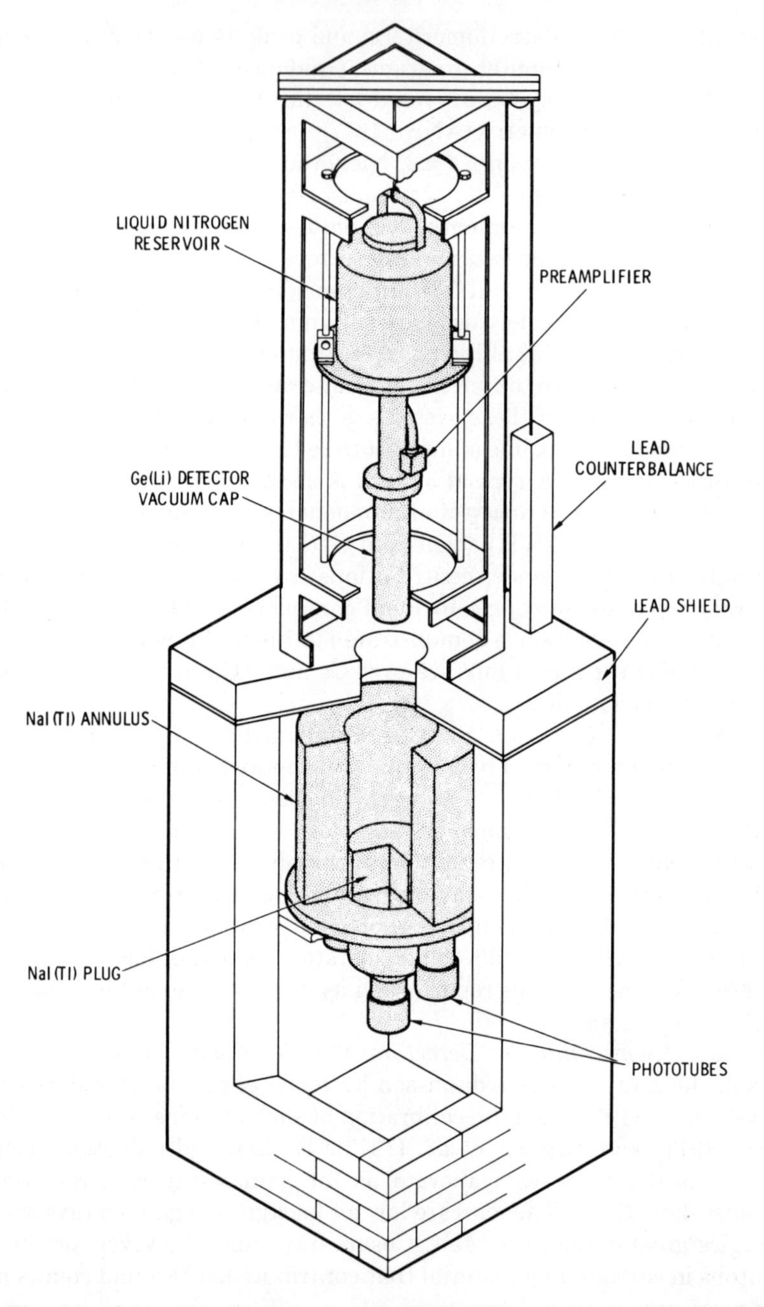

Fig. 89. Schematic diagram of a Ge(Li)-NaI(Tl) coincidence-anticoincidence gamma-ray spectrometer. (From Cooper and Perkins, 1971)

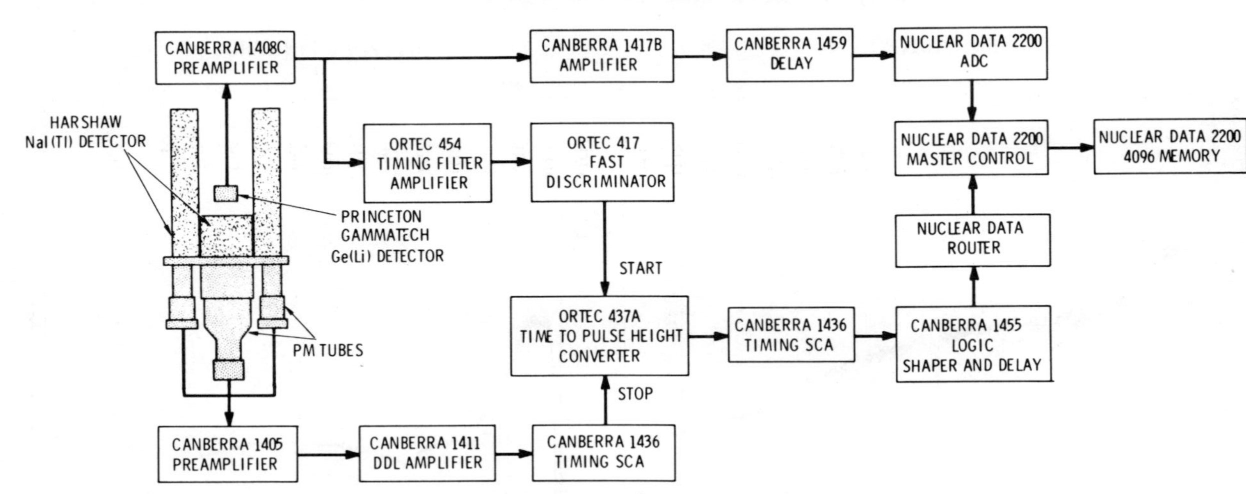

Fig. 90. Block diagram of electronic circuitry used in anticoincidence mode of operation for a NaI(Tl) shielded Ge(Li) gamma-ray spectrometer. (From Cooper and Perkins, 1971)

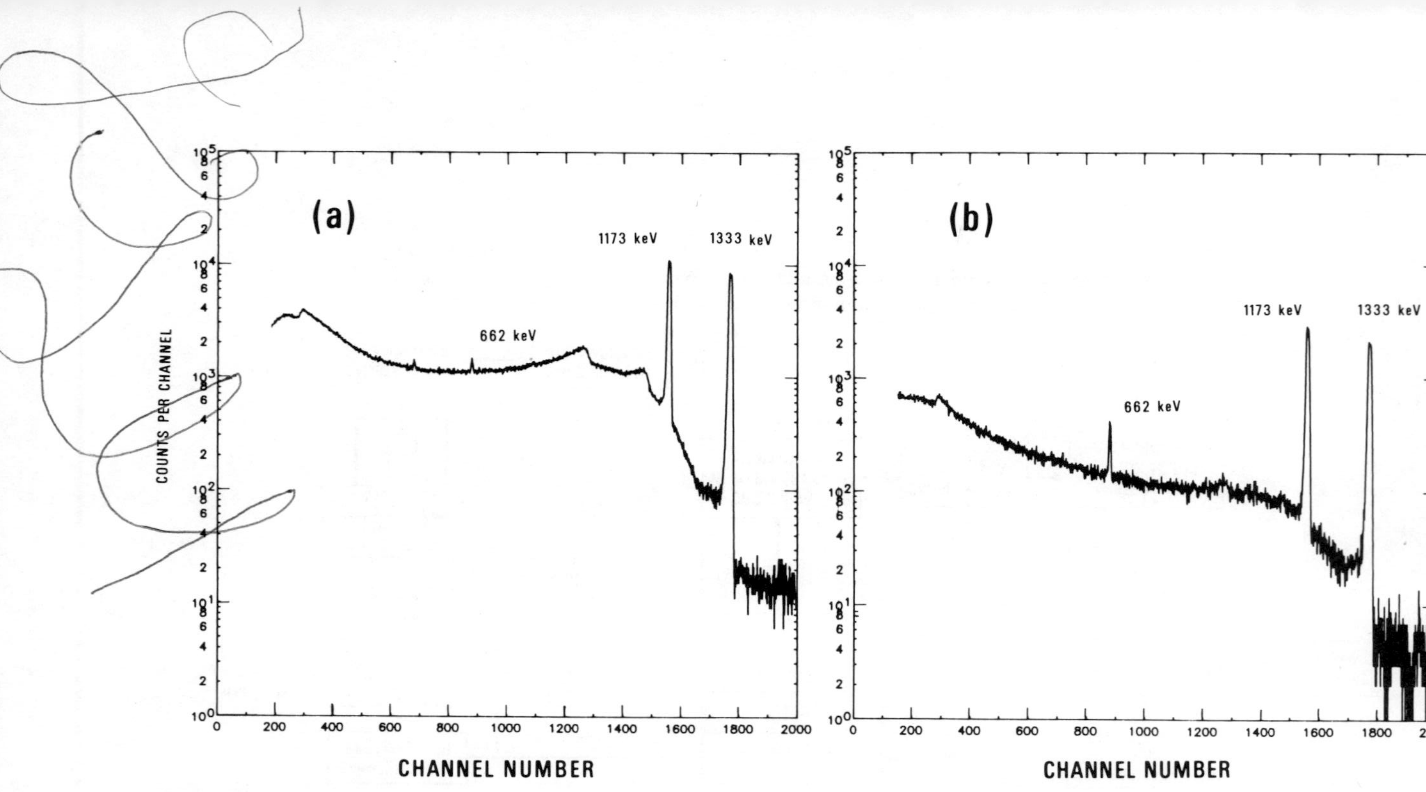

Fig. 91. Gamma-ray spectra, obtained at the National Bureau of Standards, of a mixture of ^{60}Co and ^{137}Cs-^{137m}Ba (activities 8600 s^{-1} and 93 s^{-1}, respectively), contained in 50 ml of solution in a cylindrical glass bottle, using a 60-cm^3 Ge(Li) detector within a NaI(Tl) shield. The source-to-detector distance was 1 inch, and the counting time was 10^4 s for each spectrum.

Spectrum (a) is that obtained in the normal mode (i.e., with no anticoincidence or coincidence requirement between the outputs of the two detectors). Spectrum (b) was obtained with the detectors operated in the anticoincidence mode (i.e., recording only non-coincident events). The relative enhancement of the 662-keV line due to the ^{137m}Ba is clearly seen.

measured body burdens of ^{55}Fe in reindeer and humans by electrodepositing Fe from serum onto copper planchets and subsequently assaying the ^{55}Fe using a gas proportional counter. The activity of ^{37}Ar, an electron-capture nuclide with a 35-day half life, is normally determined by internal gas proportional counting (Sections 2.2 and 6.1.3.4). Its use as a tracer in atmospheric studies has been described by Loosli *et al.* (1975) and Currie and Lindstrom (1975). Both groups assay the argon fraction condensed from large air samples.

6.1.4.4.3 *Other Techniques Used for Analysis of X-Ray Emitters.* Some other detector systems that have been used for the analysis of x-ray emitters are:

(i) x-γ coincidence counting (Brauer *et al.*, 1970),
(ii) liquid-scintillation counting (Horrocks, 1971),
(iii) xenon scintillation proportional counting (Israel *et al.*, 1971), and
(iv) neutron activation techniques (Brauer and Kaye, 1974).

It should be mentioned that in liquid-scintillation counting of electron-capture nuclides, a major fraction of the observed response may be due to the Auger electrons.

Although ^{129}I is radioactive, its specific activity is extremely low (1.57×10^7-y half life). It can, however, be readily detected by neutron activation to form ^{130}I, a gamma-ray emitter with a 12-hour half life and correspondingly higher specific activity (Edwards, 1962; Koch and Keisch, 1964; Keisch *et al.*, 1965). The ^{130}I activity is then measured by using a gamma-ray spectrometer.

6.2 Supply of Radioactive Materials

Radioactive materials are principally produced by neutron bombardment, usually in a reactor, or bombardment with charged particles in an accelerator. Some radioactive materials are also available as purified fission products from reactor fuels. Other radionuclides occur in nature and are isolated by appropriate methods.

Radionuclides are available in many physical forms as simple compounds, labeled organic compounds, radiopharmaceuticals, and radioactive sources from several commercial organizations whose catalogs should be consulted for materials and services available. The American Nuclear Society publishes an annual Buyer's Guide that contains lists of companies that supply radionuclides. The American Chemical Society publishes the Analytical Chemistry Buyer's Guide with some of the same information. Other sources of information are found in advertisements in periodicals such as Science, published by the American Association for the Advancement of Science; the Journal of Nuclear Medicine, and the Physician's Desk Reference for Radiology and Nuclear Medicine, published by the Medical Economics Company.

Standard reference materials for calibration of equipment are available from the National Bureau of Standards and many of the commercial firms that supply radioactive nuclides, sources, and pharmaceuticals. Standard reference materials are available in many geometries and as gases, liquids, or solids. Reference materials are either certified by the manufacturer as to accuracy or guaranteed by catalog statement as to accuracy. *Certified* activity standards are always higher in cost, for a given radionuclide, than *guaranteed* standards. Manufacturers' labeled amounts of regular radionuclide shipments should never be used as standard reference materials since they are not packaged or calibrated for such purposes.

Before ordering reactor-produced radionuclides above certain specified levels, the prospective user must be licensed for possession of by-product (i.e., reactor-produced) material by the U.S. Nuclear Regulatory Commission or an Agreement State. Addresses and names of regulatory personnel are available in a "Directory of Personnel Responsible for Radiological Health Programs", issued each July by the U.S. Department of Health, Education and Welfare, Bureau of Radiological Health, Rockville, Maryland 20852. Other radionuclides may also require licensing and the user should check with the appropriate state agencies prior to ordering.

6.3 Users' Responsibilities for Quality Control

6.3.1 *Check Procedures for Incoming Shipments*

Shipments of radioactive materials received by institutions fall into four general categories:

(a) Prepackaged small quantities for individual doses or tests,
(b) large quantities to be used as bulk solutions for dispensing in smaller quantities,
(c) radionuclide generators such as ^{99}Mo-^{99m}Tc generators, and
(d) sealed radioactive sources.

Most suppliers of radioactive materials have subjected their shipments to intensive quality control to prove identity, quantity, and quality. The quality-control procedures may be required and approved by regulatory agencies and, at the least, the competition of other suppliers will dictate some reasonable degree of reliability. However, the medical user, in particular, should take the precautions enumerated below before he uses a radionuclide since the ultimate responsibility for the accuracy of the dose administered to the patient rests with the physician.

The first step taken by all users should be to make a wipe test of the external shipping package and internal containers to be sure that no leakage has occurred and that there has been no compromise of primary container integrity. This practice is one that should be established by all laboratories using radioactive materials. In addition, it is a requirement for some packages of some radionuclides (see *Procedures for Picking up, Receiving, and Opening Packages*; Nuclear Regulatory Commission Regulations, Title 10, Part 20.205, Code of Federal Regulations).

For a minimum check of the product identity, the user should inspect the label attached to the primary container, to insure that this information indicates that it is the required radionuclide, and that the amount is correct. In the case of short-lived radionuclides, the calibration date should be checked and the necessary calculations or measurements should be made to determine the quantity at time of use. Technical data sheets and package inserts accompanying the radionuclide package should be read carefully for special instructions on stability, quality inspection, and chemistry of the product.

The next step beyond this visual inspection is an approximate assay of the shipment. If this assay yields results in reasonably good agreement with the value stated by the supplier, the identity of the radionuclide is unlikely to be other than expected. Gross discrepancy between the user's assay and the supplier's value should lead to the use of more accurate identification methods such as described elsewhere in this report.

If the user is using bulk solutions to prepare "in-house" pharmaceuticals or labeled materials, quality-control procedures should be instituted in order to verify activity, radionuclidic purity, and radiochemical purity by means of methods described or referenced elsewhere in this report (e.g., Section 6.4).

6.3.2 *Sterility and Pyrogenicity*

Radioactive pharmaceuticals used for *in vivo* studies must be sterile and non-pyrogenic. Materials supplied for human use by manufacturers licensed by the Food and Drug Administration are required to be sterile and non-pyrogenic and no tests for sterility and pyrogenicity are required by the user.

However, if commercial suppliers are not used, or preparation of radioactive pharmaceuticals is carried out in the hosptial pharmacy, sterility and pyrogenicity testing must be done by the user for samples to be administered intravenously. For descriptions of methods one may consult the most recent edition of the U.S. Pharmacopeia (USP, 1975).

6.4 Assay of Radiopharmaceuticals

6.4.1 *Introduction*

The requirements for radioactivity assays in a clinical laboratory vary from simple checks on the identity and quantity of preassayed shipments of commercial radiopharmaceuticals to complete chemical, biological, and radioactivity tests and measurements on products prepared within the laboratory. Radiopharmaceuticals fall into four general categories: small single-dose quantities intended for direct administration; bulk shipments from which aliquots may be taken for further processing or direct administration; short half-life radionuclide generators; and radionuclides produced on site. Radiopharmaceuticals in the first three categories are normally obtained from commercial producers and are supplied preassayed and presumably in conformity with pharmacopoeial standards. Those in the last category must be calibrated and checked by the user for both quantity and quality.

Because, however, ultimate responsibility for the accuracy of the amount of a radiopharmaceutical administered to a patient rests with the supervising physician, good practice requires that *all* individual doses be checked to minimize the possibility of error. Even commercial shipments should be routinely tested for proper quantity and identity of the radionuclide.

Although large numbers of tests, such as determinations of radioactivity concentration, radionuclidic purity, radiochemical purity, chemical purity, and purely biological criteria of sterility, pyrogenicity, isotonicity, and biological affinity are normally required to determine the suitability of a radiopharmaceutical for clinical use, only those tests primarily concerned with radioactivity measurements will be discussed here. Thus, any discussion of radiochemical purity tests will be excluded, since such determinations rely on varied and often complex chemical separations that are beyond the scope of this report. Those interested in pursuing this aspect of radiopharmaceutical testing should consult the proceedings of several comprehensive symposia on the subject (Andrews *et al.*, 1966; IAEA, 1970, 1971, 1973; Subramanian *et al.*, 1975; Tubis and Wolf, 1976).

For a detailed discussion of *in vivo* and *in vitro* radioactivity measurements and related subjects, the interested reader should consult a forthcoming NCRP Report on clinical radioactivity measurements and the references contained therein. That report also treats many of the tests for radiopharmaceutical quality not covered in this report.

The most commonly used radiopharmaceuticals are those compounded from gamma-ray-emitting nuclides, and therefore the following sections will deal mainly with the assay of such photon-emitting materials.

6.4.2 *Use of Radioactivity Standards and Reference Sources*

Radioactivity measurements in a clinical laboratory rely heavily on indirect or comparative calibration procedures. These procedures require prior calibration of instruments with calibration standards. Subsequent sample measurements must be made in geometries identical to those used for the initial calibrations in order to achieve valid comparisons. Such instrument calibrations are either performed by the user or, as is frequently the practice in nuclear medicine, by the instrument manufacturer. They should, however, be periodically checked by using such radioactivity standards as may be available.

Calibration standards are radioactive sources for which the activity of the major radionuclide and any measurable radionuclidic impurities are known within specified error limits. These can be procured from national standardizing laboratories or commercial suppliers. When these are ordered, "certified" calibration standards should be specified that conform with the certification guidelines presented in ICRU Report 12 (ICRU, 1968).

One problem that frequently arises with such calibration standards is that they come in containers that differ from the great variety of containers used for routine measurements in individual laboratories. This problem can be overcome by quantitatively transferring the certified standard, or part of it, to a laboratory container and making up to the desired volume with appropriate carrier solution (see Section 5.1). This laboratory standard is then used to calibrate the laboratory assay system. Alternatively, if it is desired to preserve the integrity of the certified standard, working standards calibrated in the same geometry by relative gamma-ray measurements can be used to calibrate the variously shaped laboratory containers.

Once an instrument or counting system has been calibrated, the performance of the instrument should be monitored by carrying out periodic measurements with a reference source of a long-lived radionuclide. These reference sources need not be calibrated and are normally mounted in rugged containers to prevent leakage. ^{226}Ra in equilibrium with its decay products is frequently selected for use as a reference source, although other long-lived radionuclides, such as ^{60}Co and ^{137}Cs-^{137m}Ba, are also used for this purpose.

Other so-called "standards" occasionally used in a clinical laboratory are simulated or mock standards. These are radioactive sources prepared from long-lived radionuclides, or mixtures of radionuclides, that approximate the electromagnetic radiation characteristics of the simulated radionuclide. Thus, a simulated standard should rarely, if ever, be used as a calibration standard for any other radionuclide. Such a "standard" when prepared from *one* radionuclide can, however, serve a useful function as a reference source to monitor the response of a measurement system in the energy region where measurements are to be made. A good example of this is the use of an ^{129}I reference source to monitor the performance of a counting system used for the routine measurement of ^{125}I samples.

6.4.3 *Radioactivity Assay Methods and Instruments*

6.4.3.1 *Introduction.* Pre- and post-administration radioactivity measurements in a clinical laboratory are primarily performed to determine the total activity or concentration of the principal radionuclide or the radionuclidic purity. The majority of these measurements are made on gamma-emitting radionuclides and, to a lesser extent, on pure beta emitters (Williams and Sutton, 1969; Gerrard, 1971). The assay of alpha-emitting radionuclides is seldom required and will, therefore, be neglected in this discussion.

Frequently, there exist prescribed approved standard methods for the assay of radiopharmaceuticals. The user should consult the current edition of the U.S. Pharmacopeia and any other relevant regulatory guidelines for the existence and description of recommended methods, such as, for example, standards developed by the American National Standards Institute. Where relevant prescribed methods exist, they should be followed, whenever possible. It should be noted, however, that the user may substitute an alternate assay method for the recommended method. When this is done, it will be the responsibility of the user to demonstrate, if called upon to do so, that this alternate method consistently yields the same measurement results as the standard method.

6.4.3.2 *Assay of Gamma-Ray-Emitting Radionuclides.* Assays of gamma-emitting radionuclides are performed to determine the total amount or concentration of the major radionuclide in a radiopharmaceutical. When the identification and assay of possible gamma-emitting impurities are also required, spectrometric methods based on such detectors as NaI(Tl) or Ge(Li) crystals must be used (see Section 4.2). When assaying a gamma-ray-emitting radiopharmaceutical for radionuclidic impurities that decay without the emission of photons, such as pure alpha or beta emitters, then alternative measurement techniques are required (see Sections 3.5, 3.6, 3.7, 3.8, 6.1.2, and 6.1.3). These techniques often involve chemical separation of the impurity from the principal radionuclide, in order to avoid interference by the radiations from the principal radionuclide. For quantitative assays of samples of established purity of relatively high activities, ionization-chamber devices, that have no intrinsic energy-discrimination capabilities, are frequently selected.

6.4.3.2.1 *NaI(Tl) Detectors.* The use of NaI(Tl) detectors for the assay of gamma-emitting radionuclides in clinical laboratories is well established. Three basic types or configurations of these detector systems are frequently employed in a clinical laboratory. The first and simplest is the integral counting system employing either a solid cylindrical or a well-type NaI(Tl) crystal and it is normally used for the routine quantitative assay of radionuclidically pure samples.

This simple system can be modified to perform pulse-height spectrometry by adding a single channel or multichannel analyzer. With this additional capability, the detector system may be used for the simultaneous quantitative assays of radiopharmaceuticals containing more than one gamma-emitting radionuclide (see Sections 2.4 and 4.2). Other relevant reviews and texts should also be consulted (e.g., Price, 1964; IAEA, 1965; Hine, 1967; Wagner, 1968; Adams and Dams, 1970; Belcher and Vetter, 1971; Cradduck, 1973; Krugers, 1973; Hine and Sorenson, 1974).

NaI(Tl) well crystals are frequently used in clinical laboratories for *in vitro* measurements. If samples are small enough to be placed at the bottom of the well, the solid angles subtended by the detector approach 4π steradians and, for low-energy photons, the total detection efficiency approaches 100 percent. Care must be exercised in the control of the sample position and size, since these detectors are very sensitive to variations in either. The effects of varying detector geometries are shown in Figure 92.

Increased source-to-detector distances can be used to extend the useful range of NaI(Tl) detectors to millicurie or greater quantities of gamma-emitters by performing sample measurements and calibrations at appropriate source-to-detector distances (Figure 92A).

The nearly 4π geometry achieved by using well crystals is shown in Figure 92C. When this geometry is used, the position and size of the sample must be carefully controlled in order to achieve reproducible results as demonstrated by the change in sensitivity of a typical well-crystal counter with variations in source position and volume (Figures 93 and 94). The energy summing of two or more photons emitted in coincidence is greatly enhanced over that observed with lower-geometry detectors. For this reason, well-crystal detector systems, calibrated in terms of the photopeak efficiency as a function of photon energy, are of limited use and should be avoided for purposes of making indirect activity determinations on radionuclides that emit two or more photons in coincidence. However, this effect can frequently be used to advantage to perform direct measurements as described in Section 3.3. Several examples of the application of this method, relevant to nuclear medicine, are described by Brinkman and Aten (1963)

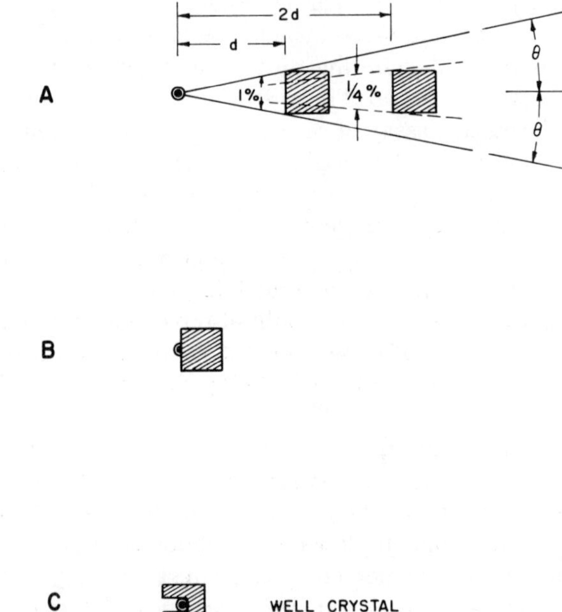

Fig. 92. Examples of different point-source-detector geometries: A, with the source positioned 2.5 times the detector diameter away, the geometric efficiency is 1 percent. At double the distance, it is 0.25 percent, using the approximate law of *inverse squares*, i.e., count rate ~1/(distance)^2. B, for small detector-to-source distances the inverse square law does not apply; however, the *cosine law* [geometric efficiency = ½ (1 − cos θ), with θ the half-angle of the cone subtended by the detector at the source] gives the geometric efficiency at all distances. C, with a well crystal the geometric efficiency approaches 100 percent (4π), provided that the diameter of the well is small and its depth large in relation to the source dimensions.

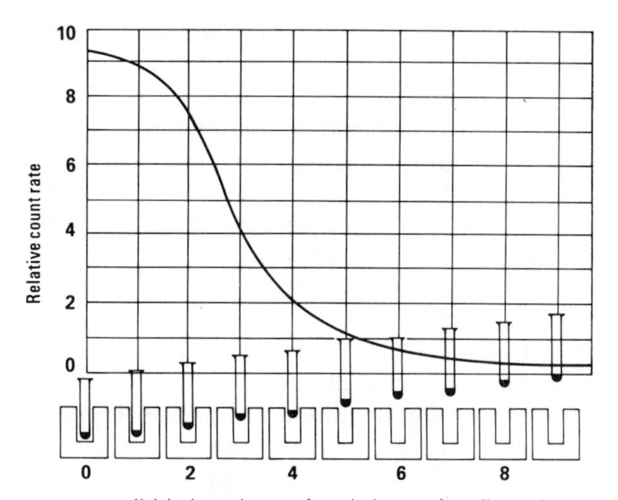

Fig. 93. Response of a typical 2″ × 2″ NaI(Tl) well-crystal as a function of detector-to-source position (in one-centimeter increments) of a 1-ml-volume source. For any well-crystal-counting system, the greatest change of response with source position occurs at intermediate source-to-detector positions, as demonstrated by the 2- to 4-cm positions in this example. (After Brucer, 1970)

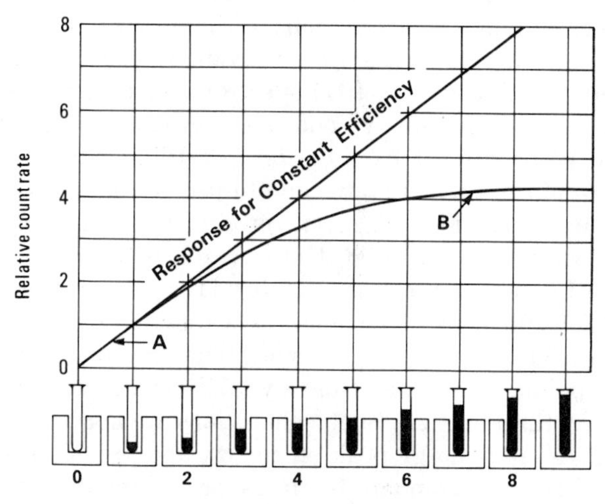

Fig. 94. Response of a typical 2″ × 2″ NaI(Tl) well-crystal detector to different volumes of a solution source of ^{57}Co (in one-milliliter increments) of constant radioactivity concentration. A, for a source volume <1 ml the maximum counting efficiency per milliliter of solution for the detector is approached. B, beyond approximately 7 ml in source volume, no appreciable increase in detector response is observed with increasing source volume. (After Brucer, 1970)

and Eldridge and Crowther (1964) for positron-emitting radionuclides and ^{125}I, respectively.

The third configuration in which NaI(Tl) detectors are typically encountered in nuclear-medicine laboratories is in imaging devices, either scanners or Anger cameras. These systems were neither designed for, nor should they be used for, performing qualitative or quantitative radioassays. Not only will much less expensive NaI(Tl)-spectrometer systems nearly always out-perform imaging devices in terms of increased sensitivity, energy resolution, and reproducibility for routine radioactivity assays, but even the temporary removal of a very expensive imaging device from service in order to perform tasks for which it is ill-suited should be avoided.

6.4.3.2.2 *Semiconductor Detectors.* The principal type of solid-state detector currently used for assaying gamma-emitting radionuclides is the lithium-drifted germanium, Ge(Li), detector. For a review of the physics of operation of these detectors and associated assay techniques, Sections 2.5 and 4.2 of this report should be consulted.

Due largely to cost considerations, these detectors have not been widely adopted in medical laboratories. However, any laboratory that routinely performs large numbers of qualitative analyses, particularly for radionuclidic purity, on gamma-emitting radionuclides should either have, or seriously consider procuring, a high-resolution Ge(Li)-spectrometer system. Determinations of the impurities present in the ^{99m}Tc eluates from ^{99}Mo generators provide an excellent example of the advantages of using a Ge(Li)-spectrometer system for radionuclidic-purity determinations (Wood and Bowen, 1971; Barrall, 1972; Bowen and Wood, 1972; Ficken *et al.*, 1973; Billinghurst *et al.*, 1974). For a review of the current status and future of semiconductor detectors in nuclear medicine, the recent proceedings of a symposium on the subject should be consulted (Hoffer *et al.*, 1971).

Both Ge(Li) spectrometers and NaI(Tl) spectrometers must be calibrated for efficiency of response as a function of photon energy, and the efficiency for photons having energies lying between the calibration points determined from an efficiency curve as described in Sections 4.2.6 through 4.2.8. Such calibrations are normally performed in terms of the full-energy peak efficiency, and provide the basis for one of the prescribed standard methods for determining the activities of gamma-emitting radiopharmaceuticals for which calibration standards are unavailable (USP, 1975).

6.4.3.2.3 *Dose Calibrators.* Ionization chamber systems, reading directly in activity, have become predominantly the instruments of choice for performing quantitative preadministration assays of radiopharmaceuticals. They are generally known as "dose calibrators". Because nuclear medicine is a multidisciplinary specialty encompassing both medicine and radiation or nuclear physics, the popularization of the term "dose calibrator" is unfortunate. The term dose has one meaning or definition when applied to radiation quantities and another when used in the medical sense. In this instance, the term dose is to be interpreted in the medical sense as the prescribed quantity of a pharmaceutical and not as the radiation quantity, absorbed dose.

One of the early forerunners of the present commercial dose calibrators is the type 1383A beta-gamma ionization chamber (Figures 72 and 95) developed for the National Physical Laboratory (Dale *et al.*, 1961). This re-entrant chamber is designed so that the sample is almost completely surrounded by the sensitive volume of the ionization chamber. This arrangement provides high geometrical efficiency that minimizes the effects of small variations in sample position and volume. The lower half of the ionization chamber has a thin window to admit beta particles, thereby permitting use of the chamber for the assay of beta emitters.

Fig. 95. Combined NPL beta-gamma reference ionization chamber of which a schematic diagram is shown in Fig. 72. (From Robinson, B. 1960)

Most of the dose calibrators available today are variations of this basic design. Typically the beta-particle ionization chamber has been deleted and many manufacturers use pressurized chambers filled with argon up to a pressure of 20 atmospheres. With the high-pressure argon filling, the ionization current from the source is up to 20 times, while the background current is about twice, that obtained in the same chamber filled with air at atmospheric pressure.

The general acceptance of these instruments in clinical laboratories for the purpose of performing comparative or indirect radioactivity assays is primarily due to their ability to permit measurements of gamma-emitting solutions in ampoules or serum vials over a range of activities from a few microcuries to many millicuries. Because of the lack of energy selectivity in these detectors, care must be exercised to ensure that the measurements are performed on radionuclidically pure sources. Otherwise, the amount and identities of any impurities present must be determined by measurements that use energy-selective systems with the overall response corrected for any contributions due to such impurities. Sections 2.1 and 4.4 give further information on the operation and use of ionization chambers in performing comparative measurements. It should be noted, however, that, recently, several manufacturers have introduced dose calibrators that rely on detectors other than ionization chambers, such as plastic scintillators in a well configuration. Much of the information presented in this section will also apply to these devices.

The feature that distinguishes dose calibrators from other ionization-chamber systems is the special electronic circuitry that permits the response of the instrument to be displayed directly in terms of activity. This is accomplished by dividing the amplified output of the detector by some unique value for each radionuclide to provide a digital or analog output equivalent to the activity of the radionuclide. Thus, a dose calibrator could be considered a simple analog-computer detector system that is programmed by setting a potentiometer, or by plugging in a fixed resistor, to convert electronically the detector output into direct activity readings, usually in sub-multiples of the curie.

The air of finality imparted by a direct reading in multiples of the curie of a given radionuclide should not mislead the user and, indeed, recent reports of the accuracy of dose calibrators support this warning (Genna *et al.*, 1972; Garfinkel and Hine, 1973; Hare *et al.*, 1974; Hauser, 1974). Some confidence in a manufacturer's calibration can be gained from good, accompanying documentation, but it behooves the operator to assure himself that the calibrations and quoted errors are applicable to the size, shape, and filling of containers routinely used. These calibrations should be checked by means of certified standards of the radionuclides of interest and, if necessary, the corrections applicable to containers of different geometry should be determined in the manner described in Section 6.4.2.

It may frequently happen that a dose calibrator may not have a calibration (variable-resistance-dial) setting, or plug-in resistor module, for a gamma-emitting radionuclide contained in a radiopharmaceutical that has been recently introduced into use. A calibration point for such a radiopharmaceutical can, however, be established by an interpolative method, provided that reliable decay-scheme data are available for the radionuclide in question.

Basically, for *any one radionuclide dial setting or module*, the readout is proportional to the ionization current flowing across the ionization chamber. Thus, by measuring a series of monoenergetic photon sources having known emission rates, at *any dose-calibrator setting*, a curve of reciprocal $K_{R,\gamma}$, as a function of photon energy, can be plotted as described in Section 4.4.3, and will be similar to that shown in Figure 96. If R is the instrument reading for the radionuclide of activity A, P_γ is the fractional probability of the radionuclide emitting a gamma ray of energy E per decay, and R_R is the instrument reading for a long-lived reference source, then

$$K_{R,\gamma} = P_\gamma A \frac{R_R}{R}, \tag{6.3}$$

and

$$A = \frac{K_{R,\gamma} R}{P_\gamma R_R}. \tag{6.4}$$

Provided that the same long-lived reference source is used, an identical curve of $1/K_{R,\gamma}$ *vs.* E will be obtained at *all other* radionuclide settings of the dose calibrator. As stated in Section 4.4.1, the reciprocal of the K factor, or of the relative factor $K_{R,\gamma}$ (as plotted in Figure 94), is equivalent to an efficiency.

After such an efficiency curve has been established, the activity of any other monoenergetic gamma-emitting nuclide can be determined by using Eq. 6.4 and an interpolated value of $1/K_{R,\gamma}$. The activity of a nuclide emitting more than one photon can be determined by weighting the "efficiency" ($1/K_{R,\gamma}$) for each photon of different energy, including x rays, by the appropriate probability per decay and by determining the activity by using Eq. 4.20, namely

$$A = \frac{R}{R_R} \cdot \sum_i^n \left(\frac{K_{R,\gamma}}{P_\gamma}\right)_i. \tag{6.5}$$

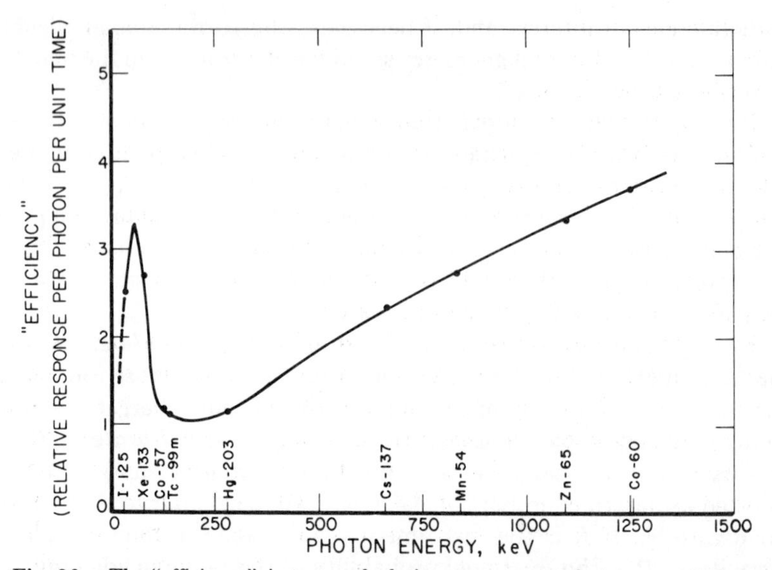

Fig. 96. The "efficiency", in terms of relative response per photon per unit time, as a function of photon energy, for a typical commercial dose calibrator (see Eq. 6.3) from data obtained by R. L. Ayres (1977). (*NOTE:* The numerical values on the vertical scale will depend upon the particular dose calibrator and the activity of the long-lived reference standard. In some cases of non-monoenergetic photon emitters, R/R_R for photons with energies in uncalibrated regions of the curve can be calculated by subtracting from the total response the response obtainable from Eq. 6.3, due to photons with energies in regions where $K_{R,\gamma}$ is known.)

After the activity has thus been determined, a calibration setting can be established for the "new" radiopharmaceutical. From the nature of the "efficiency" curve, it can be seen that serious error could arise if a preponderance of photons had energies below 200 keV, due to the difficulty in establishing the curve in this region.

One method for determining the response function of the type 1383A beta-gamma ionization chamber has been described by Dale (1961). This reference should be consulted for a description of the method itself and the physical processes that determine the observed response function. Response functions for both air and pressurized argon chambers have been determined and discussed in a review of the status of activity determinations by means of "4π"γ ionization chambers (Weiss, 1973).

Two other considerations that must be taken into account in determining or checking the calibration of a dose calibrator are its linearity as a function of activity and its geometry dependence on sample size and position. The linearity of the instrument at any given setting can be checked by preparing a source of a relatively short-lived radionuclide, such as ^{99m}Tc, and following the activity of the source as a function of time. The initial activity of the source should be equal to, or greater than, the highest activity that one would routinely expect to measure, and the decay of the source should be followed down to the lowest levels of activity that would normally be measured. By following this procedure, one would detect any nonlinearities between range changes on multi-range instruments, and the region and magnitude of ion-recombination-induced nonlinearities. An example of both of these effects is shown in Figure 97. Over the range indicated by A, a 10-percent discrepancy is observed between the 1- and 10-millicurie-range settings. In this instance it would be necessary to return the instrument to the manufacturer for correction of the problem. For rates at or above those indicated by the range B, ion recombination produces a nonlinearity. These higher rates should either be avoided, or, alterna-

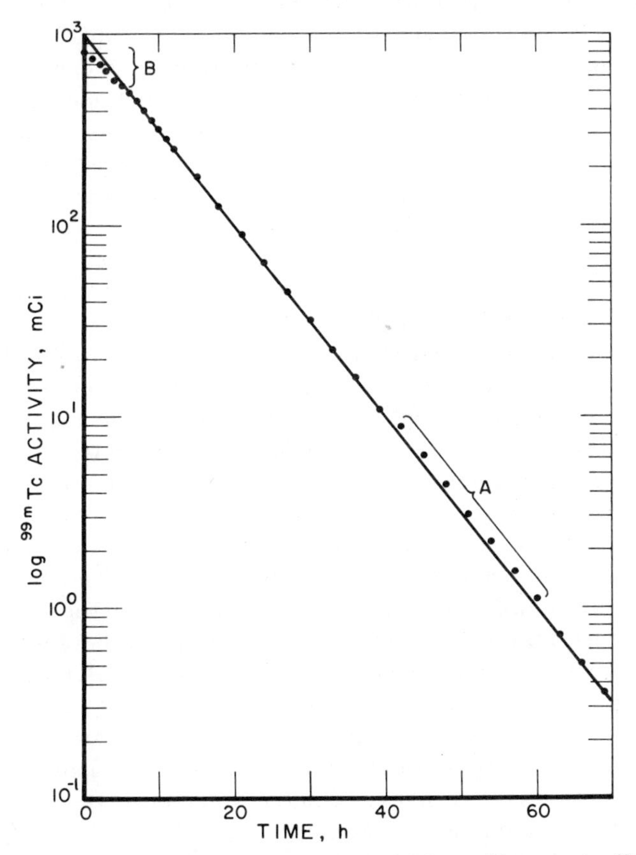

Fig. 97. Simulated linearity check of a commercial dose calibrator using ^{99m}Tc. (See Section 6.4.3.2.3 for discussion.)

tively, the magnitude of the nonlinearity as a function of activity, for a given radionuclide, can be measured and appropriate corrections applied. Test material should be free of other radionuclides in order to prevent other distortions of the half-life curve. An alternative procedure described in Section 4.4.4 using identical sources, but with different radioactivity concentrations, can be used. If measurements are made over an interval of time short compared with the half-lives of the impurities, then the method is insensitive to these photon-emitting impurities.

In order to minimize measurement errors, careful control of source geometry and positioning is required. The effects of varying source position and volume are similar to those presented in Figures 93 and 94, but are less severe because of the greater well depth of most dose calibrators. A discussion of the effects of varying source position in two typical reentrant ionization chambers is presented by Weiss (1973), and of varying source volume by Seliger and Schwebel (1954). A more severe geometry-dependent source of error, frequently encountered in the assay of radiopharmaceuticals, is due to the containers of varying size and composition used by commercial distributors of radiopharmaceuticals. In order to minimize such source-geometry-induced errors in the measurement results, the user must either adopt a standard container and content for all of his measurements or develop individual calibrations for each source geometry. The latter option would require individual calibrations for each container and for variations in the volume of radiopharmaceutical within the container.

The importance of performing daily checks with a long-lived source cannot be over-emphasized. Without such checks, significant drift or alteration of the instrument's response could escape detection and result in measurement errors. Moreover, the practice of pairing measurements of the unknown with a long-lived reference source compensates for changes in amplifier gain—or even errors in settings. That is, the *ratio* of the response per unit activity of a given radionuclide to that of a reference source should be a constant for a given instrument. This ratio is established during the initial calibration of the instrument, and can be used thereafter to calculate the activity of that radionuclide from future paired measurements with the same reference source (corrected for decay) at any arbitrary instrument setting. For example, if the ratio of the response of 1 mCi of a calibration standard of ^{99m}Tc to that of the reference source was determined to be 0.5 and, in a later measurement of a ^{99m}Tc source of unknown activity, the ratio to the same reference source was determined to be 3.6, then the activity of the unknown ^{99m}Tc source would have been equal to 7.2 mCi.

Although a dose calibrator lacks energy selectivity, useful purity and identification checks of a radiopharmaceutical may be made by measuring the half life of a sample. Significant departures from the known value of the half life indicate either the presence of impurities or that it is the wrong radionuclide. Such a check, made immediately after receipt, can be useful in cases of short-lived radionuclide generators such as ^{99}Mo-^{99m}Tc. In addition, when pre-assayed radiopharmaceuticals are being checked, close agreement between the measured and stated activities provides some assurance that the correct radiopharmaceutical has been received.

A measure of energy discrimination can be introduced by using absorbers to selectively attenuate lower-energy gamma rays from given radionuclides and thereby enhance the relative sensitivity of the instrument for radionuclidic impurities emitting higher-energy gamma rays than those of the principal radionuclide. The primary example of the application of this procedure is the use of the "molybdenum breakthrough kit" to measure the quantity of ^{99}Mo in a solution of ^{99m}Tc. Again, due to the lack of energy selectivity in the instrument itself, gross errors have been reported when other high-energy gamma-emitting impurities, other than ^{99}Mo, have been present (Briner and Harris, 1974).

6.4.3.3 *Assay of Beta-Particle-Emitting Radionuclides.* For the determination of the activity of a pure beta-emitting radionuclide, the two methods commonly employed are gas proportional counting and liquid-scintillation counting. For a description of the operation of, and measurements using, proportional counters see Sections 2.2 and 3.5. Similarly, for liquid-scintillation counting see Sections 2.4 and 4.3.

Another method that is frequently used in clinical laboratories to determine the activity of a ^{32}P source is based on the measurement of the bremsstrahlung radiation produced by the deceleration of the high-energy beta particles in the solution and source container. The bremsstrahlung radiation is measured with a photon-sensitive system such as a dose calibrator or NaI(Tl)-detector system.

In order to achieve accurate and reproducible results, great care must be exercised to minimize any variations in counting geometry. Since most of the bremsstrahlung radiation is produced through interactions of the beta particles with the solution and container, the composition and volume of the solution and the wall thickness of the container must be carefully controlled. Attempts to perform such measurements with the sample in typical molded-glass pharmaceutical serum vials will yield widely scattered and erroneous results because the wall thicknesses of these containers vary considerably. Reproducible results can be achieved only by using a "standard" container or set of containers and carefully controlling the solution volume.

Methods for the qualitative identification of pure beta emitters normally rely on determining the half life and the maximum beta-particle energy, $E_{\beta max}$, of the radionuclide. This energy is a unique index for each beta-emitting radionuclide and is most commonly determined by measuring the count rate of a source as a function of successive thicknesses of absorbers placed between the source and the detector. From these data the maximum range or mass absorption coefficient can be obtained and related to $E_{\beta max}$ (Friedlander *et al.*, 1964; Evans, 1955; Mann *et al.*, 1979; USP, 1975).

6.5 Laboratory Equipment for Radioactivity Measurements

6.5.1 *General-Purpose Equipment*

A well-equipped radioactivity measurements laboratory would have some or all of the following instrumentation available to it:

(1) A dose-calibrator or ionization-chamber type of detector for measuring millicurie quantities quickly and accurately in various geometries, and for providing some qualitative indication of radionuclidic purity (see Section 6.4).

(2) A multichannel-analysis system utilizing either
   (a) A NaI(Tl) detector (Section 2.4 and 4.2) or
   (b) A semiconductor detector (Section 2.5 and 4.2). A Ge(Li) detector will allow the user to resolve multicomponent gamma radiations from a sample for radionuclide identification.

(3) A NaI(Tl) well counter with either a ratemeter or digital display, and a variable discriminator circuit. This system is useful for routine counting and simple pulse-height analyses (see Sections 2.4 and 4.2). Automatic, multi-sample counting equipment of this type is also available.

(4) A liquid-scintillation counter for counting pure beta-emitting nuclides. This type of counter can also be used to count the x rays from electron-capture nuclides and low-energy gamma rays (and associated Auger or internal-conversion electrons). Most systems are multi-sample units with two or more channels for measurements in different energy ranges, and with some computational ability (see Sections 2.4 and 4.3).

(5) Proportional counters of various types for counting alpha, beta, or gamma radiation. Single and multi-sample units are available in a variety of configurations, some with automatic computational ability (see also Sections 2.2 and 3.6).

Table 17 indicates the instrumentation appropriate to some typical sample forms in typical ranges of activity.

The choice of equipment appropriate to a laboratory's investigations should be based on:

(1) The types of measurements and accuracies required.
(2) The particular radionuclides and the radiations to be measured.
(3) The types of samples to be measured as well as their number and variety.
(4) The funds available.

6.5.2 *Buyers' Guides*

Various publications describe the different types of equipment available to assay radioactive material. A recent book by Hendee (1973) provides a basic, easy-to-read description of the operation of various types of nuclear-detection equipment.

Several Buyers' Guides published by scientific journals provide excellent sources of information concerning the variety of commercial instruments available (e.g., the "Guide to Scientific Instruments" issue

TABLE 17—*Types of radiation-detection instrumentation*

Type of instrument	Typical activity range	Typical sample form	Data acquisition and display
	mCi		
Gas proportional counters	10^{-7} to 10^{-3}	Film disc mount, gas	Ratemeter or scaler
Liquid-scintillation counter	10^{-7} to 10^{-3}	Up to 20 ml of liquid or gel	Digital Accessories for: background subtraction, quench correction, internal standard, sample comparison
NaI (Tl) cylindrical or well crystals	10^{-6} to 10^{-3}	Liquid, solid, or contained gas <4 ml	Ratemeter Digital Discriminators for measuring various energy regions Multichannel analyzer, or computer plus analog-to-digital converter Computational accessories for full-energy-peak identification, quantification, and spectrum stripping
Ionization chambers	10^{-2} to 10^{3}	Liquid, solid, or contained gas (can be large in size)	Ionization-current measurement Digital (mCi) readout, as in dose calibrators (Section 6.4.3.2.3)
Solid-state detectors	10^{-2} to 10	Various	Multichannel analyzer or computer with various readout options

of *Science*, The "Instruments Specifier" issue of *Industrial Research*, the "Buyers' Guide" issue of *Nuclear News*, and the "Physicians' Desk Reference for Radiology and Nuclear Medicine" published by the Medical Economics Co., Oradell, New Jersey).

6.5.3 *Minimum Instrument Requirements for Some Specific Laboratories*

In general, the following list provides the minimum instrument requirements for the type of laboratory involved. This list does not include monitoring or surveying equipment for radiation protection.

(1) Therapeutic and diagnostic nuclear medicine laboratories: A dose-calibrator or ionization-chamber type of detector (a multichannel-analyzer system with a NaI(Tl) or Ge(Li) detector would be desirable; a typical gamma-ray camera or scanner with an open discriminator setting may be used for this purpose, but is not recommended because of its limitations discussed in Section 6.4.3.2.1).

(2) Clinical diagnostic laboratory for *in vitro* testing:
   - (a) NaI(Tl) well crystal with associated electronics;
   - (b) Liquid-scintillation counting system, basically for those radionuclides emitting only beta particles, that can also be used to detect low-energy photon-emitting nuclides (see Section 4.3).

(3) Biomedical research laboratory:
   Liquid-scintillation counter.

(4) Physical and chemical research laboratory:
   - (a) NaI(Tl) well crystal with associated electronics;
   - (b) Some type of multichannel analyzer with a NaI(Tl) or semiconductor detector;
   - (c) Gas-proportional-counting system appropriate to the type of radiation being measured.

(5) Teaching laboratory:
   - (a) Proportional-counting system appropriate to the type or radiation being measured;
   - (b) NaI(Tl) well crystal with digital or ratemeter output.

(6) Industrial applications or radiography laboratory:
   - (a) Survey equipment (usually of the ionization-chamber type) to measure exposure rates from radiation sources;
   - (b) A NaI(Tl)- or semiconductor-spectrometer system, if radionuclide identification is necessary.

6.5.4 *Instrument Stability*

All laboratories making radioactivity measurements utilizing the above equipment should institute a good quality-control program for their counting procedures. Laboratories carrying out clinical or diagnostic procedures may be required by regulation to have such quality control. Instrument performance should be checked daily with a suitable long-lived source to assure the reproducibility of daily measurements. Counting efficiencies for each radionuclide measured at a particular instrument setting, counting geometry, and sample composition should be determined and periodically checked. All detector systems should be characterized for the effect of variations of sample geometry, composition, and activity on counting efficiencies.

Records should be kept of all the measured data, the instrument settings, and measurement geometry. Data must be checked for statistical reliability by using techniques such as the χ^2 test for "goodness of fit" (see Section 7.3). If the instrument-monitoring data, in terms of a long-lived reference source, and the instrument background readings are plotted as a continuing function of time, long-term trends, reflected in small changes in slope or spread of points, can indicate a change in instrument performance or in instrument environment even before statistical evaluation of the data might do so. For nuclear medicine technologists, a simplified discussion on counting statistics may be found in Simmons (1970).

6.5.5 *Radiation Protection*

Every laboratory handling radioactive material should also have an active radiation-protection program that will help avoid radioactive-contamination problems and personnel overexposure. Survey equipment, such as portable Geiger-Müller counters or "Cutie-Pie"-type ionization-chamber detectors should be used to monitor the laboratory area daily for low-level radioactive contamination outside designated areas to assure that all radioactive material is adequately shielded and that no excessive radiation field exists in the facility. Wipe smears should be taken inside and outside of the designated work area to supplement the survey-instrument readings. The smears should be measured on low-level detectors, such as proportional counters or liquid-scintillation counters, to assure against unknown radioactive contamination that may not be detected with the survey equipment. For a more comprehensive treatment of the subject of radiation protection, texts such as those by Cember (1969) and Shapiro (1972) should be consulted, as well as reports by the National Council on Radiation Protection and Measurements (NCRP) and the International Commission on Radiological Protection (ICRP).

6.5.6 *Further Reading*

Information regarding the type of equipment suitable for specific purposes and the procedures that can be used can be found in handbooks available from the Oak Ridge National Laboratory (Davis and Gupton, 1970), and the National Institutes of Health (NIH, 1973) as well as a text by Shapiro (1972). Additional general references are Lapp and Andrews (1964), Price (1964), Bertolini and Coche (1968), Adams and Dams (1970), Krugers (1973), and Ouseph (1975).

7. Statistics

7.1 Treatment of Counting Data

7.1.1. *Statistics of Pulse Counting*

The process of radioactive decay is a random sequence in time. From the law of radioactive decay (see Section 1.3) the decay rate, dN/dt, is given by

$$\frac{dN}{dt} = -\lambda N \tag{7.1}$$

where λ is the decay constant and N is the number of unstable nuclei. Let n denote the number of atoms that decay during a certain period of time T. For T small compared to the half life, n will be small compared to N, so that the decay rate ρ (= dN/dt) can be treated as a constant. Then the probability of exactly k disintegrations during the period T can be approximated by the Poisson distribution:

$$\text{Probability } (n = k) = \frac{(\rho T)^k e^{-\rho T}}{k!} \tag{7.2}$$

Equation 7.2 has been derived in most treatises on probability based on either (a) axiomatic assumptions, or (b) the limit to binomial (Bernoulli) distributions (see, for example, Feller, 1968; Parzen, 1962; and Stevenson, 1966). (See also Appendix B.3.2.) The Poisson distribution is also applicable to a large number of physical phenomena in daily life, such as telephone calls and traffic accidents.

Once we accept the view that the Poisson Law adequately describes the phenomenon of radioactive decay, several useful results follow by deduction:

a. The mean and variance of the number of decays or transitions during time T are both equal to ρT. The standard deviation is $\sqrt{\rho T}$. (See Appendix B.3.5.)

b. Addition Theorem. If n_1 is the number of decays of one species of nuclide in time T with decay rate ρ_1, and n_2 the number of decays of another species of nuclide in time T with decay rate ρ_2, then the distribution of the sum $(n_1 + n_2)$ is also Poisson with mean and variance $(\rho_1 + \rho_2)T$.

c. For large numbers of decays n, the Poisson distribution is well approximated by the Gaussian, or normal, distribution, with a mean ρT and a variance ρT. Let $\xi = (n - \rho T)/\sqrt{\rho T}$ be the normalized deviate, i.e., the difference from the mean relative to the standard deviation.
Then the probability, P, of $a < \xi < b$ is

$$P(a < \xi < b) = \frac{1}{\sqrt{2\pi}} \int_a^b e^{-x^2/2}\, dx$$

d. The distribution of the time interval t between decays is the negative exponential,

$$f(t) = \rho e^{-\rho t}$$

where $0 < t < \infty$, and the probability of $t_1 < t < t_2$ is the integral

$$P(t_1 < t < t_2) = \int_{t_1}^{t_2} f(u)du.$$

The properties of Poisson and normal distributions may be found in most texts on statistics, e.g., Brownlee (1950), Haight (1967), and Cramér (1946). Tables of the individual and cumulative terms of Poisson distribution have been provided, e.g., by the General Electric Company (1962).

The decay rate ρ of a sample bears a known relationship, through the counting efficiency, to the "true" counting rate, ρ'. However, counting for a finite period of time t can only yield an estimate $r = n/t$ of the "true" counting rate ρ'. The measured r is subject to statistical fluctuations. A measure for the scatter of the measured counting rate r about the "true" counting rate ρ' can be derived from the Poisson assumption; such a measure is provided by the standard deviation.[17] The standard deviation of r is a function of t and can be expressed as

$$\sigma_r = \sqrt{\frac{\rho'}{t}}. \tag{7.3}$$

Because ρ' is unknown, only an estimate, $\hat{\sigma}_r$, of the standard deviation, σ_r, can be computed from the measured r, namely,

$$\hat{\sigma}_r = \sqrt{\frac{r}{t}} = \frac{1}{t}\sqrt{n}. \tag{7.4}$$

[17] If counting losses, or dead time, are not negligible, then the Poisson approximation does not hold. Dead time, decay, and background considerations are treated in Section 2.7.

An estimate of the standard deviation of the number of counts is

$$\hat{\sigma}_n = \sqrt{n}. \tag{7.5}$$

The standard deviation is a measure of the scatter of a set of observations around their mean value. For example, about 10,000 counts are necessary for a standard deviation in the total count of about 100 or of 1 percent. When many observations are made, approximately two-thirds of the observations ought to lie within one standard deviation from the mean and the remainder outside this limit. The following limits apply to representative multiples of the standard deviation in the theoretical distribution:

Deviation	$\pm 0.675\sigma$	$\pm\sigma$	$\pm 2\sigma$	$\pm 3\sigma$
Probability that observation lies within this deviation	0.5	0.68	0.95	0.997

The above limits allow an appraisal of the performance of a counting system to be quickly made, i.e., if a counting rate deviates more than what would be expected statistically from a previously determined rate. This is of importance when equipment is checked routinely with a performance standard. For instance, a deviation of more than 2σ from a previously established mean should be considered with suspicion, even though it may be correct, since the probability of this occurring is only about 5 percent.

The application of the general description given above can be formalized through the use of "control charts" that provide a running graphical record of the values obtained from counting a reference source or of their variability. For further details on control charts, see Natrella (1963) and Ku (1969).

7.1.2 *Background Considerations*

In most instances the radiation counter exhibits a background rate that is not negligible and that has to be subtracted from the gross counting rate. An estimate of the statistical error due to the background must be included in the final estimate of the error. On the basis that the source and background counts are additive, the standard deviation of the net counting rate is

$$\sigma = \sqrt{\sigma_T^2 + \sigma_B^2} = \sqrt{\frac{\rho_T'}{t_T} + \frac{\rho_B'}{t_B}}. \tag{7.6}$$

T refers to the counting of the sample plus background (total) and B to the counting of the background. In counting sources of reasonable activity, it is usually sufficient to determine the background counting rate only once (or perhaps before and after a series of samples) by counting long enough to make σ_B negligible compared to σ_T. The optimum division of available time between background and source counting has been derived by Loevinger and Berman (1951), and is given by

$$\frac{t_B}{t_T} = \sqrt{\frac{\rho'_B}{\rho'_T}}. \tag{7.7}$$

To approximate the optimum division, estimates of ρ'_B and ρ'_T have to be obtained from a preliminary run. The minimum combined time necessary to achieve a predetermined precision can be estimated for particular values of total counting rate relative to background counting rate. Curves published by Loevinger and Berman (1951) enable this to be accomplished for any practical case. These curves are reproduced in Figure 98. As an example of the application of these curves, suppose that it is desired to determine with a standard deviation of 5 percent the activity of a sample that has a counting rate approximately half that of the background. Then k, the ratio of total to background

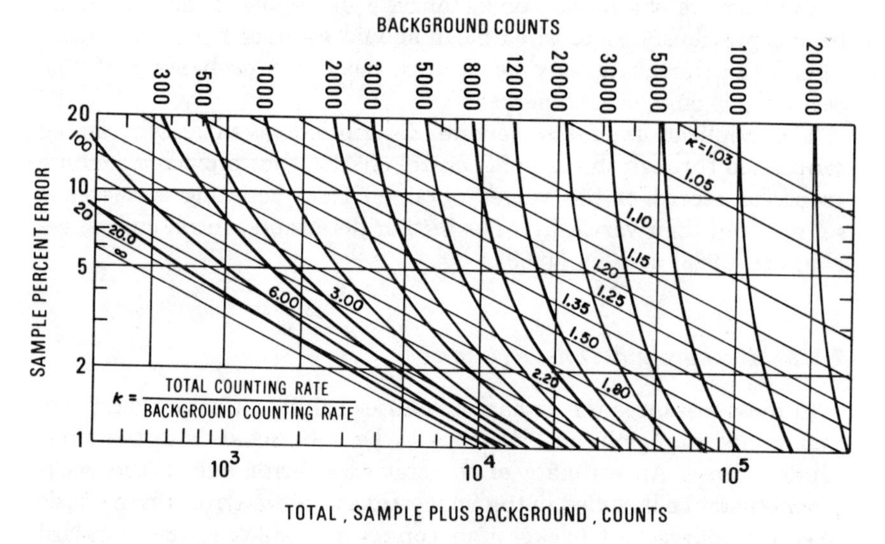

Fig. 98. Relationship between precision and number of observed counts taking account of background. For a given desired percentage error in the measurement of sample activity, the optimum distribution of counts between background and the combined background plus sample counts can be chosen so that the time spent in counting is a minimum. The ratio k is that of sample plus background counting rate to background counting rate. The curves are particularly useful when a predetermined precision is required in counting a sample of low counting rate using the pre-set count technique. (From Loevinger and Berman, 1951).

counting rate, is 1.5, and from Figure 98, the 5-percent line intersects the oblique line for k = 1.5 at a total count of 6,600. The point of intersection lies about one-third of the distance between the background-count curves for 3,000 and 5,000. Thus, the background count should be about 3,600. These curves are valuable guides whenever the preset-count technique can be used, and are especially useful when only a few measurements are to be made. They are, however, seldom used in the case of well-established measurements systems for which "running backgrounds" are maintained, as, for example, in a low-level radioactivity measurements laboratory. A preliminary rough determination of the background and the total counting rates is necessary in order to establish the value of k. The preset-count technique used with these curves then enables all sample activities to be determined with approximately the same precision.

7.1.3 *Minimum Detectable Activity*

Counting intervals cannot be made indefinitely long; therefore, time and background considerations determine minimum detectable activity. In addition, the situations in which the concept of minimal detectable activity is used also tend to give rise to similar expressions but with entirely different interpretations. Currie (1968) commented on a number of such terms and proposed three of his own that constitute successive levels of activities leading from a *decision rule* that indicates detection based on the data at hand to a preselected *detection limit* such that a signal at this level is almost sure to be detected, and then to a quantitative estimate of the signal with a particular precision. Hence, in using the term "minimum detectable activity" or similar expressions, the exact definition and meaning of the term must be given at the same time.

Moreover, since these limits are based on counting statistics alone and do not include other random errors or any systematic components of error that are always inherent in the procedures, the use of these terms is limited to that of serving as guideposts only, and not as absolute levels of activity that can or cannot be detected by a counting system. Bearing these remarks in mind, definitions of decision limit and detection limit are given as two convenient and useful terms in the interpretation of minimum detectable activity in the framework of statistical hypothesis testing, as originally conceived by Neyman and Pearson (1936), and adopted by Currie (1968), and others. Treatment of statistical hypothesis testing may be found in most statistical texts, for example, Natrella (1963, Chapter 3). Ordinarily, a rule is applied to the results observed and then a decision is made between a pair of alternatives. The two alternatives may be formally stated as follows:

(a) We shall proceed on the supposition that the observed counts (counting rate) are greater than background.

(b) We shall proceed on the supposition that the observed counts (counting rate) are not greater than background.

Since the observations are made in a finite length of time, we may sometimes make an erroneous decision. There are two ways in which we can take a wrong action:

(a) We proceed on the supposition that there is a signal when in fact there is none: Error of the first kind, or *false detection.*

(b) We proceed on the supposition that there is no signal, when a signal really exists: Error of the second kind, or *false non-detection.*

In any particular experiment, we never can be absolutely sure that the correct decision has been made. The rule of decision adopted by us, in turn, would depend on the importance we attach to each type of wrong decision, and is a compromise of (i) the time available for the experiment,[18] (ii) what proportion of false detection can be tolerated, and (iii) what proportion of false non-detection can be tolerated. A rule that works well for one laboratory, therefore, may not work at all for another laboratory with a different objective in mind.

The proportion of false detection that can be tolerated is usually denoted by α, and the proportion of false non-detection that can be tolerated is denoted by β. In general, we would like to have both α and β be small. A useful selection of the values of α and β is that $\alpha = \beta = 0.05$, i.e., we are willing to accept 5 percent of false detection and 5 percent of false non-detection in a sequence of such decisions.

If equal values of α and β have been decided upon, e.g., $\alpha = \beta = 0.05$, it can be shown that the *decision limit,* L_C and the detection limit, L_D can be expressed in terms of standardized normal deviates (normal deviates in units of σ) and the background count as follows (Currie, 1968):

$$L_C = k\sqrt{2}\sqrt{B} = 1.64\sqrt{2}\sqrt{B} = 2.32\sqrt{B} \qquad (7.8)$$

$$L_D = k^2 + 2L_C = 2.71 + 4.65\sqrt{B} \qquad (7.9)$$

where:

L_C: the net number of counts (total minus background) for reaching

[18] We note that the problem is not necessarily solved even if time can be extended, since then we will have instrument drift and a host of other interfering problems.

a decision of no detection of signal if the actual net counts are less than L_C and detection if the actual net counts are larger than L_C.

L_D: the signal level such that a signal at or above this level is likely to be detected.

B: the "true" background counts.

k = 1.64: the value of the standardized normal deviate that is exceeded with probability $\alpha = 0.05$.

For values of L_C and L_D thus computed, the probability of a false detection at a level of L_C is approximately 5 percent and the probability of non-detection at a level L_D is approximately 5 percent.

The above formulation assumes paired observations of total and background counts. Since the "true" background count B is usually not known and an estimate is used, the computed values of L_C and L_D are approximate for given α and β. For other cases, including those where the background count can no longer be assumed to be Poisson because of interference, see Currie (1968) and Donn and Wolke (1977).

The relation of the just described procedure and definitions of two terms—the decision limit L_C and the detection limit L_D—to the minimum detectable activity of a system is indicated in the plot of β against signal level in Figure 99. This figure shows the behavior of β, the probability of not detecting a signal as a function of signal level. Note that the procedure calls for a decision of no detection if the net number of counts is less than L_C. Should the signal level corresponding to L_C be called the minimum detectable activity? Most likely not, since at that level all one needs is to flip a coin, and the results based on heads or tails is as good as the result of our measurements, in the long run. This condition is unsatisfactory, because by minimum detectable activity is meant the activity level that could be detected most of the time, and not detected only occasionally.

As the signal increases from L_C toward L_D, the percentage of false non-detection continues to decrease and reaches a tolerable 5 percent error rate when the signal level is at L_D. Perhaps it could be claimed that this is the level at which the system will consistently detect the signal. However, since L_D is fixed by the background and the parameters α and β, there is no unique point corresponding to a "minimum".

In fact, fixing any two of the four parameters automatically determines the remaining two. In the above case, L_C is fixed by choosing $\alpha = 0.05$, and L_D is fixed by selecting $\beta = 0.05$ in conjunction with L_C already determined. If a low value of L_D is arbitrarily fixed, and a value for β is selected, then L_C is fixed, and the value of α corresponding to this L_C may not be acceptable for the given background counts.

Statistical hypothesis testing, and hence the decision and the detec-

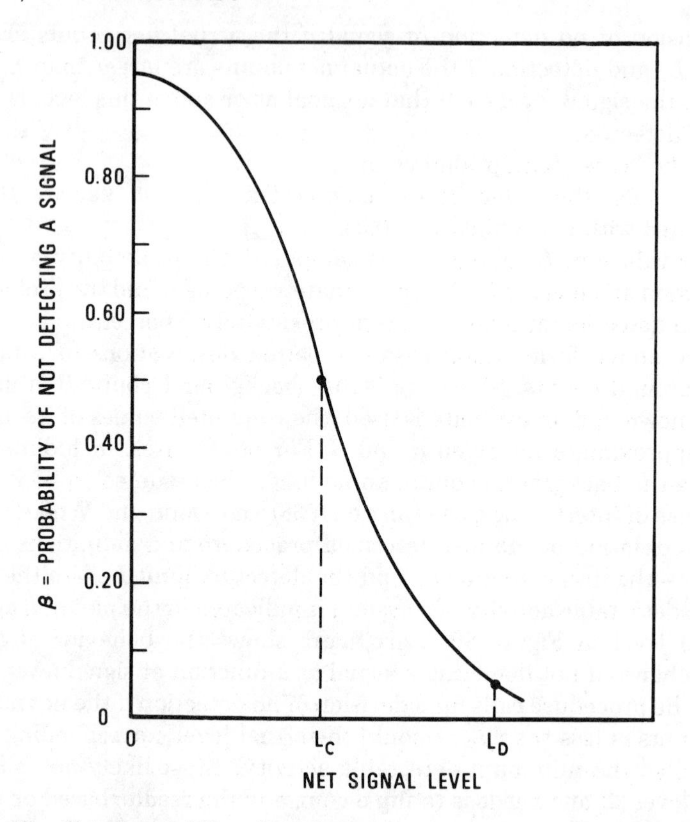

Fig. 99. Plot of probability of not detecting a signal (β) *vs* net signal level. The curve shown is one of a family of curves so that the location of L_C and L_D depends on α, β, and background. In this illustration for $\alpha = 0.05$, the probability of non-detection at zero signal level is 0.95.

tion limits, is a "yes" or "no", "go" or "no go" decision making procedure. In many instances, an "interval estimate" of the signal level could be more informative than a decision process. Confidence interval for the mean is described in Section 7.3.2.

From the above discussion, it is clear that the minimum detectable activity, however defined, does not guarantee with certainty the presence or absence of a signal. These concepts have been frequently misused in the handling of data, particularly in low-level counting. For example, in a series of data to be averaged, to consider the minimum detectable activity either as zero or at the minimum detectable activity would bias the value of the average. In such a situation, all values should be used *as measured* even if some values are below the minimum detectable activity. Negative values (below background) are as valid a measurement as positive values and should be retained in any series of data.

7.1.4 *Background Equivalent Activity*

A useful concept for expressing the intrinsic sensitivity of any detector is that of the Background Equivalent Activity (BEA) (Seliger and Schwebel, 1954; Mann and Seliger, 1958). This quantity, for any given instrument and radioactive nuclide, is, as its name implies, the activity of the nuclide that will, in the particular geometry used, produce a response of the instrument that is equal to its background reading.

7.1.5 *Statistics of Ionization Chambers*

Statistical variation in the number of particles produced in a radioactive-decay process is reflected in the variation of total charge released in an ionization chamber (or "mean-level" system; see Price, 1958). The total charge, Q, collected in a given time in the chamber will be equal to the product qn, where q is the average charge released per interacting particle and n is the number of particles interacting within the chamber in that time. Therefore the standard deviation in Q will be q times as large as the standard deviation in n, or $\sigma_Q = q\sqrt{n}$. Substituting $Q = qn$ one obtains

$$\sigma_Q = \sqrt{qQ}. \tag{7.10}$$

This treatment neglects the fluctuations in the charge released by the individual particles. Price (1958) gives the treatment for rate-type ionization-chamber instruments. For certain applications (e.g., semiconductor detectors, Section 2.5), the Fano factor (Fano, 1947) is used to correct σ_Q to agree with experimental results.

7.2 Statistical Models of a Measurement Process

The previous section has dealt with counting errors, the variability due to the decay process. The counting error is the fundamental, inescapable statistical fluctuation associated with the counting process. In addition to these counting errors, however, other sources of error may also be considered as "random" in nature. These errors in a sense are also inescapable because no two samples are exactly alike and no physical quantities can be measured without error. The standard deviation computed from counting errors can be used as a measure to judge if these other sources of error are in fact significant or negligible.

The use of a factor to correct for "geometry" of the counting instrument may introduce a systematic error in a series of results; the use of a specific calibration curve or half-life value may bias results from those of another laboratory using a different curve or a different value. Systematic errors may be the consequences of a particular measurement procedure or of particular instruments. The use of referee methods and standard reference samples helps to coordinate results of different laboratories to the same benchmark.

Another source of bias, which is important in the statistical sense, is the reduction and summary of data. The weighting of observations, including the rejection of outlying values (outliers), is forever a problem. Should the weights represent counting error alone or should the weights also include a component representing measurement errors in addition to counting error? Should a line be fitted to the data with a model that goes through the origin? These and a number of similar problems do not have readily available answers, since the appropriate technique may depend on the way data are generated, the theoretical basis for their relationship, and their distributional properties.

The measurement process and its error structure can usually be approximated by statistical models, shorthand expressions to account for and to focus attention on the variability of results of measurements. In a statistical model, a measured value is usually expressed as a function of parameters on which the measured value depends plus the errors that may be associated with various factors involved in the measurement process, including assumptions made with regard to the distribution of these errors.

For illustration, consider a set of measurements of ^{90}Sr activities in each of the soil samples collected in duplicate at 5 sites near a reactor. The measured activity, for example, of the second sample from the third site may be represented by the following statistical model:

Measured activity of the second sample of the third site
= (grand mean of samples from five sites at date of collection)
+ (effect of third site)
+ (activity difference of second sample)
+ (possible instrument drift between tests)
+ (errors in sample preparation)
+ (counting error, including background).

The activity is also influenced by other possible errors, e.g., in the:

(geometry factor),
(efficiency factor used for instrument),
(half-life value used), and
(chemical yield).

The last four items may introduce errors that are systematic in nature and that would affect equally the measured activity at different times, and thus would cancel out if the measured activity of the soil at one time period was compared to that measured at another time period by the same laboratory. These systematic errors could, however, be the main sources of discrepancy when results from one laboratory were compared to results from another.

Errors due to sample preparation and instrument drift may be assumed to be random and their standard deviations may be estimated from duplicate determinations, but the counting-error component has to be considered. The effects of different sites may or may not be a subject of interest. If the average of the five sites selected is defined to be the criterion for safe levels of radioactivity surrounding a reactor, then the sites by themselves are of no interest; but if the five sites are selected as a sample of all sites surrounding the reactor, then variability of activity from site to site may contribute substantially to the uncertainty of the reported value.[19]

The formalization of an "error account" in a statistical model such as above, for a particular measurement process, serves at least two purposes:

A. To list exhaustively the possible sources of errors that are relevant to the interpretation and use of the results, and
B. To help in planning or designing the experiment such that the results may be interpreted as unambiguously as possible and the errors due to various factors can be estimated.

In the following section a selection of statistical procedures is presented to help in explaining some of the concepts and techniques commonly used. For additional reference see Jarrett (1946), Brownlee (1960), Youden (1951, 1963), Natrella (1963), Stevenson (1966), and Ku (1969).

[19] The above example, on soil samples taken about a reactor, points to another area in which more research and standardization are needed, namely, as to how many samples (where, when, and how obtained) should be taken in order to ensure that the radioactive effluent from a reactor conforms with prescribed environmental limits. Statistical sampling methods deal with this type of problem, but a general plan for selecting sampling sites cannot be given because different situations arise in each individual case. For further information on this subject, see NCRP Report No. 50 (NCRP, 1976b).

7.3 Statistical Terms and Formulae Useful in Radioactivity Measurements

7.3.1 *Population, Mean, and Standard Deviation*

One of the basic concepts of statistics is the notion of a *population* of "things" with fixed properties, and the use of observations on a *sample* from the population to estimate such properties. Greek characters are usually used to denote the parameters of a population, and Roman characters for estimators of these parameters, or estimates of these parameters computed from a series of observations say, x_1, x_2, $\cdots$, x_n. Some of the most commonly used parameters and their estimators are given in Table 18.

The mean is a measure of location (center) of the distribution of the population, and the standard deviation is a measure of the spread (scatter) of the distribution about the mean. The square of the standard deviation, σ^2, is called the variance.

Since repeated counts for equal intervals of time follow the Poisson distribution, which has the property that the variance is equal to the mean, this property can be used to test for the presence of extraneous variability in addition to the inherent variability of the counts. Let c_1, c_2, $\cdots$, c_k be the results of repeated counts. Theoretically an estimate of the variance of each of these counts is expected to be equal to $\bar{c}$. An experimental estimate of the variance is given by:

TABLE 18—*Population parameters and corresponding estimators*

Population parameters	Corresponding estimators of parameters
μ_x (mean or first moment)	$\bar{x} = \frac{1}{n}\sum_{i=1}^{n} x_i$
σ_x^2 (variance or second central moment)	$s_x^2 = \frac{1}{n-1}\sum_{i=1}^{n}(x_i - \bar{x})^2$
$\sigma_{xy} = \sigma_{yx}$ (covariance)	$s_{xy} = s_{yx} = \frac{1}{n-1}\sum_{i=1}^{n}(x_i - \bar{x})(y_i - \bar{y})$
σ_{xy} (correlation coefficient)	$r_{xy} = r_{yx} = \frac{s_{xy}}{s_x s_y}$
σ_x (standard deviation of x about μ_x)	s_x
$\sigma_{\bar{x}} = \frac{\sigma_x}{\sqrt{n}}$ (standard deviation of the average $\bar{x}$, or standard error)	$s_{\bar{x}} = \frac{s_x}{\sqrt{n}}$
$100\frac{\sigma}{\mu}$ (coefficient of variation or relative standard deviation, expressed in per cent)	$v_x = 100\frac{s_x}{\bar{x}}$

$$s_c^2 = \frac{\sum_{i=1}^{k} (c_i - \bar{c})^2}{k - 1}. \tag{7.11}$$

In the absence of other random errors the ratio of these two quantities, $s_c^2/\bar{c}$ would be expected to be about unity. Large values of this ratio indicate excessive variation among the repeated counts, perhaps due to instrument drift, sample placements; values significantly less than one are equally suspicious and may indicate large dead-time losses, or other de-randomizing effects, that invalidate the assumption of Poisson statistics.

The quantity $\sum(c_i - \bar{c})^2/\bar{c}$ follows approximately the Chi-squared (χ^2) distribution with $(k - 1)$ degrees of freedom. A table of percentiles of the Chi-squared distribution are readily available (e.g., Natrella, 1963), and the value for $\chi^2_{0.95}$ with $k - 1$ degrees of freedom can be used as a criterion to decide if the variance computed from the k counts is larger than that expected purely due to chance. The degree of freedom is the number of independent measurements used in the estimate of the variance.

Results of experiments, however, are not always expressed in counts, or counts per unit time, but are reduced to numbers such as activities per gram, etc. These numerical values are often treated as if they are distributed normally, with mean μ and standard deviation σ. Such an assumption affords mathematical convenience, and can often be shown to be valid to a satisfactory degree of approximation.

A normal (Gaussian) distribution is completely specified by its mean and standard deviation. For distributions other than normal, other additional parameters may be needed to specify properties of the distribution, e.g., skewness (third central moment) and peakedness (fourth central moment). While it is true that, in many applications, the normal distribution can be assumed and used as an adequate approximation, caution must be exercised to check for departure of data from normality. A simple, graphical test for normality can be made using normal-probability paper (Dixon and Massey, 1957).

The median of a set of n observations can also be used as a measure of location, and is defined as the value occupying the $[(n + 1)/2]$th position for n odd if the n values are arranged in an ascending sequence with $x_1 \leq x_2 \leq x_3 \cdots \leq x_n$. For n even, the median can be defined as the value midway between the $(n/2)$th position and the $[(n/2) + 1]$th position. A large difference between median and the mean indicates nonnormality, or the presence of outlying observations.

For small sets of normally distributed data, an estimate of standard deviation can be obtained by multiplying the range R (difference between largest and smallest) of these measurements by a constant (approximately equal to $1/\sqrt{n}$). Table 19 lists these constants corresponding to the number n of measurements in the set. Clearly the use of one set of three or four measurements gives a very uncertain estimate of the standard deviation. If there are k sets of n measurements each, the average range $\bar{R}$ can be used.

In summarizing a set of data, the following three quantities should be computed and presented:

1. The average, $\bar{x}$—the quantity of interest;
2. An estimate of the standard deviation, s—a measure of the precision of the measurements; and
3. The number of measurements, n—an indication of how well $\bar{x}$ and s are known as estimates of μ and σ.

These three quantities provide a useful summary of the information in the data, unless there are very discrepant values, usually called "outliers". "Outliers" may provide information on the cause of lack of control in the measurement procedure and should never be discarded, i.e., expunged from the record, because a later review of outliers may reveal the cause of their production. An outlier should not be excluded from computation of the mean without adequate physical explanation. Outliers that recur frequently should always be included in the computation of the standard deviation to avoid an over-optimistic evaluation of the precision of the measurements.

In calculating a grand average, $\bar{X}$, of a number of groups of measurements of the same quantity, each group having a mean $\bar{x}_i$ and a variance of the mean, var $(\bar{x}_i)$, it may be desirable to take a weighted average when the variances of the mean of the different groups differ

TABLE 19—*Estimate of σ from the range*[a]

n	Multiplying factor	$1/\sqrt{n}$
2	0.886	.707
3	0.591	.577
4	0.486	.500
5	0.430	.447
6	0.395	.408
7	0.370	.378
8	0.351	.354
9	0.337	.333
10	0.325	.316
12	0.307	.289
16	0.283	.250

[a] Adapted from Table 2-1, Natrella (1963).

greatly. The reciprocal of the estimated variance is the proper weighting factor, so that the weighted average and variance are given by (Stevenson, 1966):

$$\overline{X} = \frac{\sum_i \frac{\bar{x}_i}{\text{var}\,(\bar{x}_i)}}{\sum_i \frac{1}{\text{var}\,(\bar{x}_i)}}, \tag{7.12}$$

$$\text{var}\,\overline{X} = \frac{1}{\sum_i \frac{1}{\text{var}\,(\bar{x}_i)}}.$$

This variance of $\overline{X}$ is the "internal" one, referring only to the variance of the sample due to fluctuations in decay. Stevenson (1966, p. 98) also shows how an "external" variance, the variability associated with errors other than counting, can be incorporated into a weighted average.

If it is assumed that the data are normally distributed to a close approximation, more can be said about the results. The confidence intervals for the mean and the statistical tolerance interval for a fixed percentage of the population are discussed in the next two sections.

7.3.2 *Confidence Interval for the Mean*

In practice, the exact values of the population parameters are never known; only sample estimates are available computed from one or more sets of data. It is postulated, however, that the larger the sample, the closer these estimates will approach to the true parameter values, provided there are no systematic errors in the sampled values. For the population mean in particular, an interval about the observed $\bar{x}$ can be constructed that will cover μ_x with a specified probability, 0.95 or 0.99 for instance. This interval is called the *confidence interval for the mean*, and is often used as the term expressing limits to random measurement errors in the uncertainty statement of a reported value.

To construct a confidence interval *for the mean*, three quantities must be calculated; and to be sure of the probability level, i.e., the percentage of coverage in the long run when the same experiment is repeated a large number of times, two assumptions must be verified. The three quantities are:

$\bar{x}_n$: an average computed from n observations.

s_ν: a standard deviation based on ν degrees of freedom. If the same n observations are used to compute the standard deviation, the $\nu = n-1$. If the standard deviation is pooled from k sets of data each with n observations, then $\nu = k(n-1)$.

$t_{(1-\alpha),\nu}$: a tabulated value of the "t" statistic with entries ν and $(1-\alpha)$, where $(1-\alpha)$ is the confidence coefficient (or probability level), 0.99 for example, (or confidence level 99%).

The two assumptions are:

(1) The observations $x_1, x_2, \cdots, x_n$ are independent in the statistical sense. A minimum requirement would be that a plot of these values in some meaningful order (e.g., run numbers, days, instruments, operators) does not show pattern, trend, or grouping.

(2) The observations are approximately normally distributed. A convenient way to check the validity of this assumption is to plot the observations on normal-probability paper and to see if these observations fall approximately on a straight line. (If the distribution is skewed, i.e., not symmetrical with respect to the mean, the probability level, $(1-\alpha)$, could be seriously in error.)

The confidence interval *for the mean* is computed as:

$$\bar{x}_n - t_{(1-\alpha),\nu}\left(\frac{s_\nu}{\sqrt{n}}\right) \text{ to } \bar{x}_n + t_{(1-\alpha),\nu}\left(\frac{s_\nu}{\sqrt{n}}\right). \tag{7.13}$$

Sometimes $\bar{x}_n \pm t_{(1-\alpha)\nu}\,(s_\nu/\sqrt{n})$ is used to express this interval and the endpoints are termed confidence limits.

From the above expression it is seen that "n" is associated with $\bar{x}_n$ and the denominator $\sqrt{n}$, but "ν" is associated with $t_{(1-\alpha),\nu}$ and s_ν. Since various sets of measurements could have the same precision, the standard deviations calculated from these sets can be pooled together to yield a better estimate of σ.[20] The pooled standard deviation, s_p, could be used with any average to form the confidence interval provided the average is computed from a set of data derived from the same measuring process.

If duplicate determinations are made on each of k ampoules of different but similar activities, then pooled standard deviations from the k pairs of duplicates can be computed by a simplified formula as:

$$s_p^2 = \frac{1}{2k}\sum_1^k d_i^2 \tag{7.14}$$

[20] In general, the pooled standard deviation can be computed from k sets of data as:

$$s_p^2 = (\nu_1 s_1^2 + \nu_2 s_2^2 + \cdots + \nu_k s_k^2) \,/\, (\nu_1 + \nu_2 + \cdots + \nu_k)$$

with $\nu = \nu_1 + \nu_2 + \cdots + \nu_k$ degrees of freedom.

where d_i is the difference of the duplicate readings of the ith ampoule, and s_p has k degrees of freedom. Then the confidence interval for the mean activity of the ith ampoule based on the duplicate determinations is:

$$\bar{x}_i - t_{(1-\alpha),k}\left(\frac{s_p}{\sqrt{2}}\right) \text{ to } \bar{x}_i + t_{(1-\alpha),k}\left(\frac{s_p}{\sqrt{2}}\right) \quad (7.15)$$

where $\nu = k$, but $n = 2$.

Since t is a constant and s is a number that converges to σ as a limit, the confidence interval would have zero width with n sufficiently large. That is to say, for a large number of measurements, $\bar{x}$ converges to μ, and the mean of the distribution is known almost exactly.

Many an experimenter feels uneasy with this situation where the confidence interval of his value shrinks toward an arbitrarily small interval while, in fact, he knows that his uncertainty is probably larger than these limits. This illusion, however, is probably due to any or all of the following causes:

(1) In precision measurements, it is typical that variability between sequences of measurements made on different occasions is larger than variability within a sequence of measurements. A sequence of measurements, $x_1, x_2, \cdots, x_n$, could depend on various environmental conditions, e.g., time, temperature, barometric pressure, the particular instrument setting, etc. It follows, then, that the first assumption has been violated because of correlation within sets of measurements resulting in (a) an underestimate of σ, and more important, (b) an erroneous use of the divisor $\sqrt{n}$, since n is the number of independent measurements.[21]

(2) The confidence interval relates only to random errors, and not to systematic errors. Some systematic errors may be small when compared to the standard deviation of a single measurement, s_x, and hence cannot be detected, but may show up when averages

[21] Two cases are of particular interest:

a. If σ_b^2 is the component of variance corresponding to between sequences of measurements, and σ_w^2 that within a sequence of measurements, then,

$$\sigma_{\bar{x}}^2 = \sigma_b^2 + \frac{\sigma_w^2}{n}$$

b. If successive measurements are correlated with average coefficient of correlation r, then,

$$\sigma_{\bar{x}}^2 = \frac{\sigma^2}{n}\{1 + (n-1)r\}.$$

(See page 58 of Brownlee, 1960.)

of a large number of measurements are compared. These systematic errors would then be a dominant part of the uncertainty.

(3) The homogeneity of samples measured is always a factor that must be considered, since any minor differences between samples can only add to the uncertainty of the value reported. Limits to the inhomogeneity that may be inherent in the samples can be constructed using the concept of statistical tolerance limits discussed below.

7.3.3 *Statistical Tolerance Interval for a Fixed Percentage of Population*

Frequently, a large number of units (ampoules, bottles) is prepared, of which only a fraction can be measured. From this sample of units examined, it is hoped to find out, first, whether these units are sufficiently homogeneous for the purpose intended, and secondly, if these tested units are acceptable, what can one say about the expected variability within the entire lot (population) of units.

For simplicity, assume that the measurement error is negligible in comparison with the variability of the lot of units, and that the number of units is large. Then, under the same two basic assumptions listed in Section 7.3.2, i.e., independence and normality, an interval can be constructed to cover at least p percent of all the units in the population with the desired probability $\gamma = (1 - \alpha)$. This statistical tolerance interval is computed from a random sample of n units as:

$$\bar{x}_n - Ks \text{ to } \bar{x}_n + Ks \tag{7.16}$$

where K is the tabulated value of the two-sided tolerance factor for the specified p and γ. Note that, when n is large and s approaches the standard deviation of the population, the tolerance interval approaches to an interval of constant width, as expected.

Some selected values of "t" and "K" for $\gamma = 1 - \alpha = 0.95$ and 0.99 are given in Table 20 for ready reference.

If both analytical (measurement) error and material variability are present, then standard deviations corresponding to each of the two sources of error have to be estimated from the data. In the simplest case, duplicate determinations can be made on n units selected at random from the lot. The standard deviation s_a due to analytical error can be computed from the replicate measurements as:

$$s_a^2 = \frac{1}{2n} \sum_1^n d_i^2 \tag{7.17}$$

TABLE 20—*t factors for confidence interval and K factors for tolerance interval.*[a]

Degrees of freedom	Confidence interval, t Factor[b] $1-\alpha$		Tolerance interval, K Factor[b]			
			$\gamma = 0.95$		$\gamma = 0.99$	
	0.95	0.99	p = 0.95	p = 0.99	p = 0.95	p = 0.99
1	12.706	63.657	37.674	48.430	188.491	242.300
2	4.303	9.925	9.916	12.861	22.401	29.055
3	3.182	5.841	6.370	8.299	11.150	14.527
4	2.776	4.604	5.079	6.634	7.855	10.260
5	2.571	4.032	4.414	5.775	6.345	8.301
6	2.447	3.707	4.007	5.248	5.488	7.187
7	2.365	3.499	3.732	4.891	4.936	6.468
8	2.306	3.355	3.532	4.631	4.550	5.966
9	2.262	3.250	3.379	4.433	4.265	5.594
10	2.228	3.169	3.259	4.277	4.045	5.308
15	2.131	2.947	2.903	3.812	3.421	4.492
20	2.086	2.845	2.723	3.577	3.121	4.100
25	2.060	2.787	2.612	3.432	2.941	3.865
30	2.042	2.750	2.536	3.332	2.820	3.706
60	2.000	2.660	2.329	3.061	2.500	3.285
∞	1.960	2.576	1.960	2.576	1.960	2.576

[a] Adapted from Tables A-4 and A-6, Natrella (1963).

[b] The t and K factors are for "two-sided" limits based on a normal distribution. The term "two-sided" means that both over and under limits from the average are of interest. Some tables give "one-sided" factors, in which case the use of $\gamma/2$ instead of α yields the same result. (Such t-factor tables are sometimes called "Percentiles of the t distribution.")

where d_i represents the difference between duplicate determinations of the ith pair, and that for the material variability, s_{m}, as:

$$s_{\mathrm{m}}^2 = s^2 - \frac{s_{\mathrm{a}}^2}{2} \tag{7.18}$$

where s is the standard deviation computed from the n averages of duplicates.

The value of $\bar{x}_n$ may be used in two distinct ways:

(a) to characterize the true mean of the whole lot, or

(b) to characterize the bulk (e.g., 95 or 99 percent) of measured values of units from the lot.

In the first case, the concept of confidence interval applies, i.e., if n samples are taken at random repeatedly from the lot, are measured, and $\bar{x}_n$ computed, $\bar{x}_n \pm ts/\sqrt{n}$ is expected to include the true mean of the lot at a specified confidence level, for example, 95 percent.

In the second case, the concept of tolerance interval applies, i.e., if all units were measured, then the interval $\bar{x}_n \pm Ks$ is expected to include at least p percent of all such units, say 99 percent, with confidence level of 95 percent. In either case, the number of degrees of freedom associated with s is $n - 1$, since s is computed from n values.

It is to be noted that by measuring more units, the confidence interval will decrease in size proportional to $\sqrt{n}$, but the tolerance limits will approach a constant width, representing the variability inherent in the units within the lot.

The separation of variability into its components corresponding to the various sources of possible error (e.g., the model used for ^{90}Sr activities given in Section 7.2) demands careful planning and arrangement of the measurements schedule, and can be facilitated by statistical design of experiments.

7.3.4 *Formulae for the Propagation of Errors*

Frequently one cannot measure the quantity of interest directly, but must derive the value of this quantity as a function of other measurable quantities. If W is the quantity of interest, and X and Y are the quantities to be measured, then the problem is to determine the error in the computed value of W that is due to the measurement errors in $\bar{x}$ and $\bar{y}$.

The results in the measurements of X and Y can be expressed by their averages $\bar{x}$ and $\bar{y}$, and standard deviations of these averages $s_{\bar{x}}$ and $s_{\bar{y}}$, respectively. If the estimate of W is computed from:

$$\hat{W} = F(\bar{x}, \bar{y}), \tag{7.19}$$

then the variance of $\hat{W}$ is approximately:

$$\text{var } \hat{W} = \left[\frac{\partial F}{\partial X}\right]^2 \frac{\sigma_{\bar{x}}^2}{n} + \left[\frac{\partial F}{\partial Y}\right]^2 \frac{\sigma_{\bar{y}}^2}{n} + 2\left[\frac{\partial F}{\partial X}\right]\left[\frac{\partial F}{\partial Y}\right]\frac{\sigma_{\bar{x}\bar{y}}}{n} \tag{7.20}$$

where the partial derivatives in square brackets are to be evaluated at $\bar{x}$ and $\bar{y}$. If $\bar{x}$ and $\bar{y}$ are statistically independent, $\sigma_{\bar{x}y} = 0$, and, therefore the last terms equals zero. If $\bar{x}$ and $\bar{y}$ are measured in pairs, $s_{\bar{x}\bar{y}}$ can be used as an estimate of $\sigma_{\bar{x}\bar{y}}$. Also $\sigma_{\bar{x}}^2$ and $\sigma_{\bar{y}}^2$ are not known in practice and $s_{\bar{x}}^2$ and $s_{\bar{y}}^2$ are usually used in the formula in their place.

The above formula is called the propagation of error formula and is based on the first order Taylor series approximation of the function form. Formulae for some simple functions are shown in Table 21.

In these formulae, if

(a) the partial derivatives when evaluated at the averages are small, and

(b) $\sigma_{\bar{x}}$ and $\sigma_{\bar{y}}$ are small compared to $\bar{x}$ and $\bar{y}$,

then the approximations are good and $\hat{W}$ tends to be distributed normally with var $\hat{W}$ given by Eq. 7.20.

TABLE 21—*Propagation of error formulae for some simple functions*[a]

Function form of $\hat{w}$[b]	Approx. formula for var ($\hat{w}$) (x and y are assumed to be statistically independent)	Term to be added if x and y are correlated, and a reliable estimate of $\sigma_{\bar{x}\bar{y}}$, $s_{\bar{x}\bar{y}}$, can be assumed
$A\bar{x} + B\bar{y}$	$A^2 s_{\bar{x}}^2 + B^2 s_{\bar{y}}^2$	$2AB s_{\bar{x}\bar{y}}$
$\left(\frac{\bar{x}}{\sigma_{\bar{x}}^2} + \frac{\bar{y}}{\sigma_{\bar{y}}^2}\right) \Big/ \left(\frac{1}{\sigma_{\bar{x}}^2} + \frac{1}{\sigma_{\bar{y}}^2}\right)$ [c]	$1 \Big/ \left(\frac{1}{\sigma_{\bar{x}}^2} + \frac{1}{\sigma_{\bar{y}}^2}\right)$	
$\frac{\bar{x}}{\bar{y}}$	$\left(\frac{\bar{x}}{\bar{y}}\right)^2 \left(\frac{s_{\bar{x}}^2}{\bar{x}^2} + \frac{s_{\bar{y}}^2}{\bar{y}^2}\right)$	$\left(\frac{\bar{x}}{\bar{y}}\right)^2 \left(-2\frac{s_{\bar{x}\bar{y}}}{\bar{x}\bar{y}}\right)$
$\frac{1}{\bar{y}}$	$\frac{s_{\bar{y}}^2}{\bar{y}^4}$	
$\frac{\bar{x}}{\bar{x} + \bar{y}}$	$\left(\frac{\hat{w}}{\bar{x}}\right)^4 (\bar{y}^2 s_{\bar{x}}^2 + \bar{x}^2 s_{\bar{y}}^2)$	$\left(\frac{\hat{w}}{\bar{x}}\right)^4 (-2\bar{x}\bar{y} s_{\bar{x}\bar{y}})$
$\frac{\bar{x}}{1 + \bar{x}}$	$\frac{s_{\bar{x}}^2}{(1 + \bar{x})^4}$	
$\bar{x}\bar{y}$	$(\bar{x}\bar{y})^2 \left(\frac{s_{\bar{x}}^2}{\bar{x}^2} + \frac{s_{\bar{y}}^2}{\bar{y}^2}\right)$	$(\bar{x}\bar{y})^2 \left(2\frac{s_{\bar{x}\bar{y}}}{\bar{x}\bar{y}}\right)$
$\bar{x}^2$	$4\bar{x}^2 s_{\bar{x}}^2$	
$\sqrt{\bar{x}}$	$\frac{1}{4}\frac{s_{\bar{x}}^2}{\bar{x}}$	
$\ln \bar{x}$	$\frac{s_{\bar{x}}^2}{\bar{x}^2}$	
$k\bar{x}^a\bar{y}^b$	$\hat{w}^2 \left(a^2 \frac{s_{\bar{x}}^2}{\bar{x}^2} + b^2 \frac{s_{\bar{y}}^2}{\bar{y}^2}\right)$	$(\hat{w})^2 \left(2ab\frac{s_{\bar{x}\bar{y}}}{\bar{x}\bar{y}}\right)$
$e^{\bar{x}}$ [d]	$e^{2\bar{x}} s_{\bar{x}}^2$	
$\frac{\sin\left(\frac{\bar{x}+\bar{y}}{2}\right)}{\sin\frac{\bar{x}}{2}}$	$\frac{1}{4}\left\{\frac{\cos^2\left(\frac{\bar{x}+\bar{y}}{2}\right)}{\sin^2\frac{\bar{x}}{2}} s_{\bar{y}}^2 + \frac{\sin^2\frac{\bar{y}}{2}}{\sin^4\frac{\bar{x}}{2}} s_{\bar{x}}^2\right\}$ ($s_{\bar{x}}$ and $s_{\bar{y}}$ in radians)	$-\frac{\sin\frac{\bar{y}}{2}\cos\left(\frac{\bar{x}+\bar{y}}{2}\right)}{2\sin^3\frac{\bar{x}}{2}} s_{\bar{x}\bar{y}}$
$100\frac{s_{\bar{x}}}{\bar{x}}$ (= coefficient of variation in percent)	$\frac{\hat{w}^2}{2(n-1)}$ (not directly derived from the formulae)[e]	

[a] From Ku (1966).

[b] It is assumed that the value of $\hat{w}$ is finite and real, e.g., $\bar{y} \neq 0$ for ratios with $\bar{y}$ as denominator, $\bar{x} > 0$ for $\sqrt{\bar{x}}$ and $\ln \bar{x}$.

[c] Weighted mean as a special case of $A\bar{x} + B\bar{y}$, with σ_x and σ_y considered known.

[d] Distribution of $\hat{w}$ is highly skewed and normal approximation could be seriously in error for small n.

[e] See, for example, *Statistical Theory with Engineering Applications*, p. 301, by A. Hald (John Wiley & Sons, New York, N.Y., 1952).

7.3.5 *Linear Relationships and Calibration Curves*

By use of the arithmetic mean of n measurements as an estimate of the population mean, in fact, a constant has been fitted to the data by the method of least squares, i.e., a value $\hat{m}$ has been selected for μ such that

$$\sum_{1}^{n} (y_i - \hat{m})^2 = \sum_{1}^{n} d_i^2 \quad (7.21)$$

is a minimum. The solution is $\hat{m} = \bar{y}$. The deviations $d_i = y_i - \hat{m} = y_i - \bar{y}$ are called residuals.

The measurements can be expressed in the form of a statistical model

$$y = \mu + \epsilon, \quad (7.22)$$

where y stands for the observed values, μ the population mean (a constant), and ϵ the random error (normal) of measurement with a population mean, μ_ϵ, equal to zero and with a standard deviation σ. It follows that

$$\mu_y = \mu + \mu_\epsilon = \mu \quad (7.23)$$

and

$$\sigma_y^2 = \sigma^2.$$

The method of least squares requires us to use that estimator $\hat{m}$ for μ such that the sum of squares of the residuals is a minimum. As a corollary, the method also states that the sum of squares of residuals divided by the number of measurements n less the number of fitting constants, k, will give us an estimate of σ^2, i.e.,

$$s^2 = \frac{\Sigma(y_i - \hat{m})^2}{n - k} \quad (7.24)$$

It is seen that for $k = 1$, the above agrees with the definition of s^2.

Suppose Y, the quantity measured, exhibits a linear functional relationship with a variable that can be controlled accurately, then a model can be written as:

$$y = a + bX + \epsilon, \quad (7.25)$$

where, as before, y is the quantity measured, a (the intercept) and b (the slope) are two constants to be estimated, and ϵ the measurement error with a mean equal to zero, and variance σ^2 not depending on X. X is set at x_i, and y_i is observed. For example, y_i might be the natural logarithm of counting rates observed in a series of measurements on n successive days designated as x_i. In this case the quantity of interest is the decay constant, λ, which is the slope of the line.

For any estimates of a and b, say $\hat{a}$ and $\hat{b}$, a value $\hat{y}_i$ can be computed for each x_i, thus,

$$\hat{y}_i = \hat{a} + \hat{b}x_i.$$

If the sum of squares of the residuals

$$\sum_{i=1}^{n} (y_i - \hat{y}_i)^2$$

are to be a minimum, then it can be shown that

$$\hat{b} = \frac{\sum_{i=1}^{n} (x_i - \bar{x})(y_i - \bar{y})}{\sum_{i=1}^{n} (x_i - \bar{x})^2} \tag{7.26}$$

and

$$\hat{a} = \bar{y} - \hat{b}\bar{x} \tag{7.27}$$

The variance of Y can be estimated by

$$s^2 = \frac{\Sigma (y_i - \hat{y}_i)^2}{n - 2}, \tag{7.28}$$

with $n - 2$ degrees of freedom since two constants have been estimated from the data.

The standard errors of $\hat{b}$ and $\hat{a}$ are respectively estimated by $s_{\hat{b}}$ and $s_{\hat{a}}$, where

$$s_{\hat{b}}^2 = \frac{s^2}{\Sigma (x_i - \bar{x})^2}, \tag{7.29}$$

$$s_{\hat{a}}^2 = s^2 \left[\frac{1}{n} + \frac{\bar{x}^2}{\Sigma (x_i - \bar{x})^2} \right]. \tag{7.30}$$

With these estimates and the degrees of freedom associated with s^2, confidence limits can be computed for $\hat{a}$ and $\hat{b}$ for the confidence coefficient selected if it is assumed that errors are normally distributed.

Thus, the lower and upper limits of a and b, respectively, are:

$$\begin{array}{ll} \hat{a} - ts_{\hat{a}}, & \hat{a} + ts_{\hat{a}} \\ \hat{b} - ts_{\hat{b}}, & \hat{b} + ts_{\hat{b}} \end{array} \tag{7.31}$$

for the value of t corresponding to the degree of freedom and the selected confidence coefficient.

There are many variations to the problem of fitting a function to a set of data, viz., curvilinear relationships, weighted least squares, nonlinear least squares, etc. For a comprehensive treatment on these subjects, Draper and Smith (1968) is one of the best references available.

7.4 Statistical Experiment Design[22]

The definition of an experiment as "a considered course of action aimed at answering one or more carefully framed questions" was coined by the late W. J. Youden, an expert in the theory and practice of the design of experiments. Eleven of his papers on designs useful in physical sciences, including inter-laboratory tests, are reprinted in Ku (1969).

There are certain characteristics an experiment must have in order to accomplish the defined objective. These essential requisites are:

(1) As far as possible, the effects of the factors selected for study should not be obscured by other variables. The use of an appropriate experimental pattern helps to free the comparisons of interest from the effects of variables not under experimental study, and simplifies the analysis of results.

(2) As far as possible, the experiment should be free from bias. The effects of known variables may be balanced out by planned grouping and effects of unsuspected variables may be given a chance to play their parts through randomization.

(3) An experiment should provide a measure of precision. Replicated measurements provide a measure of precision. The standard deviation computed from replicated results may be compared to that of counting statistics to find out whether the precision of experiment can be improved. Planned grouping may also aid to trace the source of undesired or excess variability.

Since each experiment has its own individual characteristics and constraints, consultation with an experienced statistician who understands the problem is perhaps the prudent way to achieving an optimal design.

The Youden Diagram, or Plot, is given below as an illustration of a simple, but highly effective, procedure used in interlaboratory tests. The name "Youden Plot" is used universally to specify not only the plotting technique, but also the experimental procedure for sampling the respective performance of the laboratories through the results obtained on paired test items. Suppose the accuracy of radiation measurement detectors used in standard laboratories or hospitals is to be checked; two samples, A and B, are prepared at different levels of activity but reasonably close to the level of interest. Identical samples of A and B are sent to each laboratory and the activities of the pair are to be measured as unknowns. The results reported by the participating laboratories are used to prepare a graph.

[22] See also Chapter 11 of Natrella (1963).

In Figure 100, the value of A is plotted in the horizontal direction and that of B in the vertical direction. Thus each laboratory is represented by one point. The known values of A and B are used as the origin, and lines through the origin parallel to the x- and y-axes divide the graph into four quadrants as shown. The points that fall in the four quadrants are, respectively,

Q_1: high in A, high in B
Q_2: high in A, low in B
Q_3: low in A, low in B
Q_4: low in A, high in B.

If there are no systematic biases among the laboratories, all points should scatter about the center and divide equally among the four quadrants. In practice, however, this ideal picture is never realized on the first try, but can be achieved approximately after various causes of troubles are removed. A majority of points usually clusters about the 45° line drawn through the origin, and for extremely precise measurements, such as those on the voltage of standard cells, the points fall almost exactly on the line. An ellipse with the major axis on the 45° line can be constructed to enclose practically all the points in most cases, indicating the existence of some systematic errors among laboratories.

Laboratory L_1 will receive a copy of the diagram with a circle around its point. Immediately, this laboratory knows that its values are unreasonably high as compared with the standards, and that corrective measures have to be taken. L_3 is assured that its detector is in order,

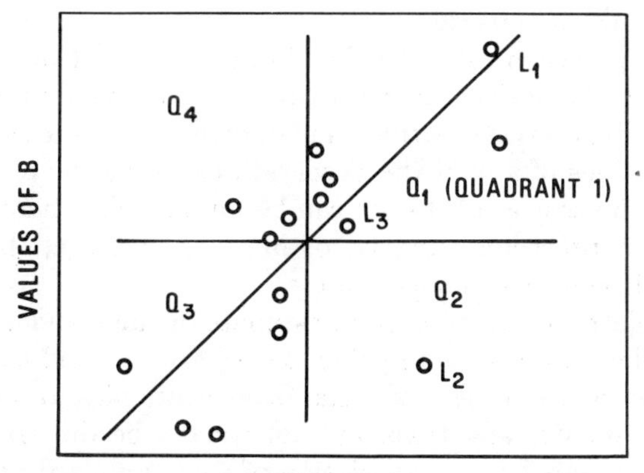

Fig. 100. A Youden Diagram. Each point represents values of samples A and B obtained by a laboratory. The results of 15 laboratories are shown here, divided into four quadrants.

whereas L_2 has to improve its measurement procedure and locate the causes of its inconsistent value.

The paper and rubber industries have adopted this procedure since 1969 for routine surveillance of their laboratory measurements procedures. The samples are prepared at NBS and computerized reports are distributed monthly to more than 100 laboratories. For simplicity in concept and execution, and for ready and meaningful interpretation, this interlaboratory test design cannot be surpassed. The original paper by Youden was published in *Industrial Quality Control,* 1959, and reprinted as paper 3.1 in Ku (1969).

Many of Youden's other designs are presented in papers reprinted in the same NBS Special Publication (Ku, 1969), including one on "Comparison of four national radium standards", and one on "Design and statistical procedures for the evaluation of an automatic gamma-ray point-source calibration", (with W. B. Mann and S. B. Garfinkel). A quick and economical procedure in locating sources of trouble, e.g., to check on seven sources in eight determinations, is given in Youden (1963).

7.5 Precision, Systematic Error, and Uncertainty

In the above sections, the terms precision and systematic error have been used, without specific definitions as to their meanings. This section brings into focus the meanings of these two terms and clarifies their connotations in usage.

The term precision refers to the closeness of a set of observations (data) among themselves, being a sample of a conceivably large sequence of such observations that can be generated by the same process. Individual values of any of the observations cannot be predicted, yet collectively the scatter of the set can be numerically characterized by a measure of precision (strictly, of imprecision), e.g., the standard deviation σ if known, or its estimate s.

The calculated value of s, as an estimate of the standard deviation σ, is a function of the sample values $x_1, x_2, \ldots, x_n$, and therefore is a statistic. For repeated sets of data, each containing n values, these calculated standard deviations will not usually be the same, but will vary as individuals from a population of such standard deviations so generated, and their properties may be predicted.

Systematic error, on the other hand, cannot be estimated or calculated from the data themselves, as each and every datum point is affected by the systematic error in the same way. The magnitude and direction of the systematic error remain the same in the average of these data points as in each individual point, and are not reduced by taking the average, as is the case with standard deviations, where the standard deviation of the average is $1/n$ times the standard deviation of individual values.

If the instrument used is suspected to be a source of systematic error, the experiment may be repeated with another instrument. If the confidence intervals constructed about the two results do not overlap, the presence of a systematic error in one instrument with respect to the other is indicated.

If the second instrument used is a standard and known to be accurate, then an estimate can be made for the systematic error of the first instrument as the difference between the two averages. If the two instruments are claimed to be equally good, the further course of action is not clear. The instruments may have to be recalibrated to see what is the systematic error in each instrument.

Suppose a number of instruments is available. Then the results obtained by using these instruments may be construed as a sample from a population of such results obtainable with the use of such instruments. Accordingly, a standard deviation can be computed from these results and a statistical tolerance interval used to give bounds to the uncertainty in the result of any one of the instruments.

What has been said about instruments applies equally well to other factors involved in any experiment: operators, reagents, time between sampling and analysis, size of sample, etc. Obviously a definitive program to establish the effects of all these factors is out of the question for most laboratories in view of the time and cost involved. Nevertheless, statistical designs are available that use a minimum number of experimental observations to check on the gross effects of these factors. Preferably, these experiments are incorporated in the course of regular routine so that if all of these effects are negligible, no effort is wasted. Youden (1963) gives an example in which a design is suggested to check seven factors each at two levels using only eight determinations.

An immediate extension of the above example would be to different laboratories, then to different test methods, to different national standards, and finally to fundamental physical constants. It follows then that the use of the terms "precision" and "systematic error" must be conditioned by the frame of reference, and these terms cannot be defined in a single way applicable to all situations. The precision of a test method [called "reproducibility" by some ASTM committees (American Society for Testing Materials)] used by many laboratories must include in its measure of variability a component due to differences among laboratories, as well as the component due to factors within each individual laboratory (called repeatability by some ASTM committees). Thus, a systematic error, peculiar to a laboratory, behaves as if it were a random error when the results of two test methods used by all of a large number of qualified laboratories are under investigation. When there are only a few laboratories, systematic differences among them should be treated as systematic error while an effort is made to remove the cause.

Proceeding in the opposite direction, the random measurement error on a particular standard calibrated by a national laboratory becomes a systematic error for the laboratory receiving the standard, to be used as if the assigned value were exact and correct. Since all related measurements will inherit the same deviation attached to that standard, the uncertainty given for the standard is a systematic error common to all the laboratory's results, although perhaps of negligible magnitude.

If a measure of the magnitude of the systematic error can be obtained or verified by an experiment, this should always be done as indicated in Section 7.4. If time and cost prohibit such an experiment, then some reasonable bounds to the systematic error will have to be estimated through experience or judgment. Since different experimenters will use different criteria, there is no way to give guidelines for such estimates. Moreover, the combined effect of two or more such sources of error may exceed the sum total of individual effects. For a detailed discussion on systematic errors, see Eisenhart (1963), reprinted in Ku (1969).

The uncertainty of a reported value refers to its likely inaccuracy in terms of credible limits, and combines both (1) components based on data that are amenable to statistical treatments and (2) components due to systematic errors that cannot be treated statistically. Depending on the circumstance in which the value is likely to be used, the uncertainty statement should be phrased accordingly so that it will be understandable and useful to the user. For example, in ICRU Report 12 (ICRU, 1968) the overall uncertainty is recommended to be expressed as $\pm\ (ts_{\bar{x}} + \delta)$ where $ts_{\bar{x}}$ is the confidence interval for the mean at a specified probability level, and δ the linear sum of estimated maximum values of conceivable systematic errors. Eisenhart (1968) provides further guidelines on the expression of the uncertainties of final results, reprinted in Ku (1969).

8. The Assurance of Accuracy and Precision in National and International Measurements of Radioactivity

8.1 General

In a sense, this is the motherhood and apple-pie (MAP) section of this report. But here it is called Measurements Assurance Programs. The ability to reproduce, whether with the help of genes or gauges, is important in all fields of human endeavor, but uniformity is as often mistaken for accuracy as it is for excellence in apple pie. To create and reproduce and then, from that base, further to create is the key to all human progress, and measurement is the foundation on which all scientific progress to the full comprehension of life and the universe is built. Cooperative endeavor and progress in practically every field of science depend on the ability to communicate experience in terms of common measure and mutual comprehension between scientific research groups throughout the world. Diverse systems of units can impede such a free exchange of ideas and information. The fact that the gallon changes in size as one goes to Canada from the United States is a little confusing, as anyone who has poured a Canadian half pint of milk into (and overflowing) a U.S. half-pint glass will, perhaps with embarrassment, appreciate. To remove impediments to world trade, standards of length and mass were defined, often different in different parts of the world, but eventually related to each other by intercomparative measurement. In science, the metric system, based on the defined meter, kilogram, and second, became the common denominator for the normalization of scientific data. And because these were defined standards, they were also absolute because they, in themselves, had no error and errors arose only in comparing other lengths, masses, or intervals of time with them. On the economic side of the world's trade, money became the measure of effort that went into the production of a kilogram of butter, or a square meter of woven cloth. It is not just by chance that in many countries the standardizing laboratories were set up under the aegis of departments or ministries of trade or commerce.

In the case of radioactivity, early standards consisted of masses of pure radium in a salt of that element, often radium chloride or radium bromide, with the mass of radium ranging from about 1 to 100 mg. The international radium standards were kept at the Bureau International des Poids et Mesures (BIPM) at Sèvres. In 1959, however, it was agreed that the relative values of a set of radium standards prepared by O. Hönigschmid in 1934 were so well known that the whole system of Hönigschmid radium standards distributed in many countries of the world could be considered to be the international standard.

In 1910 the *curie* was defined as the quantity of radon that is in equilibrium with one gram of radium. A curie of radon consists of about 0.65 ml of gas at 0 °C and 760 torr (760 mm Hg, 0.1013 MPa). Subsequently, measurements were made to determine the alpha-particle activity of a curie of radon and the first measurements yielded a result of $3.7 \times 10^{10}\ s^{-1}$. Later values tended to give a value for the curie of around $3.608 \times 10^{10}\ s^{-1}$. The best current value is $3.658 \times 10^{10}\ s^{-1}$.

With the advent of nuclear energy, it became necessary to have an activity standard applicable to any radionuclide. Accordingly in 1950, at meetings of the Joint Commission on Standards, Units, and Constants of Radioactivity of the International Unions of Pure and Applied Chemistry and Physics and at the Sixth International Congress of Radiology, it was decided to redefine the curie to be applicable to all radionuclides at a value of exactly 3.700×10^{10} transitions per second. The millicurie and the microcurie are, respectively, 3.700×10^{7} and $3.700 \times 10^{4}\ s^{-1}$.

In 1975 a new unit of radioactivity was introduced to be consistent with the Système International d'Unités (SI) (BIPM, 1973; NBS, 1977), in which the unit of choice is in terms of one kilogram, one meter, or one second. In the case of radioactive transitions, the unit of activity was named the becquerel (Bq), which is an activity of one transition per second.

In many branches of technology, however, the curie, or multiples thereof, has become so well established in the last thirty years that it is unlikely to be immediately replaced. Thus, in medicine, the millicurie is as commonplace to the nuclear-medicine practitioner as a cup of sugar is to the housewife. And in the environment, picocurie per liter seems to be becoming as well known to the lay public as it is to those who measure radioactivity around nuclear power plants or nuclear fuel processing plants. There is also a considerable literature based on the curie that has been produced over many years in several fields of science.

8.2 The Role of Standards in the Measurement of Radioactivity

The primary role of a radioactivity standard, or of any standard, is to enable the results of measurements to be communicated. That the median lethal absorbed dose of gamma rays is around 450 rad for man is information that can be comprehended by all concerned.

This point was emphasized in the opening scientific paper at the Herceg Novi, Yugoslavia, International Summer School (Cavallo *et al.*, 1973), as follows:

> "As in other fields of scientific endeavor, radioactivity standards are needed to relate observed effects in terms of a well-defined and known stimulus. Thus, if a known flux density of alpha particles, defined in terms of the standards of length and time, of known energy, defined in terms of the standards of mass, length, and time, bombard a surface of beryllium, how many neutrons are emitted in unit time per unit volume? For a given alpha-particle energy, and knowing the number of beryllium atoms per unit volume, this experiment reduces to a cross-section measurement for the reaction $^{9}Be(\alpha,n)^{12}C$. Through the use of standards, a cross-section obtained in Vinča can be compared with one measured in Chalk River, without the necessity of moving equipment, or people, from one place to another. Biological and physiological effects in response to a given applied radioactivity can also be related, one to another, through standards of that activity that are used to calibrate the measuring equipment used by the different investigators. But all of this implies a continuing availability of radioactivity standards both nationally and internationally. By and large, standards of most radionuclides of interest can be produced, but to keep a continuous supply available at all times, including standards of the shorter-lived activities is beyond the production capacity of the national and commercial laboratories in most countries. Therefore, throughout the world, the need for radioactivity standards has been met by calibrated instruments of greater or lesser degrees of complexity, all of which are normalized to the national or international radioactivity measurements systems by calibrations, by

> means of radioactivity standards, over a wide spectrum of radionuclides with widely different modes of decay. Thus, the national standard of ^{131}I in the U.S.A. is, for the periods between recalibrations, a "4π"γ ionization chamber and a sealed radium reference source at NBS, and any laboratory that wishes to check the calibration of its own measuring equipment for ^{131}I can, at any time, send a solution of that radionuclide to NBS for assay. In turn, NBS has assayed ^{131}I distributed internationally in order to be assured that the national system conforms with the international system."

It is of interest to note that, in the early days of investigating the properties of, and effects caused by, newly discovered radionuclides at the Radiation Laboratory at Berkeley, the most popular instrument used for measuring relative radioactivity in the late 1930's was the Lauritsen electroscope. Thus, the biological effects observed when a radionuclide was administered to experimental animals were often related to an activity measured as a drift rate of the electroscope, in divisions per second, for the given radionuclide in a specified geometry relative to the electroscope. The half lives of the newly discovered radionuclides were invariably measured with such instruments. The Lauritsen electroscope, the principle of which is incorporated into many pen-type dosimeters, is still a relatively inexpensive, accurate, and rugged asset to a radioactivity measurements laboratory (see Mann and Seliger, 1958; Price, 1964).

Radioactivity standards are ephemeral; unlike many other standards, they are not lumps of, say, platinum that can be stored in some safe place for future use. If a calibration error is made in the assay of, say, ^{99m}Tc or ^{198}Au, administered to a patient, there may be no evidence of this months later, in the event of legal or regulatory proceedings.

In the United States we are trying to maintain traceability of the National Radioactivity Measurements System (NRMS) to the National Bureau of Standards (NBS). For this purpose, laboratory standards prepared both by the Environmental Monitoring and Support Laboratory of the Environmental Protection Agency, Las Vegas (EPA), and by the Idaho Falls Radiological and Environmental Services Laboratory of the Department of Energy (DOE), under contract to the Nuclear Regulatory Commission (NRC), have been measured and certified by NBS as being in agreement with the NBS measurements to within, say, 1 percent or 3 percent, and so forth. The laboratory in question can then state that its measurements were traceable to within 1 percent or 3 percent to NBS and, hence, are consistent with the National Radioactivity Measurements System, for that particular radionuclide at that time.

Another method of traceability is for NBS to distribute calibrated sources of undisclosed value for measurement by key laboratories in NRMS. Such distributions take place regularly to EPA and NRC, to medical and hospital laboratories in a program organized by the College of American Pathologists (CAP), and to industrial laboratories in a research-associate program organized by the Atomic Industrial Forum (AIF) and NBS (Collé, 1975). In this NBS-AIF program NBS produces, monthly, microcurie- and millicurie-level radioactivity standards for distribution to seven manufacturers of radiopharmaceuticals and to the Food and Drug Administration. The values of the "unknowns" are, however, withheld from the participants until they demonstrate the agreement of their measurements with those of NBS, thereby achieving traceability. These values can then be used to effect any necessary corrections to the calibrations of instruments, because they are NBS certified values.

This kind of traceability is, however, *implicit* in that it establishes only the ability of a laboratory to assay a given radionuclide, or perhaps a small fraction of the many different radionuclides that it may be handling. In the long term it is the competence and reliability of a laboratory and the skill of its staff that comprise the traceability of a laboratory. Such implicit traceability also exists when a laboratory submits one or more of its standards, of stated activity or emission rate, to a key laboratory for testing.

For standards, an *explicit* form of traceability to NBS can be established when a distributing laboratory prepares a batch of radioactivity standards, from one master solution, and submits several randomly selected samples to NBS for confirmation or verification.

If a laboratory engaged in the assay of radioactivity uses NBS standards to calibrate its measuring systems, this does not necessarily, in the NBS view, constitute traceability. Only when the laboratory can measure the activity of an unknown sample and obtain a value agreeing with the NBS value, or that of another specified key laboratory in NRMS, within a certain specified range of error, is traceability achieved. This can be achieved by using a national standard, a commercial standard, or a directly-calibrated "in-house" standard.

Results obtained in one such distribution of "unknown" solution samples of ^{125}I in a "round robin" organized by the College of American Pathologists and NBS are shown in Figure 101. Further results are discussed for other CAP "round robins" by Cavallo *et al.* (1973) and by Hauser (1974). Results of an intercomparison of a multiple-energy radionuclide gamma-ray "unknown" measured by seven laboratories concerned with the monitoring of radioactive effluent from a nuclear

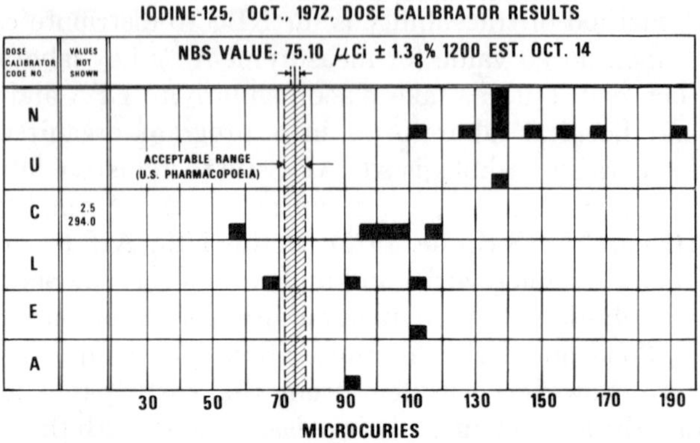

IODINE-125, OCT., 1972, DOSE CALIBRATOR RESULTS

NBS VALUE: 75.10 μCi ± 1.3₈% 1200 EST. OCT. 14

DOSE CALIBRATOR CODE NO | VALUES NOT SHOWN

N
U
C 2.5 294.0
L
E
A

ACCEPTABLE RANGE (U.S. PHARMACOPOEIA)

30 50 70 90 110 130 150 170 190

MICROCURIES

Fig. 101. Results of assays, using different dose calibrators (Section 6.4.3.2.3), of "unknown" NBS samples of ^{125}I (each having an activity of 75.10 μCi), distributed to various hospitals and medical laboratories in 1972. It is interesting to note that none of the results lies in the range of ±5 percent required by the U.S. Pharmacopoeia; that the measurements range from 2.5 to 294.0 μCi; and that results, obtained by 10 different laboratories using dose calibrators "N", are distributed around a value some 100 percent too high, which indicates an error in the "radionuclide" dial setting.

power station are also given in Cavallo *et al.* (1973), and further relevant programs have been described by Coursey *et al.* (1975).

On the other hand, it is incumbent upon the national standardizing laboratories to maintain traceability to the International Radioactivity Measurements System (IRMS) and this is attempted in a small way by international "round robins" organized by the Bureau International des Poids et Mesures (BIPM) or by bilateral intercomparisons between two or more national laboratories. BIPM and IAEA have also installed "4π"γ pressure ionization chambers for the purpose of monitoring the agreement between gamma-ray-emitting solution standards of activity prepared by different national laboratories and the consistency, over the years, of such standards produced by a given laboratory. Standards can be submitted to BIPM and IAEA for monitoring, in standard glass ampoules (Section 4.4.7). In the process, BIPM and IAEA should gradually build up very excellent calibrations, for photon-emission-rate response as a function of energy for their own "4π"γ chambers that would enable them, in turn, to calibrate solution standards for countries requiring a standard of any particular gamma-ray-emitting radionuclide.

Figure 102 illustrates how the national and international radioactivity measurements systems can be envisaged as being traceable to and consistent with each other. The national laboratories maintain traceability links with BIPM and with each other, while each of the national

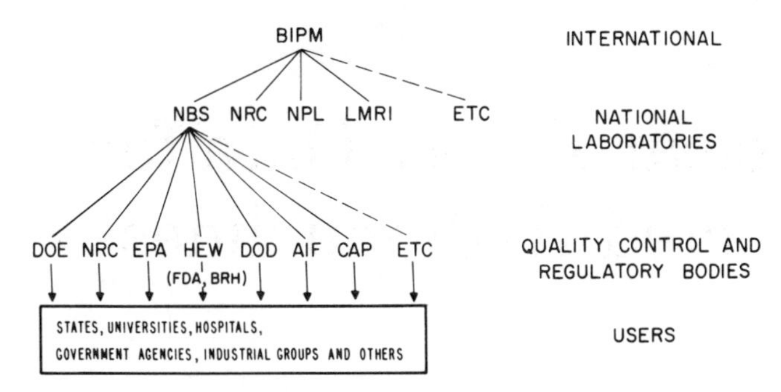

Fig. 102. The international and national radioactivity measurements traceability "tree". Most of the acronyms used in this figure have been explained in the text, but at the national standardizing laboratory level, on the national level, NRC refers to the National Research Council of Canada. On the regulatory level, NRC refers to the U.S. Nuclear Regulatory Commission.

laboratories will presumably seek to maintain traceability with the radioactivity measurements laboratories in its own jurisdiction. In the United States the NBS, in the second generation of the genealogical tree, will seek to establish traceability between itself and key laboratories in the third generation, as shown in Figure 102, namely NRC, EPA, and so forth. With an industrial production of radionuclides, including radiopharmaceuticals, approaching $100 million annually in the United States, to say nothing of radioactive waste and effluents from the use of nuclear energy, it is clearly quite impossible for traceability to be maintained, in the measurement of radioactivity, *directly* between NBS and every laboratory measuring radionuclides. NBS can only attempt to maintain traceability to the third generation organizations, but would expect occasionally to penetrate to the fourth generation of measurements laboratories, and beyond if necessary, to confirm that traceability continues to be maintained across the generation gaps.

The National Bureau of Standards seeks, therefore, to maintain strong links of traceability, renewed as frequently as possible, with every one of the key laboratories of the "Quality Control and Regulatory Bodies" shown in the third-generation level of Figure 102.

Appendix A

Nuclear-Decay Data for Selected Radionuclides

A.1 Description of the Tabulated Information[23]

A.1.1 *Introduction*

Contained in this appendix are tabulations of the atomic and nuclear radiations emitted by some 200 radioactive nuclides.[24] The nuclei included comprise most of those currently of interest in medical practice or research, health physics, industry, nuclear power, environmental impact studies, and as reference standards. Listed in tabular form are recommended values for half-lives, energies, intensities (probabilities per decay), and equilibrium absorbed-dose constants (discussed in Section A.1.4.4) for each of the atomic and nuclear radiations emitted by these radioactive atoms. The data contained here supersede those in earlier publications by Martin and Blichert-Toft (1970), and Martin (1973, 1976). Only data available before July 1978 have been used in the present evaluations.

[23] The NCRP is indebted to M. J. Martin for this contribution and for the data prepared from the Evaluated Nuclear Structure Data File of the Oak Ridge Nuclear Data Project, with support from the U.S. Department of Energy by contract to the Union Carbide Corporation. The Nuclear Data Project is comprised of W. B. Ewbank (Director); F. E. Bertrand (Deputy Director); D. J. Horen (former Director); R. L. Auble, Y. A. Ellis, R. L. Haese, B. Harmatz, H. J. Kim, D. C. Kocher, M. J. Martin, and M. R. Schmorak; and technical assistants S. J. Ball, S. H. Dockery, F. W. Hurley, M. R. McGinnis, and J. T. Miller.

[24] Many of the symbols in this Appendix, including those in the table, are nonstandard (thus D rather than d for day, etc). Their form is dictated by the symbols available on the computer print train used to prepare the table. See also Appendix A.1.4.3.

Space limitations do not permit the inclusion of references for the nuclear data appearing in this table. If documentation or the most recent nuclear data are required, they can be obtained by writing to the Nuclear Data Project.[25]

A.1.2 *Review of Radiation Processes*

A.1.2.1 *Alpha-Particle Transitions (α).* In α decay, an atom, atomic number Z and mass number A, emits an α particle (^{4}He nucleus, $Z = 2$, $A = 4$) and decays to an atom with atomic number $Z - 2$ and mass number $A - 4$, and two surplus electrons.

For α decay between ground states of parent and daughter atoms, the maximum energy available for the α particle is,

$$E_\alpha = Q_\alpha - E_\mathrm{R}$$

where Q_α is the difference, in energy units, between the mass of the parent atom and the sum of the masses of the daughter atom and a helium atom. E_R is the recoil energy of the daughter atom which, if M and M_α are the masses of the daughter atom and the α particle, respectively, is given by

$$E_\mathrm{R} = Q_\alpha \frac{M_\alpha}{M_\alpha + M}$$

The energy available will be increased or decreased, respectively, if α decay is from an excited state of the parent or to an excited state of the daughter nucleus.

A.1.2.2 *Beta-Particle Transitions (β).*

A.1.2.2.1 *β^- Decay.* In β^- decay, an antineutrino ($\bar{\nu}$) and a negative electron (β^-) are emitted from the nucleus as a result of the process $\mathrm{n} \rightarrow \mathrm{p} + \beta^- + \bar{\nu}$. The decay increases the nuclear charge by one unit.

The energy released in a single β transition is divided between the β particle and the antineutrino in a statistical manner so that, when a large number of transitions is considered, both the antineutrinos and β particles have energy distributions extending from zero up to some maximum value. For decay to a particular energy level, E_L, in the daughter nucleus, the maximum energy available is $E_\mathrm{max} = Q^- - E_\mathrm{L}$, where Q^- is the atomic mass difference, expressed in energy units,

[25] Oak Ridge National Laboratory, Post Office Box X, Oak Ridge, Tennessee 37830.

between ground states of parent and daughter nuclides. The average energy of a β particle in this transition is given by

$$E_{av} = \int_0^{E_{max}} EN(E)dE \Big/ \int_0^{E_{max}} N(E)dE$$

where $N(E)$ is the number of β particles with energy between E and $E + dE$. Thus, the β^- particles from a particular transition i, denoted in the main table by β^-i, constitute a "group" having a continuous distribution of energies with a definite E_{max} and E_{av}.

A.1.2.2.2 *β^+ Decay.* In β^+ decay, a neutrino (ν) and a positive electron (β^+) (positron) are emitted from the nucleus as a result of the process $p \rightarrow n + \beta^+ + \nu$. This decay decreases the nuclear charge by one unit.

As in β^- decay (see above), the β^+ particles emitted in a transition to a particular level in the daughter nucleus have a continuous distribution of energies with a definite E_{max} and E_{av}. For positron decay, $E_{max} = Q^+ - 2mc^2$, where Q^+ is the atomic mass difference, expressed in energy units, between ground states of parent and daughter nuclides. Note that β^+ decay to a particular energy level, E_L, cannot occur unless $Q^+ - E_L > 2mc^2$ ($2mc^2 = 1022$ keV).

A.1.2.2.3 *Electron-Capture Decay (ϵ).* In electron-capture decay, an atomic electron is captured by a nucleus and a neutrino is emitted as a result of the process $p + e^- \rightarrow n + \nu$. This decay decreases the nuclear charge by one unit and leaves the daughter nucleus with a vacancy in one of its atomic shells. *K*-shell electron capture, for example, refers to a capture process where the final-state vacancy is in the *K*-shell. For *K*-shell electron-capture to a particular energy level, E_L, the energy released is given by $Q^+ - E_L - E_K$, where E_K is the *K*-shell binding energy in the daughter atom.

The electron-capture process always competes with β^+ decay, but also can occur when the transition energy is too small to allow β^+ emission, that is, when $Q^+ - E_L < 1022$ keV.

A.1.2.3 *Gamma-Ray Transitions (γ).* Electromagnetic radiation is emitted by a nucleus in a transition from a higher- to a lower-energy state. A numeric index on γ, as in $\gamma 1$, denotes a γ ray emitted in the transition between a particular pair of nuclear levels. The energy of this γ ray, $E(\gamma 1)$, is equal to the energy difference between the two levels (except for a usually negligible amount of nuclear recoil energy, E_r (in keV) $\simeq 5.4 \times 10^{-7} E^2(\gamma)/A$, where A is the mass number and $E(\gamma)$ is in keV).

A.1.2.4 *Internal-Conversion-Electron Transitions (ce).* An atomic electron can be emitted as an alternative to γ-ray emission in the transition of a nucleus from a higher- to a lower-energy state. In the internal-conversion process, the energy difference between these states is transferred directly to a bound atomic electron which is ejected from the atom. A letter hyphenated to "ce" refers to the shell from which the atomic electron is ejected. Thus, ce-K denotes K-shell conversion and ce-MNO denotes conversion in the M, N, and O shells combined. A numeric index, as in ce-K-1, has a meaning analogous to that used above for γ rays. For a transition with energy $E(\gamma 1)$, the K-shell internal-conversion electron is emitted with energy $E(\text{ce-}K\text{-}1) = E(\gamma 1) - E_K$, where E_K is the K-shell binding energy. The energy of an electron converted in one of the other shells is given by this same expression with the replacement of E_K by the binding energy appropriate to that shell.

For a particular transition, the ratio of the probability for emission of a K-conversion electron to that for emission of a γ ray, $I(\text{ce-}K)/I(\gamma)$, is called the K-shell conversion coefficient for that transition. Conversion coefficients for the other shells are defined in an analogous manner.

A.1.2.5 *X-Ray and Auger-Electron Transitions (X, e_A).* Whenever a vacancy is produced in an inner electron shell of an atom, the filling of this vacancy is accompanied by the emission of either an x ray (X) or Auger electron (e_A). Vacancies created by the filling of the initial vacancy will in turn produce further x rays or Auger electrons. This cascade of radiations continues until all vacancies have been transferred to the outermost electron shell. Inner-shell vacancies are always produced in two types of nuclear decay, electron capture and internal conversion. Other processes that produce electron vacancies following a nuclear decay, such as electron shakeoff (the ejection of one or more atomic electrons due to a sudden change in nuclear charge) or the ejection of atomic electrons by escaping α or β particles, are not discussed here because of the low probability of their occurrence.

A.1.2.6 *Fluorescence Yield (ω).* The fluorescence yield for a particular atomic shell (ω_K, $\bar{\omega}_L$, etc.) is defined as the probability that a vacancy in that shell will result in the emission of an x ray. If n_K is the number of K-shell vacancies per transition, the number of K x rays will be $n_K\omega_K$, and the number of K Auger electrons will be $n_K(1 - \omega_K)$.

The most recent and thorough review of fluorescence yields is that of Bambynek *et al.* (1972). The K-shell fluorescence yields, ω_K, adopted in the present work are the "fitted values" from Table III · IV of the above reference. The average L-shell fluorescence yields, $\bar{\omega}_L$, with uncertainties, have been estimated by Martin (1976) from Fig. 4-34 of this same review (Bambynek *et al.*, 1972).

A.1.2.7 *X Rays*

A.1.2.7.1 X_K. An x ray emitted as the result of the filling of a K-shell vacancy by an electron from a higher shell, for example the Y shell, has an energy $E_K - E_Y$, where E_K and E_Y are the electron binding energies in the K and Y shells, respectively. This transition can be denoted by $K - Y$. In order of decreasing intensity, the most important transitions are

$$K_{\alpha_1} = K - L_3 \qquad K_{\beta_3} = K - M_2$$
$$K_{\alpha_2} = K - L_2 \qquad K_{\beta_4} = K - N_2$$
$$K_{\beta_1} = K - M_3 \qquad K_{\beta_5} = K - M_4$$
$$K_{\beta_2} = K - N_3$$

Energies and intensities for the K_{α_1} and K_{α_2} x rays and the $K_\beta = \Sigma_i K_{\beta i}$ group are given explicitly. The x-ray energies are taken from Bearden (1967), and the intensity ratios K_β/K_α and $K_{\alpha_2}/K_{\alpha_1}$ are taken from Rao *et al.* (1972).

As already mentioned, the number of K x rays per disintegration is $n_K\omega_K$. Thus, the number of K_{α_1} x rays is $n_K\omega_K/(1 + K_\beta/K_\alpha)(1 + K_{\alpha_2}/K_{\alpha_1})$ with similar expressions for K_{α_2} and K_β. The number of K-shell vacancies per disintegration is the sum of the vacancies produced by K capture and those produced by internal conversion in the K shell. Thus, $n_K = \epsilon_K + I(\text{ce-}K)$, where ϵ_K is the total number of K-capture events per disintegration, and $I(\text{ce-}K)$ is the total number of K-shell internal-conversion electrons per disintegration.

A.1.2.7.2 X_L. As in the case of the K shell, many transitions contribute to the L x-ray spectrum. However, since the relative intensities of these transitions are not known for all Z values, and since the energy differences between the strong transitions are small ($\lesssim$3 keV for $Z \lesssim 92$), the total L x radiation is treated here as a single group having the energy of the strongest transition, $L_{\alpha_1} = L_3 - M_5$. For $Z < 37$, the M_5 transitions have not been resolved experimentally from the M_4 transitions. In these cases, the energy given is that for the transition $L_{\alpha_1,\alpha_2} = L_3 - M_4, M_5$. Figures 4-8 to 4-11 of Bambynek *et al.* (1972) show examples of L x-ray spectra.

The calculation of the number of L x rays per disintegration, $n_L\omega_L$, is similar to that for K x rays except that, in addition to vacancies produced by direct L-shell capture and by L-shell conversion, those created by the transfer of L-shell electrons to K-shell vacancies must also be included. Thus,

$$n_L = \epsilon_L + I(\text{ce-}L) + n_{KL}n_K,$$

where n_{KL} is the number of L-shell vacancies created per K-shell vacancy, and the other symbols have meanings analogous to those used above for the K shell. Values of n_{KL} are taken from Bambynek *et al.* (1972).

A.1.2.8 *Auger Electrons (e_A).* The Auger process competes with the emission of x rays as a means of carrying off the energy released in the filling of an inner-shell vacancy by an electron from an outer shell. In the Auger process, the filling of an inner-shell vacancy is accompanied by the simultaneous ejection to the continuum of an outer-shell electron. The resulting atom is left with two vacancies. The Auger-electron yield per disintegration of a particular atomic shell is $n_K(1 - \omega_K)$, $n_L(1 - \overline{\omega}_L)$, etc. If the original vacancy is in the K shell and if this vacancy is filled by an electron from shell X with the ejection of an electron from shell Y, the transition is denoted by KXY. The energy of the ejected electron is $E_K - E_X - E'_Y$ where E_K and E_X are neutral atom K- and X-shell binding energies, respectively, and E'_Y is the binding energy of a Y-shell electron in an atom containing a vacancy in the X shell. The most intense K-Auger transitions are of the type KLL. Since the relative intensities of the electrons in the KLL group are not accurately known for all Z values and since the energy difference between the transitions is small ($\leq$5 keV for $Z \lesssim 92$), the K-Auger electrons are treated here as a single group having the energy of the strong transition, KL_2L_3. The energy of this transition as a function of atomic number is taken from Table 1 of Bergstrom and Nordling (1965). A K-Auger-electron spectrum is shown in Fig. 3-5 of Bambynek *et al.* (1972).

In the case of the L-Auger process, very little is known about the energies or relative intensities of the individual L-Auger electrons. For this compilation, the L-Auger electrons are treated as a single group having the energy appropriate to an $L_3M_4M_5$ transition. An example of an L-Auger electron spectrum is shown in Fig. 4-13 of Bambynek *et al.* (1972).

A.1.2.9 *Bremsstrahlung.* In addition to any monoenergetic x rays or γ rays that may be present in radioactive decay, every β or electron-capture decay produces continuous electromagnetic radiation called bremsstrahlung. Two processes contribute to this continuous spectrum:

1. "External" bremsstrahlung is produced by collisions between β's or conversion electrons and the atoms of the material surrounding the radiating atoms. Since the intensity of the external bremsstrahlung depends on the atomic number, Z, of the surrounding material, this radiation is not included in the present tables. In

many cases this radiation cannot be neglected in dose calculations. An approximate value for the average energy of the external bremsstrahlung associated with a β group of maximum energy E_β (in keV) is given, for the thick-target approximation, by Evans (1955) as

$$E_{av} \approx 1.4 \times 10^{-7}\, Z\, E_\beta^{\,2} \text{ keV per } \beta.$$

In the decay of ^{32}P, where $E_\beta = 1711$ keV, the average energy per disintegration of the external bremsstrahlung is $\approx 0.4\, Z$ keV;

2. "Internal" bremsstrahlung, originating within a decaying atom, is produced by the sudden change of nuclear charge which occurs in β^+, β^-, or ϵ decay. This radiation is not included here because of its low intensity and low average energy. The average energy of the internal bremsstrahlung associated with a β group of maximum energy E_β (in keV) is given, for $E_\beta \gg mc^2$ (511 keV) by Schopper (1966, pp. 76–84) as

$$E_{av} \approx 1.5 \times 10^{-3}\, E_\beta \log\,(0.004\, E_\beta - 2.2) \text{ keV per } \beta.$$

In electron capture, the corresponding expression for a transition with energy E_ϵ (in keV) is (Schopper, 1966, pp. 76–84)

$$E_{av} \approx 1.5 \times 10^{-7}\, E_\epsilon^{\,2} \text{ keV per capture.}$$

In the decay of ^{32}P, where $E_\beta = 1711$ keV, the average internal bremsstrahlung energy per disintegration is ≈ 1.7 keV.

A.1.3 *Computer Program, MEDLIST*[26]

Based on the radioactive decay data sets in the Evaluated Nuclear Structure Data File and computerized tabulations of the relevant Z-dependent constants (ω_K, n_{KL}, etc.) discussed in the preceding section, the computer program, MEDLIST, calculates the energies and intensities of the atomic radiations (x-ray and Auger-electron transitions). The program then combines these radiations with the nuclear radiations, sorts them according to radiation type (i.e., α, β, γ, ce $\cdots$), and, within each type, arranges and numerically labels them in order of increasing energy.

Uncertainties in all experimental quantities, including the atomic constants, are consistently carried through the calculations. A 3 per-

[26] Written by W. B. Ewbank, Nuclear Data Project, and M. J. Kowalski, Great Lakes College Association (Oak Ridge Science Semester participant from Ohio Wesleyan University).

cent uncertainty is assigned to all theoretical conversion coefficients and is propagated, within MEDLIST, along with the experimental uncertainties.

A variable low-intensity cutoff limit is built into MEDLIST. For the present tables this intensity limit is 0.1 percent. Immediately following the listing of each of the radiation types α, β, and γ, MEDLIST prints a comment giving the number and the summed intensity of the radiations omitted because of the chosen cutoff limit, provided that this summed intensity is greater than 0.01 percent.

It should be noted that the numerical labels on the γ radiations are based on all the radiations contained in the original data set on which MEDLIST operates. These labels are maintained throughout the MEDLIST calculations and are carried into the output. In particular, when low-intensity radiations are omitted, the remaining radiations are not, in the γ-ray listing, relabeled. Thus, in the γ-ray listing, there may appear the sequence $\gamma1$, $\gamma2$, $\gamma4$, $\cdots$, which tells the reader that $\gamma3$, with energy between that of $\gamma2$ and $\gamma4$, has been omitted because its intensity is below the cutoff limit. Note that there could still appear a conversion line associated with $\gamma3$, say ce-K-3, if its conversion coefficient is large enough such that the intensity of ce-K-3 exceeds the cutoff limit.

For those nuclei that decay by positron emission, MEDLIST prints a comment at the end of the γ-ray list, giving the maximum possible observable intensity of the annihilation radiation ($\gamma\pm$), calculated from $I(\gamma\pm) = 2\ \Sigma_i I(\beta_i^+)$. The actual fraction of these events observed in practice depends upon source attenuation and the detector efficiency.

A.1.4 *Explanation of Table of Nuclear-Decay Data*

A.1.4.1 *Policies Followed.* This table contains data on the following atomic and nuclear radiations: Auger (K, L shells); x ray (L, K_{α_1}, K_{α_2}, K_β); β^-; β^+; α; γ; and ce (K, L, $\cdots$ shells) emitted with intensity >0.1 per 100 disintegrations of the parent nucleus (i.e., >0.1 percent). Following the listing of α, β, and γ radiations for each data set is a comment stating the number and total intensity of the observed radiations omitted because of this intensity cutoff.

Auger electrons and x rays resulting from the filling of vacancies in the M, N, $\cdots$ shells, although sometimes occurring with large intensities, are not included. These transitions have energies ranging from $\lesssim$2 keV in ^{197}Hg decay to $\lesssim$0.6 keV in ^{113}Sn decay. Their intensities, however, are not, in general, calculable since the contribution to the total number of, say, M-shell vacancies due to initial K- or L-shell vacancies has not been measured. However, theoretical estimates of some of these quantities are available for limited Z regions. For ^{83m}Kr, based on theoretical values of the required atomic constants (Bambynek *et al.*, 1972), the yield of M-shell Auger transitions (with energy ≈0.2 keV) is estimated to be ≈300 percent. As an additional example, for ^{125}I decay, where M-shell vacancies are produced both in electron-capture decay and the subsequent internal conversion process, the yield of M-shell Auger transitions is estimated to be ≈590 percent. The total energy emitted per decay by these low-energy Auger transitions (≈0.5 keV) is ≈3 keV.

The radiations are listed in the first column by type. For an explanation of the notation employed see SYMBOLS AND DEFINITIONS (Section A.1.4.3). Particle radiations (Auger, ce, α, β) are given first, followed by the electromagnetic radiations (x ray, γ). In the case of β decay, whenever more than one group is present, a separate entry (designated "total β") is made for the composite spectrum.

The intensity given for the "total β" entry is the sum of the intensities of the individual groups. For β^- emitters with no alternate mode of decay, this intensity must be exactly 100 percent. However, it should be noted that since the intensities of the individual β^- groups are usually determined from the γ-ray intensities, which in turn are usually determined by the requirement that the total $(\beta^- + \gamma)$ feeding of the daughter ground state be 100 percent, the intensity of the total β^- contribution, as calculated from the individual β^- intensities, does not necessarily always come to exactly 100 percent. This summed value could be replaced by 100 percent, but this has not been done in the tables.

Columns 2 and 3 contain, respectively, the energy (in keV) and intensity (particles or photons per 100 disintegrations of the parent nucleus). For the β groups, both the maximum and the average energies are given.

In column 4, Δ, the mean energy emitted per nuclear decay, is tabulated in the units g rad/μCi h. For a discussion of Δ and its units, see Appendix A.1.4.4.

The nuclei are listed in order of increasing mass number, A. For nuclei with the same A, the order is by increasing Z.

Each decay mode of a given radionuclide is listed separately. In those cases where a nucleus decays by more than one mode, a cross reference is given to the alternate mode or modes. For example, ^{64}Cu decays by both β^- and β^+ branches. A separate listing is given for each of these decay modes. Following the heading "64CU B− DECAY" is the comment "SEE ALSO 64CU B+ DECAY" and following the heading "64CU B+ DECAY" is the comment "SEE ALSO 64CU B− DECAY". As another example, 154-d ^{121}Te decays by both electron-capture (EC) and isomeric transition (IT) branches. Following the heading "121TE EC DECAY" is the comment "SEE ALSO 121TE IT DECAY", and conversely. In such cases, the radiations in each data list should simply be combined to obtain all the radiations from the parent under consideration. In those cases where a parent decays to one or more radioactive descendants, the parent and each descendant are treated separately. If one of the daughters is an isomer, then for that isomer data listing, a comment is made giving the percentage feeding from the immediate parent. For example, ^{99}Mo decays to ^{99m}Tc. Following the heading "99MO B− DECAY" is the comment "SEE ALSO 99TC IT DECAY (6.02 H)" and following the heading "99TC IT DECAY" is the comment "FEEDING of 6.02-H 99TC in 99MO DECAY = 87.52% 18". To include correctly all the contributions of such radioactive chains, the user must combine the radiations from each member of the chain using the standard equations for parent-daughter activity relations discussed in Section A.1.4.2 and the feeding of each member of the chain (given in the table only for isomers; otherwise feeding is 100 percent).

No separate entry is made in the γ-ray listing for the 511-keV annihilation radiation (γ±) accompanying positron decay. Instead, the intensity of this radiation, calculated from $I(\gamma\pm) = 2 \times \Sigma_i\ I(\beta_i^+)$, is given in a comment at the end of each data set. This intensity is presented separately, and as a "maximum" value, since the annihilation process takes place external to the decaying atom in the medium surrounding the atom. The fraction of the maximum that will be of importance in a given application therefore depends upon the extent of the region surrounding the decaying atom that is being considered in that application.

The following policy on the use of "B+" and "EC" in decay type headings should be noted. "EC" as in "··· EC DECAY" is used to denote either EC or EC + B+ (if positrons are energetically allowed). "B+" as in "··· B+ DECAY" is also sometimes used, especially if the positron decay dominates electron capture, or if "B+ DECAY" is a more common descriptor of a certain nucleus. Thus, the use of "EC" as a heading does not preclude the possibility of positron decay.

A.1.4.2 *Parent-Daughter Activity Ratios.* A thorough discussion of the general case of activity relations among members of a radioactive decay chain is contained in Evans (1955). For the special case of a radioactive parent with one radioactive daughter, fed in a fraction, f, of the parent decays, the ratio of daughter activity to that of the parent (both at time t) is given by

$$\frac{f\,\lambda_d}{\lambda_d - \lambda_p}\left[1 - e^{-(\lambda_d - \lambda_p)t}\right], \qquad \text{(A.1)}$$

where λ_p and λ_d are the decay constants ($\lambda = \ln 2/T_{1/2}$) of the parent and daughter nuclides, respectively. The activity of the daughter is taken to be zero at time $t = 0$ (i.e., the parent source is initially pure).

A common case occurring in these tables is that in which the daughter half-life is shorter than the parent half-life. In such cases, the ratio of daughter-to-parent activities first increases with time and then approaches a constant value (see Figure 2 of this report). For time, t, large compared with $1/(\lambda_d - \lambda_p)$, Eq. (A1) reduces to

$$\frac{f\,\lambda_d}{\lambda_d - \lambda_p} = \frac{f\,T_{1/2,p}}{T_{1/2,p} - T_{1/2,d}} \qquad (A.2)$$

An example of this case is the decay of ^{99}Mo to ^{99m}Tc where $T_{1/2,p} = 66.0$ h 2, $T_{1/2,d} = 6.02$ h 3, and f = 0.8752 18. For large t ($t > 10\ T_{1/2,d}$ is sufficient here), the ratio of ^{99m}Tc activity to that of ^{99}Mo is, from Eq. A.2 (0.8752 18) × (1.1004 4) = 0.9631 20. Thus, to correctly account for all the radiations from a ^{99}Mo source, the radiations from ^{99m}Tc should be multiplied by 0.9631 20, and combined with those from ^{99}Mo.

Each nuclide entry in the table of nuclear-decay data (Appendix A.3) is characterized by atomic mass number, chemical symbol, and half life. The "m" which usually accompanies the mass number for decays that do not originate from the ground state, as in ^{99m}Tc and ^{110m}Ag, does not appear in the table.

A.1.4.3 *Symbols and Definitions*

Auger-K, Auger-L	*K*-shell Auger electron, *L*-shell Auger electron
avg	average
ce-K-1, ce-L-2, etc.	*K*-shell conversion electron for transition 1, *L*-shell conversion electron for transition 2, etc.
Ci	curie
D	day
H	hour
$K\alpha_1$, $K\alpha_2$, $K\beta$; L	*K* x rays; *L* x ray
M	minute
max	maximum
rad	unit of absorbed dose, equal to 6.25×10^7 MeV/g
S	second
Y	year (365.242 days)
$\alpha 1$, $\alpha 2$, ···	alpha particle corresponding to transition 1, 2, etc.

$\beta 1, \beta 2, \cdots$	beta particle corresponding to transition 1, 2, etc.
$\gamma 1, \gamma 2, \cdots$	gamma ray corresponding to transition 1, 2, etc.
$\gamma\pm$	annihilation radiation
Δ	equilibrium absorbed-dose constant (See Appendix A.1.4.4)
μCi-h	microcurie-hour
3.624 12	3.624 ± 0.012
3.6 H 12	3.6 ± 1.2 hours
2.1 E5 Y 2	$(2.1 \pm 0.2) \times 10^5$ years

A.1.4.4 *Mean Energy Emitted Per Nuclear Decay (Δ).* From a knowledge of the activity of a radiopharmaceutical administered to an individual, it is possible to estimate the absorbed dose to various organs of that individual. Let A_0 be the activity administered, and let f be the fraction of the activity taken up by some organ of a patient's body. Then $A_s = f A_0$ is the activity (at zero time) in the source organ. The mean absorbed dose, $\bar{D}$, to another organ (the target organ), due to the activity in the source organ, can be given in the form (Loevinger and Berman, 1976; ICRP, 1975):

$$\bar{D}(t \leftarrow s) = \tilde{A}_s \, \Sigma_i \, \Delta_i \, \phi_i(t \leftarrow s)/m_t. \tag{A.3}$$

In this equation $\tilde{A}_s = f \, \tilde{A}_0$ is the time integral of the activity in the source organ. $\phi_i(t \leftarrow s)$ is the fraction of the energy of the ith radiation, from the radionuclide in the source organ, that is imparted to the target organ, and is called the *absorbed fraction.* The mass of the target organ is m_t. (The quotient $\phi_i(t \leftarrow s)/m_t = \Phi_i(t \leftarrow s)$ is often called the specific absorbed fraction.)

Δ_i is the mean energy of the ith radiation emitted per nuclear decay, and can be expressed as

$$\Delta_i = k \, P_i E_i, \tag{A.4}$$

where P_i and E_i are the probability of emission per nuclear decay and the mean energy, respectively, of the ith radiation. The value of k is dependent upon the system of units used, and is equal to unity for the Système International d'Unités (SI). Then, using SI units

$$\Delta_i = P_i E_i, \tag{A.5}$$

where Δ has the units J/Bq s. (The special name of the SI unit of absorbed dose is the gray (Gy) equal to 1 J/kg.) Using traditional units, namely the microcurie for the unit of activity, the rad for absorbed dose, the kilo-electron-volt for the mean energy E_i, and the gram and hour for mass and time, respectively, Δ is given by

$$\Delta_i = 0.00213\, P_i E_i, \tag{A.6}$$

with the units g rad/μCi h. It should be noted that, while these conventional units are dimensionally correct, they are conceptually in error, since Δ is the energy *emitted* per nuclear decay and the rad is a unit of energy *imparted* per unit mass. Nevertheless, these traditional units are widely used in current practice, and are the units used in Appendix A.3 for tabulating Δ_i.

Radiations that deposit only a small fraction of the emitted energy outside the source organ are considered to be non-penetrating. In this event, the target and source organs coincide and the absorbed fraction ϕ is approximately unity. For larger organs, such as most of those in the human body, such non-penetrating radiations are typically photons of energies below 10 to 15 keV, electrons, and, of course, alpha particles. For non-penetrating radiations the mean absorbed dose can be calculated from the above equation if the administered activity A_0, the fractional uptake f, and the mass of the organ m are specified. Photons and electrons originating in very small organs, or photons of higher energy in all organs, should be considered penetrating, and it is necessary to obtain appropriate values of the absorbed fraction or the specific absorbed fraction. An extensive table of specific absorbed fractions, and references to tables of absorbed fractions, are given by the ICRP (1975).

Appendix A.2 List of Radionuclides
Appendix A.3 Table of Nuclear-Decay Data (as of July 1978)

A.2 List of Radionuclides

```
3H B- DECAY (12.35 Y 1)
11C B+ DECAY (20.38 M 2)
13N B+ DECAY (9.965 M 4)
14C B- DECAY (5730 Y 40)
15O B+ DECAY (122.24 S 16)
16N B- DECAY (7.12 S 2)
18F B+ DECAY (109.74 M 4)
22NA B+ DECAY (2.602 Y 2)
24NA B- DECAY (15.00 H 4)
26AL B+ DECAY (7.16E5 Y 32)
27MG B- DECAY (9.462 M 11)
28MG B- DECAY (20.91 H 3)
28AL B- DECAY (2.240 M 1)
32SI B- DECAY (330 Y 40)
32P B- DECAY (14.29 D 3)
33P B- DECAY (25.4 D 2)
35S B- DECAY (87.39 D 10)
38CL B- DECAY (37.21 M 4)
41AR B- DECAY (1.827 H 7)
42K B- DECAY (12.36 H 1)
43K B- DECAY (22.6 H 2)
44SC B+ DECAY (3.927 H 8)
44TI EC DECAY (47.3 Y 12)
45CA B- DECAY (163 D 1)
46SC B- DECAY (83.83 D 2)
46SC IT DECAY (18.67 S 9)
47CA B- DECAY (4.536 D 2)
47SC B- DECAY (3.351 D 2)
48V EC DECAY (15.974 D 3)
51CR EC DECAY (27.704 D 4)
52MN EC DECAY (5.591 D 3)
52MN IT DECAY (21.1 M 2)
52MN B+ DECAY (21.1 M 2)
52FE B+ DECAY (8.275 H 8)
54MN EC DECAY (312.5 D 5)
55FE EC DECAY (2.7 Y 1)
87SR EC DECAY (2.805 H 3)
87SR IT DECAY (2.805 H 3)
87Y EC DECAY (80.3 H 3)
88KR B- DECAY (2.84 H 2)
88RB B- DECAY (17.8 M 1)
88Y EC DECAY (106.64 D 8)
89RB B- DECAY (15.2 M 1)
89SR B- DECAY (50.5 D 1)
90SR B- DECAY (29.12 Y 24)
90Y B- DECAY (64.0 H 1)
94NB B- DECAY (2.03E4 Y 16)
94NB B- DECAY (6.26 M 1)
94NB IT DECAY (6.26 M 1)
97ZR B- DECAY (16.90 H 5)
97NB B- DECAY (72.1 M 7)
97NB IT DECAY (60 S 1)
97RU EC DECAY (2.9 D 1)
99MO B- DECAY (66.0 H 2)
99TC IT DECAY (6.02 H 3)
103RU B- DECAY (39.28 D 1)
103PD EC DECAY (16.96 D 2)
103RH IT DECAY (56.12 M 1)
108AG EC DECAY (2.37 M 1)
108AG B- DECAY (2.37 M 1)
108AG EC DECAY (127 Y 21)
108AG IT DECAY (127 Y )
109PD B- DECAY (13.427 H 14)
109CD EC DECAY (464 D 1)
110AG EC DECAY (24.6 S 2)
110AG B- DECAY (24.6 S 2)
110AG IT DECAY (249.9 D 1)
110AG B- DECAY (249.9 D 1)
111AG B- DECAY (7.45 D 1)
111IN EC DECAY (2.83 D 1)
113IN IT DECAY (1.658 H 1)
113SN EC DECAY (115.1 D 2)
135I B- DECAY (6.61 H 1)
135XE B- DECAY (9.09 H 1)
135XE IT DECAY (15.29 M 3)
135BA IT DECAY (28.7 H 2)
136CS B- DECAY (13.1 D 1)
137BA IT DECAY (2.552 M 2)
137CS B- DECAY (30.0 Y 2)
138XE B- DECAY (14.17 M 7)
138CS B- DECAY (32.2 M 1)
139BA B- DECAY (82.7 M 2)
139CE EC DECAY (137.66 D 2)
140BA B- DECAY (12.74 D 5)
140LA B- DECAY (40.272 H 7)
141CE B- DECAY (32.501 D 5)
142PR B- DECAY (19.13 H 4)
143CE B- DECAY (33.0 H 2)
143PR B- DECAY (13.56 D 2)
144PR B- DECAY (17.28 M 3)
145PM EC DECAY (17.7 Y 4)
147ND B- DECAY (10.98 D 1)
147PM B- DECAY (2.6234 Y 2)
148PM B- DECAY (5.37 D 1)
148PM IT DECAY (41.3 D 1)
148PM B- DECAY (41.3 D 1)
149ND B- DECAY (1.73 H 1)
149PM B- DECAY (53.08 H 5)
152EU EC DECAY (13.33 Y 4)
152EU B- DECAY (13.33 Y 4)
152EU EC DECAY (9.32 H 2)
152EU B- DECAY (9.32 H 2)
153SM B- DECAY (46.7 H 1)
153GD EC DECAY (242 D 1)
154EU B- DECAY (8.6 Y 1)
155EU B- DECAY (4.96 Y 1)
157TB EC DECAY (150 Y 30)
157DY EC DECAY (8.1 H 1)
```

```
57CO EC DECAY (270.9 D 6)
58CO B+ DECAY (70.80 D 8)
59FE B- DECAY (44.529 D 7)
60CO B- DECAY (5.271 Y 1)
63NI B- DECAY (96 Y 4)
64CU B+ DECAY (12.701 H 2)
64CU B- DECAY (12.701 H 2)
65ZN EC DECAY (243.9 D 1)
66GA B+ DECAY (9.40 H 7)
67CU B- DECAY (61.86 H 12)
67GA EC DECAY (78.26 H 3)
68GA B+ DECAY (68.0 M 2)
68GE EC DECAY (288 D 6)
69ZN B- DECAY (57 M 1)
69ZN IT DECAY (13.76 H 3)
72GA B- DECAY (14.1 H 2)
72AS EC DECAY (26.0 H 1)
73AS EC DECAY (80.30 D 6)
73SE EC DECAY (7.15 H 8)
74AS B+ DECAY (17.76 D 2)
74AS B- DECAY (17.78 D 3)
75SE EC DECAY (119.8 D 1)
76AS B- DECAY (26.32 H 7)
77GE B- DECAY (11.30 H 1)
77AS B- DECAY (38.8 H 3)
77BR EC DECAY (56 H 2)
79KR B+ DECAY (35.04 H 10)
81KR EC DECAY (2.1E5 Y 2)
82BR B- DECAY (35.30 H 3)
84RB B+ DECAY (32.77 D 4)
84RB B- DECAY (33.0 D 2)
85KR B- DECAY (10.72 Y 1)
85KR B- DECAY (4.48 H 1)
85KR IT DECAY (4.48 H 1)
85SR EC DECAY (64.84 D 3)
86RB B- DECAY (18.66 D 2)
86Y B+ DECAY (14.74 H 2)
86ZR EC DECAY (16.5 H 1)
87KR B- DECAY (76.3 M 5)
115IN B- DECAY (4.3 H 1)
116IN B- DECAY (54.15 M 6)
117SN IT DECAY (13.61 D 4)
117SB EC DECAY (2.80 H 1)
119SN IT DECAY (293.0 D 13)
121TE EC DECAY (17 D 1)
121TE EC DECAY (154 D 7)
121TE IT DECAY (154 D 7)
122SB EC DECAY (2.70 D 1)
122SB B- DECAY (2.70 D 1)
123I EC DECAY (13.2 H 1)
124SB B- DECAY (60.20 D 3)
124I B+ DECAY (4.18 D 2)
125I EC DECAY (60.14 D 11)
126I EC DECAY (13.02 D 7)
126I B- DECAY (13.02 D 7)
127XE EC DECAY (36.41 D 2)
129TE B- DECAY (69.6 M 2)
129TE IT DECAY (33.6 D 1)
129TE B- DECAY (33.6 D 1)
129I B- DECAY (1.57E7 Y 4)
129XE IT DECAY (8.0 D 2)
129CS EC DECAY (32.06 H 6)
130I B- DECAY (12.36 H 1)
131I B- DECAY (8.04 D 1)
131XE IT DECAY (11.9 D 1)
131CS EC DECAY (9.69 D 1)
131BA EC DECAY (11.8 D 2)
132TE B- DECAY (78.2 H 8)
132I B- DECAY (2.30 H 3)
132CS B+ DECAY (6.475 D 10)
132CS B- DECAY (6.475 D 10)
133I B- DECAY (20.8 H 1)
133XE B- DECAY (5.245 D 6)
133XE IT DECAY (2.19 D 1)
133BA EC DECAY (10.74 Y 5)
133BA IT DECAY (38.9 H 1)
134TE B- DECAY (41.8 M 8)
134I B- DECAY (52.6 M 5)
134CS B- DECAY (2.062 Y 5)
170TM B- DECAY (128.6 D 3)
171ER B- DECAY (7.52 H 3)
171TM B- DECAY (1.92 Y 1)
175YB B- DECAY (4.19 D 1)
177LU B- DECAY (6.71 D 1)
181HF B- DECAY (42.4 D 1)
181W EC DECAY (121.2 D 3)
182TA B- DECAY (115.0 D 2)
185W B- DECAY (75.1 D 3)
187W B- DECAY (23.9 H 1)
188W B- DECAY (69.4 D 5)
188RE B- DECAY (16.98 H 2)
191OS B- DECAY (15.4 D 2)
192IR EC DECAY (74.02 D 18)
192IR B- DECAY (74.02 D 18)
194IR B- DECAY (19.15 H 3)
195PT IT DECAY (4.02 D 1)
195AU EC DECAY (183 D 1)
195AU IT DECAY (30.5 S 2)
197PT B- DECAY (18.3 H 3)
197HG EC DECAY (64.1 H 1)
198AU B- DECAY (2.696 D 2)
199AU B- DECAY (3.139 D 7)
201TL EC DECAY (3.044 D 9)
203HG B- DECAY (46.60 D 2)
203PB EC DECAY (52.05 H 10)
206BI EC DECAY (6.243 D 3)
207BI EC DECAY (38 Y 3)
226RA A DECAY (1600 Y 7)
222RN A DECAY (3.8235 D 3)
218PO A DECAY (3.05 M )
214PB B- DECAY (26.8 M )
214BI B- DECAY (19.9 M 4)
214PO A DECAY (164.3 US 20)
210PB B- DECAY (22.3 Y 2)
210BI B- DECAY (5.012 D 9)
210PO A DECAY (138.38 D 1)
239NP B- DECAY (2.355 D 4)
242CM A DECAY (162.8 D 4)
```

A.3 Table of Nuclear-Decay Data (as of July 1978)

3H B- DECAY (12.35 Y 1) I(MIN) = 0.10%

Radiation Type	Energy (keV)	Intensity (%)	Δ(g-rad/ μCi-h)
β- 1 max	18.600 20		
avg	5.680 10	100	0.0121

11C B+ DECAY (20.38 M 2) I(MIN) = 0.10%

Radiation Type	Energy (keV)	Intensity (%)	Δ(g-rad/ μCi-h)
Auger-K	0.17	0.230 10	≈0
β+ 1 max	960.0 14		
avg	385.5 6	99.760 10	0.819

Maximum γ± intensity =199.52%

15O B+ DECAY (122.24 S 16) I(MIN) = 0.10%

Radiation Type	Energy (keV)	Intensity (%)	Δ(g-rad/ μCi-h)
Auger-K	0.38	0.107 6	≈0
β+ 1 max	1731.8 7		
avg	735.2 3	99.887 6	1.56

Maximum γ± intensity =199.77%

16N B- DECAY (7.12 S 2) I(MIN) = 0.10%

Radiation Type	Energy (keV)	Intensity (%)	Δ(g-rad/ μCi-h)
β- 1 max	1546.9 24		
avg	630.5 11	1.00 20	0.0134
β- 2 max	3299.9 24		
avg	1460.6 12	4.9 4	0.152
β- 3 max	4287.9 23		
avg	1941.0 12	68.0 20	2.81
β- 4 max	10418.6 23		
avg	4979.2 12	26.0 20	2.76
total β- avg	2695.0 15	100 3	5.74

13N B+ DECAY (9.965 M 4) I(MIN) = 0.10%

Radiation Type	Energy (keV)	Intensity (%)	Δ(g-rad/ μCi-h)
Auger-K	0.26	0.170 10	≈0
β+ 1 max	1198.5 9		
avg	491.8 4	99.820 10	1.05

Maximum γ± intensity =199.64%

14C B- DECAY (5730 Y 40) I(MIN) = 0.10%

Radiation Type	Energy (keV)	Intensity (%)	Δ(g-rad/ μCi-h)
β- 1 max	156.478 9		
avg	49.470 3	100	0.105

1 weak β's omitted (ΣIβ = 0.01%)

Radiation Type	Energy (keV)	Intensity (%)	Δ(g-rad/ μCi-h)
γ 3	1753.0 6	0.13 4	0.0049
γ 5	2741.0 6	0.76 15	0.0444
γ 8	6129.139 18	69.0 20	9.01
γ 10	7117.0 4	5.0 4	0.758

7 weak γ's omitted (ΣIγ = 0.16%)

18F B+ DECAY (109.74 M 4) I(MIN) = 0.10%

Radiation Type	Energy (keV)	Intensity (%)	Δ(g-rad/ μCi-h)
Auger-K	0.52	2.88 20	≈0
β+ 1 max	633.5 6		
avg	249.4 3	96.90 17	0.515

Maximum γ± intensity =193.80%

22NA B+ DECAY (2.602 Y 2) I(MIN) = 0.10%

Radiation Type	Energy (keV)	Intensity (%)	Δ (g-rad/ μCi-h)
Auger-K	0.82	9.20 5	0.0002
β+ 1 max	545.5 5		
avg	215.54 21	89.84 11	0.412
1 weak β's omitted (ΣIβ = 0.06%)			
K X-ray	0.84	0.12 4	≈0
γ 1	1274.540 20	99.940 20	2.71
Maximum γ± intensity = 179.80%			

24NA B- DECAY (15.00 H 4) I(MIN) = 0.10%

Radiation Type	Energy (keV)	Intensity (%)	Δ (g-rad/ μCi-h)
β- 1 max	1390.2 7		
avg	553.9 4	99.935 4	1.18
3 weak β's omitted (ΣIβ = 0.07%)			
γ 2	1368.53 5	100	2.91
γ 3	2754.09 5	99.863 5	5.86
4 weak γ's omitted (ΣIγ = 0.06%)			

28MG B- DECAY (20.91 H 3) I(MIN) = 0.10%

Radiation Type	Energy (keV)	Intensity (%)	Δ (g-rad/ μCi-h)
Auger-K	1.39	26 6	0.0008
ce-K- 1	29.080 20	27 7	0.0168
ce-L- 1	30.522 20	2.6 7	0.0017
β- 1 max	211.8 20		
avg	65.2 7	4.60 20	0.0064
β- 2 max	458.9 20		
avg	155.9 8	94.0 20	0.312
β- 3 max	859.6 20		
avg	319.3 9	1.40 20	0.0095
total β- avg	154.0 9	100.0 21	0.328
K X-ray	1.48	0.97 24	≈0
γ 1	30.640 20	66 4	0.0431
γ 2	400.690 20	36.6 10	0.312
γ 5	941.45 3	38.3 10	0.768
γ 8	1342.25 3	52.6 16	1.50
γ 9	1372.89 6	4.70 20	0.137
γ 10	1589.36 3	4.20 20	0.142
γ 11	1620.00 15	0.30 10	0.0104
4 weak γ's omitted (ΣIγ = 0.13%)			

26AL B+ DECAY (7.16E5 Y 32) I(MIN)= 0.10%

Radiation Type	Energy (keV)	Intensity (%)	Δ(g-rad/ μCi-h)
Auger-K	1.18	16.21 19	0.0004
β+ 1 max	1174.0 5		
avg	543.8 23	81.80 20	0.947
K X-ray	1.25	0.44 6	≈0
γ 1	1129.67 10	2.50 20	0.0602
γ 2	1808.65 7	99.76 4	3.84
γ 3	2938.24 13	0.240 20	0.0150

Maximum γ± intensity =163.60%

27MG B- DECAY (9.462 M 11) I(MIN)= 0.10%

Radiation Type	Energy (keV)	Intensity (%)	Δ(g-rad/ μCi-h)
β- 1 max	1594.8 12		
avg	645.8 5	29.0 4	0.399
β- 2 max	1765.5 12		
avg	724.3 6	71.0 4	1.10
total β-			
avg	701.5 6	100.0 6	1.49
γ 1	1701.686 15	0.80 10	0.0029
γ 2	843.76 3	71.8 4	1.29
γ 3	1014.44 4	28.0 4	0.605

28AL B- DECAY (2.240 M 1) I(MIN)= 0.10%

Radiation Type	Energy (keV)	Intensity (%)	Δ(g-rad/ μCi-h)
β- 1 max	2864.2 6		
avg	1247.2 3	99.990 10	2.66
γ 1	1778.85 3	100	3.79

32SI B- DECAY (330 Y 40) I(MIN)= 0.10%

Radiation Type	Energy (keV)	Intensity (%)	Δ(g-rad/ μCi-h)
β- 1 max	213 7		
avg	64.7 22	100	0.138

32P B- DECAY (14.29 D 3) I(MIN)= 0.10%

Radiation Type	Energy (keV)	Intensity (%)	Δ(g-rad/ μCi-h)
β- 1 max	1710.4 6		
avg	695.0 3	100	1.48

33P B- DECAY (25.4 D 2) I(MIN) = 0.10%

Radiation Type	Energy (keV)	Intensity (%)	Δ(g-rad/ μCi-h)
β- 1 max	249.0 20		
avg	76.6 6	100	0.163

35S B- DECAY (87.39 D 10) I(MIN) = 0.10%

Radiation Type	Energy (keV)	Intensity (%)	Δ(g-rad/ μCi-h)
β- 1 max	167.47 19		
avg	48.80 10	100	0.104

38CL B- DECAY (37.21 M 4) I(MIN) = 0.10%

Radiation Type	Energy (keV)	Intensity (%)	Δ(g-rad/ μCi-h)
β- 1 max	1107.2 9		
avg	420.40 10	32.5 6	0.291
β- 2 max	2749.6 9		
avg	1181.5 4	11.4 8	0.287

42K B- DECAY (12.36 H 1) I(MIN) = 0.10%

Radiation Type	Energy (keV)	Intensity (%)	Δ(g-rad/ μCi-h)
β- 1 max	1683.7 16		
avg	700.9 8	0.320 20	0.0048
β- 2 max	1996.4 16		
avg	822.3 8	17.5 5	0.307
β- 3 max	3521.1 16		
avg	1563.9 8	82.1 5	2.73
total β-			
avg	1429.8 9	100.0 7	3.05
2 weak β's omitted (ΣIβ = 0.12%)			
γ 1	312.75 3	0.319 17	0.0021
γ 6	1524.665 20	17.9 5	0.581
6 weak γ's omitted (ΣIγ = 0.14%)			

43K B- DECAY (22.6 H 2) I(MIN) = 0.10%

Radiation Type	Energy (keV)	Intensity (%)	Δ(g-rad/ μCi-h)
β- 1 max	422 10		
avg	137 4	2.24 9	0.0065
β- 2 max	827 10		
avg	298 4	92.1 13	0.585
β- 3 max	1224 10		
avg	469 5	3.7 4	0.0370
β- 4 max	1444 10		
avg	568 5	0.8 13	0.0097

Radiation Type		Energy (keV)	Intensity (%)	Δ(g-rad/ μCi-h)
β- 3	max	4917.2 9		
	avg	2244.1 4	56.0 5	2.68
total β-	avg	1529.5 4	99.9 12	3.25
γ 1		1642.42 6	32.5 6	1.14
γ 2		2167.51 5	44.0 5	2.03

1 weak γ's omitted (ΣIγ = 0.03%)

41AR B- DECAY (1.827 H 7) I(MIN) = 0.10%

Radiation Type		Energy (keV)	Intensity (%)	Δ(g-rad/ μCi-h)
β- 1	max	1198.1 8		
	avg	459.3 3	99.170 20	0.970
β- 2	max	2491.8 8		
	avg	1076.6 4	0.780 20	0.0179
total β-	avg	464.0 3	100.00 3	0.988

1 weak β's omitted (ΣIβ = 0.05%)

Radiation Type		Energy (keV)	Intensity (%)	Δ(g-rad/ μCi-h)
γ 1		1293.64 4	99.160 20	2.73

1 weak γ's omitted (ΣIγ = 0.05%)

Radiation Type		Energy (keV)	Intensity (%)	Δ(g-rad/ μCi-h)
β- 5	max	1817 10		
	avg	762 5	1.3 3	0.0211
total β-	avg	309 5	100.1 19	0.659
γ 1		184.00 20	0.27 6	0.0011
γ 2		220.608 18	4.11 22	0.0193
γ 3		372.763 15	87.3 5	0.693
γ 4		396.870 20	11.43 12	0.0966
γ 5		404.30 20	0.109 8	0.0009
γ 6		593.40 8	11.0 3	0.139
γ 7		617.494 25	80.5 14	1.06
γ 8		800.8 10	0.147 10	0.0025
γ 9		990.25 20	0.33 7	0.0069
γ 10		1015.1 10	0.16 7	0.0034
γ 11		1021.79 18	1.88 8	0.0409
γ 12		1394.2 7	0.102 12	0.0030

44SC B+ DECAY (3.927 H 8) I(MIN) = 0.10%

Radiation Type		Energy (keV)	Intensity (%)	Δ(g-rad/ μCi-h)
Auger-L		0.3	8.56 16	≈0
Auger-K		3.3	4.22 9	0.0003
β+ 1	max	1475.8 20		
	avg	632.7 9	94.370 20	1.27
X-ray	$K\alpha_2$	3.68809 4	0.244 25	≈0
X-ray	$K\alpha_1$	3.69168 4	0.49 5	≈0
γ 1		1157.002 11	99.884 15	2.46
γ 2		1499.451 23	0.912 20	0.0291
γ 5		2656.41 3	0.112 4	0.0064

Maximum γ± intensity = 188.74%

44TI EC DECAY (47.3 Y 12) I(MIN) = 0.10%

Radiation Type	Energy (keV)	Intensity (%)	Δ(g-rad/ μCi-h)
Auger-L	0.37	172 6	0.0014
Auger-K	3.64	83 3	0.0064
ce-K- 1	63.36 3	10.8 21	0.0146
ce-L- 1	67.35 3	1.1 3	0.0015
ce-MNO- 1	67.80 3	0.35 7	0.0005
ce-K- 2	73.91 6	2.9 5	0.0046
ce-L- 2	77.90 6	0.28 5	0.0005
X-ray L	0.4	0.34 12	≈0
X-ray $K\alpha_2$	4.08610 20	5.8 6	0.0005
X-ray $K\alpha_1$	4.09060 20	11.5 11	0.0010
X-ray $K\beta$	4.46	2.26 21	0.0002
γ 1	67.85 3	87.7 15	0.127
γ 2	78.40 6	94.7 15	0.158
γ 3	147.0 15	0.10 3	0.0003

45CA B- DECAY (163 D 1) I(MIN) = 0.10%

Radiation Type	Energy (keV)	Intensity (%)	Δ(g-rad/ μCi-h)
β- 1 max	256.9 10		
avg	77.2 4	100	0.164

47CA B- DECAY (4.536 D 2) I(MIN) = 0.10%

Radiation Type	Energy (keV)	Intensity (%)	Δ(g-rad/ μCi-h)
β- 1 max	691 3		
avg	241.2 10	82.0 20	0.421
β- 2 max	1988.3 25		
avg	817.2 12	18.0 20	0.313
total β-			
avg	344.9 13	100 3	0.735

2 weak β's omitted (ΣIβ = 0.11%)

Radiation Type	Energy (keV)	Intensity (%)	Δ(g-rad/ μCi-h)
γ 2	489.23 10	6.7 3	0.0702
γ 3	530.4 3	0.105 16	0.0012
γ 4	767.0 3	0.195 16	0.0032
γ 5	807.86 10	6.9 3	0.119
γ 6	1297.09 10	74.9 19	2.07

2 weak γ's omitted (ΣIγ = 0.03%)

47SC B- DECAY (3.351 D 2) I(MIN) = 0.10%

Radiation Type	Energy (keV)	Intensity (%)	Δ(g-rad/ μCi-h)
Auger-L	0.42	0.347 18	≈0
Auger-K	4	0.175 9	≈0
ce-K- 1	154.415 15	0.224 10	0.0007
β- 1 max	440.6 20		

46SC B- DECAY (83.83 D 2) I(MIN) = 0.10%

Radiation Type	Energy (keV)	Intensity (%)	Δ(g-rad/ μCi-h)
β- 1 max	357.3 8		
avg	112.0 3	99.9964 7	0.239
γ 1	889.25 3	99.9840 10	1.89
γ 2	1120.51 5	99.9870 10	2.39

46SC IT DECAY (18.70 S 5) I(MIN) = 0.10%

Radiation Type	Energy (keV)	Intensity (%)	Δ(g-rad/ μCi-h)
Auger-L	0.37	56 3	0.0004
Auger-K	3.64	27.1 13	0.0021
ce-K- 1	138.035 3	33.4 15	0.0983
ce-L- 1	142.028 3	3.40 15	0.0103
ce-MNO- 1	142.474 3	1.12 4	0.0034
X-ray L	0.4	0.11 4	≈0
X-ray Kα2	4.08610 20	1.88 18	0.0002
X-ray Kα1	4.09060 20	3.7 4	0.0003
X-ray Kβ	4.46	0.74 8	≈0
γ 1	142.528 3	62.0 20	0.188

avg	142.5 8	68.0 20	0.206
β- 2 max	600.0 20		
avg	203.8 8	32.0 20	0.139
total β- avg	162.1 9	100 3	0.345
γ 1	159.381 15	68.0 20	0.231

48V EC DECAY (15.974 D 3) I(MIN) = 0.10%

Radiation Type	Energy (keV)	Intensity (%)	Δ(g-rad/ μCi-h)
Auger-L	0.42	73.4 22	0.0007
Auger-K	4	34.8 10	0.0030
β+ 1 max	698 4		
avg	292.0 20	50.1 5	0.312

1 weak β's omitted (ΣIβ = 0.09%)

Radiation Type	Energy (keV)	Intensity (%)	Δ(g-rad/ μCi-h)
X-ray L	0.45	0.15 6	≈0
X-ray Kα2	4.50486 4	2.88 25	0.0003
X-ray Kα1	4.51084 4	5.7 5	0.0005
X-ray Kβ	5	1.14 10	0.0001
γ 1	802.8 13	0.140 20	0.0024
γ 2	928.28 4	0.90 10	0.0178
γ 3	944.12 3	7.9 4	0.159
γ 4	983.50 3	99.9940 20	2.09
γ 5	1312.04 3	97.39 20	2.72
γ 6	1436.80 10	0.12 3	0.0037
γ 7	2240.35 6	2.40 20	0.115

2 weak γ's omitted (ΣIγ = 0.02%)
Maximum γ± intensity = 100.37%

51CR EC DECAY (27.704 D 4) I(MIN) = 0.10%

Radiation Type	Energy (keV)	Intensity (%)	Δ(g-rad/ μCi-h)
Auger-L	0.47	143.8 12	0.0014
Auger-K	4.38	66.9 7	0.0062
X-ray L	0.5	0.33 12	≈0
X-ray $K\alpha_2$	4.94464 4	6.60 21	0.0007
X-ray $K\alpha_1$	4.95220 4	13.1 4	0.0014
X-ray $K\beta$	5.43	2.62 9	0.0003
γ 1	320.076 6	9.83 14	0.0670

52MN EC DECAY (5.591 D 3) I(MIN) = 0.10%

Radiation Type	Energy (keV)	Intensity (%)	Δ(g-rad/ μCi-h)
Auger-L	0.54	99.4 20	0.0011
Auger-K	4.78	44.7 10	0.0046
β+ 1 max	575.3 23		
avg	241.6 10	29.4 7	0.151
X-ray L	0.57	0.26 9	≈0
X-ray $K\alpha_2$	5.40551 5	5.20 18	0.0006
X-ray $K\alpha_1$	5.41472 5	10.3 4	0.0012
X-ray $K\beta$	6	2.06 8	0.0003
γ 2	346.03 3	0.980 20	0.0072
γ 4	399.56 5	0.183 8	0.0016

52MN B+ DECAY (21.1 M 2) I(MIN) = 0.10%
%(EC+B+)=98.25 5
SEE ALSO 52MN IT DECAY (21.1 M)
FEEDING OF 21.1-M 52MN IN 52FE B+
DECAY=100%

Radiation Type	Energy (keV)	Intensity (%)	Δ(g-rad/ μCi-h)
Auger-L	0.54	2.37 5	≈0
Auger-K	4.78	1.068 24	0.0001
β+ 1 max	905.2 23		
avg	383.0 10	0.164 8	0.0013
β+ 2 max	2632.8 23		
avg	1173.8 11	96.5 20	2.41
total β+			
avg	1172.1 11	96.7 20	2.41

3 weak β's omitted (ΣIβ = 0.05%)

Radiation Type	Energy (keV)	Intensity (%)	Δ(g-rad/ μCi-h)
X-ray $K\alpha_2$	5.40551 5	0.124 5	≈0
X-ray $K\alpha_1$	5.41472 5	0.246 8	≈0
γ 4	1434.060 10	98.22 5	3.00
γ 6	1727.53 7	0.216 10	0.0080

13 weak γ's omitted (ΣIγ = 0.17%)
Maximum γ± intensity =193.43%

Radiation Type	Energy (keV)	Intensity (%)	Δ(g-rad/μCi-h)
γ 5	502.05 5	0.210 20	0.0022
γ 6	600.18 4	0.390 20	0.0050
γ 7	647.50 4	0.400 20	0.0055
γ 8	744.214 11	90.0 19	1.43
γ 9	848.13 3	3.32 8	0.0600
γ 11	935.520 20	94.5 20	1.88
γ 13	1246.246 12	4.21 10	0.112
γ 14	1247.85 9	0.38 4	0.0101
γ 15	1333.615 16	5.07 11	0.144
γ 16	1434.056 16	100	3.05

9 weak γ's omitted (ΣIγ = 0.37%)
Maximum γ± intensity = 58.80%

52MN IT DECAY (21.1 M 2) I(MIN) = 0.10%
%IT=1.75 5
SEE ALSO 52MN B+ DECAY (21.1 M)
FEEDING OF 21.1-M 52MN IN 52FE B+ DECAY=100%

Radiation Type	Energy (keV)	Intensity (%)	Δ(g-rad/μCi-h)
γ 1	377.738 11	1.75 5	0.0141

52FE B+ DECAY (8.275 H 8) I(MIN) = 0.10%
SEE ALSO 52MN IT DECAY (21.1 M)
SEE ALSO 52MN B+ DECAY (21.1 M)

Radiation Type	Energy (keV)	Intensity (%)	Δ(g-rad/μCi-h)
Auger-L	0.6	61.6 23	0.0008
Auger-K	5.19	26.9 12	0.0030
β+ 1 max	804 12		
avg	340 6	56.0 13	0.406
X-ray L	0.64	0.19 7	≈0
X-ray Kα2	5.88765 3	3.6 3	0.0005
X-ray Kα1	5.89875 3	7.2 6	0.0009
X-ray Kβ	6.49	1.45 12	0.0002
γ 1	168.684 11	100	0.359

Maximum γ± intensity = 112.00%

54MN EC DECAY (312.5 D 5) I(MIN) = 0.10%

Radiation Type	Energy (keV)	Intensity (%)	Δ(g-rad/μCi-h)
Auger-L	0.54	142.1 12	0.0016
Auger-K	4.78	63.9 7	0.0065
X-ray L	0.57	0.37 13	≈0
X-ray Kα2	5.40551 5	7.43 21	0.0009
X-ray Kα1	5.41472 5	14.7 4	0.0017
X-ray Kβ	6	2.95 10	0.0004
γ 1	834.827 21	99.9760 20	1.78

55FE EC DECAY (2.7 Y 1) I(MIN) = 0.10%

Radiation Type	Energy (keV)	Intensity (%)	Δ(g-rad/ μCi-h)
Auger-L	0.6	139 4	0.0018
Auger-K	5.19	60.7 22	0.0067
X-ray L	0.64	0.42 14	≈0
X-ray $K\alpha_2$	5.88765 3	8.2 7	0.0010
X-ray $K\alpha_1$	5.89875 3	16.3 12	0.0020
X-ray Kβ	6.49	3.3 3	0.0005

56MN B- DECAY (2.5785 H 6) I(MIN) = 0.10%

Radiation Type	Energy (keV)	Intensity (%)	Δ(g-rad/ μCi-h)
β- 1 max	324.8 12		
avg	98.8 5	1.16 3	0.0024
β- 2 max	734.6 12		
avg	254.8 5	14.6 4	0.0792
β- 3 max	1037.0 12		
avg	381.5 6	27.8 8	0.226
β- 4 max	2847.7 12		
avg	1216.3 6	56.3 9	1.46
total β- avg	829.8 9	100.0 13	1.77
3 weak β's omitted (ΣIβ = 0.12%)			
γ 1	846.754 20	98.9 3	1.78
γ 4	1810.72 4	27.2 8	1.05

58CO B+ DECAY (70.80 D 8) I(MIN) = 0.10%

Radiation Type	Energy (keV)	Intensity (%)	Δ(g-rad/ μCi-h)
Auger-L	0.67	116.5 13	0.0017
Auger-K	5.62	49.4 6	0.0059
β+ 1 max	474.6 14		
avg	201.2 6	15.00 5	0.0643
X-ray L	0.7	0.36 12	≈0
X-ray $K\alpha_2$	6.39084 3	7.78 21	0.0011
X-ray $K\alpha_1$	6.40384 3	15.4 4	0.0021
X-ray Kβ	7	3.10 10	0.0005
γ 1	810.757 18	99.4 3	1.72
γ 2	863.935 18	0.676 10	0.0124
γ 3	1674.68 4	0.517 10	0.0184

Maximum γ± intensity = 30.00%

59FE B- DECAY (44.529 D 7) I(MIN) = 0.10%

Radiation Type	Energy (keV)	Intensity (%)	Δ(g-rad/ μCi-h)
β- 1 max	130.8 20		
avg	35.8 6	1.27 3	0.0010
β- 2 max	273.4 20		
avg	80.8 6	45.6 8	0.0785
β- 3 max	465.8 20		
avg	149.3 7	52.8 12	0.168

Radiation Type	Energy (keV)	Intensity (%)	Δ(g-rad/µCi-h)
γ 5	2113.05 4	14.3 4	0.645
γ 6	2522.88 6	0.99 3	0.0531
γ 8	2657.45 5	0.653 20	0.0369
γ 9	2959.77 6	0.306 10	0.0193
γ 10	3369.60 7	0.168 10	0.0121

3 weak γ's omitted (ΣIγ = 0.16%)

57CO EC DECAY (270.9 D 6) I(MIN) = 0.10%

Radiation Type	Energy (keV)	Intensity (%)	Δ(g-rad/µCi-h)
Auger-L	0.67	249 3	0.0036
Auger-K	5.62	105.5 14	0.0126
ce-K- 1	7.3007 10	69.6 4	0.0108
ce-L- 1	13.5666 6	7.79 22	0.0023
ce-MNO- 1	14.3198 10	1.15 7	0.0004
ce-K- 2	114.951 4	1.87 10	0.0046
ce-L- 2	121.217 3	0.183 10	0.0005
ce-K- 3	129.364 4	1.40 12	0.0039
ce-L- 3	135.630 3	0.140 12	0.0004
X-ray L	0.7	0.8 3	≈0
X-ray Kα2	6.39084 3	16.6 5	0.0023
X-ray Kα1	6.40384 3	32.8 8	0.0045
X-ray Kβ	7	6.63 21	0.0010
γ 1	14.4127 4	9.54 13	0.0029
γ 2	122.063 3	85.59 19	0.223
γ 3	136.476 3	10.61 18	0.0309
γ 9	692.00 3	0.160 5	0.0024

6 weak γ's omitted (ΣIγ = 0.03%)

Radiation Type	Energy (keV)	Intensity (%)	Δ(g-rad/µCi-h)
β- 4 max	1565.0 20		
avg	635.8 8	0.18 4	0.0024
total β- avg	117.4 8	99.9 15	0.250

1 weak β's omitted (ΣIβ = 0.08%)

Radiation Type	Energy (keV)	Intensity (%)	Δ(g-rad/µCi-h)
γ 1	142.648 4	1.00 3	0.0030
γ 2	192.344 6	3.00 7	0.0123
γ 3	334.80 20	0.270 10	0.0019
γ 5	1099.224 25	56.1 12	1.31
γ 6	1291.56 3	43.6 8	1.20

2 weak γ's omitted (ΣIγ = 0.08%)

60CO B- DECAY (5.271 Y 1) I(MIN) = 0.10%

Radiation Type	Energy (keV)	Intensity (%)	Δ(g-rad/µCi-h)
β- 1 max	317.87 11		
avg	95.80 10	99.920 20	0.204

2 weak β's omitted (ΣIβ = 0.08%)

Radiation Type	Energy (keV)	Intensity (%)	Δ(g-rad/µCi-h)
γ 3	1173.210 10	99.900 20	2.50
γ 4	1332.470 10	99.9824 5	2.84

4 weak γ's omitted (ΣIγ = 0.02%)

63NI B- DECAY (96 Y 4) I(MIN) = 0.10%

Radiation Type	Energy (keV)	Intensity (%)	Δ(g-rad/ μCi-h)
β- 1 max	65.87 20		
avg	17.13 5	100	0.0365

64CU B+ DECAY (12.701 H 2) I(MIN) = 0.10%
%(EC+B+)=62.8 4
SEE ALSO 64CU B- DECAY

Radiation Type	Energy (keV)	Intensity (%)	Δ(g-rad/ μCi-h)
Auger-L	0.84	59.1 21	0.0011
Auger-K	6.54	23.4 12	0.0033
β+ 1 max	652.9 8		
avg	278.1 4	17.87 18	0.106
X-ray L	0.85	0.23 8	≈0
X-ray $K\alpha_2$	7.46089 4	4.9 4	0.0008
X-ray $K\alpha_1$	7.47815 4	9.6 7	0.0015
X-ray $K\beta$	8.26	1.96 14	0.0003
γ 1	1345.9 3	0.49 4	0.0140

Maximum γ± intensity = 35.74%

66GA B+ DECAY (9.40 H 7) I(MIN) = 0.10%

Radiation Type	Energy (keV)	Intensity (%)	Δ(g-rad/ μCi-h)
Auger-L	0.99	56 3	0.0012
Auger-K	7.53	20.6 13	0.0033
β+ 1 max	361 3		
avg	157.0 13	1.00 10	0.0033
β+ 2 max	720 3		
avg	308.7 13	0.160 10	0.0011
β+ 3 max	772 3		
avg	330.9 13	0.70 10	0.0049
β+ 4 max	821 3		
avg	352.2 13	0.220 20	0.0017
β+ 5 max	924 3		
avg	397.1 13	4.10 20	0.0347
β+ 6 max	1780 3		
avg	781.6 13	0.37 3	0.0062
β+ 7 max	4153 3		
avg	1904.1 14	49.9 10	2.02
total β+			
avg	1726.3 20	56.4 11	2.08
X-ray L	1	0.28 12	≈0
X-ray $K\alpha_2$	8.61578 5	5.6 4	0.0010
X-ray $K\alpha_1$	8.63886 5	11.0 8	0.0020
X-ray $K\beta$	9.57	2.25 16	0.0005
γ 4	448.90 10	0.111 5	0.0011
γ 7	686.28 10	0.261 10	0.0038
γ 8	833.56 10	6.12 15	0.109
γ 10	856.70 10	0.123 5	0.0022
γ 11	907.0 3	0.115 8	0.0022
γ 15	1039.29 10	38.4 8	0.850
γ 18	1190.36 10	0.134 12	0.0034
γ 20	1232.9	0.538 19	0.0141

64CU B- DECAY (12.701 H 2) I(MIN) = 0.10%
%B-=37.2 4
SEE ALSO 64CU B+ DECAY

Radiation Type	Energy (keV)	Intensity (%)	Δ(g-rad/ μCi-h)
β- 1 max	578.2 15		
avg	190.3 5	37.2 4	0.151

65ZN EC DECAY (243.9 D 1) I(MIN) = 0.10%

Radiation Type	Energy (keV)	Intensity (%)	Δ(g-rad/ μCi-h)
Auger-L	0.92	126.7 18	0.0025
Auger-K	7	48.3 8	0.0072
β+ 1 max	329.9 11		
avg	143.0 5	1.460 20	0.0044
X-ray L	0.93	0.57 20	≈0
X-ray $K\alpha_2$	8.027830 10	11.5 3	0.0020
X-ray $K\alpha_1$	8.047780 10	22.6 5	0.0039
X-ray Kβ	9	4.61 13	0.0009
γ 3	1115.52 3	50.75 10	1.21

Maximum γ± intensity = 2.92%

γ	21	1333.20 20	1.25 3	0.0355
γ	22	1356.2	0.38 4	0.0109
γ	23	1356.6	0.127 20	0.0037
γ	24	1357	0.19 8	0.0055
γ	25	1418.88 10	0.645 16	0.0195
γ	27	1508.33 10	0.584 17	0.0188
γ	29	1899.18 10	0.434 15	0.0176
γ	30	1918.64 10	2.17 5	0.0887
γ	32	2173.90 20	0.12 3	0.0055
γ	33	2190.00 20	5.76 15	0.269
γ	34	2213.60 20	0.142 9	0.0067
γ	36	2393.30 20	0.253 10	0.0129
γ	37	2422.70 10	1.97 5	0.102
γ	40	2752.10 20	23.5 6	1.38
γ	41	2780.50 20	0.129 5	0.0076
γ	42	2934.30 20	0.219 6	0.0137
γ	46	3229.26 10	1.51 4	0.104
γ	48	3381.32 10	1.43 4	0.103
γ	49	3422.64 15	0.833 24	0.0607
γ	50	3433.00 15	0.284 7	0.0208
γ	53	3767.40 20	0.142 9	0.0114
γ	54	3791.47 10	1.025 25	0.0828
γ	58	4086.36 10	1.15 4	0.0999
γ	59	4295.70 20	3.53 9	0.323
γ	60	4462.01 15	0.718 17	0.0682
γ	61	4806.59 15	1.49 5	0.153

27 weak γ's omitted (ΣIγ = 1.24%)
Maximum γ± intensity =112.90%

67CU B- DECAY (61.86 H 12) I(MIN) = 0.10%

Radiation Type	Energy (keV)	Intensity (%)	Δ (g-rad/ μCi-h)
Auger-L	0.99	19.4 10	0.0004
Auger-K	7.53	7.0 5	0.0011
ce-K- 1	81.607 17	0.512 16	0.0009
ce-K- 2	83.656 13	12.1 4	0.0215
ce-L- 2	92.121 13	1.48 5	0.0029
ce-M- 2	93.179 13	0.218 8	0.0004
ce-K- 3	174.918 20	0.93 10	0.0034
β- 1 max	181 8		
avg	50.7 25	1.130 20	0.0012
β- 2 max	390 8		
avg	121 3	57.2 5	0.147
β- 3 max	482 8		
avg	154 3	21.6 4	0.0709
β- 4 max	575 8		
avg	189 3	20	0.0805
total β-			
avg	141 4	99.9 7	0.300
X-ray Kα2	8.61578 5	1.93 14	0.0004
X-ray Kα1	8.63886 5	3.8 3	0.0007
X-ray Kβ	9.57	0.77 6	0.0002
γ 1	91.266 17	7.01 11	0.0136
γ 2	93.315 13	16.12 22	0.0320
γ 3	184.577 20	48.7 5	0.191
γ 4	208.951 20	0.115 5	0.0005
γ 5	300.219 25	0.797 13	0.0051
γ 6	393.529 20	0.220 8	0.0018

68GA B+ DECAY (68.0 M 2) I(MIN) = 0.10%

Radiation Type	Energy (keV)	Intensity (%)	Δ (g-rad/ μCi-h)
Auger-L	0.99	13.8 7	0.0003
Auger-K	7.53	5.1 4	0.0008
β+ 1 max	821.7 12		
avg	352.6 6	1.08 7	0.0081
β+ 2 max	1899.1 12		
avg	836.0 6	88 3	1.57
total β+			
avg	830.1 6	89 3	1.58
X-ray Kα2	8.61578 5	1.38 10	0.0003
X-ray Kα1	8.63886 5	2.71 19	0.0005
X-ray Kβ	9.57	0.55 4	0.0001
γ 3	1077.40 10	3.30 10	0.0757
γ 6	1883.20 10	0.143 6	0.0057

6 weak γ's omitted (ΣIγ = 0.24%)
Maximum γ± intensity = 178.16%

68GE EC DECAY (288 D 6) I(MIN) = 0.10%

Radiation Type	Energy (keV)	Intensity (%)	Δ (g-rad/ μCi-h)
Auger-L	1	121.5 18	0.0028
Auger-K	8	42.4 7	0.0073
X-ray L	1	0.67 19	≈0
X-ray Kα2	9.22482 7	13.1 3	0.0026

Radiation Type	Energy (keV)	Intensity (%)	Δ(g-rad/μCi-h)
X-ray Kα$_1$	9.25174 7	25.6 5	0.0050
X-ray Kβ	10.3	5.46 14	0.0012

67GA EC DECAY (78.26 H 3) I(MIN) = 0.10%

Radiation Type	Energy (keV)	Intensity (%)	Δ(g-rad/μCi-h)
Auger-L	0.99	168 9	0.0035
Auger-K	7.53	61 4	0.0098
ce-K- 1	81.607 5	0.224 9	0.0004
ce-K- 2	83.652 5	28.7 13	0.0511
ce-L- 2	92.117 5	3.52 15	0.0069
ce-M- 2	93.175 6	0.516 22	0.0010
ce-K- 3	174.918 10	0.40 5	0.0015
X-ray L	1	0.8 4	≈0
X-ray Kα$_2$	8.61578 5	16.8 12	0.0031
X-ray Kα$_1$	8.63886 5	32.9 23	0.0061
X-ray Kβ	9.57	6.7 5	0.0014
γ 1	91.266 5	3.07 10	0.0060
γ 2	93.311 5	38.3 12	0.0760
γ 3	184.577 10	20.9 6	0.0823
γ 4	208.951 10	2.37 7	0.0105
γ 5	300.219 10	16.8 4	0.107
γ 6	393.529 10	4.70 14	0.0394
γ 10	887.693 15	0.145 5	0.0027

3 weak γ's omitted ($\Sigma I\gamma$ = 0.13%)

69ZN B- DECAY (57 M 1) I(MIN) = 0.10%

Radiation Type	Energy (keV)	Intensity (%)	Δ(g-rad/μCi-h)
β- 1 max	905 3		
avg	320.9 13	100	0.684

69ZN IT DECAY (13.76 H 3) I(MIN) = 0.10%

Radiation Type	Energy (keV)	Intensity (%)	Δ(g-rad/μCi-h)
Auger-L	0.99	6.3 4	0.0001
Auger-K	7.53	2.28 15	0.0004
ce-K- 1	428.975 18	4.38 14	0.0401
ce-L- 1	437.440 18	0.512 19	0.0048
ce-MNO- 1	438.498 18	0.171 10	0.0016
X-ray Kα$_2$	8.61578 5	0.62 5	0.0001
X-ray Kα$_1$	8.63886 5	1.23 9	0.0002
X-ray Kβ	9.57	0.250 18	≈0
γ 1	438.634 18	94.90 16	0.887

72GA B- DECAY (14.1 H 2)   I(MIN) = 0.10%

Radiation Type	Energy (keV)	Intensity (%)	Δ(g-rad/ μCi-h)
Auger-L	1.19	0.56 4	≈0
Auger-K	8.56	0.204 15	≈0
ce-K- 20	680.10 20	0.443 19	0.0064
β- 1 max	233 4		
avg	66.7 11	0.112 19	0.0002
β- 2 max	314 4		
avg	93.2 11	0.78 3	0.0015
β- 3 max	426 4		
avg	132.4 12	0.247 10	0.0007
β- 4 max	536 4		
avg	173.3 12	0.350 20	0.0013
β- 5 max	552 4		
avg	179.3 12	0.340 20	0.0013
β- 6 max	650 4		
avg	217.0 13	14.9 4	0.0689
β- 7 max	667 4		
avg	223.6 13	22.0 4	0.105
β- 8 max	956 4		
avg	341.8 13	27.4 8	0.199
β- 9 max	1048 4		
avg	380.8 14	1.96 4	0.0159
β-10 max	1477 4		
avg	569.0 14	9.10 24	0.110
β-11 max	1528 4		
avg	591.9 14	0.14 4	0.0018
β-12 max	1590 4		
avg	619.8 14	0.33 6	0.0044
β-13 max	1927 4		
avg	774.1 15	2.90 20	0.0478
β-14 max	2263 4		
avg	930.6 15	0.7 5	0.0139

72GA B- DECAY (CONTINUED)

Radiation Type	Energy (keV)	Intensity (%)	Δ(g-rad/ μCi-h)
γ 49	1567.9 4	0.20 4	0.0067
γ 50	1571.50 20	0.80 3	0.0269
γ 51	1596.65 9	4.3 4	0.146
γ 54	1680.77 8	0.99 4	0.0356
γ 55	1710.90 20	0.411 20	0.0150
γ 57	1837.80 20	0.229 10	0.0090
γ 58	1861.09 8	5.26 10	0.208
γ 59	1878.0 3	0.229 20	0.0092
γ 60	1920.2 3	0.143 20	0.0059
γ 61	1991.3 3	0.112 3	0.0047
γ 63	2109.50 10	1.07 3	0.0481
γ 64	2201.67 8	25.6 8	1.20
γ 65	2214.5 8	0.153 20	0.0072
γ 68	2490.98 8	7.93 20	0.421
γ 69	2507.80 8	12.7 4	0.679
γ 70	2515.6 5	0.239 20	0.0128
γ 73	2621.0 4	0.143 20	0.0080
γ 76	2844.10 20	0.478 20	0.0290

40 weak γ's omitted (ΣIγ = 1.30%)

72AS EC DECAY (26.0 H 1)   I(MIN) = 0.10%

Radiation Type	Energy (keV)	Intensity (%)	Δ(g-rad/ μCi-h)
Auger-L	1.19	15.9 7	0.0004
Auger-K	8.56	5.5 4	0.0010
ce-K- 7	680.10 20	0.87 9	0.0125

Radiation		Energy	Intensity	Δ
β-15	max	2528 4		
	avg	1055.0 15	9.0 9	0.202
β-16	max	3158 4		
	avg	1354.3 15	9.5 12	0.274
total β-				
	avg	492.8 20	100.0 19	1.05

8 weak β's omitted (ΣIβ = 0.23%)

Radiation		Energy	Intensity	Δ
X-ray	$K\alpha_1$	9.88642 7	0.138 9	≈0
γ	2	112.52 3	0.11 5	0.0003
γ	6	289.3 3	0.172 10	0.0011
γ	9	336.3 3	0.105 10	0.0008
γ	10	381.20 20	0.268 10	0.0022
γ	12	428.3 3	0.220 10	0.0020
γ	13	449.6 3	0.153 20	0.0015
γ	14	479.1 3	0.105 10	0.0011
γ	17	587.9 4	0.105 10	0.0013
γ	18	600.85 3	5.45 20	0.0697
γ	19	629.86 4	25.2 8	0.339
γ	21	735.90 20	0.373 10	0.0058
γ	24	786.40 10	3.26 9	0.0546
γ	25	810.24 9	2.01 9	0.0346
γ	26	834.02 3	95.59 11	1.70
γ	27	861.11 5	0.92 3	0.0168
γ	29	894.22 5	9.9 3	0.189
γ	30	924.10 20	0.143 10	0.0028
γ	32	939.40 20	0.259 7	0.0052
γ	34	970.54 6	1.09 3	0.0225
γ	36	999.86 6	0.803 20	0.0171
γ	38	1050.69 6	6.88 20	0.154
γ	42	1215.16 8	0.784 20	0.0203
γ	43	1230.86 7	1.46 4	0.0383
γ	44	1260.10 8	1.10 3	0.0295
γ	45	1276.75 8	1.558 20	0.0424
γ	47	1464.00 10	3.54 10	0.110

(CONTINUED ON NEXT COLUMN)

Radiation		Energy	Intensity	Δ
β+ 1	max	814 7		
	avg	351 3	0.450 20	0.0034
β+ 2	max	927 7		
	avg	400 3	0.167 9	0.0014
β+ 3	max	1865 7		
	avg	822 4	6.0 3	0.105
β+ 4	max	2495 7		
	avg	1115 4	64.1 12	1.52
β+ 5	max	2638 7		
	avg	1203 4	0.12 9	0.0031
β+ 6	max	3329 7		
	avg	1526 4	16.5 14	0.536
total β+				
	avg	1165 5	87.6 19	2.17

6 weak β's omitted (ΣIβ = 0.23%)

Radiation		Energy	Intensity	Δ
X-ray	$K\alpha_2$	9.85532 7	1.91 10	0.0004
X-ray	$K\alpha_1$	9.88642 7	3.73 19	0.0008
X-ray	$K\beta$	11	0.83 5	0.0002
γ	5	600.90 10	0.319 17	0.0041
γ	6	629.93 5	8.1 3	0.109
γ	9	786.43 5	0.430 18	0.0072
γ	10	834.00 3	79.7 14	1.42
γ	12	894.27 5	0.79 3	0.0150
γ	21	1050.76 5	0.98 4	0.0219
γ	25	1215.14 5	0.231 17	0.0060
γ	28	1390.44 5	0.239 9	0.0071
γ	29	1464.00 7	1.14 5	0.0355
γ	30	1475.91 7	0.502 19	0.0158
γ	32	1568.20 10	0.143 9	0.0048
γ	36	1680.70 10	0.120 9	0.0043
γ	37	1710.90 7	0.271 17	0.0099
γ	41	1991.14 8	0.383 18	0.0162
γ	43	2105.9 3	0.74 8	0.0332
γ	44	2109.8 3	0.223 25	0.0100

(CONTINUED ON NEXT PAGE)

72AS EC DECAY (CONTINUED)

Radiation Type	Energy (keV)	Intensity (%)	Δ(g-rad/ μCi-h)
γ 46	2201.74 9	0.518 19	0.0243
γ 48	2248.50 10	0.335 17	0.0160
γ 53	2507.90 10	0.335 17	0.0179
γ 58	2621.50 10	0.391 18	0.0218
γ 64	2940.10 10	0.335 17	0.0210
γ 66	2982.10 20	0.191 25	0.0121
γ 67	3094.30 20	0.143 9	0.0095

56 weak γ's omitted (ΣIγ = 1.75%)
Maximum γ± intensity = 175.13%

73AS EC DECAY (80.30 D 6) I(MIN) = 0.10%

Radiation Type	Energy (keV)	Intensity (%)	Δ(g-rad/ μCi-h)
Auger-L	1.19	321 10	0.0081
ce-K- 1	2.160 17	27.8 6	0.0013
Auger-K	8.56	88 5	0.0161
ce-L- 1	11.849 15	60.3 7	0.0152
ce-M- 1	13.083 15	8.9 3	0.0025
ce-NOP- 1	13.263 15	3.00 10	0.0008
ce-K- 2	42.334 12	76.1 6	0.0686
ce-L- 2	52.023 9	11.1 4	0.0123
ce-M- 2	53.257 9	1.73 5	0.0020
ce-NOP- 2	53.437 9	0.580 10	0.0007
X-ray L	1.19	1.9 7	≈0
X-ray Kα2	9.85532 7	30.5 16	0.0064

73SE EC DECAY (CONTINUED)

Radiation Type	Energy (keV)	Intensity (%)	Δ(g-rad/ μCi-h)
X-ray Kα1	10.54370 10	16.6 10	0.0037
X-ray Kβ	11.7	3.84 24	0.0010
γ 1	67.00 10	77.3 8	0.110
γ 2	361.10 10	96.6 5	0.743
γ 4	510	1.1 4	0.0115
γ 12	865.40 20	0.47 3	0.0087
γ 13	901.2 4	0.1159 6	0.0022
γ 14	901.2 4	0.1159 6	0.0022
γ 17	1111.0 4	0.17 5	0.0041
γ 19	1422.90 20	0.11 3	0.0032

13 weak γ's omitted (ΣIγ = 0.49%)
Maximum γ± intensity = 130.44%

74AS B+ DECAY (17.76 D 2) I(MIN) = 0.10%
%(EC+B+) = 65.8 16
SEE ALSO 74AS B- DECAY

Radiation Type	Energy (keV)	Intensity (%)	Δ(g-rad/ μCi-h)
Auger-L	1.19	43.5 19	0.0011
Auger-K	8.56	14.7 9	0.0027
β+ 1 max	944.3 18		
avg	407.9 8	25.7 7	0.223

Radiation Type	Energy (keV)	Intensity (%)	Δ(g-rad/μCi-h)
X-ray Kα1	9.88642 7	60 3	0.0125
X-ray Kβ	11	13.3 7	0.0031
γ 2	53.437 9	10.5 3	0.0120

1 weak γ's omitted (ΣIγ = 0.09%)

73SE EC DECAY (7.15 H 8) I(MIN) = 0.10%

Radiation Type	Energy (keV)	Intensity (%)	Δ(g-rad/μCi-h)
Auger-L	1.24	68 4	0.0018
Auger-K	9.1	22.1 17	0.0043
ce-K- 1	55.13 10	19.1 5	0.0224
ce-L- 1	65.47 10	2.09 7	0.0029
ce-MNO- 1	66.80 10	0.386 16	0.0005
ce-K- 2	349.23 10	1.12 4	0.0083
ce-L- 2	359.57 10	0.124 4	0.0009
β+ 1 max	1208 10		
avg	525 5	0.3 3	0.0034
β+ 2 max	1290 10		
avg	562 5	64.2 8	0.769
β+ 3 max	1651 10		
avg	725 5	0.68 20	0.0105
total β+			
avg	563 5	65.2 9	0.783

3 weak β's omitted (ΣIβ = 0.04%)

Radiation Type	Energy (keV)	Intensity (%)	Δ(g-rad/μCi-h)
X-ray L	1.28	0.48 14	≈0
X-ray Kα2	10.50800 10	8.5 6	0.0019

(CONTINUED ON NEXT COLUMN)

Radiation Type	Energy (keV)	Intensity (%)	Δ(g-rad/μCi-h)
β+ 2 max	1540.1 18		
avg	700.9 9	3.4 8	0.0508
total β+			
avg	442.1 9	29.1 11	0.274
X-ray L	1.19	0.26 9	≈0
X-ray Kα2	9.85532 7	5.1 3	0.0011
X-ray Kα1	9.88642 7	9.9 6	0.0021
X-ray Kβ	11	2.22 13	0.0005
γ 1	595.80 8	59.2 15	0.752
γ 2	608.39 8	0.53 12	0.0069
γ 9	1204.34 8	0.290 14	0.0074

8 weak γ's omitted (ΣIγ = 0.09%)
Maximum γ± intensity = 58.20%

74AS B- DECAY (17.78 D 3) I(MIN) = 0.10%
%B-=34.2 16
SEE ALSO 74AS B+ DECAY

Radiation Type	Energy (keV)	Intensity (%)	Δ(g-rad/μCi-h)
β- 1 max	718 3		
avg	242.9 11	15.4 3	0.0797
β- 2 max	1353 3		
avg	530.9 12	18.8 16	0.213
total β-			
avg	400.8 14	34.2 17	0.292

1 weak β's omitted (ΣIβ = 0.04%)

Radiation Type	Energy (keV)	Intensity (%)	Δ(g-rad/μCi-h)
γ 1	634.78 8	15.4 8	0.208
γ 2	635.0 20	0.142 7	0.0019

75SE EC DECAY (119.8 D 1) I(MIN) = 0.10%

Radiation Type	Energy (keV)	Intensity (%)	Δ(g-rad/μCi-h)
Auger-L	1.24	136 8	0.0036
Auger-K	9.1	44 4	0.0085
ce-K- 1	12.53 20	4.3 4	0.0011
ce-L- 1	22.87 20	0.84 7	0.0004
ce-M- 1	24.20 20	0.134 11	≈0
ce-K- 2	54.183 10	0.39 3	0.0004
ce-K- 4	84.8663 22	2.73 17	0.0049
ce-L- 4	95.2065 22	0.364 22	0.0007
ce-K- 5	109.248 3	0.66 3	0.0015
ce-K- 6	124.133 5	1.59 7	0.0042
ce-L- 6	134.473 5	0.165 7	0.0005
ce-K- 10	252.784 8	0.380 17	0.0020
ce-K- 11	267.661 8	0.179 14	0.0010
X-ray L	1.28	1.0 3	≈0
X-ray $K\alpha_2$	10.50800 10	16.9 11	0.0038
X-ray $K\alpha_1$	10.54370 10	32.9 21	0.0074
X-ray $K\beta$	11.7	7.6 5	0.0019
γ 2	66.050 10	1.10 6	0.0016
γ 4	96.7330 20	3.50 18	0.0072
γ 5	121.115 3	17.7 6	0.0457
γ 6	136.000 5	60.6 19	0.176
γ 7	198.596 7	1.50 5	0.0064
γ 10	264.651 8	59.4 19	0.335
γ 11	279.528 8	25.2 8	0.150
γ 13	303.910 11	1.313 21	0.0085
γ 16	400.646 9	11.35 21	0.0968

15 weak γ's omitted ($\Sigma I\gamma$ = 0.09%)

76AS B- DECAY (CONTINUED)

Radiation Type	Energy (keV)	Intensity (%)	Δ(g-rad/μCi-h)
γ 23	867.63 8	0.125 7	0.0023
γ 29	1129.87 7	0.143 11	0.0034
γ 31	1212.72 18	1.63 12	0.0420
γ 32	1216.02 7	3.84 25	0.0996
γ 33	1228.52 8	1.39 8	0.0363
γ 35	1439.13 8	0.326 20	0.0100
γ 36	1453.60 8	0.130 11	0.0040
γ 42	1787.67 8	0.331 23	0.0126
γ 46	2096.33 14	0.66 5	0.0295
γ 47	2110.79 15	0.393 25	0.0177

34 weak γ's omitted ($\Sigma I\gamma$ = 0.65%)

77GE B- DECAY (11.30 H 1) I(MIN) = 0.10%

Radiation Type	Energy (keV)	Intensity (%)	Δ(g-rad/μCi-h)
β- 1 max	158 3		
avg	43.2 9	0.101 6	≈0
β- 2 max	188 3		
avg	52.5 10	0.221 8	0.0002
β- 3 max	277 3		
avg	80.8 10	0.156 6	0.0003
β- 4 max	347 3		
avg	104.6 11	1.04 4	0.0023

76AS B- DECAY (26.32 H 7) I(MIN) = 0.10%

Radiation Type		Energy (keV)	Intensity (%)	Δ(g-rad/ μCi-h)
β- 1	max	299.2 19		
	avg	88.0 7	0.64 4	0.0012
β- 2	max	313.6 19		
	avg	92.8 7	1.19 7	0.0024
β- 3	max	540.1 19		
	avg	173.8 8	1.89 12	0.0070
β- 4	max	1181.2 19		
	avg	436.4 9	2.02 12	0.0188
β- 5	max	1752.8 19		
	avg	691.7 9	7.6 5	0.112
β- 6	max	1846.6 19		
	avg	749.3 9	0.78 6	0.0124
β- 7	max	2409.8 19		
	avg	996.5 9	34.6 15	0.734
β- 8	max	2968.9 19		
	avg	1267.1 9	51.0 20	1.38
total β-				
	avg	1064.1 12	100 3	2.27

9 weak β's omitted (ΣIβ = 0.30%)

Radiation Type	Energy (keV)	Intensity (%)	Δ(g-rad/ μCi-h)
γ 7	559.10 5	44.7 19	0.532
γ 8	563.23 8	1.17 6	0.0140
γ 9	571.30 20	0.139 8	0.0017
γ 12	657.03 5	6.1 4	0.0851
γ 14	665.31 7	0.39 4	0.0056
γ 17	740.12 8	0.116 7	0.0018
γ 19	771.76 8	0.116 11	0.0019

(CONTINUED ON NEXT COLUMN)

Radiation Type		Energy (keV)	Intensity (%)	Δ(g-rad/ μCi-h)
β- 5	max	360 3		
	avg	108.8 11	2.28 8	0.0053
β- 6	max	591 3		
	avg	193.5 12	2.25 8	0.0093
β- 7	max	701 3		
	avg	236.7 12	8.1 5	0.0408
β- 8	max	730 3		
	avg	248.3 12	2.52 8	0.0133
β- 9	max	1128 3		
	avg	414.1 13	1.87 6	0.0165
β-10	max	1141 3		
	avg	419.8 13	7.18 23	0.0642
β-11	max	1163 3		
	avg	429.2 13	0.19 15	0.0017
β-12	max	1173 3		
	avg	433.8 13	0.91 3	0.0084
β-13	max	1244 3		
	avg	464.6 14	4.0 4	0.0396
β-14	max	1303 3		
	avg	490.6 14	1.77 7	0.0185
β-15	max	1356 3		
	avg	514.2 14	0.153 5	0.0017
β-16	max	1382 3		
	avg	525.5 14	0.43 15	0.0048
β-17	max	1512 3		
	avg	583.5 14	19.3 7	0.240
β-18	max	1536 3		
	avg	594.6 14	0.140 5	0.0018
β-19	max	1643 3		
	avg	642.7 14	0.176 9	0.0024
β-20	max	1812 3		
	avg	720.1 14	0.307 19	0.0047
β-21	max	1826 3		
	avg	726.5 14	0.270 18	0.0042
β-22	max	1917 3		
	avg	768.1 14	0.69 7	0.0113

(CONTINUED ON NEXT PAGE)

77GE B- DECAY (CONTINUED)

Radiation Type	Energy (keV)	Intensity (%)	Δ (g-rad/ μCi-h)
β-23 max	2067 3		
avg	837.7 14	0.98 4	0.0175
β-24 max	2070 3		
avg	838.9 14	20.8 9	0.372
β-25 max	2087 3		
avg	846.9 14	0.83 7	0.0150
β-26 max	2226 3		
avg	911.7 14	15.4 9	0.299
β-27 max	2437 3		
avg	1010.7 15	2.5 8	0.0538
β-28 max	2486 3		
avg	1033.7 15	5.3 8	0.117
β-29 max	2507 3		
avg	1043.5 15	0.47 6	0.0104
total β- avg	643.0 18	100.5 20	1.38

5 weak β's omitted (ΣIβ = 0.20%)

Radiation Type	Energy (keV)	Intensity (%)	Δ (g-rad/ μCi-h)
γ 2	156.36 3	0.81 5	0.0027
γ 3	159.11 15	0.232 11	0.0008
γ 4	177.28 3	0.181 8	0.0007
γ 5	194.762 20	1.79 8	0.0074
γ 6	208.98 6	0.95 4	0.0042
γ 7	211.031 19	31.1 13	0.140
γ 8	215.505 22	28.9 12	0.133
γ 9	219.1 3	0.15 15	0.0007
γ 10	254.74 16	0.213 8	0.0012
γ 11	264.440 17	54.4 17	0.306
γ 12	268.10 22	0.3 3	0.0017
γ 14	337.63 6	0.234 10	0.0017

77GE B- DECAY (CONTINUED)

Radiation Type	Energy (keV)	Intensity (%)	Δ (g-rad/ μCi-h)
γ 81	923.143 20	0.696 22	0.0137
γ 83	925.473	0.72 8	0.0143
γ 85	928.853 12	1.05 4	0.0208
γ 86	939.350 15	0.287 10	0.0057
γ 93	996.55 3	0.106 4	0.0023
γ 100	1061.699 23	0.152 5	0.0034
γ 101	1080.82 8	0.243 8	0.0056
γ 102	1085.188 13	6.10 20	0.141
γ 104	1114.80 4	0.104 4	0.0025
γ 105	1124.99 3	0.119 4	0.0028
γ 108	1151.837 25	0.197 7	0.0048
γ 112	1193.263 13	2.59 9	0.0659
γ 114	1215.418 25	0.128 4	0.0033
γ 116	1242.183 15	0.401 13	0.0106
γ 117	1263.862 15	0.86 3	0.0230
γ 118	1279.957 20	0.175 6	0.0048
γ 121	1309.271 16	0.491 16	0.0137
γ 122	1312.802 16	0.361 12	0.0101
γ 123	1319.662 17	0.304 10	0.0085
γ 130	1368	3.4 4	0.0983
γ 134	1452.59 4	0.122 4	0.0038
γ 138	1476.524 22	0.244 8	0.0077
γ 139	1479	0.128 14	0.0040
γ 141	1495.597 17	0.503 16	0.0160
γ 143	1538.763 20	0.144 5	0.0047
γ 146	1573.688 20	0.664 21	0.0222
γ 151	1709.812 25	0.310 10	0.0113
γ 152	1719.656 22	0.402 13	0.0147
γ 154	1727.18 3	0.149 5	0.0055
γ 162	1846.41 3	0.172 6	0.0068
γ 168	2000.10 3	0.565 18	0.0241
γ 170	2077.20 3	0.235 8	0.0104

γ 15	338.66 4	0.67 3	0.0049
γ 18	367.397 16	14.1 5	0.111
γ 20	416.328 14	22.0 8	0.195
γ 21	419.75 3	1.24 4	0.0111
γ 23	439.438 20	0.204 7	0.0019
γ 25	461.378 15	1.28 5	0.0126
γ 27	475.433 17	1.00 4	0.0101
γ 29	520	0.30 4	0.0033
γ 33	558.018 13	16.2 6	0.193
γ 36	582.537 14	0.787 25	0.0098
γ 38	614.39	0.51 6	0.0067
γ 40	624.76 9	0.184 6	0.0024
γ 41	631.823 13	7.03 23	0.0946
γ 42	634.389 15	2.10 7	0.0284
γ 47	673	0.54 6	0.0077
γ 48	673	0.134 15	0.0019
γ 52	698.538 25	0.231 8	0.0034
γ 53	705.24 8	0.108 4	0.0016
γ 54	712.35 4	0.84 3	0.0127
γ 55	714.345 12	7.21 23	0.110
γ 57	743.649 25	0.179 6	0.0028
γ 58	745.748 12	0.97 3	0.0155
γ 59	749.861 12	0.89 3	0.0142
γ 60	766.715 13	0.79 3	0.0129
γ 62	781.261 13	1.02 4	0.0170
γ 63	784.770 12	1.33 5	0.0222
γ 65	794.328 18	0.279 9	0.0047
γ 68	810.352 12	2.29 8	0.0395
γ 69	813.36 8	0.133 5	0.0023
γ 70	823.13 4	0.606 20	0.0106
γ 71	843.173 17	0.210 7	0.0038
γ 73	875.191 17	0.789 25	0.0147
γ 76	896.51 5	0.123 4	0.0024
γ 77	900.97 11	0.122 4	0.0023
γ 78	906.986 13	0.96 3	0.0185
γ 79	913.805 20	0.369 12	0.0072

(CONTINUED ON NEXT COLUMN)

γ 171	2089.60 3	0.241 8	0.0107
γ 172	2126.15 4	0.206 7	0.0093
γ 176	2341.63 4	0.476 15	0.0237

94 weak γ's omitted (ΣIγ = 3.08%)

77AS B- DECAY (38.8 H 3) I(MIN) = 0.10%

Radiation Type	Energy (keV)	Intensity (%)	Δ(g-rad/ μCi-h)
Auger-L	1.32	0.11 3	≈0
β- 1 max	170 4		
avg	46.8 12	0.70 20	0.0007
β- 2 max	441 4		
avg	137.1 14	0.70 20	0.0020
β- 3 max	451 4		
avg	141.0 15	1.6 4	0.0048
β- 4 max	690 4		
avg	231.8 16	97.1 8	0.479
total β- avg	228.4 17	100.1 10	0.487

2 weak β's omitted (ΣIβ = 0.03%)

γ 3	87.876 20	0.21 6	0.0004
γ 5	161.933 20	0.13 4	0.0004
γ 7	238.999 20	1.6 4	0.0081
γ 8	249.790 20	0.43 11	0.0023
γ 12	520.652 20	0.62 16	0.0069

7 weak γ's omitted (ΣIγ = 0.10%)

77BR EC DECAY (56 H 2) I(MIN) = 0.10%

Radiation Type	Energy (keV)	Intensity (%)	Δ(g-rad/ μCi-h)
Auger-L	1.32	115 7	0.0032
Auger-K	9.67	36 3	0.0074
ce-K- 3	75.14 10	0.146 6	0.0002
ce-K- 8	149.17 8	0.79 4	0.0025
ce-L- 8	160.18 8	0.134 6	0.0005
ce-K- 14	226.32 7	0.213 8	0.0010
β+ 1 max	343 3		
avg	151.6 3	0.73 3	0.0024
X-ray L	1.38	1.05 24	≈0
X-ray Kα₂	11.18140 20	15.4 10	0.0037
X-ray Kα₁	11.22240 20	30.0 18	0.0072
X-ray Kβ	12.5	7.2 5	0.0019
γ 3	87.80 10	1.40 4	0.0026
γ 5	138.95 9	0.129 6	0.0004
γ 8	161.83 8	1.10 3	0.0038
γ 9	180.68 7	0.284 10	0.0011
γ 12	200.40 7	1.21 6	0.0052
γ 14	238.98 7	23.1 5	0.118
γ 16	249.77 7	2.98 10	0.0159
γ 17	270.83 7	0.321 14	0.0019
γ 19	281.65 7	2.29 7	0.0137
γ 20	297.23 8	4.16 21	0.0263
γ 21	303.76 9	1.18 4	0.0076
γ 27	384.99 8	0.84 3	0.0069
γ 32	439.47 6	1.56 5	0.0146
γ 34	484.57 7	1.00 4	0.0103
γ 36	517.9 4	0.16 5	0.0018
γ 37	520.69 6	22.4 6	0.249
γ 39	565.91 19	0.427 17	0.0052
γ 40	567.90 8	0.86 3	0.0104

79KR B+ DECAY (CONTINUED)

Radiation Type	Energy (keV)	Intensity (%)	Δ(g-rad/ μCi-h)
X-ray Kα₁	11.92420 20	29.0 22	0.0074
X-ray Kβ	13.3	7.2 6	0.0020
γ 1	44.2 4	0.210 20	0.0002
γ 2	135.99 10	1.00 10	0.0029
γ 3	180.21 15	0.10 5	0.0004
γ 4	208.45 10	0.78 4	0.0035
γ 5	217.02 10	2.40 10	0.0111
γ 6	261.26 10	12.7 4	0.0707
γ 8	299.51 10	1.57 7	0.0100
γ 9	306.31 15	2.60 10	0.0170
γ 10	344.70 10	0.240 10	0.0018
γ 11	389.00 10	1.52 7	0.0126
γ 12	397.56 10	9.5 3	0.0804
γ 16	522.98 20	0.250 10	0.0028
γ 17	525.32 15	0.430 20	0.0048
γ 19	606.07 10	8.10 20	0.105
γ 25	832.04 10	1.26 6	0.0223
γ 30	934.81 15	0.126 7	0.0025
γ 31	1025.70 10	0.156 9	0.0034
γ 35	1115.1 3	0.370 20	0.0088
γ 39	1332.13 10	0.44 3	0.0125

22 weak γ's omitted (ΣIγ = 0.52%)
Maximum γ± intensity = 14.21%

Radiation Type	Energy (keV)	Intensity (%)	Δ(g-rad/μCi-h)
γ 41	574.64 8	1.19 4	0.0145
γ 42	578.91 7	2.96 10	0.0365
γ 43	585.48 7	1.57 5	0.0196
γ 48	755.35 7	1.67 5	0.0268
γ 51	817.79 6	2.08 7	0.0362
γ 59	1005.05 6	0.92 3	0.0198

37 weak γ's omitted (ΣIγ = 0.77%)
Maximum γ± intensity = 1.46%

79KR B+ DECAY (35.04 H 10) I(MIN) = 0.10%

Radiation Type	Energy (keV)	Intensity (%)	Δ(g-rad/μCi-h)
Auger-L	1.4	105 7	0.0032
Auger-K	10.2	31 3	0.0067
ce-K- 1	30.7 4	0.214 22	0.0001
β+ 1 max	348 8		
avg	154 4	0.180 17	0.0006
β+ 2 max	609 8		
avg	265 4	6.9 7	0.0389
total β+			
avg	262 4	7.1 7	0.0396

2 weak β's omitted (ΣIβ = 0.03%)

Radiation Type	Energy (keV)	Intensity (%)	Δ(g-rad/μCi-h)
X-ray L	1.48	1.1 4	≈0
X-ray Kα2	11.87760 20	14.9 11	0.0038

(CONTINUED ON NEXT COLUMN)

81KR EC DECAY (2.1E5 Y 2) I(MIN) = 0.10%

Radiation Type	Energy (keV)	Intensity (%)	Δ(g-rad/μCi-h)
Auger-L	1.4	111 6	0.0033
Auger-K	10.2	31 3	0.0068
X-ray L	1.48	1.1 4	≈0
X-ray Kα2	11.87760 20	15.1 9	0.0038
X-ray Kα1	11.92420 20	29.3 17	0.0075
X-ray Kβ	13.3	7.2 5	0.0021
γ 1	275.990 11	2.0 20	0.0118

82BR B- DECAY (35.30 H 3) I(MIN) = 0.10%

Radiation Type	Energy (keV)	Intensity (%)	Δ(g-rad/μCi-h)
β- 1 max	264.6 15		
avg	75.6 5	1.360 20	0.0022
β- 2 max	444.3 15		
avg	137.6 5	98.6 9	0.289
total β-			
avg	136.8 5	100.0 9	0.291
γ 1	92.184 8	0.72 4	0.0014
γ 4	137.40 20	0.14 3	0.0004
γ 6	221.45 3	2.26 7	0.0107

(CONTINUED ON NEXT PAGE)

82BR B- DECAY (CONTINUED)

Radiation Type	Energy (keV)	Intensity (%)	Δ(g-rad/ μCi-h)
γ 7	273.45 3	0.80 3	0.0047
γ 12	554.320 20	70.8 7	0.836
γ 14	606.30 10	1.17 9	0.0151
γ 15	619.070 20	43.5 5	0.573
γ 17	698.330 20	28.5 3	0.424
γ 19	776.49 3	83.6 9	1.38
γ 20	827.81 3	24.04 25	0.424
γ 23	952.10 20	0.37 3	0.0075
γ 24	1007.57 9	1.272 13	0.0273
γ 25	1043.97 3	27.2 3	0.605
γ 27	1081.40 20	0.63 4	0.0144
γ 30	1317.47 5	26.5 3	0.744
γ 32	1426	0.11 5	0.0033
γ 33	1474.82 8	16.32 17	0.513
γ 34	1650.30 10	0.743 8	0.0261
γ 35	1779.60 20	0.1136 17	0.0043

18 weak γ's omitted (ΣIγ = 0.76%)

84RB B- DECAY (33.0 D 2) I(MIN) = 0.10%
%B-=3.6 6
SEE ALSO 84RB B+ DECAY

Radiation Type	Energy (keV)	Intensity (%)	Δ(g-rad/ μCi-h)
β- 1 max	890 4		
avg	331.2 17	3.6 6	0.0254

85KR B- DECAY (10.72 Y 1) I(MIN) = 0.10%

Radiation Type	Energy (keV)	Intensity (%)	Δ(g-rad/ μCi-h)
β- 1 max	173.0 20		
avg	47.5 6	0.430 10	0.0004
β- 2 max	687.0 20		
avg	251.4 8	99.570 10	0.533
total β- avg	250.5 8	100.000 15	0.534
γ 1	513.990 10	0.430 10	0.0047

84RB B+ DECAY (32.77 D 4) I(MIN) = 0.10%
%(EC+B+)=96.4 6
SEE ALSO 84RB B- DECAY

Radiation Type	Energy (keV)	Intensity (%)	Δ(g-rad/ μCi-h)
Auger-L	1.5	77 4	0.0025
Auger-K	10.8	21.7 19	0.0050
β+ 1 max	777 3		
avg	338.5 13	14.4 3	0.104
β+ 2 max	1658 3		
avg	756.3 13	11.9 4	0.192
total β+			
avg	527.5 16	26.3 5	0.296
X-ray L	1.59	1.0 4	≈0
X-ray Kα2	12.5980 20	11.6 6	0.0031
X-ray Kα1	12.6490 20	22.4 11	0.0060
X-ray Kβ	14	5.7 3	0.0017
γ 1	881.50 10	71.0 20	1.33
γ 2	1015.90 10	0.43 7	0.0093
γ 3	1897.00 20	0.78 8	0.0315

Maximum γ± intensity = 52.60%

85KR B- DECAY (4.48 H 1) I(MIN) = 0.10%
%B-=78.9 6
SEE ALSO 85KR IT DECAY (4.48 H)

Radiation Type	Energy (keV)	Intensity (%)	Δ(g-rad/ μCi-h)
Auger-L	1.68	3.69 10	0.0001
Auger-K	11.4	1.00 4	0.0002
ce-K- 2	135.980 10	3.02 7	0.0087
ce-L- 2	149.115 10	0.340 16	0.0011
β- 1 max	710.8 20		
avg	238.3 7	0.290 10	0.0015
β- 2 max	840.7 20		
avg	290.4 7	78.6 6	0.486
total β-			
avg	290.2 7	78.9 6	0.488

1 weak β's omitted (ΣIβ = 0.02%)

Radiation Type	Energy (keV)	Intensity (%)	Δ(g-rad/ μCi-h)
X-ray Kα2	13.33580 20	0.587 18	0.0002
X-ray Kα1	13.39530 20	1.14 3	0.0003
X-ray Kβ	15	0.298 10	≈0
γ 1	129.850 20	0.300 11	0.0008
γ 2	151.180 10	75.5 9	0.243

2 weak γ's omitted (ΣIγ = 0.02%)

85KR IT DECAY (4.48 H 1) I(MIN) = 0.10%
%IT=21.1 6
SEE ALSO 85KR B- DECAY (4.48 H)

Radiation Type	Energy (keV)	Intensity (%)	Δ(g-rad/ μCi-h)
Auger-L	1.5	7.8 5	0.0002
Auger-K	10.8	2.14 21	0.0005
ce-K- 1	290.544 20	6.0 3	0.0374
ce-L- 1	302.949 20	0.90 5	0.0058
ce-MNO- 1	304.582 20	0.182 8	0.0012
X-ray L	1.59	0.10 4	≈0
X-ray $K\alpha_2$	12.5980 20	1.14 8	0.0003
X-ray $K\alpha_1$	12.6490 20	2.21 16	0.0006
X-ray $K\beta$	14	0.56 4	0.0002
γ 1	304.870 20	14.0 6	0.0909

85SR EC DECAY (64.84 D 3) I(MIN) = 0.10%

Radiation Type	Energy (keV)	Intensity (%)	Δ(g-rad/ μCi-h)
Auger-L	1.68	108.2 23	0.0039
Auger-K	11.4	29.1 9	0.0071
ce-K- 1	498.790 10	0.617 20	0.0066
X-ray L	1.69	1.6 6	≈0
X-ray $K\alpha_2$	13.33580 20	17.1 4	0.0048
X-ray $K\alpha_1$	13.39530 20	33.0 7	0.0094

86Y B+ DECAY (CONTINUED)

Radiation Type	Energy (keV)	Intensity (%)	Δ(g-rad/ μCi-h)
β+ 2 max	485 10		
avg	215 4	0.330 20	0.0015
β+ 3 max	606 10		
avg	267 5	0.37 3	0.0021
β+ 4 max	889 10		
avg	389 4	0.200 20	0.0017
β+ 5 max	933 10		
avg	408 5	1.3 3	0.0113
β+ 6 max	1066 10		
avg	467 5	2.0 5	0.0199
β+ 7 max	1195 10		
avg	524 4	1.40 20	0.0156
β+ 8 max	1254 10		
avg	550 4	12.4 5	0.145
β+ 9 max	1373 10		
avg	603 4	0.69 17	0.0089
β+10 max	1578 10		
avg	691 9	5.6 5	0.0824
β+11 max	1769 10		
avg	783 6	1.7 10	0.0284
β+12 max	2021 10		
avg	898 5	3.6 10	0.0689
β+13 max	2397 10		
avg	1093 5	0.9 9	0.0210
β+14 max	3174 10		
avg	1452 5	1.8 11	0.0557
total β+			
avg	666 6	32.9 23	0.466

6 weak β's omitted ($\Sigma I\beta$ = 0.30%)

X-ray	Kβ	15	8.66 24	0.0028
γ	1	513.990 10	98.0 10	1.07

1 weak γ's omitted (ΣIγ = 0.01%)

86RB B- DECAY (18.66 D 2) I(MIN) = 0.10%

Radiation Type		Energy (keV)	Intensity (%)	Δ(g-rad/ µCi-h)
β- 1	max	697.8 19		
	avg	232.4 7	8.76 8	0.0434
β- 2	max	1774.4 19		
	avg	709.2 8	91.24 8	1.38
total β-				
	avg	667.4 9	100.00 12	1.42
γ	1	1076.63 10	8.81 8	0.202

86Y B+ DECAY (14.74 H 2) I(MIN) = 0.10%

Radiation Type		Energy (keV)	Intensity (%)	Δ(g-rad/ µCi-h)
Auger-L		1.79	69 4	0.0026
Auger-K		12	18.0 16	0.0046
β+ 1	max	420 10		
	avg	187 4	0.30 4	0.0012

(CONTINUED ON NEXT COLUMN)

X-ray	L	1.8	1.1 4	≈0
X-ray	$K\alpha_2$	14.09790 20	11.7 6	0.0035
X-ray	$K\alpha_1$	14.16500 20	22.5 11	0.0068
X-ray	Kβ	15.8	6.1 3	0.0020
γ	1	132.34 10	0.165 9	0.0005
γ	3	182.34 20	0.11 4	0.0004
γ	4	187.87 13	1.26 5	0.0051
γ	5	190.80 13	1.01 4	0.0041
γ	6	209.80 23	0.396 17	0.0018
γ	7	235.37 23	0.396 17	0.0020
γ	8	237.9 3	0.132 25	0.0007
γ	9	252.05 13	0.371 17	0.0020
γ	11	264.53 13	0.536 25	0.0030
γ	12	307.00 10	3.46 9	0.0227
γ	13	331.08 23	0.83 3	0.0059
γ	15	370.28 17	0.82 5	0.0065
γ	16	380.4 3	0.45 4	0.0037
γ	17	382.86 23	3.63 12	0.0296
γ	18	425.97 23	0.305 17	0.0028
γ	19	439.5 3	0.20 7	0.0019
γ	20	443.13 10	16.9 5	0.160
γ	21	444.18 23	0.64 17	0.0061
γ	23	469.24 25	0.297 25	0.0030
γ	26	515.18 20	4.89 15	0.0537
γ	27	580.57 10	4.78 15	0.0592
γ	28	608.29 10	2.01 15	0.0261
γ	29	618.2 4	0.21 4	0.0028
γ	30	627.72 10	32.6 10	0.436
γ	32	644.82	2.2 4	0.0300
γ	33	645.87	9.2 11	0.126
γ	35	689.29 25	0.17 4	0.0025
γ	36	702.2 6	0.25 9	0.0037
γ	37	703.33 10	15.4 5	0.231
γ	38	709.90 10	2.62 8	0.0397
γ	39	719.17 23	0.22 4	0.0034
γ	40	740.81 13	1.36 5	0.0215

(CONTINUED ON NEXT PAGE)

86Y B+ DECAY (CONTINUED)

Radiation Type	Energy (keV)	Intensity (%)	Δ (g-rad/ μCi-h)
γ 41	767.63 13	2.4 4	0.0391
γ 42	768.25	0.32 11	0.0053
γ 43	777.37 10	22.4 6	0.372
γ 44	783.6 3	0.26 4	0.0044
γ 45	826.02 13	3.30 9	0.0581
γ 46	833.72	1.5 4	0.0264
γ 47	835.67	4.4 6	0.0778
γ 48	882.96 17	0.25 9	0.0047
γ 49	887.40 17	0.44 5	0.0083
γ 50	955.35 20	1.04 5	0.0212
γ 51	971.43 18	0.27 4	0.0056
γ 52	1017.93 23	0.18 12	0.0039
γ 53	1024.04 10	3.79 17	0.0828
γ 54	1076.63 10	82.5 4	1.89
γ 56	1092.68 13	0.69 5	0.0161
γ 57	1102.02 23	0.198 25	0.0046
γ 58	1133.3 10	0.297 25	0.0072
γ 61	1153.05 10	30.5 10	0.750
γ 63	1163.03 10	1.18 5	0.0292
γ 64	1253.11 10	1.53 5	0.0410
γ 65	1270.16 13	0.65 10	0.0176
γ 66	1283.96 13	0.29 11	0.0079
γ 67	1294.9 3	0.29 9	0.0080
γ 68	1296.03 23	0.54 4	0.0150
γ 70	1349.15 10	2.95 10	0.0846
γ 71	1404.8 4	0.18 5	0.0054
γ 72	1415.20 23	0.33 9	0.0099
γ 73	1507.86 10	0.35 5	0.0114
γ 74	1533.19 13	0.22 4	0.0073
γ 75	1535.67 13	0.12 4	0.0038
γ 76	1564.4 5	0.18 5	0.0060
γ 77	1696.25 13	0.635 17	0.0230

86ZR EC DECAY (CONTINUED)

Radiation Type	Energy (keV)	Intensity (%)	Δ (g-rad/ μCi-h)
ce-MNO- 1	27.6 5	2.00 10	0.0012
ce-K- 2	226.0 10	3.55 11	0.0171
ce-L- 2	240.6 10	0.440 10	0.0023
ce-MNO- 2	242.6 10	0.18	0.0009
X-ray L	2	4.3 14	0.0002
X-ray $K\alpha_2$	14.88290 20	33.1 19	0.0105
X-ray $K\alpha_1$	14.95840 20	64 4	0.0203
X-ray Kβ	16.7	17.5 10	0.0062
γ 1	28.0 5	19.4 6	0.0116
γ 2	243.0 10	95.80 10	0.496
γ 3	612.0 20	5.2 8	0.0678

87KR B- DECAY (76.3 M 5) I(MIN) = 0.10%

Radiation Type	Energy (keV)	Intensity (%)	Δ (g-rad/ μCi-h)
Auger-L	1.68	0.218 10	≈0
ce-K- 2	387.378 20	0.178 8	0.0015
β− 1 max	579 7		
avg	188 3	0.50 3	0.0020
β− 2 max	833 7		
avg	287 3	0.110 10	0.0007
β− 3 max	927 7		

γ 78	1711.6 7	0.17 4	0.0063
γ 79	1724.15 10	0.55 5	0.0203
γ 80	1790.90 10	1.00 5	0.0381
γ 81	1801.70 10	1.65 5	0.0633
γ 82	1854.38 13	17.2 5	0.678
γ 83	1920.72 13	20.8 7	0.851
γ 85	2017.1 6	0.132 17	0.0057
γ 86	2088.09 25	0.247 25	0.0110
γ 89	2291.8 5	0.124 9	0.0060
γ 90	2482.08 17	0.115 9	0.0061
γ 92	2567.97 18	2.25 11	0.123
γ 93	2610.11 20	1.24 8	0.0688
γ 94	2641.9 4	0.16 5	0.0093
γ 96	2794.9 4	0.206 17	0.0123
γ 99	2865.9 3	0.38 7	0.0232
γ 101	3069.7 4	0.115 17	0.0076
γ 102	3334.0 5	0.124 17	0.0088

23 weak γ's omitted (ΣIγ = 1.03%)
Maximum γ± intensity = 65.77%

86ZR EC DECAY (16.5 H 1) I(MIN)= 0.10%

Radiation Type	Energy (keV)	Intensity (%)	Δ(g-rad/ μCi-h)
Auger-L	2	189 12	0.0077
ce-K- 1	11.0 5	69.9 4	0.0163
Auger-K	12.7	46 6	0.0126
ce-L- 1	25.6 5	8.7 3	0.0047

(CONTINUED ON NEXT COLUMN)

avg	326 3	4.4 3	0.0306
β- 4 max	1077 7		
avg	389 3	0.58 3	0.0048
β- 5 max	1333 7		
avg	500 3	9.5 3	0.101
β- 6 max	1474 7		
avg	562 3	5.60 20	0.0670
β- 7 max	1510 7		
avg	579 3	0.42 6	0.0052
β- 8 max	2147 7		
avg	870 3	0.67 10	0.0124
β- 9 max	2310 7		
avg	946 3	0.16 6	0.0032
β-10 max	2498 7		
avg	1029 7	0.120 10	0.0026
β-11 max	3043 7		
avg	1294 4	6.9 4	0.190
β-12 max	3485 7		
avg	1501 3	40.8 14	1.30
β-13 max	3888 7		
avg	1694 3	30.5 22	1.10
total β- avg	1323 4	100 3	2.82
γ 2	402.578 20	49.5 16	0.424
γ 6	673.87 4	1.91 10	0.0274
γ 7	814.25 7	0.168 12	0.0029
γ 8	836.37 6	0.75 4	0.0134
γ 9	845.43 4	7.3 4	0.131
γ 12	946.64 15	0.139 11	0.0028
γ 15	1175.40 8	1.12 6	0.0281
γ 16	1337.96 8	0.65 4	0.0185
γ 17	1382.53 7	0.287 18	0.0085
γ 18	1389.91 16	0.124 11	0.0037
γ 20	1531.2 4	0.36 6	0.0116
γ 21	1577.99 14	0.129 11	0.0043

(CONTINUED ON NEXT PAGE)

87KR B- DECAY (CONTINUED)

Radiation Type	Energy (keV)	Intensity (%)	Δ(g-rad/ μCi-h)
γ 22	1611.16 16	0.104 20	0.0036
γ 23	1740.52 8	2.05 10	0.0760
γ 24	1842.61 24	0.139 11	0.0054
γ 25	2011.88 12	2.90 14	0.124
γ 27	2408.50 20	0.213 17	0.0109
γ 28	2554.80 20	9.3 6	0.506
γ 29	2558.10 20	3.9 3	0.213
γ 31	2811.40 20	0.317 23	0.0190
γ 34	3308.50 20	0.450 25	0.0317

13 weak γ's omitted (ΣIγ = 0.65%)

87SR EC DECAY (2.805 H 3) I(MIN) = 0.10%
%EC=0.30 8
FEEDING OF 2.805-H 87SR IN 87Y DECAY=100%
SEE ALSO 87SR IT DECAY (2.805 H)

Radiation Type	Energy (keV)	Intensity (%)	Δ(g-rad/ μCi-h)
Auger-L	1.68	0.32 8	≈0

87Y EC DECAY (CONTINUED)

Radiation Type	Energy (keV)	Intensity (%)	Δ(g-rad/ μCi-h)
β+ 1 max	450.6 14		
avg	200.0 6	0.160 20	0.0007
X-ray L	1.8	1.7 6	≈0
X-ray Kα₂	14.09790 20	17.5 8	0.0053
X-ray Kα₁	14.16500 20	33.8 14	0.0102
X-ray Kβ	15.8	9.1 4	0.0031
γ 2	484.70 20	92.2 10	0.952

Maximum γ± intensity = 0.32%

88KR B- DECAY (2.84 H 2) I(MIN) = 0.10%

Radiation Type	Energy (keV)	Intensity (%)	Δ(g-rad/ μCi-h)
Auger-L	1.68	14.8 7	0.0005
Auger-K	11.4	4.00 20	0.0010
ce-K- 1	12.313 14	10.7 6	0.0028
ce-L- 1	25.448 14	1.23 8	0.0007
ce-MNO- 1	27.191 14	0.214 18	0.0001
ce-K- 4	150.78 4	0.21 3	0.0007
ce-K- 7	181.120 15	1.14 12	0.0044
ce-L- 7	194.255 15	0.135 17	0.0006
β- 1 max	142 17		
avg	38 5	0.350 20	0.0003

87SR IT DECAY (2.805 H 3) I(MIN) = 0.10%
%IT=99.70 8
FEEDING OF 2.805-H 87SR IN 87Y DECAY=100%
SEE ALSO 87SR EC DECAY (2.805 H)

Radiation Type		Energy (keV)	Intensity (%)	Δ(g-rad/ μCi-h)
Auger-L		1.79	18.1 10	0.0007
Auger-K		12	4.5 4	0.0012
ce-K- 1		372.30 8	14.6 5	0.115
ce-L- 1		386.18 8	2.39 8	0.0196
ce-MNO- 1		388.04 8	0.4113 9	0.0034
X-ray	L	1.8	0.29 10	≈0
X-ray	$K\alpha_2$	14.09790 20	2.92 15	0.0009
X-ray	$K\alpha_1$	14.16500 20	5.6 3	0.0017
X-ray	Kβ	15.8	1.51 8	0.0005
γ 1		388.40 8	82.26 17	0.681

87Y EC DECAY (80.3 H 3) I(MIN) = 0.10%
SEE ALSO 87SR IT DECAY (2.805 H)
SEE ALSO 87SR EC DECAY (2.805 H)

Radiation Type		Energy (keV)	Intensity (%)	Δ(g-rad/ μCi-h)
Auger-L		1.79	105 6	0.0040
Auger-K		12	27.0 23	0.0070
ce-K- 2		468.60 20	0.240 8	0.0024

(CONTINUED ON NEXT COLUMN)

Radiation Type		Energy (keV)	Intensity (%)	Δ(g-rad/ μCi-h)
β− 2	max	365 17		
	avg	109 6	2.65 16	0.0062
β− 3	max	521 17		
	avg	165 7	67 3	0.235
β− 4	max	681 17		
	avg	227 7	9.1 5	0.0440
β− 5	max	824 17		
	avg	284 7	0.14 3	0.0008
β− 6	max	997 17		
	avg	355 8	0.200 20	0.0015
β− 7	max	1198 17		
	avg	441 8	1.92 11	0.0180
β− 8	max	1252 17		
	avg	464 8	0.23 4	0.0023
β− 9	max	1471 17		
	avg	561 8	0.22 3	0.0026
β−10	max	1731 17		
	avg	678 8	0.90 6	0.0130
β−11	max	1772 17		
	avg	697 8	0.10 6	0.0015
β−12	max	2051 17		
	avg	825 8	1.31 25	0.0230
β−13	max	2551 17		
	avg	1058 8	0.14 14	0.0032
β−14	max	2717 17		
	avg	1136 8	1.80 25	0.0436
β−15	max	2913 17		
	avg	1233 8	14 4	0.368
total β−				
	avg	358 13	100 5	0.766

6 weak β's omitted (ΣIβ = 0.25%)

Radiation Type		Energy (keV)	Intensity (%)	Δ(g-rad/ μCi-h)
X-ray	L	1.69	0.23 8	≈0
X-ray	$K\alpha_2$	13.33580 20	2.35 12	0.0007
X-ray	$K\alpha_1$	13.39530 20	4.54 21	0.0013

(CONTINUED ON NEXT PAGE)

88KR B- DECAY (CONTINUED)

Radiation Type	Energy (keV)	Intensity (%)	Δ(g-rad/ μCi-h)
X-ray Kβ	15	1.19 6	0.0004
γ 1	27.513 14	1.94 17	0.0011
γ 3	122.27 6	0.197 12	0.0005
γ 4	165.98 4	3.10 15	0.0110
γ 7	196.320 15	26.0 13	0.109
γ 8	240.71 4	0.253 14	0.0013
γ 9	311.69 3	0.107 9	0.0007
γ 10	334.71 3	0.145 10	0.0010
γ 12	362.226 13	2.25 12	0.0174
γ 14	390.543 11	0.64 6	0.0054
γ 16	421.70 18	0.128 25	0.0011
γ 17	471.80 3	0.73 4	0.0073
γ 25	677.34 5	0.235 18	0.0034
γ 30	788.28 4	0.53 3	0.0089
γ 31	790.32 7	0.125 12	0.0021
γ 34	834.830 3	13.0 7	0.231
γ 35	850.34 5	0.173 14	0.0031
γ 36	862.327 19	0.67 4	0.0123
γ 39	944.92 4	0.294 20	0.0059
γ 42	985.780 16	1.31 7	0.0276
γ 43	990.09 9	0.142 19	0.0030
γ 44	1039.59 3	0.48 3	0.0107
γ 45	1049.48 12	0.142 13	0.0032
γ 48	1141.33 6	1.28 7	0.0312
γ 49	1179.51 3	1.00 5	0.0250
γ 50	1184.95 4	0.69 5	0.0174
γ 51	1209.84 8	0.14 3	0.0037
γ 52	1212.73 17	0.14 5	0.0036
γ 53	1245.22 4	0.363 25	0.0096
γ 54	1250.67 4	1.12 6	0.0299
γ 57	1324.98 4	0.16 4	0.0045
γ 59	1352.32 11	0.159 22	0.0046

88RB B- DECAY (CONTINUED)

Radiation Type	Energy (keV)	Intensity (%)	Δ(g-rad/ μCi-h)
β- 4 max	795 11		
avg	271 5	2.13 11	0.0123
β- 5 max	895 11		
avg	312 5	0.211 17	0.0014
β- 6 max	2091 11		
avg	842 5	0.98 6	0.0176
β- 7 max	2575 11		
avg	1068 6	13.3 7	0.303
β- 8 max	3473 11		
avg	1494 6	4.2 3	0.134
β- 9 max	5309 11		
avg	2370 6	77.4 14	3.91
total β-			
avg	2067 8	99.5 16	4.38

4 weak β's omitted (ΣIβ = 0.06%)

Radiation Type	Energy (keV)	Intensity (%)	Δ(g-rad/ μCi-h)
γ 5	898.03 4	14.0 8	0.269
γ 8	1366.26 12	0.103 14	0.0030
γ 9	1382.39 5	0.74 5	0.0219
γ 11	1779.83 7	0.216 18	0.0082
γ 13	1836.00 5	21.4 13	0.837
γ 14	2111.22 12	0.118 13	0.0053
γ 17	2577.72 6	0.180 14	0.0099
γ 18	2677.86 5	1.96 12	0.112
γ 19	2734.03 7	0.109 9	0.0064
γ 20	3009.43 7	0.244 17	0.0156
γ 22	3218.48 8	0.214 14	0.0147
γ 23	3486.46 9	0.131 9	0.0097
γ 26	4742.69 11	0.143 11	0.0145

14 weak γ's omitted (ΣIγ = 0.38%)

Radiation Type	Energy (keV)	Intensity (%)	Δ(g-rad/μCi-h)
γ 60	1369.50 20	1.48 9	0.0431
γ 61	1406.94 10	0.218 20	0.0065
γ 62	1464.84 9	0.114 15	0.0036
γ 63	1518.39 3	2.15 12	0.0696
γ 64	1529.77 3	10.9 6	0.356
γ 65	1603.79 5	0.46 4	0.0156
γ 68	1685.6 4	0.66 8	0.0239
γ 72	1892.76 13	0.14 3	0.0056
γ 73	1908.7 4	0.100 15	0.0041
γ 74	2029.84 3	4.53 23	0.196
γ 75	2035.411 18	3.74 21	0.162
γ 76	2186.5 3	0.29 6	0.0134
γ 77	2195.842 7	13.2 7	0.617
γ 78	2231.772 21	3.39 17	0.161
γ 80	2352.08 4	0.73 4	0.0366
γ 82	2392.11 4	34.6 16	1.76
γ 83	2408.91 7	0.104 12	0.0053
γ 85	2548.40 3	0.62 3	0.0338
γ 86	2771.02 5	0.149 10	0.0088

36 weak γ's omitted (ΣIγ = 1.89%)

88RB B- DECAY (17.8 M 1) I(MIN) = 0.10%

Radiation Type	Energy (keV)	Intensity (%)	Δ(g-rad/μCi-h)
β- 1 max	456 11		
avg	141 4	0.67 4	0.0020
β- 2 max	464 11		
avg	144 4	0.370 20	0.0011
β- 3 max	566 11		
avg	182 5	0.144 10	0.0006

(CONTINUED ON NEXT COLUMN)

88Y EC DECAY (106.64 D 8) I(MIN) = 0.10%

Radiation Type	Energy (keV)	Intensity (%)	Δ(g-rad/μCi-h)
Auger-L	1.79	105 6	0.0040
Auger-K	12	27.1 23	0.0070
β+ 1 max	755 4		
avg	355.1 15	0.22 3	0.0017
X-ray L	1.8	1.7 6	≈0
X-ray Kα2	14.09790 20	17.6 8	0.0053
X-ray Kα1	14.16500 20	33.9 14	0.0102
X-ray Kβ	15.8	9.1 4	0.0031
γ 2	898.020 20	93.4 7	1.79
γ 4	1836.040 20	99.35 3	3.89
γ 5	2734.03 7	0.64 3	0.0370

3 weak γ's omitted (ΣIγ = 0.10%)
Maximum γ± intensity = 0.44%

89RB B- DECAY (15.2 M 1) I(MIN) = 0.10%

Radiation Type	Energy (keV)	Intensity (%)	Δ(g-rad/μCi-h)
β- 1 max	977 12		
avg	346 5	1.52 13	0.0112
β- 2 max	1258 12		
avg	466 6	33 3	0.328

(CONTINUED ON NEXT PAGE)

89RB B- DECAY (CONTINUED)

Radiation Type	Energy (keV)	Intensity (%)	Δ(g-rad/ µCi-h)
β- 3 max	1779 12		
avg	711 6	2.19 13	0.0332
β- 4 max	1916 12		
avg	761 6	3.0 9	0.0486
β- 5 max	2206 12		
avg	895 6	34 4	0.648
β- 6 max	2429 12		
avg	999 6	0.49 13	0.0104
β- 7 max	2478 12		
avg	1022 6	0.46 21	0.0100
β- 8 max	2546 12		
avg	1054 6	0.22 4	0.0049
β- 9 max	3013 12		
avg	1277 6	0.25 5	0.0068
β-10 max	3454 12		
avg	1485 6	1 4	0.0316
β-11 max	4486 12		
avg	1979 6	25 5	1.05
total β-			
avg	1012 8	102 9	2.19
7 weak β's omitted (ΣIβ = 0.38%)			
γ 3	272.45 10	1.42 12	0.0082
γ 4	289.76 10	0.54 10	0.0033
γ 8	657.71 7	10.0 9	0.140
γ 10	766.79 15	0.162 21	0.0027
γ 14	947.69 7	9.2 8	0.186
γ 16	1025.3 5	0.23 9	0.0049
γ 17	1031.88 7	58 5	1.27
γ 23	1220.32 10	0.220 24	0.0057
γ 24	1228.40 15	0.122 20	0.0032

90Y B- DECAY (64.0 H 1) I(MIN) = 0.10%

Radiation Type	Energy (keV)	Intensity (%)	Δ(g-rad/ µCi-h)
β- 1 max	2284 4		
avg	934.8 18	99.984 3	1.99

1 weak β's omitted (ΣIβ = 0.02%)

94NB B- DECAY (2.03E4 Y 16) I(MIN) = 0.10%

Radiation Type	Energy (keV)	Intensity (%)	Δ(g-rad/ µCi-h)
β- 1 max	471 3		
avg	145.6 10	100	0.310
γ 1	702.627 19	100	1.50
γ 2	871.099 18	100	1.86

Radiation Type	Energy (keV)	Intensity (%)	Δ(g-rad/ μCi-h)
γ 26	1248.10 7	43 4	1.13
γ 29	1473.22 20	0.35 4	0.0111
γ 30	1501.07 20	0.197 22	0.0063
γ 31	1538.08 10	2.55 25	0.0836
γ 35	1940.2 3	0.33 4	0.0137
γ 37	2007.54 10	2.38 24	0.102
γ 38	2058.0 11	0.23 9	0.0102
γ 40	2196.00 15	13.3 13	0.624
γ 42	2280.06 10	0.180 22	0.0087
γ 45	2570.14 10	9.9 9	0.540
γ 48	2707.20 10	2.03 19	0.117
γ 57	3508.84 25	1.15 11	0.0858

41 weak γ's omitted (ΣIγ = 1.18%)

89SR B- DECAY (50.5 D 1) I(MIN) = 0.10%

Radiation Type	Energy (keV)	Intensity (%)	Δ(g-rad/ μCi-h)
β- 1 max	1492 4		
avg	583.1 13	99.9910 10	1.24

90SR B- DECAY (29.12 Y 24) I(MIN) = 0.10%

Radiation Type	Energy (keV)	Intensity (%)	Δ(g-rad/ μCi-h)
β- 1 max	546.0 20		
avg	195.8 8	100	0.417

94NB B- DECAY (6.26 M 1) I(MIN) = 0.10%
%B-=0.48 10
SEE ALSO 94NB IT DECAY (6.26 M)

Radiation Type	Energy (keV)	Intensity (%)	Δ(g-rad/ μCi-h)
β- 1 max	1214 3		
avg	443.7 12	0.48 10	0.0045
γ 2	871.099 18	0.48 10	0.0089

94NB IT DECAY (6.26 M 1) I(MIN) = 0.10%
%IT=99.52 10
SEE ALSO 94NB B- DECAY (6.26 M)

Radiation Type	Energy (keV)	Intensity (%)	Δ(g-rad/ μCi-h)
Auger-L	2.15	91 4	0.0042
Auger-K	14	14.5 19	0.0043
ce-K- 1	22.5 5	57.62 6	0.0276
ce-L- 1	38.8 5	33.34 4	0.0276
ce-M- 1	41.0 5	6.369 7	0.0056
ce-NOP- 1	41.4 5	2.0899 21	0.0018
X-ray L	2.17	2.8 10	0.0001
X-ray $K\alpha_2$	16.52100 20	12.4 6	0.0044
X-ray $K\alpha_1$	16.61510 20	23.8 11	0.0084
X-ray Kβ	18.6	6.9 4	0.0027

1 weak γ's omitted (ΣIγ = 0.08%)

97ZR B- DECAY (16.90 H 5) I(MIN) = 0.10%
SEE ALSO 97NB IT DECAY

Radiation Type	Energy (keV)	Intensity (%)	Δ(g-rad/ μCi-h)
β- 1 max	418 16		
avg	127 6	0.41 4	0.0011
β- 2 max	559 16		
avg	178 6	5.5 3	0.0209
β- 3 max	901 16		
avg	312 6	2.10 20	0.0140
β- 4 max	915 16		
avg	318 7	0.4 3	0.0027
β- 5 max	1012 16		
avg	359 7	0.120 20	0.0009
β- 6 max	1117 16		
avg	403 7	0.45 7	0.0039
β- 7 max	1117 16		
avg	403 7	0.65 7	0.0056
β- 8 max	1389 16		
avg	521 7	0.21 11	0.0023
β- 9 max	1414 16		
avg	532 7	4.1 5	0.0465
β-10 max	1922 16		
avg	760 8	86.0 7	1.39
total β-			
avg	700 9	99.9 10	1.49
γ 3	218.87 20	0.176 19	0.0008
γ 4	254.15 20	1.25 14	0.0068
γ 5	272.27 20	0.25 4	0.0015
γ 8	330.43 20	0.11 3	0.0008
γ 9	355.39 10	2.27 24	0.0172
γ 10	400.39 20	0.32 5	0.0028
γ 11	507.63 10	5.1 6	0.0547
γ 12	513.38 20	0.56 10	0.0061

97NB B- DECAY (CONTINUED)

Radiation Type	Energy (keV)	Intensity (%)	Δ(g-rad/ μCi-h)
β- 3 max	665 16		
avg	218 6	0.210 20	0.0010
β- 4 max	910 16		
avg	315 6	1.10 10	0.0074
β- 5 max	1276 16		
avg	470 7	98.30 10	0.984
total β-			
avg	467 7	99.99 15	0.994

1 weak β's omitted (ΣIβ = 0.09%)

Radiation Type	Energy (keV)	Intensity (%)	Δ(g-rad/ μCi-h)
γ 3	480.9	0.148 20	0.0015
γ 5	657.92 10	98.34 11	1.38
γ 12	1268.63 10	0.157 20	0.0043
γ 13	1515.64 20	0.118 20	0.0038

10 weak γ's omitted (ΣIγ = 0.49%)

97NB IT DECAY (60 S 1) I(MIN) = 0.10%
FEEDING OF 60-S 97NB IN 97ZR DECAY=94.6% 9

Radiation Type	Energy (keV)	Intensity (%)	Δ(g-rad/ μCi-h)
Auger-L	2.15	1.99 12	≈0
Auger-K	14	0.44 6	0.0001
ce-K- 1	724.37 10	1.74 5	0.0268
ce-L- 1	740.66 10	0.225 7	0.0035

Radiation Type	Energy (keV)	Intensity (%)	Δ(g-rad/ μCi-h)
γ 13	602.41 20	1.39 14	0.0179
γ 14	690.63 20	0.25 4	0.0037
γ 15	699.2 3	0.121 19	0.0018
γ 16	703.80 10	0.93 10	0.0139
γ 18	804.53 10	0.65 7	0.0111
γ 19	829.80 10	0.223 19	0.0039
γ 20	854.90 10	0.33 4	0.0061
γ 21	971.39 10	0.29 3	0.0059
γ 22	1021.3 3	1.34 14	0.0293
γ 23	1110.45 20	0.111 19	0.0026
γ 24	1147.95 10	2.6 3	0.0646
γ 25	1276.09 10	0.97 10	0.0265
γ 26	1362.66 10	1.34 14	0.0390
γ 27	1750.46 10	1.34 14	0.0501
γ 28	1851.55 10	0.35 4	0.0139

5 weak γ's omitted (ΣIγ = 0.21%)

97NB B- DECAY (72.1 M 7) I(MIN) = 0.10%

Radiation Type	Energy (keV)	Intensity (%)	Δ(g-rad/ μCi-h)
Auger-L	2.27	0.180 13	≈0
ce-K- 5	637.92 10	0.182 6	0.0025
β- 1 max	305 16		
avg	89 6	0.120 20	0.0002
β- 2 max	418 16		
avg	127 6	0.170 20	0.0005

(CONTINUED ON NEXT COLUMN)

Radiation Type	Energy (keV)	Intensity (%)	Δ(g-rad/ μCi-h)
X-ray Kα2	16.52100 20	0.375 20	0.0001
X-ray Kα1	16.61510 20	0.72 4	0.0003
X-ray Kβ	18.6	0.207 12	≈0
γ 1	743.36 10	97.95 6	1.55

97RU EC DECAY (2.9 D 1) I(MIN) = 0.10%

Radiation Type	Energy (keV)	Intensity (%)	Δ(g-rad/ μCi-h)
Auger-L	2.17	97 6	0.0045
Auger-K	15.5	20 3	0.0065
ce-K- 5	194.64 4	2.94 8	0.0122
ce-L- 5	212.64 4	0.361 18	0.0016
ce-K- 6	303.51 7	0.173 9	0.0011
X-ray L	2.42	4.6 16	0.0002
X-ray Kα2	18.2508 8	20.0 9	0.0078
X-ray Kα1	18.3671 8	38.3 17	0.0150
X-ray Kβ	20.6	11.5 6	0.0050
γ 2	108.80 4	0.108 12	0.0003
γ 5	215.68 4	86.0 4	0.395
γ 6	324.55 7	10.2 4	0.0705
γ 7	460.55 5	0.117 5	0.0011
γ 11	569.27 5	0.89 4	0.0108

14 weak γ's omitted (ΣIγ = 0.34%)

99MO B- DECAY (66.0 H 2)   I(MIN) = 0.10%
SEE ALSO 99TC IT DECAY (6.02 H)

Radiation Type	Energy (keV)	Intensity (%)	Δ(g-rad/μCi-h)
Auger-L	2.17	5.4 3	0.0002
Auger-K	15.5	1.11 17	0.0004
ce-K- 2	19.543 15	3.77 7	0.0016
ce-L- 2	37.544 15	0.457 16	0.0004
ce-MNO- 2	40.043 15	0.110 6	≈0
ce-K- 4	119.422 15	0.491 24	0.0012
ce-K- 8	160.013 15	0.764 25	0.0026
ce-L- 8	178.014 15	0.114 4	0.0004
β- 1 max	214.9 10		
avg	59.9 3	0.111 3	0.0001
β- 2 max	352.7 10		
avg	104.3 4	0.134 4	0.0003
β- 3 max	436.1 10		
avg	133.0 4	16.55 7	0.0469
β- 4 max	847.6 10		
avg	289.6 4	1.17 3	0.0072
β- 5 max	1214.1 10		
avg	442.7 5	81.96 18	0.773
total β-			
avg	388.7 6	99.94 20	0.827

2 weak β's omitted (ΣIβ = 0.01%)

Radiation Type	Energy (keV)	Intensity (%)	Δ(g-rad/μCi-h)
X-ray L	2.42	0.25 9	≈0
X-ray Kα2	18.2508 8	1.12 6	0.0004
X-ray Kα1	18.3671 8	2.15 10	0.0008
X-ray Kβ	20.6	0.64 3	0.0003
γ 2	40.587 15	1.15 4	0.0010
γ 4	140.466 15	4.95 9	0.0148
γ 8	181.057 15	6.06 8	0.0234

103RU B- DECAY (39.28 D 1)   I(MIN) = 0.10%
SEE ALSO 103RH IT DECAY (56.12 M)

Radiation Type	Energy (keV)	Intensity (%)	Δ(g-rad/μCi-h)
Auger-L	2.39	1.03 10	≈0
Auger-K	17	0.21 4	≈0
ce-K- 3	30.065 5	0.69 7	0.0004
ce-K- 13	473.860 10	0.414 13	0.0042
β- 1 max	112 4		
avg	29.5 10	6.40 10	0.0040
β- 2 max	225 4		
avg	62.8 11	87.1 15	0.117
β- 3 max	467 4		
avg	143.4 13	0.231 10	0.0007
β- 4 max	722 4		
avg	238.7 14	6 3	0.0325
total β-			
avg	72.1 13	100 4	0.154

4 weak β's omitted (ΣIβ = 0.10%)

Radiation Type	Energy (keV)	Intensity (%)	Δ(g-rad/μCi-h)
X-ray Kα2	20.07370 20	0.255 20	0.0001
X-ray Kα1	20.21610 20	0.48 4	0.0002
X-ray Kβ	22.7	0.151 12	≈0
γ 3	53.285 5	0.38 4	0.0004
γ 9	294.980 10	0.252 10	0.0016
γ 12	443.800 10	0.324 10	0.0031
γ 13	497.080 10	90.1 5	0.954
γ 15	557.040 20	0.82 3	0.0097
γ 17	610.330 10	5.59 12	0.0726

13 weak γ's omitted (ΣIγ = 0.13%)

Radiation Type	Energy (keV)	Intensity (%)	Δ(g-rad/μCi-h)
γ 11	366.421 15	1.193 24	0.0093
γ 25	739.500 15	12.194 17	0.192
γ 27	777.921 20	4.32 7	0.0715
γ 28	822.972 15	0.133 4	0.0023

25 weak γ's omitted (ΣIγ = 0.27%)

99TC IT DECAY (6.02 H 3) I(MIN) = 0.10%
FEEDING OF 6.02-H 99TC IN 99MO DECAY=87.52% 18

Radiation Type	Energy (keV)	Intensity (%)	Δ(g-rad/μCi-h)
ce-M- 1	1.630 5	86.84 16	0.0030
ce-NOP- 1	2.106 5	12.18 20	0.0005
Auger-L	2.17	10.4 7	0.0005
Auger-K	15.5	2.1 4	0.0007
ce-K- 2	119.422 15	8.8 4	0.0224
ce-K- 3	121.63 3	0.69 4	0.0018
ce-L- 2	137.423 15	1.06 4	0.0031
ce-L- 3	139.63 3	0.215 14	0.0006
ce-M- 2	139.922 15	0.198 10	0.0006
X-ray L	2.42	0.49 17	≈0
X-ray Kα2	18.2508 8	2.12 13	0.0008
X-ray Kα1	18.3671 8	4.06 24	0.0016
X-ray Kβ	20.6	1.22 8	0.0005
γ 2	140.466 15	88.97 24	0.266

2 weak γ's omitted (ΣIγ = 0.02%)

103PD EC DECAY (16.96 D 2) I(MIN) = 0.10%
SEE AISO 103RH IT DECAY (56.12 M)

Radiation Type	Energy (keV)	Intensity (%)	Δ(g-rad/μCi-h)
Auger-L	2.39	91 6	0.0046
Auger-K	17	17 3	0.0060
X-ray L	2.7	4.8 16	0.0003
X-ray Kα2	20.07370 20	19.8 9	0.0085
X-ray Kα1	20.21610 20	37.7 15	0.0162
X-ray Kβ	22.7	11.7 5	0.0057

9 weak γ's omitted (ΣIγ = 0.03%)

103RH IT DECAY (56.12 M 1) I(MIN) = 0.10%
FEEDING OF 56.12-M 103RH IN 103RU DECAY=99.75% 1
FEEDING OF 56.12-M 103RH IN 103PD DECAY=99.975% 1

Radiation Type	Energy (keV)	Intensity (%)	Δ(g-rad/μCi-h)
Auger-L	2.39	76.6 15	0.0039
ce-K- 1	16.530 7	9.5 3	0.0033
Auger-K	17	1.8 3	0.0007
ce-L- 1	36.338 7	71.290 10	0.0552
ce-M- 1	39.123 7	14.4 4	0.0120
ce-NOP- 1	39.669 7	4.70 20	0.0040

(CONTINUED ON NEXT PAGE)

103RH IT DECAY (CONTINUED)

Radiation Type		Energy (keV)	Intensity (%)	Δ(g-rad/ μCi-h)
X-ray	L	2.7	4.0 13	0.0002
X-ray	$K\alpha_2$	20.07370 20	2.19 12	0.0009
X-ray	$K\alpha_1$	20.21610 20	4.17 21	0.0018
X-ray	$K\beta$	22.7	1.30 7	0.0006

1 weak γ's omitted (ΣIγ = 0.07%)

108AG EC DECAY (2.37 M 1) I(MIN) = 0.10%
%(EC+B+)=2.35 32
SEE ALSO 108AG B- DECAY (2.37 M)

Radiation Type		Energy (keV)	Intensity (%)	Δ(g-rad/ μCi-h)
Auger-L		2.5	2.00 16	0.0001
Auger-K		17.7	0.35 7	0.0001
β+ 1	max	899 7		
	avg	400 3	0.244 19	0.0021
X-ray	L	2.84	0.12 5	≈0
X-ray	$K\alpha_2$	21.02010 20	0.46 4	0.0002
X-ray	$K\alpha_1$	21.17710 20	0.87 6	0.0004
X-ray	$K\beta$	23.8	0.275 20	0.0001
γ 3		433.927 9	0.50 9	0.0046
γ 6		618.86 5	0.26 5	0.0035

10 weak γ's omitted (ΣIγ = 0.03%)
Maximum γ± intensity = 0.49%

108AG EC DECAY (CONTINUED)

Radiation Type		Energy (keV)	Intensity (%)	Δ(g-rad/ μCi-h)
X-ray	L	2.84	4.8 17	0.0003
X-ray	$K\alpha_2$	21.02010 20	18.3 8	0.0082
X-ray	$K\alpha_1$	21.17710 20	34.7 14	0.0157
X-ray	$K\beta$	23.8	11.0 5	0.0056
γ 1		433.927 9	90.3 6	0.834
γ 2		614.37 10	90.8 6	1.19
γ 3		722.95 8	90.9 6	1.40

108AG IT DECAY (127 Y) I(MIN) = 0.10%
%IT=8.9 6
SEE ALSO 108AG EC DECAY (127 Y)

Radiation Type		Energy (keV)	Intensity (%)	Δ(g-rad/ μCi-h)
Auger-L		2.6	8.2 5	0.0005
ce-K- 1		4.87 6	0.231 18	≈0
Auger-K		18.5	0.35 6	0.0001
ce-L- 1		26.57 6	6.5 5	0.0037
ce-MNO- 1		29.66 6	2.16 16	0.0014
ce-K- 2		53.69 5	1.84 14	0.0021
ce-L- 2		75.39 5	0.231 18	0.0004
X-ray	L	3	0.55 19	≈0
X-ray	$K\alpha_2$	21.9903 3	0.49 4	0.0002
X-ray	$K\alpha_1$	22.16290 10	0.93 7	0.0004
X-ray	$K\beta$	24.9	0.300 22	0.0002
γ 2		79.20 5	6.8 5	0.0114

108AG B- DECAY (2.37 M 1) I(MIN) = 0.10%
%B-=97.65 32
SEE ALSO 108AG EC DECAY (2.37 M)

Radiation Type	Energy (keV)	Intensity (%)	Δ(g-rad/μCi-h)
β- 1 max	1017 8		
avg	356 4	1.75 10	0.0133
β- 2 max	1650 8		
avg	629 4	95.9 3	1.28
total β- avg	624 4	97.6 4	1.30
γ 1	632.98 5	1.75 17	0.0236

108AG EC DECAY (127 Y 21) I(MIN) = 0.10%
%EC=91.1 6
SEE ALSO 108AG IT DECAY (127 Y)

Radiation Type	Energy (keV)	Intensity (%)	Δ(g-rad/μCi-h)
Auger-L	2.5	82 5	0.0043
Auger-K	17.7	14.2 24	0.0053
ce-K- 1	409.577 9	0.713 22	0.0062
ce-L- 1	430.323 9	0.144 5	0.0013
ce-K- 2	590.02 10	0.263 8	0.0033
ce-K- 3	698.60 8	0.173 6	0.0026

(CONTINUED ON NEXT COLUMN)

109PD B- DECAY (13.427 H 14) I(MIN) = 0.10%

Radiation Type	Energy (keV)	Intensity (%)	Δ(g-rad/μCi-h)
Auger-L	2.6	79 3	0.0044
Auger-K	18.5	7.1 11	0.0028
ce-K- 2	62.5180 21	41.8 4	0.0556
ce-L- 2	84.2262 21	44.4 6	0.0796
ce-M- 2	87.3145 21	8.7 4	0.0162
ce-NOP- 2	87.9368 21	1.49 15	0.0028
β- 1 max	1028.0 20		
avg	361.0 9	99.900 10	0.768

11 weak β's omitted (ΣIβ = 0.11%)

Radiation Type	Energy (keV)	Intensity (%)	Δ(g-rad/μCi-h)
X-ray L	3	5.3 18	0.0003
X-ray $K\alpha_2$	21.9903 3	9.9 4	0.0046
X-ray $K\alpha_1$	22.16290 10	18.7 7	0.0088
X-ray Kβ	24.9	6.04 23	0.0032
γ 2	88.0320 20	3.7 4	0.0070

29 weak γ's omitted (ΣIγ = 0.13%)

109CD EC DECAY (464 D 1) I(MIN) = 0.10%

Radiation Type	Energy (keV)	Intensity (%)	Δ(g-rad/ μCi-h)
Auger-L	2.6	166 7	0.0092
Auger-K	18.5	21 3	0.0082
ce-K- 1	62.5180 21	41.8 4	0.0557
ce-L- 1	84.2262 21	44.4 6	0.0797
ce-M- 1	87.3145 21	8.7 4	0.0162
ce-NOP- 1	87.9368 21	1.49 15	0.0028
X-ray L	3	11 4	0.0007
X-ray $K\alpha_2$	21.9903 3	29.1 10	0.0136
X-ray $K\alpha_1$	22.16290 10	55.1 18	0.0260
X-ray $K\beta$	24.9	17.8 7	0.0094
γ 1	88.0320 20	3.73 6	0.0070

110AG EC DECAY (24.6 S 2) I(MIN) = 0.10%
%EC=0.30 6
SEE ALSO 110AG B- DECAY (24.6 S)

Radiation Type	Energy (keV)	Intensity (%)	Δ(g-rad/ μCi-h)
Auger-L	2.5	0.23 5	≈0
X-ray $K\alpha_1$	21.17710 20	0.115 24	≈0

110AG IT DECAY (CONTINUED)

Radiation Type	Energy (keV)	Intensity (%)	Δ(g-rad/ μCi-h)
Auger-K	18.5	0.141 24	≈0
ce-K- 2	90.97 5	0.83 7	0.0016
ce-L- 2	112.67 5	0.39 3	0.0009
γ 1	1.28 10	0.1	≈0
X-ray $K\alpha_2$	21.9903 3	0.196 17	≈0
X-ray $K\alpha_1$	22.16290 10	0.37 3	0.0002
X-ray $K\beta$	24.9	0.120 10	≈0

110AG B- DECAY (249.9 D 1) I(MIN) = 0.10%
%B-=98.67 10
SEE ALSO 110AG IT DECAY (249.9 D)

Radiation Type	Energy (keV)	Intensity (%)	Δ(g-rad/ μCi-h)
Auger-L	2.72	0.337 25	≈0
ce-K- 26	631.038 10	0.250 10	0.0034
β^- 1 max	83.9 19		
avg	21.5 6	67.5 9	0.0309
β^- 2 max	133.8 19		
avg	35.3 6	0.408 12	0.0003

110AG B- DECAY (24.6 S 2) I(MIN) = 0.10%
%B-=99.70 6
SEE ALSO 110AG EC DECAY (24.6 S)

Radiation Type	Energy (keV)	Intensity (%)	Δ(g-rad/ μCi-h)
β- 1 max	2235.0 19		
avg	894.1 9	4.4 3	0.0838
β- 2 max	2892.8 19		
avg	1199.3 9	95.2 3	2.43
total β- avg	1185.1 9	99.7 5	2.52

9 weak β's omitted (ΣIβ = 0.09%)

Radiation Type	Energy (keV)	Intensity (%)	Δ(g-rad/ μCi-h)
γ 2	657.749 10	4.50 23	0.0630

12 weak γ's omitted (ΣIγ = 0.10%)

110AG IT DECAY (249.9 D 1) I(MIN) = 0.10%
%IT=1.33 10
SEE ALSO 110AG B- DECAY (249.9 D)

Radiation Type	Energy (keV)	Intensity (%)	Δ(g-rad/ μCi-h)
ce-M- 1	0.56 10	1.00 8	≈0
ce-NCP- 1	1.18 10	0.332 25	≈0
Auger-L	2.6	1.11 8	≈0

(CONTINUED ON NEXT COLUMN)

Radiation Type	Energy (keV)	Intensity (%)	Δ(g-rad/ μCi-h)
β- 3 max	530.7 19		
avg	165.1 7	30.6 4	0.108
total β- avg	66.2 14	98.7 10	0.139

4 weak β's omitted (ΣIβ = 0.18%)

Radiation Type	Energy (keV)	Intensity (%)	Δ(g-rad/ μCi-h)
X-ray Kα1	23.17360 20	0.172 8	≈0
γ 12	365.441 15	0.106 9	0.0008
γ 17	446.797 8	3.66 4	0.0348
γ 23	620.346 11	2.78 3	0.0367
γ 24	626.246 10	0.235 7	0.0031
γ 26	657.749 10	94.7 10	1.33
γ 27	676.60 10	0.142 19	0.0020
γ 28	677.606 11	10.72 11	0.155
γ 29	686.988 11	6.49 7	0.0950
γ 30	706.670 13	16.74 12	0.252
γ 31	708.115 20	0.28 10	0.0043
γ 32	744.260 13	4.66 5	0.0739
γ 33	763.928 13	22.36 23	0.364
γ 35	818.016 12	7.32 8	0.128
γ 36	884.667 13	72.9 8	1.37
γ 37	937.478 13	34.3 4	0.685
γ 39	997.233 18	0.125 5	0.0027
γ 49	1334.304 17	0.133 10	0.0038
γ 50	1384.270 13	24.35 25	0.718
γ 52	1475.759 22	3.99 4	0.125
γ 53	1505.001 21	13.11 14	0.420
γ 54	1562.266 22	1.184 13	0.0394

40 weak γ's omitted (ΣIγ = 0.92%)

111AG B- DECAY (7.45 D 1) I(MIN) = 0.10%

Radiation Type		Energy (keV)	Intensity (%)	Δ(g-rad/ µCi-h)
Auger-L		2.72	0.200 17	≈0
ce-K- 4		315.419 20	0.105 8	0.0007
β^- 1	max	692.9 20		
	avg	226.2 8	6.3 16	0.0304
β^- 2	max	789.6 20		
	avg	281.6 8	1.0 3	0.0060
β^- 3	max	1035.0 20		
	avg	363.3 9	92.7 16	0.717
total β^-	avg	353.7 9	100.0 23	0.754

3 weak β's omitted ($\Sigma I\beta$ = 0.04%)

Radiation Type		Energy (keV)	Intensity (%)	Δ(g-rad/ µCi-h)
X-ray	$K\alpha_1$	23.17360 20	0.102 6	≈0
γ 1		96.750 20	0.121 13	0.0002
γ 2		245.400 20	1.23 6	0.0064
γ 4		342.130 20	6.7 4	0.0488

7 weak γ's omitted ($\Sigma I\gamma$ = 0.05%)

111IN EC DECAY (2.83 D 1) I(MIN) = 0.10%

Radiation Type	Energy (keV)	Intensity (%)	Δ(g-rad/ µCi-h)
Auger-L	2.72	100 6	0.0058
Auger-K	19.3	16 3	0.0064
ce-K- 1	144.57 3	7.89 23	0.0243

113SN EC DECAY (115.1 D 2) I(MIN) = 0.10%
SEE ALSO 113IN IT DECAY (1.658 H)

Radiation Type		Energy (keV)	Intensity (%)	Δ(g-rad/ µCi-h)
Auger-L		2.84	85 6	0.0052
Auger-K		20	12.8 25	0.0055
X-ray	L	3.29	6.7 23	0.0005
X-ray	$K\alpha_2$	24.00200 20	20.7 8	0.0106
X-ray	$K\alpha_1$	24.20970 20	39.0 14	0.0201
X-ray	$K\beta$	27.3	13.0 5	0.0075
γ 1		255.115 15	1.85 6	0.0101

115IN IT DECAY (4.3 H 1) I(MIN) = 0.10%
%IT=96.3 8
SEE ALSO 115IN B- DECAY (4.3 H)

Radiation Type		Energy (keV)	Intensity (%)	Δ(g-rad/ µCi-h)
Auger-L		2.84	43 3	0.0026
Auger-K		20	6.0 12	0.0026
ce-K- 1		308.30 3	39.9 7	0.262
ce-L- 1		332.00 3	8.38 21	0.0592
ce-M- 1		335.42 3	1.70 5	0.0122
ce-NOP- 1		336.12 3	0.376 11	0.0027
X-ray	L	3.29	3.3 12	0.0002
X-ray	$K\alpha_2$	24.00200 20	9.7 4	0.0049
X-ray	$K\alpha_1$	24.20970 20	18.2 7	0.0094
X-ray	$K\beta$	27.3	6.0 3	0.0035
γ 1		336.241 25	46.7 6	0.335

ce-L- 1	167.26 3	0.96 4	0.0034
ce-MNO- 1	170.51 3	0.222 11	0.0008
ce-K- 2	218.68 3	4.94 11	0.0230
ce-L- 2	241.37 3	0.725 14	0.0037
ce-MNO- 2	244.62 3	0.158 8	0.0008
X-ray L	3.13	7.0 24	0.0005
X-ray Kα2	22.98410 20	23.5 9	0.0115
X-ray Kα1	23.17360 20	44.3 16	0.0219
X-ray Kβ	26	14.5 6	0.0081
γ 1	171.28 3	90.93 22	0.332
γ 2	245.39 3	94.17 11	0.492

113IN IT DECAY (1.658 H 1) I(MIN) = 0.10%
FEEDING OF 1.658-H 113IN IN 113SN DECAY=100%

Radiation Type	Energy (keV)	Intensity (%)	Δ(g-rad/ μCi-h)
Auger-L	2.84	29.7 18	0.0018
Auger-K	20	4.2 9	0.0018
ce-K- 1	363.748 10	28.2 3	0.218
ce-L- 1	387.450 10	5.48 7	0.0452
ce-M- 1	390.862 10	1.100 10	0.0092
ce-NOP- 1	391.566 10	0.245 3	0.0020
X-ray L	3.29	2.3 8	0.0002
X-ray Kα2	24.00200 20	6.8 3	0.0035
X-ray Kα1	24.20970 20	12.9 5	0.0066
X-ray Kβ	27.3	4.27 18	0.0025
γ 1	391.688 10	64.90 20	0.541

115IN B- DECAY (4.3 H 1) I(MIN) = 0.10%
%B-=3.7 8
SEE ALSO 115IN IT DECAY (4.3 H)

Radiation Type	Energy (keV)	Intensity (%)	Δ(g-rad/ μCi-h)
β- 1 max	859 9		
avg	290 3	3.7 8	0.0229

1 weak β's omitted (ΣIβ = 0.04%)

116IN B- DECAY (54.15 M 6) I(MIN) = 0.10%

Radiation Type	Energy (keV)	Intensity (%)	Δ(g-rad/ μCi-h)
Auger-L	3	0.73 10	≈0
Auger-K	21	0.12 3	≈0
ce-K- 4	109.126 8	0.86 11	0.0020
β- 1 max	302 8		
avg	87 3	0.33 4	0.0006
β- 2 max	352 8		
avg	103 3	2.71 10	0.0059
β- 3 max	597 8		
avg	189 3	10.2 4	0.0411
β- 4 max	869 8		
avg	294 4	33.6 15	0.210

(CONTINUED ON NEXT PAGE)

116IN B- DECAY (CONTINUED)

Radiation Type	Energy (keV)	Intensity (%)	Δ(g-rad/ μCi-h)
β- 5 max	1007 8		
avg	351 4	51.8 11	0.387
total β-			
avg	307 4	98.6 19	0.645
X-ray Kα2	25.04400 20	0.21 3	0.0001
X-ray Kα1	25.27130 20	0.39 5	0.0002
X-ray Kβ	28.5	0.133 17	≈0
γ 4	138.326 8	3.29 12	0.0097
γ 8	262.95 8	0.12 3	0.0007
γ 10	278.49 8	0.143 17	0.0009
γ 11	303.80 7	0.118 17	0.0008
γ 13	355.36 4	0.83 5	0.0063
γ 14	416.86 3	29.2 15	0.259
γ 17	463.14 12	0.83 5	0.0082
γ 22	655.7 4	0.11 5	0.0015
γ 24	689.0 3	0.16 3	0.0024
γ 25	705.7 3	0.17 3	0.0025
γ 28	781.1 8	0.110 21	0.0018
γ 29	818.70 20	11.5 5	0.200
γ 32	972.55 3	0.454 17	0.0094
γ 34	1097.30 20	56.2 12	1.31
γ 37	1293.54 4	84.4 18	2.33
γ 38	1507.40 20	10.0 4	0.320
γ 40	1753.8 6	2.46 8	0.0917
γ 42	2112.1 4	15.5 5	0.699

25 weak γ's omitted (ΣIγ = 1.02%)

117SB EC DECAY (CONTINUED)

Radiation Type	Energy (keV)	Intensity (%)	Δ(g-rad/ μCi-h)
ce-L- 1	154.097 15	1.46 5	0.0048
ce-MNO- 1	157.678 15	0.4896 23	0.0016
β+ 1 max	564 18		
avg	258 8	1.7 3	0.0093
X-ray L	3.44	8 3	0.0006
X-ray Kα2	25.04400 20	23.5 9	0.0125
X-ray Kα1	25.27130 20	44.1 16	0.0238
X-ray Kβ	28.5	14.9 6	0.0090
γ 1	158.562 15	85.9 4	0.290
γ 4	861.35 5	0.31 4	0.0057
γ 5	1004.51 15	0.21 3	0.0044
γ 6	1020.6 5	0.103 18	0.0022
γ 7	1021.0 5	0.112 18	0.0024

7 weak γ's omitted (ΣIγ = 0.26%)
Maximum γ± intensity = 3.40%

119SN IT DECAY (293.0 D 13) I(MIN) = 0.10%

Radiation Type	Energy (keV)	Intensity (%)	Δ(g-rad/ μCi-h)
Auger-L	3	137 5	0.0086
ce-L- 1	19.410 10	67.4 7	0.0279
Auger-K	21	4.5 9	0.0020

117SN IT DECAY (13.61 D 4) I(MIN) = 0.10%

Radiation Type	Energy (keV)	Intensity (%)	Δ(g-rad/ μCi-h)
Auger-L	3	91 5	0.0057
Auger-K	21	10.8 22	0.0048
ce-K- 1	126.82 3	64.8 7	0.175
ce-K- 2	129.362 15	11.7 3	0.0322
ce-L- 1	151.56 3	26.1 11	0.0843
ce-L- 2	154.097 15	1.48 4	0.0049
ce-M- 1	155.14 3	5.7 3	0.0188
ce-NOP- 1	155.88 3	1.35 6	0.0045
ce-MNO- 2	157.678 15	0.350 10	0.0012
X-ray L	3.44	8 3	0.0006
X-ray $K\alpha_2$	25.04400 20	18.7 7	0.0100
X-ray $K\alpha_1$	25.27130 20	35.1 13	0.0189
X-ray $K\beta$	28.5	11.8 5	0.0072
γ 1	156.02 3	2.110 10	0.0070
γ 2	158.562 15	86.4 4	0.292

117SB EC DECAY (2.80 H 1) I(MIN) = 0.10%

Radiation Type	Energy (keV)	Intensity (%)	Δ(g-rad/ μCi-h)
Auger-L	3	94 6	0.0059
Auger-K	21	14 3	0.0061
ce-K- 1	129.362 15	11.7 4	0.0322
ce-M- 1	22.991 10	13.2 4	0.0065
ce-NOP- 1	23.738 10	3.40 10	0.0017
ce-K- 2	36.460 10	32.0 6	0.0249
ce-L- 2	61.195 10	52.20 10	0.0680
ce-M- 2	64.776 10	12.8 3	0.0177
ce-NOP- 2	65.523 10	3.00 10	0.0042
X-ray L	3.44	12 4	0.0009
γ 1	23.875 10	16.2 4	0.0082
X-ray $K\alpha_2$	25.04400 20	7.8 4	0.0042
X-ray $K\alpha_1$	25.27130 20	14.7 6	0.0079
X-ray $K\beta$	28.5	4.96 21	0.0030

(CONTINUED ON NEXT COLUMN)

1 weak γ's omitted ($\Sigma I\gamma$ = 0.06%)

121TE EC DECAY (17 D 1) I(MIN) = 0.10%

Radiation Type	Energy (keV)	Intensity (%)	Δ(g-rad/ μCi-h)
Auger-L	3	84 5	0.0055
ce-K- 1	6.647 10	1.120 12	0.0002
Auger-K	21.8	11.6 25	0.0054
ce-I- 1	32.440 10	0.145 5	0.0001
ce-K- 2	35.057 13	0.49 3	0.0004
ce-K- 4	477.100 11	0.127 6	0.0013
ce-K- 5	542.648 11	0.369 14	0.0043
X-ray L	3.6	8 3	0.0006
X-ray $K\alpha_2$	26.11080 20	21.4 8	0.0119

(CONTINUED ON NEXT PAGE)

121TE EC DECAY (CONTINUED)

Radiation Type	Energy (keV)	Intensity (%)	Δ(g-rad/μCi-h)
X-ray $K\alpha_1$	26.35910 20	40.2 14	0.0225
X-ray $K\beta$	29.7	13.7 6	0.0087
γ 1	37.138 10	0.117 4	≈0
γ 2	65.548 13	0.259 9	0.0004
γ 3	470.472 13	1.41 4	0.0141
γ 4	507.591 11	17.7 5	0.191
γ 5	573.139 11	80.3 18	0.981

121TE EC DECAY (154 D 7) I(MIN) = 0.10%
%EC=11.4 11
SEE ALSO 121TE IT DECAY (154 D)

Radiation Type	Energy (keV)	Intensity (%)	Δ(g-rad/μCi-h)
Auger-L	3	18.0 17	0.0012
ce-K- 1	6.647 10	9.0 13	0.0013
Auger-K	21.8	2.5 6	0.0011
ce-L- 1	32.440 10	1.17 17	0.0008
ce-M- 1	36.194 10	0.23 4	0.0002
X-ray L	3.6	1.7 6	0.0001
X-ray $K\alpha_2$	26.11080 20	4.6 5	0.0026
X-ray $K\alpha_1$	26.35910 20	8.6 8	0.0048
X-ray $K\beta$	29.7	2.9 3	0.0019
γ 1	37.138 10	0.94 14	0.0007
γ 8	1102.149 18	2.5 3	0.0586

8 weak γ's omitted (ΣIγ = 0.16%)

122SB EC DECAY (2.70 D 1) I(MIN) = 0.10%
%(EC+B+)=2.38 13
SEE ALSO 122SB B- DECAY

Radiation Type	Energy (keV)	Intensity (%)	Δ(g-rad/μCi-h)
Auger-L	3	2.00 15	0.0001
Auger-K	21	0.29 6	0.0001
X-ray L	3.44	0.17 6	≈0
X-ray $K\alpha_2$	25.04400 20	0.50 4	0.0003
X-ray $K\alpha_1$	25.27130 20	0.93 6	0.0005
X-ray $K\beta$	28.5	0.315 21	0.0002
γ 1	1140.67 4	0.74 6	0.0180

122SB B- DECAY (2.70 D 1) I(MIN) = 0.10%
%B-=97.62 13
SEE ALSO 122SB EC DECAY

Radiation Type	Energy (keV)	Intensity (%)	Δ(g-rad/μCi-h)
Auger-L	3.19	0.290 21	≈0
ce-K- 1	532.43 4	0.350 11	0.0040
β- 1 max	724 4		
avg	236.6 15	4.64 13	0.0234
β- 2 max	1417 4		
avg	522.3 17	66.6 3	0.741

121TE IT DECAY (154 D 7) I(MIN) = 0.10%
%IT=88.6 11
SEE ALSO 121TE EC DECAY (154 D)

Radiation Type	Energy (keV)	Intensity (%)	Δ(g-rad/ μCi-h)
Auger-L	3.19	72 4	0.0049
Auger-K	22.7	5.1 12	0.0025
ce-K- 1	49.974 15	34.5 8	0.0367
ce-L- 1	76.849 15	41.6 9	0.0680
ce-M- 1	80.782 15	9.9 3	0.0171
ce-NOP- 1	81.620 15	2.65 9	0.0046
ce-K- 2	180.38 3	6.12 19	0.0235
ce-L- 2	207.25 3	0.81 3	0.0036
ce-MNO- 2	211.18 3	0.198 7	0.0009
X-ray L	3.77	7.1 24	0.0006
X-ray $K\alpha_2$	27.20170 20	10.1 4	0.0059
X-ray $K\alpha_1$	27.47230 20	18.9 8	0.0110
X-ray $K\beta$	31	6.5 3	0.0043
γ 2	212.19 3	81.5 11	0.368

1 weak γ's omitted ($\Sigma I\gamma$ = 0.05%)

β^- 3 max	1981 4		
avg	772.1 17	26.40 20	0.434
total β^- avg	576.1 18	97.7 4	1.20

3 weak β's omitted ($\Sigma I\beta$ = 0.04%)

X-ray $K\alpha_1$	27.47230 20	0.163 8	≈0
γ 1	564.24 4	70.04 22	0.842
γ 3	692.65 4	3.82 13	0.0563
γ 6	1256.93 4	0.81 5	0.0216

4 weak γ's omitted ($\Sigma I\gamma$ = 0.04%)

123I EC DECAY (13.2 H 1) I(MIN) = 0.10%

Radiation Type	Energy (keV)	Intensity (%)	Δ(g-rad/ μCi-h)
Auger-L	3.19	95 6	0.0064
Auger-K	22.7	12 3	0.0060
ce-K- 2	127.16 5	14.1 4	0.0381
ce-L- 2	154.03 5	1.90 8	0.0062
ce-MNO- 2	157.96 5	0.41 5	0.0014
X-ray L	3.77	9 4	0.0008
X-ray $K\alpha_2$	27.20170 20	24.7 9	0.0143
X-ray $K\alpha_1$	27.47230 20	46.2 16	0.0270
X-ray $K\beta$	31	16.0 6	0.0106
γ 2	158.97 5	82.8 5	0.280

(CONTINUED ON NEXT PAGE)

123I EC DECAY (CONTINUED)

Radiation Type	Energy (keV)	Intensity (%)	Δ(g-rad/ μCi-h)
γ 22	346.35 5	0.125 5	0.0009
γ 25	440.02 5	0.425 15	0.0040
γ 28	505.33 5	0.314 10	0.0034
γ 29	528.96 5	1.38 5	0.0156
γ 30	538.54 5	0.379 13	0.0043

42 weak γ's omitted (ΣIγ = 0.48%)

124SB B- DECAY (60.20 D 3) I(MIN) = 0.10%

Radiation Type	Energy (keV)	Intensity (%)	Δ(g-rad/ μCi-h)
Auger-L	3.19	0.341 24	≈0
ce-K- 12	570.91 4	0.411 13	0.0050
β- 1 max	131.0 19		
avg	33.1 5	0.530 10	0.0004
β- 2 max	204.3 19		
avg	54.4 6	0.560 10	0.0006
β- 3 max	212.2 19		
avg	56.7 6	8.80 20	0.0106
β- 4 max	384.5 19		
avg	111.6 6	0.130 10	0.0003
β- 5 max	422.6 19		
avg	124.5 7	0.670 10	0.0018

124SB B- DECAY (CONTINUED)

Radiation Type	Energy (keV)	Intensity (%)	Δ(g-rad/ μCi-h)
γ 22	790.78 6	0.744 10	0.0125
γ 26	968.25 6	1.83 5	0.0378
γ 29	1045.24 5	1.84 4	0.0410
γ 35	1325.49 7	1.41 3	0.0398
γ 36	1355.17 8	0.93 3	0.0269
γ 37	1368.23 6	2.35 5	0.0685
γ 38	1376.10 20	0.431 10	0.0126
γ 40	1436.66 9	1.02 3	0.0312
γ 41	1445.25 20	0.206 10	0.0063
γ 43	1489.03 20	0.55 3	0.0174
γ 44	1526.33 20	0.392 10	0.0127
γ 45	1579.90 20	0.196 10	0.0066
γ 47	1691.02 4	48.8 5	1.76
γ 53	2091.0 3	5.58 10	0.249

39 weak γ's omitted (ΣIγ = 1.23%)

124I B+ DECAY (4.18 D 2) I(MIN) = 0.10%

Radiation Type	Energy (keV)	Intensity (%)	Δ(g-rad/ μCi-h)
Auger-L	3.19	64 4	0.0043
Auger-K	22.7	8.4 19	0.0040
ce-K- 9	570.91 4	0.256 23	0.0031
β+ 1 max	810 4		
avg	365.7 18	0.295 17	0.0023

β^- 6 max	612.2 19		
avg	192.1 7	52.0 10	0.213
β^- 7 max	681.5 19		
avg	218.0 7	0.110 10	0.0005
β^- 8 max	723.3 19		
avg	234.0 7	0.260 10	0.0013
β^- 9 max	814.1 19		
avg	269.2 7	0.680 10	0.0039
β^-10 max	866.6 19		
avg	290.0 8	3.63 7	0.0224
β^-11 max	948.1 19		
avg	322.6 8	2.0 4	0.0137
β^-12 max	1580.4 19		
avg	591.1 8	5.39 11	0.0679
β^-13 max	1657.4 19		
avg	625.0 8	2.45 5	0.0326
β^-14 max	2303.2 19		
avg	916.3 9	22.6 5	0.441
total β^-			
avg	380.5 14	100.0 13	0.811

7 weak β's omitted ($\Sigma I\beta$ = 0.21%)

X-ray $K\alpha_2$	27.20170 20	0.102 5	≈0
X-ray $K\alpha_1$	27.47230 20	0.191 9	0.0001
γ 5	400.03 6	0.53 4	0.0045
γ 6	443.99 5	0.35 3	0.0033
γ 11	525.50 10	0.313 20	0.0035
γ 12	602.72 4	97.92 5	1.26
γ 13	632.36 10	0.157 10	0.0021
γ 14	645.82 4	7.21 22	0.0991
γ 16	709.31 5	1.42 5	0.0215
γ 17	713.82 4	2.39 12	0.0363
γ 18	722.78 4	11.26 16	0.173
γ 19	735.85 15	0.127 10	0.0020

(CONTINUED ON NEXT COLUMN)

β^+ 2 max	1532 4		
avg	685.9 18	11.3 6	0.165
β^+ 3 max	2135 4		
avg	973.6 18	11.3 6	0.234
total β^+			
avg	823.7 19	22.9 9	0.402
X-ray L	3.77	6.3 22	0.0005
X-ray $K\alpha_2$	27.20170 20	16.7 7	0.0097
X-ray $K\alpha_1$	27.47230 20	31.1 13	0.0182
X-ray $K\beta$	31	10.8 5	0.0071
γ 6	541.20 10	0.189 17	0.0022
γ 9	602.72 4	61 5	0.783
γ 11	645.82 4	0.95 9	0.0131
γ 13	695.0 10	0.122 10	0.0018
γ 16	713.80 20	0.110 16	0.0017
γ 17	722.78 4	10.1 9	0.155
γ 27	968.22 8	0.41 4	0.0086
γ 28	976.32 14	0.104 15	0.0022
γ 31	1045.00 10	0.43 5	0.0095
γ 32	1054.00 20	0.122 12	0.0027
γ 40	1325.50 4	1.45 12	0.0408
γ 42	1368.20 6	0.29 3	0.0084
γ 43	1376.00 10	1.68 14	0.0492
γ 48	1488.90 10	0.183 17	0.0058
γ 49	1509.49 4	3.01 25	0.0969
γ 50	1559.80 20	0.16 3	0.0055
γ 52	1637.7 5	0.195 21	0.0068
γ 54	1675.8 4	0.11 3	0.0039
γ 55	1691.02 4	10.5 9	0.378
γ 56	1720.37 14	0.171 19	0.0063
γ 58	1851.4 4	0.21 3	0.0082
γ 59	1918.58 4	0.159 23	0.0065
γ 60	2038.3 3	0.34 3	0.0149
γ 61	2078.86 7	0.35 4	0.0154
γ 62	2091.00 10	0.57 5	0.0255

(CONTINUED ON NEXT PAGE)

124I B+ DECAY (CONTINUED)

Radiation Type	Energy (keV)	Intensity (%)	Δ(g-rad/ μCi-h)
γ 64	2099.09 9	0.140 13	0.0063
γ 65	2144.320 10	0.110 11	0.0050
γ 66	2232.25 7	0.57 5	0.0273
γ 67	2283.25 8	0.66 7	0.0323
γ 75	2746.90 10	0.46 5	0.0271

48 weak γ's omitted (ΣIγ = 1.24%)
Maximum γ± intensity = 45.79%

125I EC DECAY (60.14 D 11) I(MIN) = 0.10%

Radiation Type	Energy (keV)	Intensity (%)	Δ(g-rad/ μCi-h)
Auger-L	3.19	156 9	0.0106
ce-K- 1	3.6781 6	80.0 6	0.0063
Auger-K	22.7	20 5	0.0097
ce-L- 1	30.5527 6	10.5 3	0.0068
ce-M- 1	34.4859 6	2.20 20	0.0016
ce-NOP- 1	35.3236 6	0.70 20	0.0005
X-ray L	3.77	15 6	0.0012
X-ray Kα2	27.20170 20	39.8 14	0.0230
X-ray Kα1	27.47230 20	74.2 25	0.0434
X-ray Kβ	31	25.8 10	0.0170
γ 1	35.4919 5	6.67 22	0.0050

126I B- DECAY (13.02 D 7) I(MIN) = 0.10%
%B-=43.7 20
SEE ALSO 126I EC DECAY

Radiation Type	Energy (keV)	Intensity (%)	Δ(g-rad/ μCi-h)
Auger-L	3.43	0.51 5	≈0
ce-K- 1	354.072 11	0.53 5	0.0040
β- 1 max	371 5		
avg	108.9 17	3.6 3	0.0084
β- 2 max	862 5		
avg	289.7 20	32 3	0.197
β- 3 max	1251 5		
avg	449.5 22	8 3	0.0766
total β- avg	304.1 22	44 5	0.282
X-ray Kα2	29.4580 10	0.134 12	≈0
X-ray Kα1	29.7790 10	0.249 22	0.0002
γ 1	388.633 11	34 3	0.282
γ 2	491.243 11	2.85 22	0.0298
γ 3	879.876 13	0.75 6	0.0141

127XE EC DECAY (36.41 D 2) I(MIN) = 0.10%

Radiation Type	Energy (keV)	Intensity (%)	Δ(g-rad/ μCi-h)
Auger-L	3.3	96 6	0.0068
Auger-K	23.6	12 3	0.0059
ce-K- 1	24.431 20	4.28 18	0.0022

126I EC DECAY (13.02 D 7) I(MIN) = 0.10%
% (EC+B+)=56.3 20
SEE ALSO 126I B- DECAY

Radiation Type	Energy (keV)	Intensity (%)	Δ(g-rad/ μCi-h)
Auger-L	3.19	43 3	0.0029
Auger-K	22.7	5.7 13	0.0027
ce-K- 1	634.517 12	0.108 9	0.0015
β^+ 1 max	1134 5		
avg	508.4 23	3.34 22	0.0362
X-ray L	3.77	4.3 15	0.0003
X-ray $K\alpha_2$	27.20170 20	11.3 7	0.0065
X-ray $K\alpha_1$	27.47230 20	21.1 13	0.0123
X-ray $K\beta$	31	7.3 5	0.0048
γ 1	666.331 12	33.1 25	0.470
γ 3	753.819 13	4.2 4	0.0669
γ 6	1420.19 3	0.295 23	0.0089

Maximum $\gamma\pm$ intensity = 6.68%

Radiation Type	Energy (keV)	Intensity (%)	Δ(g-rad/ μCi-h)
ce-L- 1	52.412 20	0.61 3	0.0007
ce-M- 1	56.528 20	0.123 6	0.0001
ce-K- 2	112.05 3	1.54 7	0.0037
ce-K- 3	138.93 3	3.65 16	0.0108
ce-L- 2	140.03 3	0.391 17	0.0012
ce-L- 3	166.91 3	0.475 15	0.0017
ce-K- 4	169.67 3	6.63 7	0.0240
ce-L- 4	197.65 3	0.98 3	0.0041
ce-M- 4	201.77 3	0.198 3	0.0009
ce-K- 5	341.79 5	0.289 13	0.0021
X-ray L	4	10 4	0.0008
X-ray $K\alpha_2$	28.3172 4	25.1 10	0.0151
X-ray $K\alpha_1$	28.6120 3	46.7 17	0.0285
X-ray $K\beta$	32.3	16.4 7	0.0113
γ 1	57.600 20	1.33 6	0.0016
γ 2	145.22 3	4.29 14	0.0133
γ 3	172.10 3	25.5 8	0.0936
γ 4	202.84 3	68.3 4	0.295
γ 5	374.96 5	17.2 6	0.137

1 weak γ's omitted ($\Sigma I\gamma$ = 0.01%)

129TE B- DECAY (69.6 M 2) I(MIN) = 0.10%

Radiation Type	Energy (keV)	Intensity (%)	Δ(g-rad/ μCi-h)
Auger-L	3.3	59 7	0.0041
ce-L- 1	22.62 4	65 7	0.0313
ce-M- 1	26.74 4	13.1 14	0.0075
ce-NOP- 1	27.62 4	4.4 5	0.0026

(CONTINUED ON NEXT PAGE)

129TE B- DECAY (CONTINUED)

Radiation Type	Energy (keV)	Intensity (%)	Δ(g-rad/ μCi-h)
β- 1 max	386 4		
avg	114.1 14	0.84 3	0.0020
β- 2 max	668 4		
avg	214.9 15	0.204 9	0.0009
β- 3 max	938 4		
avg	320.6 17	0.240 10	0.0016
β- 4 max	1011 4		
avg	350.0 17	8.9 4	0.0663
β- 5 max	1220 4		
avg	437.0 17	0.54 3	0.0050
β- 6 max	1470 4		
avg	544.5 18	89.3 9	1.04
total β-			
avg	521.5 19	100.1 10	1.11

4 weak β's omitted (ΣIβ = 0.06%)

Radiation Type	Energy (keV)	Intensity (%)	Δ(g-rad/ μCi-h)
X-ray L	4	6.2 22	0.0005
γ 1	27.81 4	15.6 17	0.0092
γ 2	208.96 5	0.172 8	0.0008
γ 5	250.62 5	0.366 16	0.0020
γ 7	278.43 5	0.542 23	0.0032
γ 8	281.26 5	0.158 7	0.0009
γ 14	459.60 5	7.4 4	0.0721
γ 16	487.39 6	1.35 6	0.0141
γ 32	802.10 6	0.183 8	0.0031
γ 42	1083.85 6	0.471 21	0.0109
γ 43	1111.64 6	0.183 10	0.0043

37 weak γ's omitted (ΣIγ = 0.39%)

129TE B- DECAY (CONTINUED)

Radiation Type	Energy (keV)	Intensity (%)	Δ(g-rad/ μCi-h)
β- 2 max	874 4		
avg	294.8 16	0.72 8	0.0045
β- 3 max	908 4		
avg	308.3 16	3.1 4	0.0204
β- 4 max	1604 4		
avg	602.8 18	31 6	0.398
total β-			
avg	567.2 19	35 6	0.423

5 weak β's omitted (ΣIβ = 0.08%)

Radiation Type	Energy (keV)	Intensity (%)	Δ(g-rad/ μCi-h)
γ 16	556.65	0.121 21	0.0014
γ 19	695.88 6	3.1 6	0.0453
γ 23	729.57 6	0.71 13	0.0111

36 weak γ's omitted (ΣIγ = 0.29%)

129I B- DECAY (1.57E7 Y 4) I(MIN) = 0.10%

Radiation Type	Energy (keV)	Intensity (%)	Δ(g-rad/ μCi-h)
Auger-L	3.43	74 4	0.0054
ce-K- 1	5.02 3	79.10 20	0.0085
Auger-K	24.6	8.8 16	0.0046
ce-L- 1	34.13 3	10.6 3	0.0077

129TE IT DECAY (33.6 D 1) I(MIN) = 0.10%
%IT=65 6
SEE ALSO 129TE B- DECAY (33.6 D)

Radiation Type	Energy (keV)	Intensity (%)	Δ(g-rad/ μCi-h)
Auger-L	3.19	50 4	0.0034
Auger-K	22.7	4.1 10	0.0020
ce-K- 1	73.69 5	33 3	0.0513
ce-L- 1	100.56 5	24.9 24	0.0533
ce-MNO- 1	104.49 5	7.3 7	0.0162
X-ray L	3.77	4.9 17	0.0004
X-ray $K\alpha_2$	27.20170 20	8.1 9	0.0047
X-ray $K\alpha_1$	27.47230 20	15.2 15	0.0089
X-ray $K\beta$	31	5.3 6	0.0035
γ 1	105.50 5	0.149 16	0.0003

129TE B- DECAY (33.6 D 1) I(MIN) = 0.10%
%B-=35 6
SEE ALSO 129TE IT DECAY (33.6 D)

Radiation Type	Energy (keV)	Intensity (%)	Δ(g-rad/ μCi-h)
ce-L- 1	22 3	0.110 19	≈0
β- 1 max	202 4		
avg	55.5 12	0.156 17	0.0002
ce-M- 1	38.44 3	2.10 10	0.0017
ce-NOP- 1	39.37 3	0.70 10	0.0006
β- 1 max	150 5		
avg	40 5	100	0.0852
X-ray L	4.1	8.2 25	0.0007
X-ray $K\alpha_2$	29.4580 10	20.0 6	0.0126
X-ray $K\alpha_1$	29.7790 10	37.1 9	0.0235
X-ray $K\beta$	33.6	13.2 4	0.0094
γ 1	39.58 3	7.50 20	0.0063

(CONTINUED ON NEXT COLUMN)

129XE IT DECAY (8.0 D 2) I(MIN) = 0.10%

Radiation Type	Energy (keV)	Intensity (%)	Δ(g-rad/ μCi-h)
Auger-L	3.43	147 8	0.0108
ce-K- 1	5.02 3	79.10 20	0.0085
Auger-K	24.6	16 3	0.0083
ce-L- 1	34.13 3	10.6 3	0.0077
ce-M- 1	38.44 3	2.10 10	0.0017
ce-NOP- 1	39.37 3	0.70 10	0.0006
ce-K- 2	162.00 3	63.90 10	0.220
ce-L- 2	191.11 3	24.4 7	0.0993
ce-M- 2	195.42 3	5.50 20	0.0229
ce-NOP- 2	196.35 3	1.50 10	0.0063
X-ray L	4.1	16 5	0.0014
X-ray $K\alpha_2$	29.4580 10	36.2 10	0.0227
X-ray $K\alpha_1$	29.7790 10	67.1 17	0.0426
X-ray $K\beta$	33.6	23.9 7	0.0171
γ 1	39.58 3	7.50 20	0.0063
γ 2	196.56 3	4.70 20	0.0197

129CS EC DECAY (32.06 H 6)   I(MIN)= 0.10%

Radiation Type	Energy (keV)	Intensity (%)	Δ(g-rad/μCi-h)
Auger-L	3.43	110 9	0.0080
ce-K- 1	5.020 15	31.4 22	0.0034
Auger-K	24.6	13 3	0.0068
ce-L- 1	34.128 15	4.2 3	0.0030
ce-MNO- 1	38.439 15	1.02 7	0.0008
ce-K- 3	58.768 4	0.57 4	0.0007
ce-K- 14	337.3566 23	0.62 4	0.0044
ce-K- 16	376.9286 23	0.359 22	0.0029
X-ray L	4.1	12 4	0.0011
X-ray $K\alpha_2$	29.4580 10	29.7 21	0.0186
X-ray $K\alpha_1$	29.7790 10	55 4	0.0350
X-ray $K\beta$	33.6	19.6 14	0.0140
γ 1	39.581 15	2.99 18	0.0025
γ 3	93.329 3	0.66 4	0.0013
γ 4	177.036 10	0.271 15	0.0010
γ 5	266.820 7	0.274 15	0.0016
γ 6	270.352 5	0.214 12	0.0012
γ 7	278.614 4	1.33 8	0.0079
γ 8	282.131 6	0.243 13	0.0015
γ 10	318.1800 20	2.46 14	0.0167
γ 14	371.9180 20	30.8 16	0.244
γ 16	411.4900 20	22.5 12	0.197
γ 21	548.945 8	3.42 18	0.0400
γ 27	588.549 8	0.61 4	0.0076
γ 32	906.425 6	0.221 12	0.0043

20 weak γ's omitted (ΣIγ = 0.37%)

130I B- DECAY (CONTINUED)

Radiation Type	Energy (keV)	Intensity (%)	Δ(g-rad/μCi-h)
γ 11	457.720 20	0.237 16	0.0023
γ 12	510.350 20	0.85 3	0.0093
γ 13	536.090 20	99.0 20	1.13
γ 14	539.10 3	1.40 3	0.0160
γ 15	553.900 10	0.662 21	0.0078
γ 16	586.050 20	1.69 4	0.0211
γ 17	603.530 10	0.615 25	0.0079
γ 19	668.540 10	96.1 21	1.37
γ 20	685.990 10	1.07 3	0.0156
γ 22	739.480 20	82.3 18	1.30
γ 25	800.23 3	0.101 6	0.0017
γ 26	808.290 20	0.236 8	0.0041
γ 31	877.35 4	0.191 9	0.0036
γ 34	967.020 20	0.877 22	0.0181
γ 38	1096.48 3	0.552 15	0.0129
γ 39	1122.15 3	0.253 10	0.0061
γ 40	1157.470 10	11.31 25	0.279
γ 41	1222.56 3	0.179 7	0.0047
γ 42	1272.120 20	0.748 19	0.0203
γ 44	1403.900 20	0.345 14	0.0103

31 weak γ's omitted (ΣIγ = 0.82%)

130I B- DECAY (12.36 H 1) I(MIN) = 0.10%

Radiation Type		Energy (keV)	Intensity (%)	Δ(g-rad/ μCi-h)
Auger-L		3.43	0.51 4	≈0
ce-K- 13		501.529 20	0.624 23	0.0067
β- 1	max	234 10		
	avg	65 3	0.320 20	0.0004
β- 2	max	357 10		
	avg	104 4	0.330 20	0.0007
β- 3	max	378 10		
	avg	111 4	0.500 20	0.0012
β- 4	max	559 10		
	avg	174 4	0.185 6	0.0007
β- 5	max	624 10		
	avg	198 4	46.0 10	0.194
β- 6	max	814 10		
	avg	271 4	2.14 5	0.0124
β- 7	max	904 10		
	avg	306 4	0.175 16	0.0011
β- 8	max	1042 10		
	avg	362 4	47.7 12	0.368
β- 9	max	1178 10		
	avg	419 4	1.45 4	0.0129
total β-				
	avg	280 5	99.1 16	0.592

6 weak β's omitted (ΣIβ = 0.27%)

Radiation Type		Energy (keV)	Intensity (%)	Δ(g-rad/ μCi-h)
X-ray	$K\alpha_2$	29.4580 10	0.158 7	≈0
X-ray	$K\alpha_1$	29.7790 10	0.293 13	0.0002
X-ray	Kβ	33.6	0.104 5	≈0
γ 9		418.010 20	34.2 7	0.304

(CONTINUED ON NEXT COLUMN)

131I B- DECAY (8.04 D 1) I(MIN) = 0.10%
SEE ALSO 131XE IT DECAY

Radiation Type		Energy (keV)	Intensity (%)	Δ(g-rad/ μCi-h)
Auger-L		3.43	5.1 3	0.0004
Auger-K		24.6	0.60 11	0.0003
ce-K- 1		45.622 10	3.54 13	0.0034
ce-L- 1		74.730 10	0.472 17	0.0008
ce-MNO- 1		79.041 10	0.1310 24	0.0002
ce-K- 7		249.737 11	0.248 9	0.0013
ce-K- 14		329.919 11	1.54 6	0.0108
ce-L- 14		359.027 11	0.244 8	0.0019
β- 1	max	247.9 6		
	avg	69.40 20	2.13 3	0.0031
β- 2	max	303.9 6		
	avg	87.00 20	0.620 20	0.0011
β- 3	max	333.8 6		
	avg	96.60 20	7.36 10	0.0151
β- 4	max	606.3 6		
	avg	191.6 3	89.4 10	0.365
β- 5	max	806.9 6		
	avg	283.2 3	0.420 20	0.0025
total β-				
	avg	181.7 3	100.0 10	0.387

1 weak β's omitted (ΣIβ = 0.06%)

Radiation Type		Energy (keV)	Intensity (%)	Δ(g-rad/ μCi-h)
X-ray	L	4.1	0.56 18	≈0
X-ray	$K\alpha_2$	29.4580 10	1.37 5	0.0009
X-ray	$K\alpha_1$	29.7790 10	2.54 9	0.0016
X-ray	Kβ	33.6	0.90 4	0.0006
γ 1		80.183 10	2.62 5	0.0045

(CONTINUED ON NEXT PAGE)

131I B- DECAY (CONTINUED)

Radiation Type	Energy (keV)	Intensity (%)	Δ(g-rad/μCi-h)
γ 4	177.210 10	0.265 4	0.0010
γ 7	284.298 11	6.06 9	0.0367
γ 12	325.781 11	0.251 6	0.0017
γ 14	364.480 11	81.2 12	0.631
γ 16	502.991 11	0.361 7	0.0039
γ 17	636.973 10	7.27 11	0.0986
γ 18	642.703 11	0.220 4	0.0030
γ 19	722.893 10	1.80 3	0.0278

10 weak γ's omitted ($\Sigma I\gamma$ = 0.23%)

131XE IT DECAY (11.9 D 1) I(
FEEDING OF 11.9-D 131XE IN 131I DECAY=1.086% 13

Radiation Type	Energy (keV)	Intensity (%)	Δ(g-rad/μCi-h)
Auger-L	3.43	75 4	0.0055
Auger-K	24.6	6.8 13	0.0036
ce-K- 1	129.369 8	61.2 7	0.169
ce-L- 1	158.477 8	28.6 6	0.0965
ce-M- 1	162.788 8	6.50 20	0.0225
ce-NOP- 1	163.722 8	1.78 6	0.0062
X-ray L	4.1	8 3	0.0007
X-ray $K\alpha_2$	29.4580 10	15.5 5	0.0097

131BA EC DECAY (CONTINUED)

Radiation Type	Energy (keV)	Intensity (%)	Δ(g-rad/μCi-h)
ce-MNO- 1	53.74 6	0.220 15	0.0003
ce-K- 4	56.316 11	0.61 7	0.0007
ce-L- 2	73.041 13	0.153 12	0.0002
ce-K- 5	87.818 12	18.1 8	0.0339
ce-K- 7	97.622 14	0.81 6	0.0017
ce-L- 5	118.089 12	5.99 25	0.0151
ce-MNO- 5	122.586 12	1.66 5	0.0043
ce-L- 7	127.893 14	0.151 13	0.0004
ce-K- 10	180.11 3	1.79 6	0.0069
ce-K- 11	203.65 3	0.169 18	0.0007
ce-L- 10	210.38 3	0.239 21	0.0011
ce-K- 13	213.46 3	0.171 18	0.0008
ce-K- 18	337.27 3	0.27 4	0.0019
ce-K- 26	460.30 3	0.48 7	0.0047
X-ray L	4.29	13 4	0.0012
X-ray $K\alpha_2$	30.6251 3	27.7 17	0.0181
X-ray $K\alpha_1$	30.9728 3	51 3	0.0338
X-ray $K\beta$	35	18.4 12	0.0137
γ 2	78.755 13	0.75 3	0.0013
γ 4	92.301 11	0.64 5	0.0013
γ 5	123.803 12	29.1 9	0.0766
γ 7	133.607 14	2.19 9	0.0062
γ 9	157.15 3	0.199 21	0.0007
γ 10	216.09 3	19.9 4	0.0916
γ 11	239.63 3	2.41 8	0.0123
γ 12	246.92 6	0.60 5	0.0031
γ 13	249.44 3	2.81 10	0.0149
γ 14	294.54 4	0.159 21	0.0010
γ 16	351.15 4	0.119 20	0.0009
γ 18	373.25 3	13.3 15	0.106
γ 19	404.04 3	1.29 9	0.0111

X-ray Kα1	29.7790 10	28.7 8	0.0182
X-ray Kβ	33.6	10.2 4	0.0073
γ 1	163.930 8	1.96 6	0.0068

131CS EC DECAY (9.69 D 1) I(MIN) = 0.10%

Radiation Type	Energy (keV)	Intensity (%)	Δ(g-rad/ μCi-h)
Auger-L	3.43	79 4	0.0058
Auger-K	24.6	9.3 17	0.0049
X-ray L	4.1	9 3	0.0008
X-ray Kα2	29.4580 10	21.1 6	0.0132
X-ray Kα1	29.7790 10	39.1 10	0.0248
X-ray Kβ	33.6	13.9 4	0.0100

131BA EC DECAY (11.8 D 2) I(MIN) = 0.10%

Radiation Type	Energy (keV)	Intensity (%)	Δ(g-rad/ μCi-h)
Auger-L	3.55	104 7	0.0078
ce-K- 1	18.98 6	0.61 7	0.0002
Auger-K	25.5	11.4 15	0.0062
ce-K- 2	42.770 13	1.23 9	0.0011
ce-L- 1	49.25 6	0.89 10	0.0009

(CONTINUED ON NEXT COLUMN)

γ 24	480.38 4	0.34 4	0.0035
γ 25	486.48 4	1.89 21	0.0196
γ 26	496.28 3	44 4	0.463
γ 29	572.66 4	0.159 11	0.0019
γ 30	585.02 3	1.23 9	0.0154
γ 31	620.05 3	1.57 9	0.0208
γ 32	674.41 4	0.129 9	0.0019
γ 33	696.46 4	0.147 11	0.0022
γ 37	831.63 5	0.219 21	0.0039
γ 40	923.86 4	0.70 7	0.0137
γ 43	1046.9	0.199 4	0.0044
γ 44	1047.56 4	1.194 24	0.0266

23 weak γ's omitted (ΣIγ = 0.68%)

132TE B- DECAY (78.2 H 8) I(MIN) = 0.10%

Radiation Type	Energy (keV)	Intensity (%)	Δ(g-rad/ μCi-h)
Auger-L	3.3	75 7	0.0053
ce-K- 1	16.551 10	70 6	0.0248
Auger-K	23.6	9.4 23	0.0047
ce-L- 1	44.532 10	9.3 7	0.0088
ce-MNO- 1	48.648 10	2.44 17	0.0025
ce-K- 2	78.59 8	0.91 16	0.0015
ce-K- 3	83.13 8	0.89 16	0.0016
ce-L- 2	106.57 8	0.26 5	0.0006
ce-L- 3	111.11 8	0.19 4	0.0005
ce-K- 4	194.99 6	7.14 22	0.0297

(CONTINUED ON NEXT PAGE)

132TE B- DECAY (CONTINUED)

Radiation Type	Energy (keV)	Intensity (%)	Δ(g-rad/ μCi-h)
ce-L- 4	222.97 6	1.34 4	0.0064
ce-MNO- 4	227.09 6	0.2734 13	0.0013
β- 1 max	215 4		
avg	59.4 12	100	0.127
X-ray L	4	8 3	0.0007
X-ray $K\alpha_2$	28.3172 4	19.9 15	0.0120
X-ray $K\alpha_1$	28.6120 3	37 3	0.0226
X-ray Kβ	32.3	13.0 10	0.0089
γ 1	49.720 10	14.4 10	0.0152
γ 2	111.76 8	1.85 18	0.0044
γ 3	116.30 8	1.94 18	0.0048
γ 4	228.16 6	88.2 4	0.429

132I B- DECAY (2.30 H 3) I(MIN) = 0.10%

Radiation Type	Energy (keV)	Intensity (%)	Δ(g-rad/ μCi-h)
Auger-L	3.43	0.89 7	≈0
Auger-K	24.6	0.118 22	≈0
ce-K- 24	488.09 9	0.127 14	0.0013
ce-K- 34	633.13 8	0.35 3	0.0047
ce-K- 41	738.04 8	0.206 24	0.0032
β- 1 max	226 20		
avg	63 7	0.12 6	0.0002

132I B- DECAY (CONTINUED)

Radiation Type	Energy (keV)	Intensity (%)	Δ(g-rad/ μCi-h)
β-20 max	2140 20		
avg	841 9	17.6 22	0.315
total β-			
avg	490 11	98 3	1.02
14 weak β's omitted (ΣIβ = 0.44%)			
X-ray $K\alpha_2$	29.4580 10	0.269 15	0.0002
X-ray $K\alpha_1$	29.7790 10	0.50 3	0.0003
X-ray Kβ	33.6	0.177 10	0.0001
γ 2	147.20 10	0.237 20	0.0007
γ 3	183.3	0.1579 4	0.0006
γ 4	254.80 20	0.19 3	0.0010
γ 5	262.70 10	1.44 9	0.0081
γ 7	284.80 10	0.79 7	0.0048
γ 9	306.6 4	0.1085 2	0.0007
γ 10	306.6 4	0.11 4	0.0007
γ 12	316.5 4	0.16 4	0.0011
γ 15	363.5 4	0.49 10	0.0038
γ 16	387.8 4	0.17 3	0.0014
γ 17	416.8 4	0.46 9	0.0041
γ 18	431.9 4	0.45 9	0.0042
γ 19	446.0 4	0.67 8	0.0064
γ 20	473.4 7	0.27 5	0.0027
γ 22	487.5 7	0.18 5	0.0018
γ 23	505.90 15	5.03 20	0.0542
γ 24	522.65 9	16.1 6	0.179
γ 25	535.5 4	0.52 8	0.0060
γ 26	547.10 20	1.25 9	0.0146
γ 30	620.8 10	0.3948 8	0.0052
γ 31	621.2 10	1.579 4	0.0209
γ 32	630.22 9	13.7 6	0.184

Radiation		Energy (keV)		Intensity (%)		Δ
β− 2	max	320	20			
	avg	92	7	0.261	3	0.0005
β− 3	max	353	20			
	avg	103	7	0.12	5	0.0003
β− 4	max	366	20			
	avg	107	7	< 0.19		< 0.0004
β− 5	max	425	20			
	avg	127	7	0.21	4	0.0006
β− 6	max	504	20			
	avg	154	8	0.24	6	0.0008
β− 7	max	522	20			
	avg	161	8	0.34	7	0.0012
β− 8	max	740	20			
	avg	242	8	1.81	10	0.0093
β− 9	max	741	20			
	avg	242	8	12.8	8	0.0660
β−10	max	826	20			
	avg	275	8	0.37	6	0.0022
β−11	max	910	20			
	avg	309	8	3.60	20	0.0237
β−12	max	967	20			
	avg	331	9	8.1	4	0.0571
β−13	max	996	20			
	avg	344	9	3.79	16	0.0278
β−14	max	1155	20			
	avg	409	9	2.11	7	0.0184
β−15	max	1185	20			
	avg	422	9	19.0	7	0.171
β−16	max	1413	20			
	avg	519	9	1.7	6	0.0188
β−17	max	1470	20			
	avg	543	9	10.2	10	0.118
β−18	max	1468	20			
	avg	543	9	1.9	8	0.0220
β−19	max	1617	20			
	avg	608	9	12.7	7	0.164

(CONTINUED ON NEXT COLUMN)

Radiation		Energy (keV)		Intensity (%)		Δ
γ	33	650.60	20	2.66	20	0.0369
γ	34	667.69	8	98.70	20	1.40
γ	35	669.8	3	4.9	8	0.0704
γ	36	671.6	3	5.2	4	0.0748
γ	37	727		2.2	6	0.0336
γ	38	727.2	10	3.2	6	0.0489
γ	39	729.5	4	1.1	3	0.0169
γ	41	772.60	8	76.2	18	1.25
γ	42	780.2	3	1.23	6	0.0205
γ	43	784.5	4	0.42	5	0.0071
γ	45	809.80	20	2.9	3	0.0494
γ	46	812.20	20	5.6	5	0.0973
γ	47	863.30	20	0.59	5	0.0109
γ	48	876.80	20	1.08	5	0.0201
γ	50	910.30	20	0.92	5	0.0178
γ	51	927.6	3	0.44	8	0.0088
γ	53	954.55	9	18.1	6	0.367
γ	54	984.50	20	0.56	6	0.0118
γ	58	1034.70	20	0.57	5	0.0126
γ	64	1136.03	12	2.96	20	0.0716
γ	65	1138		0.2961	6	0.0072
γ	66	1143.40	20	1.38	10	0.0337
γ	67	1148.2	7	0.21	5	0.0051
γ	68	1173.20	20	1.09	10	0.0271
γ	71	1272.7	4	0.15	3	0.0040
γ	72	1290.7	3	1.14	6	0.0312
γ	73	1295.3	3	1.97	10	0.0545
γ	74	1298.2	5	0.89	10	0.0246
γ	76	1317.1	7	0.118	20	0.0033
γ	77	1372.07	13	2.47	10	0.0721
γ	78	1398.57	10	7.1	3	0.212
γ	80	1442.56	10	1.42	6	0.0437
γ	82	1476.80	20	0.138	20	0.0043
γ	94	1757.50	20	0.38	3	0.0140
γ	101	1921.08	12	1.18	9	0.0485
γ	103	2002.30	12	1.09	10	0.0463

(CONTINUED ON NEXT PAGE)

132I B- DECAY (CONTINUED)

Radiation Type	Energy (keV)	Intensity (%)	Δ(g-rad/ μCi-h)
γ 104	2086.82 15	0.25 4	0.0110
γ 105	2172.68 15	0.19 3	0.0087
γ 107	2223.17 15	0.118 20	0.0056
γ 110	2390.48 15	0.168 20	0.0085

58 weak γ's omitted (ΣIγ = 2.23%)

132CS B+ DECAY (6.475 D 10) I(MIN) = 0.10%
%(EC+B+)=98.0 1
SEE ALSO 132CS B- DECAY

Radiation Type	Energy (keV)	Intensity (%)	Δ(g-rad/ μCi-h)
Auger-L	3.43	78 4	0.0057
Auger-K	24.6	9.2 17	0.0048
β+ 1 max	403 23		
avg	190 10	0.30 8	0.0012
X-ray L	4.1	9 3	0.0008
X-ray $K\alpha_2$	29.4580 10	21.0 6	0.0132
X-ray $K\alpha_1$	29.7790 10	39.0 10	0.0248
X-ray Kβ	33.6	13.9 4	0.0099
γ 2	505.90 15	0.80 10	0.0086
γ 3	630.22 9	1.01 8	0.0136

133I B- DECAY (20.8 H 1) I(MIN) = 0.10%
SEE ALSO 133XE IT DECAY (2.25 D)

Radiation Type	Energy (keV)	Intensity (%)	Δ(g-rad/ μCi-h)
β- 1 max	170 30		
avg	46 9	0.414 14	0.0004
β- 2 max	370 30		
avg	110 10	1.24 4	0.0029
β- 3 max	410 30		
avg	122 11	0.396 10	0.0010
β- 4 max	460 30		
avg	140 11	3.74 6	0.0112
β- 5 max	520 30		
avg	162 11	3.12 6	0.0108
β- 6 max	710 30		
avg	230 12	0.541 19	0.0027
β- 7 max	880 30		
avg	299 12	4.15 11	0.0264
β- 8 max	1020 30		
avg	352 13	1.81 5	0.0136
β- 9 max	1230 30		
avg	441 13	83.5 18	0.784
β-10 max	1530 30		
avg	573 13	1.07 5	0.0131
total β- avg	407 15	100.0 18	0.866
γ 5	262.702 12	0.356 10	0.0020
γ 6	267.173 22	0.117 6	0.0007
γ 7	345.43 5	0.104 18	0.0008
γ 8	361.09 6	0.11 4	0.0009
γ 12	417.556	0.153 11	0.0014
γ 13	422.910 16	0.309 9	0.0028
γ 15	510.530 11	1.81 5	0.0197
γ 18	529.872 11	86.3 18	0.974

γ	4	667.67 6	97.47 11	1.39
γ	6	1136.03 12	0.51 4	0.0123
γ	8	1317.80 20	0.58 5	0.0164

4 weak γ's omitted (ΣIγ = 0.27%)
Maximum γ± intensity = 0.60%

132CS B- DECAY (6.475 D 10) I(MIN) = 0.10%
%B-=2.0 1
SEE ALSO 132CS E+ DECAY

Radiation Type	Energy (keV)	Intensity (%)	Δ(g-rad/ μCi-h)
β- 1 max	240 30		
avg	67 8	0.37 3	0.0005
β- 2 max	810 30		
avg	267 11	1.57 12	0.0089
total β-			
avg	223 13	2.00 13	0.0095

1 weak β's omitted (ΣIβ = 0.06%)

γ	1	464.55 6	1.87 14	0.0185
γ	2	567.14 3	0.24 4	0.0029
γ	4	1031.70 3	0.122 12	0.0027

1 weak γ's omitted (ΣIγ = 0.06%)

γ	22	617.974 17	0.539 13	0.0071
γ	26	680.247 15	0.645 16	0.0093
γ	27	706.578 13	1.49 4	0.0225
γ	28	768.382 18	0.457 13	0.0075
γ	30	820.506 24	0.154 6	0.0027
γ	31	856.278 12	1.23 4	0.0225
γ	32	875.329 11	4.47 10	0.0834
γ	33	909.67 3	0.212 8	0.0041
γ	36	1052.296 21	0.552 13	0.0124
γ	37	1060.07 6	0.137 6	0.0031
γ	39	1236.411 12	1.49 4	0.0393
γ	40	1298.223 11	2.33 6	0.0644
γ	42	1350.38 3	0.148 5	0.0043

23 weak γ's omitted (ΣIγ = 0.62%)

133XE B- DECAY (5.245 D 6) I(MIN) = 0.10%

Radiation Type	Energy (keV)	Intensity (%)	Δ(g-rad/ μCi-h)
Auger-L	3.55	49.0 20	0.0037
Auger-K	25.5	5.5 7	0.0030
ce-K- 1	43.636 11	0.30 6	0.0003
ce-K- 2	45.0124 4	52.0 3	0.0498
ce-L- 2	75.2827 4	8.49 20	0.0136
ce-MNO- 2	79.7799 4	2.3 3	0.0039
β- 1 max	266 3		
avg	75.0 10	0.66 10	0.0011

(CONTINUED ON NEXT PAGE)

133XE B- DECAY (CONTINUED)

Radiation Type	Energy (keV)	Intensity (%)	Δ(g-rad/ μCi-h)
β- 2 max	346 3		
avg	100.5 10	99.34 10	0.213
total β- avg	100.3 10	100.01 15	0.214
X-ray L	4.29	6.1 17	0.0006
X-ray $K\alpha_2$	30.6251 3	13.3 3	0.0087
X-ray $K\alpha_1$	30.9728 3	24.6 5	0.0163
X-ray $K\beta$	35	8.84 20	0.0066
γ 1	79.621 11	0.22 6	0.0004
γ 2	81	37.1 4	0.0640

4 weak γ's omitted (ΣIγ = 0.07%)

133XE IT DECAY (2.19 D 1) I(MIN) = 0.10%
FEEDING OF 2.188-D 133XE IN 133I DECAY=2.883% 23

Radiation Type	Energy (keV)	Intensity (%)	Δ(g-rad/ μCi-h)
Auger-L	3.43	70 4	0.0051
Auger-K	24.6	7.1 13	0.0037
ce-K- 1	198.62 4	63.7 8	0.269
ce-L- 1	227.73 4	20.8 6	0.101
ce-M- 1	232.04 4	4.16 13	0.0206
ce-NOP- 1	232.97 4	1.04 17	0.0052
X-ray L	4.1	7.8 24	0.0007
X-ray $K\alpha_2$	29.4580 10	16.1 5	0.0101

133BA EC DECAY (CONTINUED)

Radiation Type	Energy (keV)	Intensity (%)	Δ(g-rad/ μCi-h)
X-ray $K\alpha_1$	30.9728 3	62.9 12	0.0415
X-ray $K\beta$	35	22.6 6	0.0168
γ 1	53.155 16	2.17 4	0.0025
γ 2	79.621 11	2.66 8	0.0045
γ 3	80.997 5	33.5 5	0.0578
γ 4	160.605 15	0.62 4	0.0021
γ 5	223.25 3	0.460 13	0.0022
γ 6	276.397 12	7.09 13	0.0417
γ 7	302.851 15	18.40 20	0.119
γ 8	356.005 17	62.1 7	0.471
γ 9	383.851 15	8.91 10	0.0729

133BA IT DECAY (38.9 H 1) I(MIN) = 0.10%

Radiation Type	Energy (keV)	Intensity (%)	Δ(g-rad/ μCi-h)
Auger-L	3.67	130 6	0.0102
ce-L- 1	6.30 4	77.5916 5	0.0104
ce-M- 1	11.00 4	15.9 5	0.0037
ce-NOP- 1	12.04 4	5.20 20	0.0013
Auger-K	26.4	5.9 16	0.0033
ce-K- 2	238.65 15	59.2 3	0.301
ce-L- 2	270.10 15	18.0 6	0.104
ce-M- 2	274.80 15	4.00 10	0.0234
ce-NOP- 2	275.84 15	1.15 3	0.0068

Radiation Type	Energy (keV)	Intensity (%)	Δ(g-rad/ μCi-h)
X-ray Kα₁	29.7790 10	29.9 8	0.0190
X-ray Kβ	33.6	10.6 4	0.0076
γ 1	233.18 4	10.3 3	0.0512

133BA EC DECAY (10.74 Y 5) I(MIN) = 0.10%

Radiation Type	Energy (keV)	Intensity (%)	Δ(g-rad/ μCi-h)
Auger-L	3.55	135 6	0.0102
ce-K- 1	17.170 16	10.5 4	0.0038
Auger-K	25.5	14.0 16	0.0076
ce-K- 2	43.636 11	3.72 15	0.0035
ce-K- 3	45.012 5	46.9 10	0.0450
ce-L- 1	47.441 16	1.43 20	0.0014
ce-MNO- 1	51.938 16	0.43 20	0.0005
ce-L- 2	73.907 11	0.59 11	0.0009
ce-L- 3	75.283 5	7.64 24	0.0122
ce-MNO- 2	78.404 11	0.194 6	0.0003
ce-M- 3	79.780 5	1.78 14	0.0030
ce-NOP- 3	80.766 5	0.32 4	0.0005
ce-K- 4	124.620 15	0.123 9	0.0003
ce-K- 6	240.412 12	0.327 12	0.0017
ce-K- 7	266.866 15	0.70 6	0.0040
ce-L- 7	297.137 15	0.103 15	0.0007
ce-K- 8	320.020 17	1.31 5	0.0089
ce-K- 9	347.866 15	0.153 5	0.0011
ce-L- 8	350.291 17	0.218 7	0.0016
X-ray L	4.29	17 5	0.0015
X-ray Kα₂	30.6251 3	34.0 8	0.0222
X-ray L	4.47	18 5	0.0017
γ 1	12.29 4	1.39 4	0.0004
X-ray Kα₂	31.8171 3	15.2 5	0.0103
X-ray Kα₁	32.1936 3	28.0 9	0.0192
X-ray Kβ	36.4	10.2 4	0.0079
γ 2	276.09 15	18.0 5	0.106

(CONTINUED ON NEXT COLUMN)

134TE B- DECAY (41.8 M 8) I(MIN) = 0.10%

Radiation Type	Energy (keV)	Intensity (%)	Δ(g-rad/ μCi-h)
Auger-L	3.3	32.3 22	0.0023
Auger-K	23.6	4.0 10	0.0020
ce-K- 3	46.281 10	26.7 10	0.0263
ce-K- 4	68.25 3	0.30 12	0.0004
ce-L- 3	74.262 10	4.0 6	0.0063
ce-M- 3	78.378 10	0.79 11	0.0013
ce-NOP- 3	79.264 10	0.20 3	0.0003
ce-K- 6	147.72 3	3.21 20	0.0101
ce-K- 8	168.07 3	0.98 20	0.0035
ce-L- 6	175.70 3	0.69 5	0.0026
ce-K- 9	177.301 20	2.1 3	0.0080
ce-MNO- 6	179.82 3	0.175 10	0.0007
ce-K- 11	244.781 20	0.8 3	0.0041
β- 1 max	600 300		
avg	190 120	14.5 8	0.0587
β- 2 max	800 300		
avg	260 120	42.6 17	0.236

(CONTINUED ON NEXT PAGE)

134TE B- DECAY (CONTINUED)

Radiation Type	Energy (keV)	Intensity (%)	Δ(g-rad/ μCi-h)
β- 3 max	900 300		
avg	290 130	43.5 14	0.269
total β- avg	260 130	100.6 24	0.563
X-ray L	4	3.4 12	0.0003
X-ray Kα2	28.3172 4	8.6 4	0.0052
X-ray Kα1	28.6120 3	16.0 8	0.0097
X-ray Kβ	32.3	5.6 3	0.0038
γ 2	76.230 20	0.29 3	0.0005
γ 3	79.450 10	20.9 8	0.0353
γ 4	101.42 3	0.34 7	0.0007
γ 5	131.05 20	0.15 3	0.0004
γ 6	180.89 3	18.4 11	0.0710
γ 7	183.05 13	0.6 3	0.0024
γ 8	201.24 3	8.9 7	0.0382
γ 9	210.470 20	22.4 11	0.100
γ 10	259.59 15	0.49 10	0.0027
γ 11	277.950 20	21.8 11	0.129
γ 12	435.06 5	19.0 13	0.176
γ 13	460.99 3	11.1 7	0.109
γ 14	464.64 6	5.22 25	0.0517
γ 15	565.99 3	19.3 11	0.233
γ 16	636.26 9	2.1 4	0.0291
γ 17	665.85	1.044 24	0.0148
γ 18	712.97 6	4.21 18	0.0639
γ 19	742.59 3	15.0 7	0.238
γ 20	767.20 3	30.7 10	0.502
γ 21	844.06 6	1.2 6	0.0221
γ 22	896.02 9	0.18 7	0.0035
γ 23	925.55 8	1.69 19	0.0333
γ 24	1027.00 10	0.46 13	0.0101

134I B- DECAY (CONTINUED)

Radiation Type	Energy (keV)	Intensity (%)	Δ(g-rad/ μCi-h)
β-13 max	1880 60		
avg	720 30	0.7 7	0.0107
β-14 max	2230 60		
avg	880 30	3.7 9	0.0694
β-15 max	2420 60		
avg	970 30	11.9 15	0.246
total β- avg	610 40	100.5 24	1.31
5 weak β's omitted (ΣIβ = 0.23%)			
X-ray L	4.1	0.13 5	≈0
X-ray Kα2	29.4580 10	0.32 6	0.0002
X-ray Kα1	29.7790 10	0.60 10	0.0004
X-ray Kβ	33.6	0.21 4	0.0002
γ 1	135.399 22	3.76 22	0.0108
γ 2	139.03 3	0.69 5	0.0020
γ 3	151.98 15	0.106 12	0.0003
γ 4	162.48 7	0.26 3	0.0009
γ 5	188.47 4	0.70 4	0.0028
γ 6	217.00 20	0.25 3	0.0011
γ 7	235.47 3	1.98 16	0.0100
γ 9	278.80 15	0.131 15	0.0008
γ 10	319.81 6	0.52 5	0.0035
γ 11	351.08 10	0.50 6	0.0037
γ 12	405.451 20	7.3 4	0.0634
γ 13	411.00 8	0.61 6	0.0053
γ 14	433.35 3	4.19 24	0.0387
γ 15	458.92 6	1.30 9	0.0127
γ 16	465.50 10	0.36 4	0.0036
γ 17	488.88 4	1.41 9	0.0147

134I B- DECAY (52.6 M 5) I(MIN) = 0.10%

Radiation Type	Energy (keV)	Intensity (%)	Δ(g-rad/ μCi-h)
Auger-L	3.43	1.14 18	≈0
Auger-K	24.6	0.14 4	≈0
ce-K- 1	100.838 22	0.98 20	0.0021
ce-L- 1	129.946 22	0.12 3	0.0003
ce-K- 31	812.46 3	0.17 3	0.0030
ce-K- 34	849.53 3	0.12 3	0.0022
β- 1 max	770 60		
avg	255 24	1.48 9	0.0080
β- 2 max	790 60		
avg	261 24	0.33 4	0.0018
β- 3 max	840 60		
avg	279 24	0.153 19	0.0009
β- 4 max	1070 60		
avg	372 25	1.22 7	0.0097
β- 5 max	1280 60		
avg	460 30	31.5 8	0.309
β- 6 max	1380 60		
avg	500 30	0.53 4	0.0056
β- 7 max	1500 60		
avg	550 30	7.6 5	0.0890
β- 8 max	1560 60		
avg	580 30	16.2 5	0.200
β- 9 max	1600 60		
avg	600 30	3.67 17	0.0469
β-10 max	1740 60		
avg	660 30	7.6 5	0.107
β-11 max	1800 60		
avg	690 30	11.2 7	0.165
β-12 max	1850 60		
avg	710 30	2.50 15	0.0378

(CONTINUED ON NEXT COLUMN)

Radiation Type	Energy (keV)	Intensity (%)	Δ(g-rad/ μCi-h)
γ 18	514.40 3	2.34 14	0.0256
γ 19	540.825 25	7.8 5	0.0901
γ 20	565.52 4	0.88 6	0.0106
γ 21	570.75 15	0.21 3	0.0026
γ 22	595.362 20	11.4 6	0.144
γ 23	621.790 25	10.6 6	0.140
γ 24	627.96 3	2.37 14	0.0316
γ 25	677.34 3	8.5 5	0.122
γ 26	706.65 10	0.83 6	0.0125
γ 27	730.74 4	1.91 12	0.0297
γ 28	739.18 8	0.76 8	0.0120
γ 29	766.68 4	4.1 3	0.0670
γ 30	816.38 7	0.52 5	0.0091
γ 31	847.025 25	95.4 3	1.72
γ 32	857.28 3	6.96 20	0.127
γ 33	864.0 3	0.19 3	0.0035
γ 34	884.09 3	65.3 10	1.23
γ 35	922.6 3	0.14 3	0.0028
γ 36	947.86 4	4.04 20	0.0815
γ 37	966.90 5	0.35 4	0.0073
γ 38	974.67 4	4.7 4	0.0970
γ 39	1040.25 10	1.91 19	0.0423
γ 43	1072.55 3	15.3 8	0.349
γ 45	1100.07 12	0.69 6	0.0161
γ 46	1103.18 12	0.73 6	0.0170
γ 47	1136.16 4	9.7 5	0.235
γ 48	1159.10 8	0.35 3	0.0087
γ 49	1164.0 3	0.13 3	0.0033
γ 51	1190.03 8	0.35 3	0.0089
γ 53	1239.0 3	0.21 6	0.0055
γ 55	1269.49 5	0.56 4	0.0152
γ 56	1322.4 3	0.10 4	0.0030
γ 57	1336.00 20	0.14 3	0.0041
γ 58	1352.62 8	0.45 4	0.0129
γ 61	1414.3 5	0.22 6	0.0066
γ 62	1428.2 3	0.17 4	0.0052

(CONTINUED ON NEXT PAGE)

134I B- DECAY (CONTINUED)

Radiation Type	Energy (keV)	Intensity (%)	Δ(g-rad/ μCi-h)
γ 63	1431.35 25	0.17 4	0.0052
γ 64	1455.24 5	2.29 15	0.0710
γ 65	1470.00 7	0.77 5	0.0242
γ 66	1505.5 4	0.11 4	0.0037
γ 67	1541.51 7	0.51 4	0.0166
γ 68	1613.80 5	4.36 24	0.150
γ 69	1629.24 8	0.26 4	0.0089
γ 70	1644.25 7	0.40 5	0.0140
γ 71	1655.19 10	0.23 3	0.0081
γ 72	1741.49 5	2.67 19	0.0991
γ 73	1806.84 4	5.7 4	0.220
γ 76	1925.88 10	0.181 19	0.0074
γ 78	2020.6 3	0.172 19	0.0074
γ 79	2159.9 3	0.21 3	0.0097
γ 82	2312.40 20	0.24 3	0.0117
γ 85	2467.4 3	0.153 19	0.0080

22 weak γ's omitted (ΣIγ = 1.33%)

134CS B- DECAY (2.062 Y 5) I(MIN) = 0.10%

Radiation Type	Energy (keV)	Intensity (%)	Δ(g-rad/ μCi-h)
Auger-L	3.67	0.65 5	≈0
ce-K- 5	531.874 15	0.120 7	0.0014
ce-K- 6	567.258 15	0.488 15	0.0059
ce-K- 7	758.404 22	0.224 11	0.0036

135I B- DECAY (6.61 H 1) I(MIN) = 0.10%
SEE ALSO 135XE IT DECAY

Radiation Type	Energy (keV)	Intensity (%)	Δ(g-rad/ μCi-h)
β- 1 max	240 30		
avg	66 10	0.14 3	0.0002
β- 2 max	240 30		
avg	68 10	0.130 20	0.0002
β- 3 max	260 30		
avg	74 10	0.14 3	0.0002
β- 4 max	300 30		
avg	86 10	1.03 5	0.0019
β- 5 max	340 30		
avg	98 10	0.91 4	0.0019
β- 6 max	350 30		
avg	103 10	1.39 5	0.0030
β- 7 max	460 30		
avg	138 11	4.73 15	0.0139
β- 8 max	480 30		
avg	145 11	7.33 20	0.0226
β- 9 max	620 30		
avg	196 12	1.57 7	0.0066
β-10 max	670 30		
avg	213 12	1.10 5	0.0050
β-11 max	740 30		
avg	243 12	7.9 3	0.0409
β-12 max	820 30		
avg	272 12	0.61 3	0.0035
β-13 max	920 30		
avg	313 12	8.7 3	0.0580
β-14 max	1030 30		
avg	359 13	21.5 5	0.164
β-15 max	1150 30		
avg	405 13	8.0 3	0.0690

β- 1	max	88.6	4			
	avg	23.09	11	27.40	20	0.0135
β- 2	max	415.2	4			
	avg	123.43	14	2.47	5	0.0065
β- 3	max	658.0	4			
	avg	210.15	15	70.1	5	0.314
total β-						
	avg	156.8	3	100.0	6	0.334

2 weak β's omitted (ΣIβ = 0.05%)

X-ray	$K\alpha_2$	31.8171	3	0.213	9	0.0001
X-ray	$K\alpha_1$	32.1936	3	0.393	15	0.0003
X-ray	Kβ	36.4		0.143	6	0.0001
γ 3		475.35	5	1.46	4	0.0148
γ 4		563.227	15	8.38	5	0.101
γ 5		569.315	15	15.43	11	0.187
γ 6		604.699	15	97.6	3	1.26
γ 7		795.845	22	85.4	4	1.45
γ 8		801.932	22	8.73	4	0.149
γ 9		1038.57	3	1.000	10	0.0221
γ 10		1167.94	3	1.80	3	0.0448
γ 11		1365.15	3	3.04	4	0.0884

2 weak γ's omitted (ΣIγ = 0.04%)

β-16	max	1250	30			
	avg	451	13	7.6	4	0.0730
β-17	max	1260	30			
	avg	454	13	0.11	5	0.0011
β-18	max	1450	30			
	avg	535	13	23.8	12	0.271
β-19	max	1580	30			
	avg	591	14	1.1	9	0.0138
β-20	max	2180	30			
	avg	858	14	0.8	6	0.0146
total β-						
	avg	364	16	98.7	19	0.766

4 weak β's omitted (ΣIβ = 0.15%)

γ 7	220.502	15	1.75	6	0.0082
γ 8	229.72	3	0.232	9	0.0011
γ 11	264.26	9	0.184	7	0.0010
γ 12	288.451	16	3.09	12	0.0190
γ 13	290.27	4	0.303	21	0.0019
γ 20	361.85	14	0.19	3	0.0014
γ 21	403.03	4	0.232	9	0.0020
γ 22	414.83	3	0.301	18	0.0027
γ 23	417.63	3	3.52	12	0.0313
γ 24	429.93	3	0.303	24	0.0028
γ 25	433.741	19	0.553	24	0.0051
γ 26	451.63	3	0.315	18	0.0030
γ 29	546.557	16	7.13	25	0.0830
γ 30	575.97	8	0.129	23	0.0016
γ 33	649.85	4	0.46	3	0.0063
γ 37	690.13	6	0.129	15	0.0019
γ 38	707.92	5	0.66	6	0.0099
γ 39	785.48	5	0.152	21	0.0025
γ 41	797.71	8	0.17	3	0.0029
γ 43	836.804	16	6.67	24	0.119
γ 45	961.46		0.15	3	0.0030

(CONTINUED ON NEXT PAGE)

135I B- DECAY (CONTINUED)

Radiation Type	Energy (keV)	Intensity (%)	Δ(g-rad/ μCi-h)
γ 46	972	0.89 6	0.0184
γ 47	972.6	1.20 6	0.0249
γ 48	995.09 10	0.15 3	0.0033
γ 49	1038.760 21	7.9 3	0.175
γ 51	1101.58 4	1.60 6	0.0376
γ 52	1124.00 4	3.61 12	0.0864
γ 53	1131.511 18	22.5 8	0.543
γ 55	1159.90 20	0.103 23	0.0025
γ 56	1169.04 4	0.87 4	0.0217
γ 59	1240.470 20	0.90 4	0.0238
γ 61	1260.409 17	28.6 10	0.769
γ 67	1367.89 4	0.61 4	0.0177
γ 69	1448.35 10	0.31 3	0.0097
γ 70	1457.56 3	8.6 3	0.268
γ 71	1502.79 4	1.07 5	0.0344
γ 73	1566.41 3	1.29 6	0.0430
γ 74	1678.03 3	9.5 4	0.341
γ 75	1706.46 3	4.09 18	0.149
γ 77	1791.20 3	7.70 25	0.294
γ 78	1830.69 4	0.58 3	0.0226
γ 80	1927.30 3	0.295 15	0.0121
γ 83	2045.88 4	0.87 4	0.0379
γ 88	2255.46 3	0.61 3	0.0294
γ 90	2408.65 4	0.95 5	0.0489

47 weak γ's omitted (ΣIγ = 1.50%)

135XE IT DECAY (15.29 M 3) I(MIN)= 0.10%
FEEDING OF 15.29-M 135XE IN 135I DECAY=15.8% 5

Radiation Type	Energy (keV)	Intensity (%)	Δ(g-rad/ μCi-h)
Auger-L	3.43	14.9 8	0.0011
Auger-K	24.6	1.7 3	0.0009
ce-K- 1	492.010 17	15.2 3	0.159
ce-L- 1	521.118 17	2.88 9	0.0320
ce-M- 1	525.429 17	0.610 20	0.0068
ce-NOP- 1	526.363 17	0.200 10	0.0022
X-ray L	4.1	1.7 5	0.0001
X-ray $K\alpha_2$	29.4580 10	3.84 13	0.0024
X-ray $K\alpha_1$	29.7790 10	7.13 23	0.0045
X-ray $K\beta$	33.6	2.54 9	0.0018
γ 1	526.571 17	81.2 5	0.911

135BA IT DECAY (28.7 H 2) I(MIN)= 0.10%

Radiation Type	Energy (keV)	Intensity (%)	Δ(g-rad/ μCi-h)
Auger-L	3.67	63 4	0.0050
Auger-K	26.4	5.9 16	0.0033
ce-K- 1	230.797 10	59.9 7	0.294
ce-L- 1	262.249 10	18.7 5	0.104
ce-M- 1	266.945 10	4.19 12	0.0238
ce-NOP- 1	267.985 10	1.20 3	0.0068

135XE B- DECAY (9.09 H 1) I(MIN) = 0.10%

Radiation Type	Energy (keV)	Intensity (%)	Δ(g-rad/ μCi-h)
Auger-L	3.55	5.12 25	0.0004
Auger-K	25.5	0.59 7	0.0003
ce-K- 3	213.809 15	5.63 17	0.0257
ce-L- 3	244.080 15	0.739 23	0.0038
β- 1 max	96 9		
avg	25.0 25	0.123 4	≈0
β- 2 max	550 9		
avg	171 4	3.12 10	0.0114
β- 3 max	750 9		
avg	245 4	0.68 3	0.0035
β- 4 max	908 9		
avg	307 4	95.98 3	0.628
total β- avg	302 4	99.98 11	0.643

1 weak β's omitted (ΣIβ = 0.08%)

Radiation Type	Energy (keV)	Intensity (%)	Δ(g-rad/ μCi-h)
X-ray L	4.29	0.63 18	≈0
X-ray $K\alpha_2$	30.6251 3	1.43 6	0.0009
X-ray $K\alpha_1$	30.9728 3	2.66 9	0.0018
X-ray Kβ	35	0.95 4	0.0007
γ 1	158.197 18	0.289 10	0.0010
γ 3	249.794 15	90.13 8	0.480
γ 4	358.39 4	0.221 9	0.0017
γ 6	407.990 20	0.359 13	0.0031
γ 9	608.185 16	2.90 9	0.0376

8 weak γ's omitted (ΣIγ = 0.21%)

Radiation Type	Energy (keV)	Intensity (%)	Δ(g-rad/ μCi-h)
X-ray L	4.47	8.6 22	0.0008
X-ray $K\alpha_2$	31.8171 3	15.3 6	0.0104
X-ray $K\alpha_1$	32.1936 3	28.3 10	0.0194
X-ray Kβ	36.4	10.3 4	0.0080
γ 1	268.238 10	16.0 5	0.0914

136CS B- DECAY (13.1 D 1) I(MIN) = 0.10%

Radiation Type	Energy (keV)	Intensity (%)	Δ(g-rad/ μCi-h)
Auger-L	3.67	20.8 13	0.0016
Auger-K	26.4	1.9 6	0.0011
ce-K- 1	29.47 5	7.3 7	0.0046
ce-K- 2	48.85 5	1.85 12	0.0019
ce-L- 1	60.92 5	1.05 10	0.0014
ce-MNO- 1	65.62 5	0.212 19	0.0003
ce-K- 3	72.22 10	0.38 3	0.0006
ce-L- 2	80.30 5	0.253 17	0.0004
ce-K- 4	115.78 5	2.38 12	0.0059
ce-K- 5	126.45 5	5.17 25	0.0139
ce-K- 6	129.09 10	0.145 21	0.0004
ce-K- 7	139.12 5	0.55 3	0.0016
ce-L- 4	147.23 5	0.67 4	0.0021
ce-MNO- 4	151.93 5	0.144 6	0.0005
ce-L- 5	157.90 5	4.11 20	0.0138
ce-M- 5	162.60 5	0.92 5	0.0032
ce-NOP- 5	163.64 5	0.240 12	0.0008
ce-K- 10	236.21 4	0.159 8	0.0008
ce-K- 14	303.13 5	1.17 6	0.0075

(CONTINUED ON NEXT PAGE)

136CS B- DECAY (CONTINUED)

Radiation Type	Energy (keV)	Intensity (%)	Δ(g-rad/ µCi-h)
ce-L- 14	334.58 5	0.201 9	0.0014
ce-K- 18	781.06 4	0.241 11	0.0040
ce-K- 19	1010.63 7	0.112 5	0.0024
β- 1 max	173.8 20		
avg	47.1 6	2.29 10	0.0023
β- 2 max	191.0 20		
avg	52.1 6	0.21 3	0.0002
β- 3 max	340.4 20		
avg	98.6 7	94 3	0.197
β- 4 max	407.3 20		
avg	120.8 7	0.13 13	0.0003
β- 5 max	493.6 20		
avg	150.5 7	0.8 8	0.0026
β- 6 max	517.0 20		
avg	175.4 8	1.8 8	0.0067
β- 7 max	681.0 20		
avg	218.8 8	2.0 20	0.0093
total β- avg	101.5 8	101 4	0.219
X-ray L	4.47	2.8 8	0.0003
X-ray $K\alpha_2$	31.8171 3	4.99 25	0.0034
X-ray $K\alpha_1$	32.1936 3	9.2 5	0.0063
X-ray $K\beta$	36.4	3.35 17	0.0026
γ 1	66.91 5	12.5 11	0.0178
γ 2	86.29 5	6.3 4	0.0116

137CS B- DECAY (30.0 Y 2) I(MIN) = 0.10%
SEE ALSO 137BA IT DECAY

Radiation Type	Energy (keV)	Intensity (%)	Δ(g-rad/ µCi-h)
β- 1 max	511.553 9		
avg	173.5 3	94.6 3	0.350
β- 2 max	1173.2		
avg	415.4 3	5.4 3	0.0478
total β- avg	186.6 4	100.0 5	0.397

138XE B- DECAY (14.17 M 7) I(MIN) = 0.10%

Radiation Type	Energy (keV)	Intensity (%)	Δ(g-rad/ µCi-h)
Auger-L	3.55	50.0 24	0.0038
ce-M- 1	3.63 5	31 5	0.0024
ce-NOP- 1	4.62 5	10.5 15	0.0010
ce-L- 2	5.14 5	52.0 19	0.0057
ce-M- 2	9.63 5	10.4 4	0.0021
ce-NOP- 2	10.62 5	2.55 9	0.0006
Auger-K	25.5	0.42 6	0.0002
ce-K- 5	117.77 3	1.61 25	0.0040
ce-L- 5	148.04 3	0.36 18	0.0011
ce-K- 7	206.58 5	0.241 11	0.0011
ce-K- 8	222.33 5	1.80 10	0.0085
ce-L- 8	252.60 5	0.28 7	0.0015
ce-K- 14	360.45 5	0.107 14	0.0008
ce-K- 17	398.51 5	0.26 5	0.0022

Radiation Type	Energy (keV)	Intensity (%)	Δ(g-rad/μCi-h)
γ 3	109.66 10	0.41 3	0.0010
γ 4	153.22 5	7.5 3	0.0244
γ 5	163.89 5	4.62 17	0.0161
γ 6	166.53 10	0.63 4	0.0022
γ 7	176.56 5	13.6 5	0.0511
γ 8	187.25 10	0.60 6	0.0024
γ 10	273.65 4	12.7 5	0.0740
γ 13	319.87 10	0.60 6	0.0041
γ 14	340.57 5	46.8 15	0.339
γ 16	507.21 10	0.98 6	0.0106
γ 18	818.50 4	100 3	1.74
γ 19	1048.07 7	79.7 25	1.78
γ 20	1235.34 5	19.8 7	0.520

8 weak γ's omitted (ΣIγ = 0.39%)

137BA IT DECAY (2.552 M 2) I(MIN)= 0.10%
FEEDING OF 2.552-M 137BA IN 137CS DECAY=94.6% 3

Radiation Type	Energy (keV)	Intensity (%)	Δ(g-rad/μCi-h)
Auger-L	3.67	7.8 5	0.0006
Auger-K	26.4	0.82 22	0.0005
ce-K- 1	624.204 9	8.24 5	0.110
ce-L- 1	655.656 9	1.50 3	0.0209
ce-MNO- 1	660.352 9	0.500 20	0.0070
X-ray L	4.47	1.1 3	0.0001
X-ray $K\alpha_2$	31.8171 3	2.11 7	0.0014
X-ray $K\alpha_1$	32.1936 3	3.90 12	0.0027
X-ray Kβ	36.4	1.42 5	0.0011
γ 1	661.645 9	89.9 4	1.27

Radiation Type		Energy (keV)	Intensity (%)	Δ(g-rad/μCi-h)
β- 1	max	320 80		
	avg	90 30	0.23 17	0.0004
β- 2	max	340 80		
	avg	100 30	0.46 14	0.0010
β- 3	max	490 80		
	avg	150 30	3.07 13	0.0098
β- 4	max	570 80		
	avg	180 30	9.5 4	0.0364
β- 5	max	810 80		
	avg	270 40	0.28 3	0.0016
β- 6	max	800 80		
	avg	270 40	32.7 13	0.188
β- 7	max	1040 80		
	avg	360 40	0.23 3	0.0018
β- 8	max	1460 80		
	avg	540 40	0.16 3	0.0018
β- 9	max	1620 80		
	avg	610 40	0.19 3	0.0025
β-10	max	1880 80		
	avg	720 40	0.27 4	0.0041
β-11	max	2380 80		
	avg	950 40	20.1 7	0.407
β-12	max	2420 80		
	avg	970 40	13.4 5	0.277
β-13	max	2570 80		
	avg	1040 40	5.6 21	0.124
β-14	max	2810 80		
	avg	1150 40	9 6	0.220
β-15	max	2820 80		
	avg	1150 40	5 5	0.122
total β-				
	avg	660 70	100 9	1.40

5 weak β's omitted (ΣIβ = 0.28%)

(CONTINUED ON NEXT PAGE)

138XE B- DECAY (CONTINUED)

Radiation Type	Energy (keV)	Intensity (%)	Δ(g-rad/ μCi-h)
X-ray L	4.29	6.2 17	0.0006
γ 1	4.85 5	0.19 3	≈0
γ 2	10.85 5	0.699 25	0.0002
X-ray $K\alpha_2$	30.6251 3	1.02 8	0.0007
X-ray $K\alpha_1$	30.9728 3	1.89 13	0.0012
X-ray Kβ	35	0.68 5	0.0005
γ 5	153.75 3	5.95 25	0.0195
γ 7	242.56 5	3.50 14	0.0181
γ 8	258.31 5	31.5 13	0.173
γ 9	282.51 6	0.428 20	0.0026
γ 12	335.28 9	0.107 11	0.0008
γ 13	371.44 5	0.50 3	0.0040
γ 14	396.43 5	6.3 3	0.0532
γ 15	401.36 5	2.17 13	0.0186
γ 17	434.49 5	20.3 9	0.188
γ 18	500.22 6	0.362 18	0.0039
γ 19	530.07 7	0.252 16	0.0028
γ 21	537.76 13	0.117 17	0.0013
γ 23	555.95 9	0.117 14	0.0014
γ 24	568.53 6	0.306 19	0.0037
γ 27	588.84 8	0.123 11	0.0015
γ 30	654.08 8	0.145 14	0.0020
γ 47	665.82 7	0.296 22	0.0055
γ 49	869.35 6	0.62 4	0.0115
γ 50	896.87 12	0.132 14	0.0025
γ 52	912.51 7	0.328 22	0.0064
γ 53	917.13 6	0.92 4	0.0180
γ 54	936.36 11	0.135 14	0.0027
γ 55	941.25 8	0.230 18	0.0046
γ 60	1093.87 9	0.41 3	0.0095
γ 61	1098.77 11	0.214 18	0.0050
γ 62	1102.24 17	0.107 14	0.0025

138CS B- DECAY (CONTINUED)

Radiation Type	Energy (keV)	Intensity (%)	Δ(g-rad/ μCi-h)
β- 4 max	1210 70		
avg	430 30	0.18 3	0.0016
β- 5 max	1350 70		
avg	490 30	0.47 6	0.0049
β- 6 max	1370 70		
avg	500 30	0.21 3	0.0022
β- 7 max	1600 70		
avg	600 30	0.30 3	0.0038
β- 8 max	1640 70		
avg	620 30	0.43 8	0.0057
β- 9 max	1920 70		
avg	740 40	0.230 20	0.0036
β-10 max	1950 70		
avg	750 40	0.170 20	0.0027
β-11 max	2050 70		
avg	800 40	0.270 20	0.0046
β-12 max	2130 70		
avg	830 40	0.34 3	0.0060
β-13 max	2240 70		
avg	880 40	0.17 3	0.0032
β-14 max	2300 70		
avg	910 40	0.64 4	0.0124
β-15 max	2360 70		
avg	940 40	0.21 4	0.0042
β-16 max	2410 70		
avg	960 40	0.37 20	0.0076
β-17 max	2440 70		
avg	970 40	0.20 5	0.0041
β-18 max	2510 70		
avg	1010 40	1.59 7	0.0342
β-19 max	2650 70		
avg	1070 40	8.70 23	0.198

Radiation Type	Energy (keV)	Intensity (%)	Δ(g-rad/μCi-h)
γ 63	1114.29 10	1.47 9	0.0350
γ 64	1141.64 9	0.51 4	0.0125
γ 65	1145.44 18	0.132 20	0.0032
γ 80	1571.84 16	0.26 3	0.0089
γ 82	1614.57 18	0.24 3	0.0081
γ 84	1768.26 13	16.7 7	0.630
γ 87	1812.54 18	0.180 20	0.0069
γ 88	1850.86 13	1.42 7	0.0561
γ 90	1925.36 14	0.56 4	0.0231
γ 91	2004.75 14	5.35 23	0.229
γ 92	2015.82 14	12.3 5	0.526
γ 94	2079.17 14	1.44 7	0.0639
γ 95	2252.26 14	2.29 11	0.110
γ 97	2321.90 16	0.62 4	0.0307
γ 99	2475.26 16	0.312 20	0.0164
γ 101	2497.56 17	0.173 14	0.0092

57 weak γ's omitted (ΣIγ = 2.66%)

138CS B- DECAY (32.2 M 1) I(MIN) = 0.10%

Radiation Type	Energy (keV)	Intensity (%)	Δ(g-rad/μCi-h)
β- 1 max	660 70		
avg	210 30	0.26 3	0.0012
β- 2 max	780 70		
avg	260 30	0.160 20	0.0009
β- 3 max	1050 70		
avg	360 30	0.100 10	0.0008

(CONTINUED ON NEXT COLUMN)

Radiation Type	Energy (keV)	Intensity (%)	Δ(g-rad/μCi-h)
β-20 max	2710 70		
avg	1100 40	1.67 8	0.0391
β-21 max	2840 70		
avg	1160 40	43.0 7	1.06
β-22 max	2870 70		
avg	1170 40	0.63 7	0.0157
β-23 max	2980 70		
avg	1220 40	7.90 20	0.205
β-24 max	3070 70		
avg	1270 40	13.0 3	0.352
β-25 max	3390 70		
avg	1410 40	13.7 7	0.411
β-26 max	3850 70		
avg	1630 40	4.6 18	0.160
total β-			
avg	1200 40	99.8 22	2.55

6 weak β's omitted (ΣIβ = 0.29%)

Radiation Type	Energy (keV)	Intensity (%)	Δ(g-rad/μCi-h)
γ 1	112.60 13	0.130 23	0.0003
γ 2	138.10 6	1.49 9	0.0044
γ 3	191.96 6	0.50 4	0.0021
γ 4	193.89 8	0.328 23	0.0014
γ 5	212.32 8	0.175 14	0.0008
γ 6	227.76 6	1.51 4	0.0073
γ 7	324.90 8	0.290 19	0.0020
γ 9	363.93 8	0.244 23	0.0019
γ 10	365.29 13	0.191 23	0.0015
γ 12	408.98 6	4.66 10	0.0406
γ 13	421.59 7	0.427 23	0.0038
γ 14	462.79 7	30.7 7	0.303
γ 15	516.74 12	0.43 5	0.0047
γ 16	546.94 7	10.76 24	0.125
γ 19	683.59 15	0.108 14	0.0016
γ 23	766.10 12	0.146 15	0.0024
γ 24	773.31 10	0.233 19	0.0038

(CONTINUED ON NEXT PAGE)

138CS B- DECAY (CONTINUED)

Radiation Type	Energy (keV)	Intensity (%)	Δ(g-rad/ μCi-h)
γ 25	782.08 9	0.33 3	0.0055
γ 31	871.80 8	5.11 14	0.0949
γ 32	880.8 3	0.11 3	0.0021
γ 33	935.03 12	0.181 16	0.0036
γ 36	1009.78 8	29.8 7	0.642
γ 38	1054.32 15	0.159 19	0.0036
γ 39	1147.22 9	1.24 7	0.0304
γ 40	1199.15 24	0.17 3	0.0043
γ 41	1203.69 13	0.40 4	0.0102
γ 42	1264.94 16	0.137 17	0.0037
γ 43	1343.59 9	1.14 6	0.0328
γ 46	1415.68 13	0.37 3	0.0110
γ 47	1435.86 9	76.3 16	2.33
γ 48	1445.04 25	0.97 19	0.0298
γ 49	1495.63 23	0.18 4	0.0058
γ 50	1555.31 10	0.366 23	0.0121
γ 51	1614.09 20	0.137 23	0.0047
γ 52	1717.1 3	0.107 23	0.0039
γ 53	1727.68 18	0.111 13	0.0041
γ 55	1778.25 23	0.137 23	0.0052
γ 60	2023.93 20	0.118 16	0.0051
γ 61	2062.34 17	0.111 12	0.0049
γ 64	2210.7 4	0.21 7	0.0101
γ 65	2218.00 10	15.2 4	0.717
γ 67	2499.4 3	0.17 5	0.0089
γ 69	2583.15 13	0.239 16	0.0131
γ 71	2639.59 13	7.63 23	0.429
γ 72	2731.12 15	0.120 8	0.0070
γ 78	3339.01 25	0.151 10	0.0107
γ 80	3366.98 25	0.227 13	0.0163

39 weak γ's omitted (ΣI_γ = 1.62%)

139CE EC DECAY (137.66 D 2) I(MIN) = 0.10%

Radiation Type	Energy (keV)	Intensity (%)	Δ(g-rad/ μCi-h)
Auger-L	3.8	89 5	0.0072
Auger-K	27.4	8.4 24	0.0049
ce-K- 1	126.928 7	17.05 16	0.0461
ce-L- 1	159.587 7	2.30 8	0.0078
ce-MNO- 1	164.492 7	0.61 3	0.0021
X-ray L	4.65	13 4	0.0013
X-ray $K\alpha_2$	33.03410 20	23.1 8	0.0163
X-ray $K\alpha_1$	33.44180 20	42.6 14	0.0304
X-ray Kβ	37.8	15.6 6	0.0126
γ 1	165.853 7	79.94 13	0.282

140BA B- DECAY (12.74 D 5) I(MIN) = 0.10%

Radiation Type	Energy (keV)	Intensity (%)	Δ(g-rad/ μCi-h)
Auger-L	3.8	100 5	0.0081
ce-L- 1	7.58 5	52 4	0.0083
ce-M- 1	12.49 5	10.7 7	0.0028
ce-NOP- 1	13.58 5	3.6 3	0.0010
ce-L- 2	23.70 5	61.0 7	0.0308
Auger-K	27.4	0.18 5	0.0001
ce-M- 2	28.61 5	12.6 3	0.0077
ce-NOP- 2	29.70 5	4.14 10	0.0026
ce-K- 5	123.684 20	1.48 5	0.0039

139BA B- DECAY (82.7 M 2) I(MIN)= 0.10%

Radiation Type	Energy (keV)	Intensity (%)	Δ(g-rad/ μCi-h)
Auger-L	3.8	4.1 3	0.0003
Auger-K	27.4	0.44 13	0.0003
ce-K- 1	126.925 20	4.69 15	0.0127
ce-L- 1	159.584 20	0.605 19	0.0021
ce-MNO- 1	164.489 20	0.1606 8	0.0006
β- 1 max	886 5		
avg	297.8 20	0.28 3	0.0018
β- 2 max	2141 5		
avg	837.5 23	27.40 10	0.489
β- 3 max	2307 5		
avg	912.5 23	72.20 10	1.40
total β-			
avg	889.9 23	99.94 15	1.89

14 weak β's omitted (ΣIβ = 0.06%)

Radiation Type	Energy (keV)	Intensity (%)	Δ(g-rad/ μCi-h)
X-ray L	4.65	0.61 15	≈0
X-ray $K\alpha_2$	33.03410 20	1.21 6	0.0008
X-ray $K\alpha_1$	33.44180 20	2.22 10	0.0016
X-ray Kβ	37.8	0.82 4	0.0007
γ 1	165.850 20	22.00 10	0.0777
γ 12	1420.50 20	0.25 3	0.0076

26 weak γ's omitted (ΣIγ = 0.09%)

Radiation Type	Energy (keV)	Intensity (%)	Δ(g-rad/ μCi-h)
ce-L- 5	156.343 20	0.199 7	0.0007
ce-K- 6	265.925 10	0.189 6	0.0011
ce-K- 10	498.349 20	0.254 8	0.0027
β- 1 max	454 10		
avg	136 4	24.69 22	0.0715
β- 2 max	567 10		
avg	177 4	9.86 8	0.0372
β- 3 max	872 10		
avg	292 4	3.81 12	0.0237
β- 4 max	991 10		
avg	340 4	39 4	0.282
β- 5 max	1005 10		
avg	357 4	23 4	0.175
total β-			
avg	276 5	100 6	0.590
X-ray L	4.65	15 4	0.0015
γ 1	13.85 5	1.20 8	0.0004
γ 2	29.97 5	13.7 4	0.0087
X-ray $K\alpha_2$	33.03410 20	0.495 21	0.0003
X-ray $K\alpha_1$	33.44180 20	0.91 4	0.0006
X-ray Kβ	37.8	0.335 14	0.0003
γ 4	132.67 8	0.202 20	0.0006
γ 5	162.609 20	6.21 8	0.0215
γ 6	304.850 10	4.30 5	0.0279
γ 7	423.722 12	3.15 4	0.0284
γ 8	437.575 20	1.93 4	0.0180
γ 9	467.57 5	0.147 15	0.0015
γ 10	537.274 20	24.39 21	0.279

1 weak γ's omitted (ΣIγ = 0.05%)

140LA B- DECAY (40.272 H 7) I(MIN) = 0.10%

Radiation Type	Energy (keV)	Intensity (%)	Δ(g-rad/μCi-h)
Auger-L	4	1.77 12	0.0001
ce-L- 1	18.046 4	0.401 24	0.0002
Auger-K	28.4	0.17 5	0.0001
ce-K- 3	28.473 6	0.19 3	0.0001
ce-K- 4	68.974 6	0.158 13	0.0002
ce-K- 5	90.678 8	0.263 18	0.0005
ce-K- 10	288.325 12	0.81 4	0.0050
ce-K- 15	446.586 19	0.443 14	0.0042
β- 1 max	1213.0 20		
avg	430.3 9	0.643 20	0.0059
β- 2 max	1238.8 20		
avg	441.1 9	11.09 13	0.104
β- 3 max	1244.4 20		
avg	443.5 9	5.74 5	0.0542
β- 4 max	1279.3 20		
avg	458.2 9	1.11 7	0.0108
β- 5 max	1296.2 20		
avg	465.3 9	5.61 5	0.0556
β- 6 max	1348.2 20		
avg	487.4 9	43.7 3	0.454
β- 7 max	1410.5 20		
avg	513.9 9	0.16 6	0.0018
β- 8 max	1412.3 20		
avg	514.7 9	5.11 5	0.0560
β- 9 max	1677.0 20		
avg	629.5 9	21.6 5	0.290
β-10 max	2164.0 20		
avg	846.2 9	5.0 5	0.0901
total β- avg	527.7 10	99.9 8	1.12

4 weak β's omitted (ΣIβ = 0.16%)

141CE B- DECAY (CONTINUED)

Radiation Type	Energy (keV)	Intensity (%)	Δ(g-rad/μCi-h)
ce-K- 1	103.449 10	18.8 3	0.0414
ce-L- 1	138.605 10	2.57 6	0.0076
ce-MNO- 1	143.929 10	0.73 5	0.0022
β- 1 max	434.6 15		
avg	129.6 6	70.5 7	0.195
β- 2 max	580.0 15		
avg	180.7 6	29.5 7	0.114
total β- avg	144.7 7	100.0 10	0.308
X-ray L	5	2.6 4	0.0003
X-ray $K\alpha_2$	35.55020 20	4.89 17	0.0037
X-ray $K\alpha_1$	36.02630 20	8.9 3	0.0069
X-ray Kβ	40.7	3.36 12	0.0029
γ 1	145.440 10	48.4 4	0.150

Radiation Type	Energy (keV)	Intensity (%)	Δ(g-rad/ μCi-h)
X-ray L	4.84	0.27 7	≈0
X-ray Kα2	34.27890 20	0.482 21	0.0004
X-ray Kα1	34.71970 20	0.88 4	0.0007
X-ray Kβ	39.3	0.329 15	0.0003
γ 4	109.417 6	0.200 15	0.0005
γ 5	131.121 8	0.49 3	0.0014
γ 6	173.544 11	0.130 20	0.0005
γ 7	241.966 12	0.47 3	0.0024
γ 8	266.551 14	0.452 25	0.0026
γ 10	328.768 12	20.74 18	0.145
γ 12	432.520 20	2.99 4	0.0275
γ 15	487.029 19	45.9 4	0.476
γ 17	751.83 8	4.41 4	0.0706
γ 18	815.78 3	23.64 17	0.411
γ 19	867.84 4	5.59 5	0.103
γ 20	919.54 4	2.68 3	0.0525
γ 21	925.19 4	7.05 8	0.139
γ 23	951.00 6	0.539 19	0.0109
γ 24	1596.17 6	95.40 8	3.24
γ 26	2347.80 6	0.846 17	0.0423
γ 28	2521.32 6	3.43 8	0.184
γ 30	2547.14 6	0.104 3	0.0056

15 weak γ's omitted (ΣIγ = 0.43%)

141CE B- DECAY (32.501 D 5)   I(MIN)= 0.10%

Radiation Type	Energy (keV)	Intensity (%)	Δ(g-rad/ μCi-h)
Auger-L	4	16.3 8	0.0014
Auger-K	29.4	1.6 5	0.0010

(CONTINUED ON NEXT COLUMN)

142PR B- DECAY (19.13 H 4)   I(MIN)= 0.10%

Radiation Type	Energy (keV)	Intensity (%)	Δ(g-rad/ μCi-h)
β- 1 max	583 3		
avg	181.5 10	3.7 5	0.0143
β- 2 max	2159 3		
avg	832.8 12	96.3 5	1.71
total β- avg	808.5 13	100.0 7	1.72

1 weak β's omitted (ΣIβ = 0.02%)

Radiation Type	Energy (keV)	Intensity (%)	Δ(g-rad/ μCi-h)
γ 2	1575.75 5	3.7 5	0.124

1 weak γ's omitted (ΣIγ = 0.02%)

143CE B- DECAY (33.0 H 2)   I(MIN)= 0.10%

Radiation Type	Energy (keV)	Intensity (%)	Δ(g-rad/ μCi-h)
Auger-L	4	59 5	0.0051
ce-K- 1	15.364 9	65 6	0.0214
Auger-K	29.4	5.7 18	0.0036
ce-L- 1	50.520 9	9.2 8	0.0099
ce-M- 1	55.844 9	1.95 16	0.0023
ce-NOP- 1	57.050 9	0.53 5	0.0006
ce-K- 6	251.270 15	2.13 17	0.0114

(CONTINUED ON NEXT PAGE)

143CE B- DECAY (CONTINUED)

Radiation Type	Energy (keV)	Intensity (%)	Δ(g-rad/ μCi-h)
ce-L- 6	286.426 15	0.352 21	0.0021
ce-MNO- 6	291.750 15	0.117 6	0.0007
β- 1 max	294 4		
avg	83.6 12	0.50 3	0.0009
β- 2 max	395 4		
avg	116.2 12	0.152 10	0.0004
β- 3 max	517 4		
avg	158.2 13	1.41 8	0.0048
β- 4 max	733 4		
avg	237.4 14	13.5 7	0.0683
β- 5 max	1104 4		
avg	384.6 15	49 3	0.401
β- 6 max	1398 4		
avg	507.5 16	35 5	0.378
total β-			
avg	402.1 17	100 6	0.855

9 weak β's omitted (ΣIβ = 0.24%)

Radiation Type	Energy (keV)	Intensity (%)	Δ(g-rad/ μCi-h)
X-ray L	5	9.6 16	0.0010
X-ray Kα₂	35.55020 20	17.6 15	0.0133
X-ray Kα₁	36.02630 20	32 3	0.0247
X-ray Kβ	40.7	12.1 10	0.0105
γ 1	57.355 9	11.7 9	0.0143
γ 5	231.563 6	2.09 12	0.0103
γ 6	293.261 15	43.4 22	0.271
γ 7	350.610 16	3.25 18	0.0243
γ 11	432.987 16	0.161 11	0.0015
γ 15	490.356 16	2.16 13	0.0225
γ 20	587.181 22	0.269 15	0.0034
γ 22	664.554 16	5.8 4	0.0814

144PR B- DECAY (CONTINUED)

Radiation Type	Energy (keV)	Intensity (%)	Δ(g-rad/ μCi-h)
β- 3 max	2997 3		
avg	1221.8 14	97.75 10	2.54
total β-			
avg	1207.6 15	100.01 13	2.57
γ 3	696.490 20	1.48 6	0.0220
γ 7	1489.15 5	0.300 13	0.0095
γ 9	2185.70 6	0.77 4	0.0360

7 weak γ's omitted (ΣIγ = 0.02%)

145PM EC DECAY (17.7 Y 4) I(MIN) = 0.10%

Radiation Type	Energy (keV)	Intensity (%)	Δ(g-rad/ μCi-h)
Auger-L	4.23	81 4	0.0073
ce-K- 1	23.63 10	2.19 10	0.0011
ce-K- 2	28.83 10	6.7 5	0.0041
Auger-K	30.5	6.5 20	0.0042
ce-L- 1	60.07 10	3.15 14	0.0040
ce-L- 2	65.27 10	1.78 16	0.0025
ce-M- 1	65.62 10	0.72 4	0.0010
ce-NOP- 1	66.88 10	0.193 9	0.0003
ce-M- 2	70.82 10	0.37 9	0.0006
ce-NOP- 2	72.08 10	0.111 24	0.0002

Radiation Type	Energy (keV)	Intensity (%)	Δ(g-rad/µCi-h)
γ 27	721.911 16	5.5 3	0.0841
γ 35	880.439 16	1.05 6	0.0198
γ 45	1103.178 25	0.423 24	0.0099

37 weak γ's omitted (ΣIγ = 0.68%)

143PR B- DECAY (13.56 D 2) I(MIN) = 0.10%

Radiation Type	Energy (keV)	Intensity (%)	Δ(g-rad/µCi-h)
β- 1 max	932.0 20		
avg	314.3 8	100	0.669

144PR B- DECAY (17.28 M 3) I(MIN) = 0.10%

Radiation Type	Energy (keV)	Intensity (%)	Δ(g-rad/µCi-h)
β- 1 max	811 3		
avg	267.0 12	1.08 5	0.0061
β- 2 max	2301 3		
avg	894.8 14	1.17 5	0.0223

(CONTINUED ON NEXT COLUMN)

Radiation Type	Energy (keV)	Intensity (%)	Δ(g-rad/µCi-h)
X-ray L	5.23	14.4 20	0.0016
X-ray Kα2	36.8474 3	21.1 7	0.0166
X-ray Kα1	37.3610 3	38.6 12	0.0307
X-ray Kβ	42.3	14.7 5	0.0132
γ 1	67.20 10	0.65 3	0.0009
γ 2	72.40 10	2.18 15	0.0034

147ND B- DECAY (10.98 D 1) I(MIN) = 0.10%

Radiation Type	Energy (keV)	Intensity (%)	Δ(g-rad/µCi-h)
Auger-L	4.38	41.0 23	0.0038
Auger-K	31.5	3.7 12	0.0025
ce-K- 1	45.9210 22	48.3 17	0.0472
ce-K- 2	75.30 5	0.31 4	0.0005
ce-L- 1	83.6771 22	7.09 25	0.0126
ce-MNO- 1	89.456 6	2.37 5	0.0045
β- 1 max	209.9 9		
avg	57.6 3	2.20 10	0.0027
β- 2 max	364.8 9		
avg	106.1 3	15.30 10	0.0346
β- 3 max	406.5 9		
avg	119.9 3	0.80 10	0.0020
β- 4 max	485.3 9		
avg	146.7 4	0.60 20	0.0019
β- 5 max	804.7 9		
avg	264.0 4	81.0 10	0.455
total β- avg	233.3 5	100.0 11	0.497

2 weak β's omitted (ΣIβ = 0.07%)

(CONTINUED ON NEXT PAGE)

147ND B- DECAY (CONTINUED)

Radiation Type	Energy (keV)	Intensity (%)	Δ(g-rad/ μCi-h)
X-ray L	5.43	7.8 11	0.0009
X-ray $K\alpha_2$	38.1712 5	12.8 6	0.0104
X-ray $K\alpha_1$	38.7247 5	23.2 11	0.0191
X-ray $K\beta$	43.8	8.9 5	0.0083
γ 1	91.1050 20	27.9 5	0.0541
γ 2	120.48 5	0.40 5	0.0010
γ 4	196.64 4	0.204 18	0.0009
γ 7	275.374 15	0.80 6	0.0047
γ 9	319.411 18	1.95 12	0.0133
γ 11	398.155 20	0.87 6	0.0074
γ 12	410.48 3	0.139 9	0.0012
γ 13	439.895 22	1.20 9	0.0112
γ 14	489.24 3	0.153 9	0.0016
γ 15	531.016 22	13.1 8	0.148
γ 17	594.803 3	0.265 18	0.0034
γ 19	685.90 4	0.81 6	0.0119

7 weak γ's omitted (ΣIγ = 0.07%)

147PM B- DECAY (2.6234 Y 2) I(MIN) = 0.10%

Radiation Type	Energy (keV)	Intensity (%)	Δ(g-rad/ μCi-h)
β- 1 max	224.6 6		
avg	61.96 12	99.9940 10	0.132

148PM IT DECAY (41.3 D 1) I(MIN) = 0.10%
%IT=4.9 6
SEE ALSO 148PM B- DECAY (41.3 D)

Radiation Type	Energy (keV)	Intensity (%)	Δ(g-rad/ μCi-h)
Auger-L	4.38	5.7 5	0.0005
ce-K- 2	30.52 10	3.2 4	0.0021
Auger-K	31.5	0.25 9	0.0002
ce-L- 1	54.07 10	3.6 5	0.0041
ce-M- 1	59.85 10	1.02 13	0.0013
ce-NOP- 1	61.17 10	0.28 4	0.0004
ce-L- 2	68.27 10	0.46 6	0.0007
X-ray L	5.43	1.09 17	0.0001
X-ray $K\alpha_2$	38.1712 5	0.85 11	0.0007
X-ray $K\alpha_1$	38.7247 5	1.55 20	0.0013
X-ray $K\beta$	43.8	0.60 8	0.0006
γ 2	75.70 10	1.09 14	0.0018

148PM B- DECAY (41.3 D 1) I(MIN) = 0.10%
%B-=95.1 6
SEE ALSO 148PM IT DECAY (41.3 D)

Radiation Type	Energy (keV)	Intensity (%)	Δ(g-rad/ μCi-h)
Auger-L	4.53	4.4 3	0.0004
Auger-K	32.6	0.40 13	0.0003
ce-K- 1	51.65 3	3.71 14	0.0041
ce-L- 1	90.74 3	0.599 22	0.0012
ce-MNO- 1	96.76 3	0.167 4	0.0003

148PM B- DECAY (5.37 D 1) I(MIN) = 0.10%

Radiation Type	Energy (keV)	Intensity (%)	Δ(g-rad/ μCi-h)
Auger-L	4.53	0.130 9	≈0
ce-K- 4	503.44 3	0.182 7	0.0019
β- 1 max	406 9		
avg	120 3	1.36 4	0.0035
β- 2 max	999 9		
avg	340 4	33.3 8	0.241
β- 3 max	1040 9		
avg	356 4	0.235 9	0.0018
β- 4 max	1914 9		
avg	728 4	9.4 3	0.146
β- 5 max	2464 9		
avg	975 4	55.5 11	1.15
total β- avg	725 6	100.0 14	1.55

5 weak β's omitted (ΣIβ = 0.23%)

Radiation Type	Energy (keV)	Intensity (%)	Δ(g-rad/ μCi-h)
γ 4	550.27 3	22.0 6	0.258
γ 5	592.83 3	0.353 11	0.0045
γ 6	611.26 3	1.02 3	0.0133
γ 8	874.18 3	0.235 9	0.0044
γ 9	896.42 3	0.981 24	0.0187
γ 11	914.85 3	11.5 3	0.223
γ 16	1465.12 3	22.2 5	0.693

15 weak γ's omitted (ΣIγ = 0.31%)

Radiation Type	Energy (keV)	Intensity (%)	Δ(g-rad/ μCi-h)
ce-K- 2	142.80 3	0.21 4	0.0006
ce-K- 3	241.28 3	0.80 17	0.0041
ce-K- 11	503.44 3	0.782 25	0.0084
β- 1 max	407 9		
avg	120 3	53.7 5	0.137
β- 2 max	5C6 9		
avg	154 4	18.4 3	0.0604
β- 3 max	695 9		
avg	222 4	21.7 4	0.103
β- 4 max	1007 9		
avg	343 4	0.9 3	0.0066
total β- avg	152 4	94.7 8	0.307
X-ray L	5.64	0.91 12	0.0001
X-ray Kα2	39.5224 3	1.45 7	0.0012
X-ray Kα1	40.1181 3	2.63 13	0.0022
X-ray Kβ	45.4	1.02 5	0.0010
γ 1	98.48 3	2.46 5	0.0052
γ 2	189.63 3	1.10 3	0.0044
γ 3	288.11 3	12.47 12	0.0765
γ 5	311.63 3	3.89 5	0.0258
γ 6	362.09 3	0.177 18	0.0014
γ 7	414.07 3	18.53 19	0.163
γ 8	432.78 3	5.31 7	0.0489
γ 9	460.57 3	0.415 18	0.0041
γ 10	501.26 3	6.70 9	0.0715
γ 11	550.27 3	94.2 10	1.10
γ 12	553.24 3	0.40 4	0.0047
γ 13	571.95 3	0.212 9	0.0026
γ 14	599.74 3	12.45 14	0.159
γ 15	611.26 3	5.44 10	0.0709
γ 16	629.97 3	88.3 6	1.19
γ 18	725.70 3	32.6 4	0.504
γ 19	915.33 3	17.04 21	0.332
γ 20	1013.81 3	20.13 21	0.435

3 weak γ's omitted (ΣIγ = 0.19%)

149ND B- DECAY (1.73 H 1) I(MIN) = 0.10%

Radiation Type	Energy (keV)	Intensity (%)	Δ(g-rad/ μCi-h)
Auger-L	4.38	34.2 21	0.0032
ce-K- 2	13.699 20	1.42 8	0.0004
ce-L- 1	22.57 3	4.8 5	0.0023
ce-M- 1	28.35 3	1.10 11	0.0007
ce-K- 4	29.15 3	3.7 6	0.0023
ce-K- 5	29.48 10	0.49 16	0.0003
ce-NOP- 1	29.67 3	0.36 4	0.0002
ce-K- 6	30.56 4	0.9 4	0.0006
Auger-K	31.5	2.6 9	0.0018
ce-L- 2	51.455 20	0.224 13	0.0002
ce-K- 8	51.823 15	2.22 20	0.0024
ce-L- 4	66.90 3	1.7 4	0.0025
ce-K- 10	69.137 14	17.1 19	0.0252
ce-M- 4	72.68 3	0.39 8	0.0006
ce-NOP- 4	74.00 3	0.105 22	0.0002
ce-I- 8	89.579 15	0.35 5	0.0007
ce-K- 15	94.026 14	0.23 3	0.0005
ce-L- 10	106.893 14	2.6 3	0.0059
ce-K- 16	110.692 10	0.41 5	0.0010
ce-M- 10	112.672 15	0.55 6	0.0013
ce-NOP-10	113.991 15	0.158 17	0.0004
ce-K- 19	143.456 9	0.36 5	0.0011
ce-K- 21	153.744 9	0.28 3	0.0009
ce-K- 22	162.964 10	0.44 8	0.0015
ce-K- 23	166.123 8	4.3 3	0.0154
ce-L- 19	181.212 9	0.108 13	0.0004
ce-K- 27	195.034 8	2.18 21	0.0090
ce-I- 23	203.879 8	0.67 5	0.0029
ce-MNO-23	209.658 10	0.183 13	0.0008
ce-K- 32	222.508 9	0.49 6	0.0023
ce-K- 33	224.981 8	0.170 18	0.0008
ce-L- 27	232.790 8	0.39 4	0.0019

149ND B- DECAY (CONTINUED)

Radiation Type	Energy (keV)	Intensity (%)	Δ(g-rad/ μCi-h)
γ 4	74.33 3	1.26 20	0.0020
γ 5	74.66 10	1.0 3	0.0016
γ 6	75.74 4	0.33 11	0.0005
γ 8	97.007 15	1.53 13	0.0032
γ 10	114.321 14	18.8 20	0.0457
γ 11	116.93 3	0.117 23	0.0003
γ 12	122.416 14	0.232 24	0.0006
γ 13	126.630 19	0.115 12	0.0003
γ 15	139.210 14	0.48 5	0.0014
γ 16	155.876 10	6.1 6	0.0201
γ 17	177.831 19	0.164 13	0.0006
γ 19	188.640 9	1.99 24	0.0080
γ 20	192.027 9	0.60 6	0.0024
γ 21	198.928 9	1.46 16	0.0062
γ 22	208.148 10	2.9 4	0.0130
γ 23	211.307 8	27.3 18	0.123
γ 24	213.946 17	0.41 5	0.0019
γ 25	226.846 21	0.16 4	0.0008
γ 27	240.218 8	4.0 4	0.0203
γ 28	245.699 8	1.04 11	0.0054
γ 31	258.064 14	0.38 4	0.0021
γ 32	267.692 9	6.1 6	0.0346
γ 33	270.165 8	10.7 11	0.0617
γ 34	273.25 4	0.23 5	0.0014
γ 35	275.445 13	0.60 12	0.0035
γ 36	276.960 19	0.32 7	0.0019
γ 37	282.455 11	0.62 6	0.0037
γ 38	288.192 12	0.68 7	0.0042
γ 39	294.807 12	0.58 6	0.0037
γ 40	301.133 16	0.38 4	0.0025
γ 41	310.982 14	0.52 5	0.0034
γ 42	326.556 11	4.7 5	0.0325

β- 1 max	377 4		
avg	110.0 14	0.154 10	0.0004
β- 2 max	455 4		
avg	136.2 14	0.92 5	0.0027
β- 3 max	1034 4		
avg	354.7 17	19.5 13	0.147
β- 4 max	1151 4		
avg	402.4 17	24.1 18	0.207
β- 5 max	1227 4		
avg	433.8 17	0.6 10	0.0055
β- 6 max	1264 4		
avg	449.2 17	0.91 8	0.0087
β- 7 max	1292 4		
avg	461.1 17	3.8 5	0.0373
β- 8 max	1301 4		
avg	465.0 17	0.5 5	0.0050
β- 9 max	1329 4		
avg	476.6 17	0.53 13	0.0054
β-10 max	1419 4		
avg	514.7 17	19.1 16	0.209
β-11 max	1478 4		
avg	539.8 18	26.5 21	0.305
β-12 max	1500 4		
avg	549.5 18	2.9 13	0.0339
β-13 max	1575 4		
avg	581.5 18	≈ 1	≈ 0.0136
total β- avg	457.4 18	101 4	0.980

3 weak β's omitted (ΣIβ = 0.02%)

X-ray L	5.43	6.5 9	0.0008
X-ray $K\alpha_2$	38.1712 5	9.1 6	0.0074
X-ray $K\alpha_1$	38.7247 5	16.6 11	0.0137
X-ray Kβ	43.8	6.4 5	0.0059
γ 2	58.883 20	1.52 7	0.0019

(CONTINUED ON NEXT COLUMN)

γ 44	347.833 23	0.19 9	0.0014
γ 45	349.233 10	1.48 16	0.0110
γ 48	360.055 20	0.164 13	0.0013
γ 49	366.637 15	0.66 8	0.0052
γ 51	384.691 18	0.34 4	0.0028
γ 53	423.554 10	9.4 10	0.0852
γ 55	443.550 12	1.50 15	0.0142
γ 63	538.15 7	0.11 4	0.0013
γ 64	540.510 10	7.7 8	0.0886
γ 65	556.43 5	1.2 5	0.0139
γ 70	630.238 21	0.221 24	0.0030
γ 71	635.482 25	0.112 12	0.0015
γ 72	654.831 14	7.3 8	0.102
γ 77	686.933 25	0.104 12	0.0015
γ 78	696.266 25	0.172 18	0.0026
γ 92	808.834 22	0.169 18	0.0029
γ 107	923.876 25	0.115 12	0.0023
γ 116	979.02 4	0.112 12	0.0023
γ 119	1022.78 3	0.120 12	0.0026
γ 139	1234.12 5	0.292 25	0.0077

100 weak γ's omitted (ΣIγ = 1.59%)

149PM B- DECAY (53.08 H 5) I(MIN) = 0.10%

Radiation Type	Energy (keV)	Intensity (%)	Δ(g-rad/ μCi-h)
Auger-L	4.53	0.158 16	≈0
ce-K- 9	239.07 5	0.208 20	0.0011
β- 1 max	190.6 21		
avg	51.8 6	0.116 16	0.0001

(CONTINUED ON NEXT PAGE)

149PM B- DECAY (CONTINUED)

Radiation Type	Energy (keV)	Intensity (%)	Δ(g-rad/ μCi-h)
β- 2 max	786.5 20		
avg	256.6 8	3.1 3	0.0169
β- 3 max	1072.4 20		
avg	369.6 9	96.7 3	0.761
total β-			
avg	365.2 9	100.1 5	0.779

7 weak β's omitted (ΣIβ = 0.23%)

Radiation Type	Energy (keV)	Intensity (%)	Δ(g-rad/ μCi-h)
γ 9	285.90 5	2.8 3	0.0173

27 weak γ's omitted (ΣIγ = 0.36%)

152EU EC DECAY (13.33 Y 4) I(MIN) = 0.10%
%(EC+B+)=72.08 19
SEE AISO 152EU B- DECAY (13.6 Y)

Radiation Type	Energy (keV)	Intensity (%)	Δ(g-rad/ μCi-h)
Auger-L	4.53	74 4	0.0071
Auger-K	32.6	5.7 19	0.0040
ce-K- 1	74.9482 12	19.4 6	0.0310
ce-L- 1	114.0456 12	10.6 4	0.0258
ce-M- 1	120.0596 14	2.43 8	0.0062
ce-NCP- 1	121.4367 15	0.667 21	0.0017

152EU EC DECAY (CONTINUED)

Radiation Type	Energy (keV)	Intensity (%)	Δ(g-rad/ μCi-h)
γ 82	1292.784 19	0.102 7	0.0028
γ 85	1408.011 14	20.85 9	0.625
γ 86	1457.628 15	0.494 11	0.0153
γ 87	1528.115 19	0.265 9	0.0086

60 weak γ's omitted (ΣIγ = 0.93%)

152EU B- DECAY (13.33 Y 4) I(MIN) = 0.10%
%B-=27.92 19
SEE AISO 152EU EC DECAY (13.33 Y)

Radiation Type	Energy (keV)	Intensity (%)	Δ(g-rad/ μCi-h)
Auger-L	4.84	0.78 4	≈0
ce-K- 6	294.036 4	0.83 3	0.0052
ce-L- 6	335.899 4	0.181 6	0.0013
β- 1 max	176 4		
avg	47.5 10	1.820 20	0.0018
β- 2 max	385 4		
avg	112.5 11	2.37 3	0.0057
β- 3 max	696 4		
avg	221.8 13	13.80 10	0.0652
β- 4 max	710 4		
avg	227.0 13	0.240 10	0.0012

Radiation		Energy (keV)	Intensity (%)	Δ (g-rad/μCi-h)
ce-K-	9	197.8647 12	0.610 19	0.0026
ce-I-	9	236.9621 12	0.156 5	0.0008

2 weak β's omitted (ΣIβ = 0.03%)

Radiation		Energy (keV)	Intensity (%)	Δ (g-rad/μCi-h)
X-ray	L	5.64	15.2 19	0.0018
X-ray	$K\alpha_2$	39.5224 3	21.0 7	0.0177
X-ray	$K\alpha_1$	40.1181 3	38.1 11	0.0325
X-ray	Kβ	45.4	14.8 5	0.0143
γ 1		121.7824 11	28.38 23	0.0736
γ 9		244.6989 11	7.51 7	0.0391
γ 14		295.939 8	0.430 11	0.0027
γ 17		329.433 17	0.121 3	0.0008
γ 26		416.052 6	0.111 3	0.0010
γ 29		443.983 7	0.319 15	0.0030
γ 30		444	2.80 3	0.0265
γ 32		488.66 4	0.407 5	0.0042
γ 39		564.021 8	0.467 17	0.0056
γ 41		566.421 8	0.129 5	0.0016
γ 45		656.484 12	0.144 5	0.0020
γ 48		674.678 3	0.186 7	0.0027
γ 51		688.678 6	0.853 11	0.0125
γ 52		719.353	0.267 13	0.0041
γ 59		810.459 7	0.317 7	0.0055
γ 62		841.586 8	0.163 5	0.0029
γ 64		867.388 8	4.21 5	0.0778
γ 67		919.401 8	0.436 13	0.0085
γ 68		926.324 15	0.265 11	0.0052
γ 70		964.131 9	14.49 6	0.298
γ 71		964.13	0.134 4	0.0027
γ 74		1005.279 17	0.649 21	0.0139
γ 75		1084.0 10	0.24 3	0.0054
γ 76		1085.914 13	10.16 5	0.235
γ 77		1112.116 17	13.56 6	0.321
γ 80		1212.950 12	1.40 4	0.0361
γ 81		1249.946 13	0.184 13	0.0049

(CONTINUED ON NEXT COLUMN)

Radiation		Energy (keV)	Intensity (%)	Δ (g-rad/μCi-h)
β- 5	max	889 4		
	avg	295.2 13	0.320 20	0.0020
β- 6	max	1064 4		
	avg	364.8 14	0.900 20	0.0070
β- 7	max	1475 4		
	avg	535.6 14	8.30 20	0.0947
total β-				
	avg	298.9 19	27.95 23	0.178

5 weak β's omitted (ΣIβ = 0.20%)

Radiation		Energy (keV)	Intensity (%)	Δ (g-rad/μCi-h)
X-ray	L	6	0.184 21	≈0
X-ray	$K\alpha_2$	42.3089 3	0.241 10	0.0002
X-ray	$K\alpha_1$	42.9962 3	0.434 17	0.0004
X-ray	Kβ	48.7	0.172 7	0.0002
γ 6		344.275 4	26.58 19	0.195
γ 8		367.788 4	0.863 11	0.0068
γ 9		411.115 5	2.233 13	0.0196
γ 12		503.387 5	0.152 5	0.0016
γ 17		586.294 6	0.459 9	0.0057
γ 20		678.623	0.471 7	0.0068
γ 23		764.905 8	0.175 9	0.0029
γ 24		778.910 10	12.96 7	0.215
γ 30		1089.700 15	1.658 23	0.0385
γ 31		1109.180 12	0.183 5	0.0043
γ 34		1299.124 12	1.626 17	0.0450

25 weak γ's omitted (ΣIγ = 0.68%)

152EU EC DECAY (9.32 H 2) I(MIN) = 0.10%
%(EC+B+)=28 3
SEE ALSO 152EU B- DECAY (9.32 H)

Radiation Type		Energy (keV)	Intensity (%)	Δ(g-rad/ μCi-h)
Auger-L		4.53	25.6 23	0.0025
Auger-K		32.6	2.1 7	0.0014
ce-K- 1		74.9482 12	4.9 6	0.0079
ce-L- 1		114.0456 12	2.7 4	0.0065
ce-M- 1		120.0596 14	0.61 7	0.0016
ce-NOP- 1		121.4367 15	0.166 20	0.0004
X-ray	L	5.64	5.2 8	0.0006
X-ray	$K\alpha_2$	39.5224 3	7.5 8	0.0063
X-ray	$K\alpha_1$	40.1181 3	13.6 13	0.0117
X-ray	$K\beta$	45.4	5.3 5	0.0051
γ 1		121.7824 11	7.2 8	0.0187
γ 13		562.920 10	0.226 25	0.0027
γ 22		841.63 6	14.6 16	0.262
γ 25		961.06 22	0.204 25	0.0042
γ 26		963.34 6	12.0 13	0.246
γ 33		1389.000 10	0.77 9	0.0228

32 weak γ's omitted ($\Sigma I\gamma$ = 0.40%)

153SM B- DECAY (CONTINUED)

Radiation Type		Energy (keV)	Intensity (%)	Δ(g-rad/ μCi-h)
Auger-K		33.7	4.5 10	0.0032
ce-K- 6		34.848 3	0.50 8	0.0004
ce-K- 7		40.965 4	0.37 5	0.0003
ce-K- 9		48.911 3	0.190 8	0.0002
ce-K- 10		54.6600 21	40.8 15	0.0474
ce-L- 4		61.6200 21	3.83 22	0.0050
ce-M- 4		67.8720 21	0.83 5	0.0012
ce-NOP- 4		69.3118 22	0.278 16	0.0004
ce-L- 6		75.315 3	0.24 4	0.0004
ce-L- 10		95.1270 21	6.17 23	0.0125
ce-MNO-10		101.3790 21	1.78 4	0.0038
β- 1	max	644 4		
	avg	203.0 15	34.7 16	0.150
β- 2	max	714 4		
	avg	228.8 15	43 3	0.210
β- 3	max	720 4		
	avg	230.9 15	0.61 5	0.0030
β- 4	max	817 4		
	avg	267.9 16	20.9 17	0.119
total β-	avg	227.6 16	99 4	0.482

11 weak β's omitted ($\Sigma I\beta$ = 0.25%)

Radiation Type		Energy (keV)	Intensity (%)	Δ(g-rad/ μCi-h)
X-ray	L	5.85	11.9 14	0.0015
X-ray	$K\alpha_2$	40.9019 3	17.3 7	0.0151
X-ray	$K\alpha_1$	41.5422 3	31.3 12	0.0277
X-ray	$K\beta$	47	12.2 5	0.0123
γ 4		69.6720 20	5.25 25	0.0078
γ 5		75.421 3	0.190 20	0.0003
γ 6		83.367 3	0.21 3	0.0004

152EU B- DECAY (9.32 H 2) I(MIN) = 0.10%
%B-=72 3
SEE ALSO 152EU EC DECAY (9.32 H)

Radiation Type	Energy (keV)	Intensity (%)	Δ(g-rad/ μCi-h)
β- 1 max	550 4		
avg	189.3 4	1.60 20	0.0065
β- 2 max	1521 4		
avg	579.5 5	1.70 20	0.0210
β- 3 max	1865 4		
avg	729.2 5	69 3	1.07
total β-			
avg	712.7 6	72 3	1.10

3 weak β's omitted (ΣIβ = 0.15%)

Radiation Type	Energy (keV)	Intensity (%)	Δ(g-rad/ μCi-h)
γ 5	344.275 4	2.4 3	0.0179
γ 16	970.38 3	0.60 7	0.0125
γ 19	1314.670 10	0.96 11	0.0268

19 weak γ's omitted (ΣIγ = 0.31%)

153SM B- DECAY (46.7 H 1) I(MIN) = 0.10%

Radiation Type	Energy (keV)	Intensity (%)	Δ(g-rad/ μCi-h)
Auger-L	4.69	54 3	0.0054
ce-L- 1	11.768 20	0.32	≈0
ce-K- 4	21.1530 21	23.5 14	0.0106
γ 7	89.484 4	0.170 20	0.0003
γ 9	97.430 3	0.730 20	0.0015
γ 10	103.1790 20	28.3 6	0.0622

(CONTINUED ON NEXT COLUMN)

59 weak γ's omitted (ΣIγ = 0.32%)

153GD EC DECAY (242 D 1) I(MIN) = 0.10%

Radiation Type	Energy (keV)	Intensity (%)	Δ(g-rad/ μCi-h)
Auger-L	4.69	110 5	0.0109
ce-L- 1	6.0180 4	0.146 5	≈0
ce-L- 2	11.768 20	0.298 9	≈0
ce-K- 3	21.1530 21	10.8 7	0.0049
Auger-K	33.7	9.0 20	0.0065
ce-K- 5	34.848 3	0.49 6	0.0004
ce-K- 6	40.965 4	0.15 5	0.0001
ce-K- 7	48.911 3	7.6 4	0.0079
ce-K- 8	54.6600 21	30.4 14	0.0354
ce-L- 3	61.6200 21	1.77 10	0.0023
ce-MNO- 3	67.8720 21	0.508 25	0.0007
ce-L- 5	75.315 3	0.23 3	0.0004
ce-L- 7	89.378 3	1.13 5	0.0022
ce-L- 8	95.1270 21	4.60 22	0.0093
ce-MNO- 7	95.630 3	0.322 10	0.0007
ce-MNO- 8	101.3790 21	1.33 5	0.0029
X-ray L	5.85	24 3	0.0030
X-ray $K\alpha_2$	40.9019 3	34.6 12	0.0301

(CONTINUED ON NEXT PAGE)

153GD EC DECAY (CONTINUED)

Radiation Type	Energy (keV)	Intensity (%)	Δ(g-rad/ μCi-h)
X-ray Kα₁	41.5422 3	62.5 20	0.0553
X-ray Kβ	47	24.5 9	0.0245
γ 3	69.6720 20	2.42 12	0.0036
γ 5	83.367 3	0.206 22	0.0004
γ 7	97.430 3	29.5 9	0.0612
γ 8	103.1790 20	21.1 8	0.0464

5 weak γ's omitted (ΣIγ = 0.19%)

154EU B- DECAY (8.6 Y 1) I(MIN) = 0.10%

Radiation Type	Energy (keV)	Intensity (%)	Δ(g-rad/ μCi-h)
Auger-L	4.84	32.5 16	0.0034
Auger-K	34.9	1.8 6	0.0013
ce-K- 3	72.90 4	26.8 12	0.0416
ce-L- 3	114.76 4	16.8 8	0.0410
ce-M- 3	121.26 4	3.89 17	0.0100
ce-NOP- 3	122.76 4	1.09 5	0.0029
ce-K- 25	197.80 4	0.536 24	0.0023
ce-L- 25	239.66 4	0.149 7	0.0008
β- 1 max	255 4		
avg	71.1 11	27.6 8	0.0418
β- 2 max	314 4		

154EU B- DECAY (CONTINUED)

Radiation Type	Energy (keV)	Intensity (%)	Δ(g-rad/ μCi-h)
γ 3	123.14 4	40.5 13	0.106
γ 19	188.22 5	0.228 11	0.0009
γ 25	248.04 4	6.59 21	0.0348
γ 51	401.30 5	0.210 9	0.0018
γ 58	444.40 5	0.503 20	0.0048
γ 62	478.26 5	0.215 9	0.0022
γ 75	557.56 5	0.254 10	0.0030
γ 78	582.00 5	0.84 4	0.0104
γ 79	591.74 5	4.84 16	0.0610
γ 87	625.22 5	0.310 13	0.0041
γ 94	676.59 5	0.140 6	0.0020
γ 96	692.41 5	1.70 6	0.0251
γ 99	715.76 5	0.175 8	0.0027
γ 100	723.30 4	19.7 7	0.304
γ 103	756.87 5	4.34 14	0.0700
γ 108	815.55 5	0.465 18	0.0081
γ 110	845.39 5	0.550 22	0.0099
γ 111	850.64 5	0.230 9	0.0042
γ 112	873.19 5	11.5 4	0.214
γ 114	892.73 5	0.460 18	0.0087
γ 116	904.05 5	0.82 4	0.0159
γ 123	996.32 4	10.3 4	0.219
γ 124	1004.76 4	17.4 6	0.372
γ 132	1118.50 10	0.103 4	0.0025
γ 134	1128.40 10	0.267 11	0.0064
γ 136	1140.90 10	0.216 9	0.0052
γ 143	1241.60 20	0.130 5	0.0034
γ 144	1246.60 20	0.70 3	0.0185
γ 145	1274.45 9	35.5 11	0.964

	avg	89.3 12	0.79 3	0.0015
β- 3	max	329 4		
	avg	94.1 12	0.147 5	0.0003
β- 4	max	357 4		
	avg	103.2 12	1.36 5	0.0030
β- 5	max	415 4		
	avg	122.3 12	0.127 5	0.0003
β- 6	max	443 4		
	avg	131.8 12	0.280 16	0.0008
β- 7	max	556 4		
	avg	171.0 13	0.111 5	0.0004
β- 8	max	577 4		
	avg	178.4 13	36.5 12	0.139
β- 9	max	711 4		
	avg	227.3 13	0.614 22	0.0030
β-10	max	723 4		
	avg	231.8 14	0.263 12	0.0013
β-11	max	847 4		
	avg	279.0 14	16.8 6	0.0998
β-12	max	978 4		
	avg	330.5 14	2.0 5	0.0141
β-13	max	1159 4		
	avg	403.5 15	0.18 6	0.0015
β-14	max	1604 4		
	avg	590.6 15	0.54 18	0.0068
β-15	max	1851 4		
	avg	698.3 16	12.3 20	0.183
total β-				
	avg	233.7 20	100 3	0.497

11 weak β's omitted (ΣIβ = 0.24%)

X-ray	L	6	7.6 9	0.0010
X-ray	$K\alpha_2$	42.3089 3	7.3 4	0.0065
X-ray	$K\alpha_1$	42.9962 3	13.1 7	0.0120
X-ray	Kβ	48.7	5.2 3	0.0054

(CONTINUED ON NEXT COLUMN)

γ 159	1494.60 20	0.65 3	0.0207
γ 165	1596.48 20	1.67 6	0.0568

140 weak γ's omitted (ΣIγ = 1.76%)

155EU B- DECAY (4.96 Y 1) I(MIN) = 0.10%

Radiation Type	Energy (keV)	Intensity (%)	Δ(g-rad/ µCi-h)
ce-L- 1	2.024 20	1.21 14	≈0
Auger-L	4.84	34 3	0.0035
ce-M- 1	8.519 20	0.26 3	≈0
ce-K- 9	9.7709 21	8.5 9	0.0018
ce-L- 3	10.40 4	14 3	0.0031
ce-L- 4	12.644 20	0.95 8	0.0003
ce-M- 3	16.90 4	3.1 7	0.0011
ce-L- 5	18.137 21	0.49 4	0.0002
ce-NOP- 3	18.40 4	1.03 20	0.0004
ce-M- 4	19.139 20	0.220 19	≈0
ce-L- 6	23.05 5	0.40 14	0.0002
ce-M- 5	24.632 21	0.108 9	≈0
Auger-K	34.9	1.6 6	0.0012
ce-K- 10	35.823 5	0.40 5	0.0003
ce-K- 11	36.3039 21	11.2 8	0.0087
ce-L- 7	36.9224 21	0.44 4	0.0003
ce-L- 9	51.6344 21	1.71 18	0.0019
ce-K- 12	55.069 3	4.4 4	0.0052
ce-M- 9	58.1292 21	0.38 4	0.0005
ce-NOP- 9	59.6342 22	0.107 12	0.0001

(CONTINUED ON NEXT PAGE)

155EU B- DECAY (CONTINUED)

Radiation Type	Energy (keV)	Intensity (%)	Δ(g-rad/ μCi-h)
ce-L- 11	78.1674 21	1.72 13	0.0029
ce-M- 11	84.6622 21	0.37 3	0.0007
ce-NOP-11	86.1672 22	0.105 8	0.0002
ce-L- 12	96.932 3	0.66 5	0.0014
ce-M- 12	103.427 3	0.142 11	0.0003
β- 1 max	101 3		
avg	26.3 8	0.69 8	0.0004
β- 2 max	129 3		
avg	34.0 9	2.2 3	0.0016
β- 3 max	141 3		
avg	37.6 9	46 5	0.0368
β- 4 max	160 3		
avg	42.9 9	26 4	0.0238
β- 5 max	187 3		
avg	50.6 9	8.3 11	0.0089
β- 6 max	247 3		
avg	68.5 9	18 5	0.0263
total β-			
avg	45.4 10	101 9	0.0978
X-ray L	6	8.0 11	0.0010
γ 5	26.513 21	0.318 24	0.0002
X-ray $K\alpha_2$	42.3089 3	6.5 4	0.0059
X-ray $K\alpha_1$	42.9962 3	11.8 7	0.0108
γ 7	45.2980 20	1.28 10	0.0012
X-ray Kβ	48.7	4.6 3	0.0048
γ 9	60.0100 20	1.14 12	0.0015
γ 10	86.062 5	0.151 19	0.0003

157DY EC DECAY (CONTINUED)

Radiation Type	Energy (keV)	Intensity (%)	Δ(g-rad/ μCi-h)
ce-K- 2	31.01 4	1.7 3	0.0011
Auger-K	36	5.5 20	0.0043
ce-L- 1	52.11 7	0.35 14	0.0004
ce-L- 2	74.30 4	0.25 4	0.0004
ce-K- 7	274.16 20	1.09 4	0.0064
ce-L- 7	317.45 20	0.151 5	0.0010
X-ray L	6.27	17.7 20	0.0024
X-ray $K\alpha_2$	43.7441 3	23.5 8	0.0219
X-ray $K\alpha_1$	44.4816 3	42.2 13	0.0400
X-ray Kβ	50.4	16.8 6	0.0180
γ 1	60.82 7	0.27 11	0.0003
γ 2	83.01 4	0.53 8	0.0009
γ 4	182.20 20	1.95 13	0.0076
γ 5	265.340 20	0.19 4	0.0011
γ 7	326.16 20	95.2 12	0.661

21 weak γ's omitted ($\Sigma I\gamma$ = 0.48%)

Radiation Type	Energy (keV)	Intensity (%)	Δ(g-rad/ μCi-h)
γ 11	86.5430 20	30.9 20	0.0570
γ 12	105.308 3	20.6 14	0.0462

7 weak γ's omitted (ΣIγ = 0.20%)

157TB EC DECAY (150 Y 30) I(MIN) = 0.10%

Radiation Type	Energy (keV)	Intensity (%)	Δ(g-rad/ μCi-h)
Auger-L	4.84	60 5	0.0061
Auger-K	34.9	0.7 4	0.0005
X-ray L	6	14.0 19	0.0018
X-ray $K\alpha_2$	42.3089 3	2.7 14	0.0024
X-ray $K\alpha_1$	42.9962 3	4.8 24	0.0044
X-ray $K\beta$	48.7	1.9 10	0.0020

157DY EC DECAY (8.1 H 1) I(MIN) = 0.10%

Radiation Type	Energy (keV)	Intensity (%)	Δ(g-rad/ μCi-h)
Auger-L	5	71 4	0.0076
ce-K- 1	8.82 7	2.1 9	0.0004

(CONTINUED ON NEXT COLUMN)

169YB EC DECAY (32.01 D 2) I(MIN) = 0.10%

Radiation Type	Energy (keV)	Intensity (%)	Δ(g-rad/ μCi-h)
ce-K- 3	3.729 7	38.0 25	0.0030
Auger-L	5.67	162 10	0.0196
ce-M- 1	6.094 8	76 3	0.0099
ce-NOP- 1	7.929 8	18.4 7	0.0031
ce-L- 2	10.63 5	10.4 7	0.0024
ce-M- 2	18.44 5	2.31 16	0.0009
ce-NOP- 2	20.28 5	0.77 6	0.0003
ce-K- 4	34.223 7	8.0 6	0.0059
Auger-K	40.9	10 4	0.0089
ce-K- 5	50.387 7	35.0 18	0.0376
ce-L- 3	53.003 7	6.7 5	0.0076
ce-K- 7	58.797 7	1.35 7	0.0017
ce-M- 3	60.812 7	1.51 10	0.0020
ce-NOP- 3	62.647 7	0.41 3	0.0005
ce-K- 8	71.130 7	6.3 4	0.0095
ce-L- 4	83.497 7	1.38 9	0.0024
ce-M- 4	91.306 7	0.311 22	0.0006
ce-L- 5	99.661 7	5.64 23	0.0120
ce-M- 5	107.470 7	1.26 5	0.0029
ce-L- 7	108.071 7	1.39 7	0.0032
ce-NOP- 5	109.305 7	0.367 19	0.0009
ce-M- 7	115.880 7	0.337 17	0.0008
ce-K- 10	117.820 7	10.7 6	0.0269
ce-L- 8	120.404 7	5.4 3	0.0138
ce-M- 8	128.213 7	1.30 7	0.0035
ce-NOP- 8	130.048 7	0.361 18	0.0010
ce-K- 11	138.563 7	13.2 7	0.0389
ce-L- 10	167.094 7	1.92 10	0.0068
ce-M- 10	174.903 7	0.433 22	0.0016
ce-NOP-10	176.738 7	0.125 7	0.0005

(CONTINUED ON NEXT PAGE)

169YB EC DECAY (CONTINUED)

Radiation Type	Energy (keV)	Intensity (%)	Δ(g-rad/ μCi-h)
ce-L- 11	187.837 7	2.15 11	0.0086
ce-M- 11	195.646 7	0.481 24	0.0020
ce-NOP-11	197.481 7	0.137 7	0.0006
ce-K- 14	248.340 7	0.478 24	0.0025
ce-L- 14	297.614 7	0.138 7	0.0009
X-ray L	7.18	51 7	0.0078
γ 1	8.401 8	0.35 3	≈0
γ 2	20.75 5	0.237 16	0.0001
X-ray Kα2	49.7726 4	53.2 23	0.0564
X-ray Kα1	50.7416 4	94 4	0.102
X-ray Kβ	57.5	38.6 17	0.0473
γ 3	63.119 7	41.6 24	0.0560
γ 4	93.613 7	2.55 16	0.0051
γ 5	109.777 7	17.4 7	0.0407
γ 7	118.187 7	1.91 8	0.0048
γ 8	130.520 7	11.5 5	0.0319
γ 10	177.210 7	22.3 9	0.0843
γ 11	197.953 7	35.9 14	0.151
γ 12	240.4 7	0.122 8	0.0006
γ 13	261.072 7	1.68 7	0.0094
γ 14	307.730 7	9.9 4	0.0647

24 weak γ's omitted (ΣIγ = 0.08%)

171ER B- DECAY (7.52 H 3) I(MIN) = 0.10%

Radiation Type	Energy (keV)	Intensity (%)	Δ(g-rad/ μCi-h)
ce-L- 2	2.269 8	6.4 10	0.0003
ce-M- 1	2.718 6	70 4	0.0040
ce-NOP- 1	4.553 6	23.3 14	0.0023
Auger-L	5.67	47 3	0.0056
ce-M- 2	10.078 8	1.42 22	0.0003
ce-NOP- 2	11.913 8	0.48 8	0.0001
Auger-K	40.9	2.6 10	0.0022
ce-K- 4	52.231 4	40 3	0.0445
ce-K- 5	57.266 6	1.72 10	0.0021
ce-K- 6	64.627 4	5.8 4	0.0080
ce-L- 4	101.505 4	6.5 5	0.0140
ce-I- 5	106.540 6	1.82 11	0.0041
ce-M- 4	109.314 4	1.45 10	0.0034
ce-NOP- 4	111.149 4	0.42 3	0.0010
ce-L- 6	113.901 4	5.4 4	0.0132
ce-M- 5	114.349 6	0.44 3	0.0011
ce-NOP- 5	116.184 6	0.122 7	0.0003
ce-M- 6	121.710 4	1.32 8	0.0034
ce-NOP- 6	123.545 4	0.372 23	0.0010
ce-K- 15	236.511 14	0.62 4	0.0031
ce-K- 16	248.901 18	1.06 6	0.0056
ce-L- 16	298.175 18	0.152 8	0.0010
β- 1 max	205.3 12		
avg	56.0 4	0.342 19	0.0004
β- 2 max	491.7 12		
avg	147.6 5	0.53 3	0.0017
β- 3 max	577.3 12		
avg	177.4 5	2.17 11	0.0082
β- 4 max	814.4 12		
avg	264.4 5	0.181 18	0.0010

170TM B- DECAY (128.6 D 3) I(MIN) = 0.10%
%B-=99.854 2

Radiation Type	Energy (keV)	Intensity (%)	Δ(g-rad/ μCi-h)
Auger-L	5.84	12.1 8	0.0015
ce-K- 1	22.9245 16	4.7 3	0.0023
Auger-K	42.2	0.23 10	0.0002
ce-L- 1	73.7704 16	12.2 7	0.0192
ce-M- 1	81.8587 16	3.00 18	0.0052
ce-NOP- 1	83.7696 17	0.83 5	0.0015
β- 1 max	883.7 10		
avg	290.5 4	24.0 10	0.149
β- 2 max	968.0 10		
avg	323.1 4	76.0 10	0.523
total β-			
avg	315.3 4	100.0 15	0.672
X-ray L	7.42	4.0 6	0.0006
X-ray Kα2	51.3540 5	1.27 8	0.0014
X-ray Kα1	52.3889 5	2.25 14	0.0025
X-ray Kβ	59.4	0.93 6	0.0012
γ 1	84.2568 15	3.26 16	0.0058

Radiation Type	Energy (keV)	Intensity (%)	Δ(g-rad/ μCi-h)
β- 5 max	1065.4 12		
avg	362.2 5	97 5	0.748
β- 6 max	1485.3 12		
avg	534.7 5	2.30 20	0.0262
total β-			
avg	359.5 6	103 5	0.786

6 weak β's omitted (ΣIβ = 0.19%)

Radiation Type	Energy (keV)	Intensity (%)	Δ(g-rad/ μCi-h)
γ 1	5.025 6	0.66 3	≈0
X-ray L	7.18	14.7 20	0.0022
γ 2	12.385 8	0.30 5	≈0
X-ray Kα2	49.7726 4	13.4 8	0.0142
X-ray Kα1	50.7416 4	23.6 14	0.0255
X-ray Kβ	57.5	9.7 6	0.0119
γ 4	111.621 4	21.0 12	0.0498
γ 5	116.656 6	2.35 12	0.0058
γ 6	124.017 4	9.3 5	0.0246
γ 10	210.60 3	0.66 4	0.0029
γ 11	237.14 4	0.309 17	0.0016
γ 13	277.43 5	0.59 4	0.0035
γ 15	295.901 14	29.5 15	0.186
γ 16	308.291 18	66 4	0.432
γ 18	371.96 9	0.263 15	0.0021
γ 34	670.70 20	0.258 12	0.0037
γ 36	676.1 3	0.291 14	0.0042
γ 42	784.10 20	0.245 12	0.0041
γ 43	796.60 20	0.65 3	0.0111
γ 48	907.7 4	0.65 3	0.0125

49 weak γ's omitted (ΣIγ = 1.04%)

171TM B- DECAY (1.92 Y 1) I(MIN) = 0.10%

Radiation Type	Energy (keV)	Intensity (%)	Δ(g-rad/ μCi-h)
ce-K- 1	5.386 7	1.02 20	0.0001
Auger-L	5.84	1.17 17	0.0001
ce-L- 1	56.232 7	0.71 14	0.0008
ce-MNO- 1	64.320 7	0.22 5	0.0003
β- 1 max	30.0 10		
avg	7.6 3	2.1 4	0.0003
β- 2 max	96.7 10		
avg	25.2 3	97.9 4	0.0525
total β- avg	24.8 3	100.0 6	0.0529
X-ray L	7.42	0.39 8	≈0
X-ray Kα2	51.3540 5	0.28 6	0.0003
X-ray Kα1	52.3889 5	0.49 10	0.0005
X-ray Kβ	59.4	0.20 4	0.0003
γ 1	66.718 7	0.15 3	0.0002

175YB B- DECAY (4.19 D 1) I(MIN) = 0.10%

Radiation Type	Energy (keV)	Intensity (%)	Δ(g-rad/ μCi-h)
Auger-L	6	3.1 4	0.0004
Auger-K	43.5	0.19 9	0.0002
ce-K- 1	50.489 4	3.7 5	0.0039

177LU B- DECAY (6.71 D 1) I(MIN) = 0.10%

Radiation Type	Energy (keV)	Intensity (%)	Δ(g-rad/ μCi-h)
Auger-L	6.18	8.9 6	0.0012
ce-K- 1	6.2952 21	0.111 9	≈0
Auger-K	44.8	0.28 12	0.0003
ce-K- 2	47.601 3	5.2 4	0.0053
ce-L- 2	101.681 3	7.0 5	0.0152
ce-M- 2	110.351 3	1.74 12	0.0041
ce-NOP- 2	112.414 3	0.50 4	0.0012
ce-K- 4	143.008 10	0.59 8	0.0018
ce-L- 4	197.088 10	0.100 18	0.0004
β- 1 max	175.8 10		
avg	47.3 3	12.2 7	0.0123
β- 2 max	384.1 10		
avg	111.3 4	9.1 12	0.0216
β- 3 max	497.1 10		
avg	148.9 4	78.6 10	0.249
total β- avg	133.0 5	99.9 18	0.283

1 weak β's omitted (ΣIβ = 0.05%)

Radiation Type	Energy (keV)	Intensity (%)	Δ(g-rad/ μCi-h)
X-ray L	7.9	3.3 4	0.0006
X-ray Kα2	54.6114 8	1.64 12	0.0019
X-ray Kα1	55.7902 8	2.87 20	0.0034
X-ray Kβ	63.2	1.20 9	0.0016
γ 1	71.6460 20	0.154 10	0.0002
γ 2	112.952 3	6.4 4	0.0154
γ 4	208.359 10	11.0 8	0.0488
γ 5	249.686 25	0.212 14	0.0011
γ 6	321.33 4	0.219 14	0.0015

1 weak γ's omitted (ΣIγ = 0.05%)

Radiation Type	Energy (keV)	Intensity (%)	Δ(g-rad/μCi-h)
ce-K- 2	74.342 6	0.131 19	0.0002
ce-L- 1	102.933 4	0.88 12	0.0019
ce-M- 1	111.312 4	0.20 3	0.0005
ce-K- 6	333.008 20	0.24 4	0.0017
β- 1 max	71.4 13		
avg	18.4 4	10.3 4	0.0040
β- 2 max	216.2 13		
avg	68.1 5	0.80 12	0.0012
β- 3 max	353.9 13		
avg	101.6 5	3.30 20	0.0071
β- 4 max	467.7 13		
avg	139.1 5	86.5 17	0.256
total β-			
avg	125.0 7	100.9 18	0.269
X-ray L	7.66	1.10 18	0.0002
X-ray $K\alpha_2$	52.9650 5	1.10 14	0.0012
X-ray $K\alpha_1$	54.0698 5	1.93 24	0.0022
X-ray Kβ	61.3	0.80 10	0.0010
γ 1	113.803 4	1.91 25	0.0046
γ 2	137.656 6	0.117 16	0.0003
γ 3	144.861 5	0.33 5	0.0010
γ 5	282.517 14	3.1 4	0.0184
γ 6	396.322 20	6.5 8	0.0549

1 weak γ's omitted ($\Sigma I\gamma = 0.08\%$)

181HF B- DECAY (42.4 D 1) I(MIN) = 0.10%

Radiation Type	Energy (keV)	Intensity (%)	Δ(g-rad/μCi-h)
ce-M- 1	1.19 10	2.64 14	≈0
ce-NOP- 1	3.33 10	1.36 7	≈0
ce-NOP- 2	5.64 3	0.120 12	≈0
Auger-L	6.35	40.3 24	0.0055
Auger-K	46.2	1.5 7	0.0015
ce-K- 3	65.604 20	21.4 13	0.0299
ce-K- 4	68.834 20	8.5 6	0.0124
ce-K- 5	69.44 4	2.5 4	0.0037
ce-L- 3	121.338 20	25.0 15	0.0647
ce-L- 4	124.568 20	1.71 12	0.0045
ce-L- 5	125.18 4	0.49 10	0.0013
ce-M- 3	130.312 20	6.2 4	0.0173
ce-NOP- 3	132.454 20	1.81 11	0.0051
ce-M- 4	133.542 20	0.40 3	0.0011
ce-M- 5	134.15 4	0.113 24	0.0003
ce-NOP- 4	135.684 20	0.118 8	0.0003
ce-K- 6	278.43 20	0.54 4	0.0032
ce-L- 6	334.17 20	0.169 10	0.0012
ce-K- 8	414.58 20	1.54 10	0.0136
ce-L- 8	470.32 20	0.353 21	0.0035
β- 1 max	404 4		
avg	117.7 14	7 3	0.0175
β- 2 max	408 4		
avg	118.9 14	93 3	0.236
total β-			
avg	118.8 14	100 5	0.253
X-ray L	8.15	15.7 18	0.0027
X-ray $K\alpha_2$	56.2770 10	9.5 5	0.0113

(CONTINUED ON NEXT PAGE)

181HF B- DECAY (CONTINUED)

Radiation Type	Energy (keV)	Intensity (%)	Δ(g-rad/ µCi-h)
ce-M- 1	1.19 10	2.64 14	≈0
ce-NOP- 1	3.33 10	1.36 7	≈0
X-ray $K\alpha_1$	57.5320 10	16.5 8	0.0202
X-ray $K\beta$	65.2	7.0 4	0.0097
γ 3	133.020 20	43.0 22	0.122
γ 4	136.250 20	6.1 3	0.0177
γ 5	136.86 4	1.80 9	0.0052
γ 6	345.85 20	14.0 7	0.103
γ 7	476.00 20	0.43 10	0.0044
γ 8	482.00 20	86 5	0.883
γ 9	615.5 5	0.250 13	0.0033

3 weak γ's omitted (ΣIγ = 0.04%)

181W EC DECAY (121.2 D 3) I(MIN) = 0.10%

Radiation Type	Energy (keV)	Intensity (%)	Δ(g-rad/ µCi-h)
ce-M- 1	3.50 3	25 6	0.0018
ce-NOP- 1	5.64 3	8.2 19	0.0010
Auger-L	6.35	58 6	0.0078
Auger-K	46.2	3.0 14	0.0030
ce-K- 3	85.1 3	0.107 12	0.0002
γ 1	6.21 3	0.97 22	0.0001

182TA B- DECAY (CONTINUED)

Radiation Type	Energy (keV)	Intensity (%)	Δ(g-rad/ µCi-h)
ce-K- 12	86.8567 13	0.267 12	0.0005
ce-L- 6	88.0066 5	32.4 14	0.0608
ce-M- 6	97.2868 5	8.1 4	0.0169
ce-NOP- 6	99.5114 5	2.39 11	0.0051
ce-L- 8	101.5672 13	C.99 7	0.0021
ce-K- 13	109.8654 19	1.63 12	0.0038
ce-M- 8	110.8474 13	C.230 15	0.0005
ce-K- 14	128.8227 23	0.267 16	0.0007
ce-L- 11	140.3283 15	0.122 5	0.0004
ce-K- 15	152.5760 21	0.302 14	0.0010
ce-K- 16	159.791 3	0.428 20	0.0015
ce-L- 13	167.2906 19	0.452 20	0.0016
ce-M- 13	176.5708 19	0.108 5	0.0004
ce-L- 14	186.2479 23	0.170 10	0.0007
ce-K- 17	194.546 5	0.293 14	0.0012
ce-L- 16	217.216 3	0.221 10	0.0010
ce-L- 17	251.971 5	0.127 6	0.0007
ce-K- 25	1051.75 3	0.105 4	0.0023
β- 1 max	258 3		
avg	71.5 9	28.6 10	0.0436
β- 2 max	301 3		
avg	84.7 10	0.13 2	0.0002
β- 3 max	324 3		
avg	91.8 10	2.70 20	0.0053
β- 4 max	368 3		
avg	105.9 10	0.66 4	0.0015
β- 5 max	437 3		
avg	128.5 10	20 3	0.0547
β- 6 max	480 3		
avg	142.8 11	2.3	0.0070
β- 7 max	522 3		

X-ray L	8.15	22 3	0.0039
X-ray Kα2	56.2770 10	19.0 20	0.0228
X-ray Kα1	57.5320 10	33 4	0.0407
X-ray Kβ	65.2	14.0 15	0.0194

2 weak γ's omitted (ΣIγ = 0.14%)

182TA B- DECAY (115.0 D 2) I(MIN)= 0.10%

Radiation Type	Energy (keV)	Intensity (%)	Δ(g-rad/ μCi-h)
Auger-L	6.53	58 3	0.0081
ce-K- 5	15.1555 8	15.8 10	0.0051
ce-L- 1	19.6372 5	0.80 3	0.0003
ce-M- 1	28.9174 6	0.185 7	0.0001
ce-K- 6	30.5814 5	12.5 6	0.0082
ce-L- 2	30.6145 8	0.138 10	≈0
ce-K- 8	44.1420 13	5.1 4	0.0048
Auger-K	45.7	1.6 8	0.0016
ce-L- 3	53.6219 4	6.5 4	0.0075
ce-L- 4	55.6497 4	6.5 4	0.0077
ce-M- 3	62.9021 5	1.49 9	0.0020
ce-M- 4	64.9299 5	1.48 10	0.0021
ce-NOP- 3	65.1267 5	0.45 3	0.0006
ce-NOP- 4	67.1545 5	0.42 3	0.0006
ce-L- 5	72.5807 8	3.93 25	0.0061
ce-M- 5	81.8609 8	0.93 6	0.0016
ce-K- 11	82.9031 15	0.75 3	0.0013
ce-NOP- 5	84.0855 8	0.279 18	0.0005

(CONTINUED ON NEXT COLUMN)

avg	157.1 11	40 5	0.134
β- 8 max	554 3		
avg	168.0 11	≈ 0.5	≈ 0.0018
β- 9 max	590 3		
avg	180.6 11	5.0 20	0.0192
total β- avg	126.2 12	100 7	0.269

4 weak β's omitted (ΣIβ = 0.15%)

X-ray L	8.4	25 3	0.0045
γ 1	31.7370 4	0.628 11	0.0004
γ 2	42.7143 7	0.244 15	0.0002
X-ray Kα2	57.9817 5	10.3 5	0.0127
X-ray Kα1	59.31820 10	17.9 7	0.0226
γ 3	65.72170 20	2.79 15	0.0039
X-ray Kβ	67.2	7.6 4	0.0109
γ 4	67.74950 20	41.2 23	0.0594
γ 5	84.6805 7	2.65 15	0.0048
γ 6	100.1064 3	14.0 5	0.0299
γ 8	113.6670 12	1.92 11	0.0046
γ 9	116.4149 17	0.44 3	0.0011
γ 11	152.4281 14	7.15 19	0.0232
γ 12	156.3817 12	2.72 9	0.0091
γ 13	179.3904 18	3.14 12	0.0120
γ 14	198.3477 22	1.54 8	0.0065
γ 15	222.1010 20	7.54 25	0.0357
γ 16	229.316 3	3.63 13	0.0177
γ 17	264.071 5	3.63 13	0.0204
γ 20	927.99 7	0.62 4	0.0123
γ 21	959.74 7	0.356 22	0.0073
γ 22	1001.68 7	2.09 11	0.0445
γ 23	1044.43 9	0.24 3	0.0054
γ 24	1113.38 10	0.39 4	0.0094
γ 25	1121.28 3	34.9 6	0.834
γ 26	1157.5	0.35 10	0.0085

(CONTINUED ON NEXT PAGE)

182TA B- DECAY (CONTINUED)

Radiation Type	Energy (keV)	Intensity (%)	Δ(g-rad/µCi-h)
γ 27	1157.5	0.64 13	0.0158
γ 29	1189.04 4	16.4 4	0.415
γ 30	1221.418 25	27.3 6	0.711
γ 31	1223.2	0.21 4	0.0055
γ 32	1230.97 3	11.55 25	0.303
γ 33	1257.47 5	1.51 4	0.0405
γ 34	1273.75 6	0.663 18	0.0180
γ 35	1289.17 7	1.41 4	0.0387
γ 36	1342.72 6	0.262 9	0.0075
γ 37	1373.80 7	0.230 8	0.0067

9 weak γ's omitted (ΣIγ = 0.42%)

185W B- DECAY (75.1 D 3) I(MIN)= 0.10%

Radiation Type	Energy (keV)	Intensity (%)	Δ(g-rad/µCi-h)
β- 1 max	432.5 10		
avg	126.8 4	100	0.270

1 weak β's omitted (ΣIβ = 0.08%)

1 weak γ's omitted (ΣIγ = 0.02%)

187W B- DECAY (CONTINUED)

Radiation Type	Energy (keV)	Intensity (%)	Δ(g-rad/µCi-h)
β- 7 max	694.5 18		
avg	217.8 7	7.2 3	0.0334
β- 8 max	801.1 18		
avg	257.0 7	0.15 4	0.0008
β- 9 max	1178.7 18		
avg	401.9 7	1.7 10	0.0146
β-10 max	1312.9 18		
avg	457.3 8	25.1 24	0.244
total β- avg	263.2 9	100 4	0.562

7 weak β's omitted (ΣIβ = 0.02%)

Radiation Type	Energy (keV)	Intensity (%)	Δ(g-rad/µCi-h)
X-ray L	8.65	8.6 10	0.0016
X-ray $K\alpha_2$	59.7179 6	7.7 4	0.0098
X-ray $K\alpha_1$	61.1403 6	13.3 6	0.0173
X-ray Kβ	69.3	5.69 24	0.0084
γ 7	72.060 10	11.9 5	0.0183
γ 15	134.220 10	9.4 4	0.0270
γ 21	206.28 3	0.153 7	0.0007
γ 24	246.180 10	0.128 7	0.0007
γ 30	479.530 10	23.4 10	0.239
γ 33	511.760 10	0.69 3	0.0075
γ 34	551.550 10	5.45 22	0.0641
γ 39	589.09 3	0.131 6	0.0016
γ 41	618.370 10	6.7 3	0.0887
γ 42	625.520 10	1.17 5	0.0155
γ 47	685.810 10	29.3 12	0.427
γ 50	745.210 20	0.319 14	0.0051
γ 52	772.880 20	4.42 18	0.0727

187W B- DECAY (23.9 H 1) I(MIN) = 0.10%

Radiation Type	Energy (keV)	Intensity (%)	Δ(g-rad/μCi-h)
ce-K- 7	0.384 10	9.1 6	≈0
ce-M- 1	4.2183 4	1.67 22	0.0001
ce-NOP- 1	6.5250 4	0.56 8	≈0
Auger-L	6.7	19.2 13	0.0027
ce-L- 3	16.70 3	0.15 6	≈0
ce-L- 4	23.8533 4	0.11 4	≈0
Auger-K	47	1.1 5	0.0011
ce-L- 7	59.533 10	1.65 9	0.0021
ce-K- 15	62.544 10	17.6 7	0.0234
ce-M- 7	69.128 10	0.375 20	0.0006
ce-NOP- 7	71.435 10	0.109 6	0.0002
ce-L- 15	121.693 10	2.99 15	0.0078
ce-M- 15	131.288 10	0.69 4	0.0019
ce-NOP-15	133.555 10	0.208 11	0.0006
ce-K- 21	134.60 3	0.397 22	0.0011
ce-K- 30	407.854 10	0.429 22	0.0037
ce-K- 41	546.694 10	0.202 11	0.0024
β- 1 max	433.5 18		
avg	127.1 6	0.478 19	0.0013
β- 2 max	448.4 18		
avg	132.1 6	0.603 24	0.0017
β- 3 max	496.3 18		
avg	148.2 7	0.20 9	0.0006
β- 4 max	540.0 18		
avg	163.1 7	4.33 20	0.0150
β- 5 max	627.1 18		
avg	193.6 7	58.9 22	0.243
β- 6 max	687.4 18		
avg	215.2 7	1.56 7	0.0072
γ 57	864.540 10	0.360 15	0.0066
γ 58	879.43 5	0.152 7	0.0028

(CONTINUED ON NEXT COLUMN)

50 weak γ's omitted (ΣIγ = 0.45%)

188W B- DECAY (69.4 D 5) I(MIN) = 0.10%

Radiation Type	Energy (keV)	Intensity (%)	Δ(g-rad/μCi-h)
Auger-L	6.7	0.20 4	≈0
ce-L- 1	51.05 3	0.29 5	0.0003
β- 1 max	58 3		
avg	14.9 8	0.85 5	0.0003
β- 2 max	285 3		
avg	89.9 10	0.13 8	0.0002
β- 3 max	349 3		
avg	99.7 10	99.02 10	0.210
total β- avg	99.0 11	100.01 14	0.211
γ 1	63.58 3	0.108 18	0.0001
γ 6	227.090 20	0.221 14	0.0011
γ 7	290.669 13	0.401 25	0.0025

4 weak γ's omitted (ΣIγ = 0.02%)

188RE β- DECAY (16.98 H 2) I(MIN) = 0.10%

Radiation Type		Energy (keV)	Intensity (%)	Δ(g-rad/ μCi-h)
Auger-L		6.88	6.5 5	0.0009
Auger-K		48.3	0.19 9	0.0002
ce-K- 1		81.169 20	4.9 3	0.0084
ce-L- 1		142.072 20	5.5 4	0.0167
ce-M- 1		151.991 20	1.40 9	0.0045
ce-NOP- 1		154.386 20	0.420 25	0.0014
β- 1	max	178.3 9		
	avg	48.0 3	0.107 7	0.0001
β- 2	max	354.0 9		
	avg	101.2 3	0.186 10	0.0004
β- 3	max	656.9 9		
	avg	203.9 4	0.52 3	0.0023
β- 4	max	1033.0 9		
	avg	345.0 4	0.64 3	0.0047
β- 5	max	1486.4 9		
	avg	527.4 4	1.60 14	0.0180
β- 6	max	1964.4 9		
	avg	728.5 4	25.1 13	0.389
β- 7	max	2119.4 9		
	avg	795.0 4	71.6 15	1.21
total β-	avg	764.3 5	100.0 20	1.63
10 weak β's omitted ($\Sigma I\beta$ = 0.25%)				
X-ray	L	9	3.0 4	0.0006
X-ray	$K\alpha_2$	61.4867 7	1.35 9	0.0018
X-ray	$K\alpha_1$	63.0005 7	2.34 15	0.0031
X-ray	$K\beta$	71.4	1.00 7	0.0015
γ 1		155.040 20	14.9 8	0.0491
γ 5		477.96 3	1.04 6	0.0106

192IR EC DECAY (74.02 D 18) I(MIN) = 0.10%
%EC=4.70 10
SEE ALSO 192IR β- DECAY (74.02 D)

Radiation Type		Energy (keV)	Intensity (%)	Δ(g-rad/ μCi-h)
Auger-L		6.88	3.13 19	0.0005
Auger-K		48.3	0.16 8	0.0002
ce-K- 3		131.920 7	0.64 3	0.0018
ce-L- 3		192.823 7	0.444 20	0.0018
ce-MNO- 3		202.742 7	0.145 5	0.0006
X-ray	L	9	1.47 15	0.0003
X-ray	$K\alpha_2$	61.4867 7	1.17 3	0.0015
X-ray	$K\alpha_1$	63.0005 7	2.02 5	0.0027
X-ray	$K\beta$	71.4	0.866 25	0.0013
γ 2		201.306 7	0.466 20	0.0020
γ 3		205.791 7	3.29 11	0.0144
γ 5		283.255 20	0.261 13	0.0016
γ 8		374.476 7	0.729 19	0.0058
γ 11		484.565 8	3.16 8	0.0326
γ 12		489.06 3	0.397 12	0.0041

7 weak γ's omitted ($\Sigma I\gamma$ = 0.17%)

Radiation Type	Energy (keV)	Intensity (%)	Δ(g-rad/μCi-h)
γ 10	633.03 3	1.25 13	0.0168
γ 11	635.00 4	0.15 5	0.0020
γ 12	672.51 3	0.110 6	0.0016
γ 15	829.51 3	0.408 21	0.0072
γ 17	931.32 3	0.561 25	0.0111

34 weak γ's omitted (ΣIγ = 0.72%)

191OS B- DECAY (15.4 D 2) I(MIN) = 0.10%

Radiation Type	Energy (keV)	Intensity (%)	Δ(g-rad/μCi-h)
Auger-L	7	101 9	0.0152
ce-L- 1	28.431 10	71 8	0.0428
ce-L- 2	33.63 3	0.29 4	0.0002
ce-MNO- 1	38.676 10	29 3	0.0237
ce-MNO- 2	43.88 3	0.100 12	≈0
Auger-K	49.6	3.0 15	0.0032
ce-K- 4	53.289 7	78 10	0.0890
ce-L- 4	115.981 7	16.6 20	0.0409
ce-MNO- 4	126.226 7	5.1 6	0.0137
β- 1 max	142 3		
avg	37.6 7	100	0.0801
X-ray L	9.18	50 6	0.0098
X-ray Kα2	63.2867 7	22 3	0.0294
X-ray Kα1	64.8956 7	38 5	0.0519
X-ray Kβ	73.6	16.2 20	0.0254
γ 4	129.400 7	35 4	0.0965

3 weak γ's omitted (ΣIγ = 0.03%)

192IR B- DECAY (74.02 D 18) I(MIN) = 0.10%
%B-=95.30 10
SEE ALSO 192IR EC DECAY

Radiation Type	Energy (keV)	Intensity (%)	Δ(g-rad/μCi-h)
Auger-L	7.24	7.5 5	0.0012
Auger-K	51	0.35 13	0.0004
ce-K- 1	57.945 7	0.112 7	0.0001
ce-K- 2	217.556 7	1.920 10	0.0089
ce-K- 3	230.052 8	1.786 25	0.0088
ce-K- 4	238.105 7	4.47 14	0.0226
ce-L- 2	282.071 7	0.88 3	0.0053
ce-M- 2	292.655 7	0.219 7	0.0014
ce-L- 3	294.567 8	0.772 24	0.0048
ce-L- 4	302.620 7	1.95 6	0.0125
ce-M- 3	305.151 8	0.192 6	0.0012
ce-M- 4	313.204 7	0.484 15	0.0032
ce-NOP- 4	315.778 7	0.148 5	0.0010
ce-K- 6	389.665 7	1.02 4	0.0085
ce-L- 6	454.180 7	0.293 10	0.0028
ce-K- 10	526.003 8	0.167 6	0.0019
β- 1 max	256 4		
avg	70.8 12	5.59 8	0.0084
β- 2 max	536 4		
avg	161.2 14	41.35 19	0.142
β- 3 max	672 4		
avg	208.9 15	48.0 6	0.214
total β- avg	179.8 16	95.0 7	0.364

2 weak β's omitted (ΣIβ = 0.10%)

(CONTINUED ON NEXT PAGE)

192IR B- DECAY (CONTINUED)

Radiation Type	Energy (keV)	Intensity (%)	Δ(g-rad/ μCi-h)
X-ray L	9.44	4.0 5	0.0008
X-ray Kα2	65.1220 20	2.63 7	0.0037
X-ray Kα1	66.8320 20	4.52 10	0.0064
X-ray Kβ	75.7	1.97 6	0.0032
γ 1	136.340 7	0.181 9	0.0005
γ 2	295.951 7	28.96 12	0.183
γ 3	308.447 8	29.67 9	0.195
γ 4	316.500 7	82.84 9	0.558
γ 5	416.460 8	0.662 10	0.0059
γ 6	468.060 7	47.8 6	0.477
γ 8	588.573 8	4.52 8	0.0567
γ 10	604.398 8	8.18 16	0.105
γ 11	612.451 8	5.33 10	0.0696
γ 12	884.523 18	0.302 6	0.0057

5 weak γ's omitted (ΣIγ = 0.10%)

194IR B- DECAY (19.15 H 3) I(MIN) = 0.10%
SEE ANALOGOUS EC DECAY TO LOW-SPIN STATES IN
194PT WITH COMPARABLE G-BRANCHINGS

Radiation Type	Energy (keV)	Intensity (%)	Δ(g-rad/ μCi-h)
Auger-L	7.24	0.68 8	0.0001
ce-K- 4	215.146 14	0.17 3	0.0008

194IR B- DECAY (CONTINUED)

Radiation Type	Energy (keV)	Intensity (%)	Δ(g-rad/ μCi-h)
γ 23	938.69 3	0.59 10	0.0119
γ 27	1150.75 5	0.59 10	0.0145
γ 30	1183.49 5	0.30 5	0.0076
γ 43	1468.91 7	0.19 3	0.0059

61 weak γ's omitted (ΣIγ = 0.84%)

195PT IT DECAY (4.02 D 1) I(MIN) = 0.10%

Radiation Type	Energy (keV)	Intensity (%)	Δ(g-rad/ μCi-h)
ce-L- 1	6.06 21	0.135 5	≈0
Auger-L	7.24	137 10	0.0211
ce-L- 3	16.996 6	70 4	0.0252
ce-K- 4	20.462 10	66 4	0.0287
ce-M- 3	27.580 6	15.9 10	0.0093
ce-NOP- 3	30.154 6	4.1 3	0.0026
Auger-K	51	3.0 11	0.0032
ce-K- 5	51.11 20	13.4 4	0.0146
ce-K- 6	51.340 10	1.33 8	0.0015
ce-L- 4	84.977 10	11.6 7	0.0209
ce-M- 4	95.561 10	2.61 14	0.0053
ce-NOP- 4	98.135 10	0.85 7	0.0018
ce-L- 5	115.62 20	60.8 8	0.150
ce-L- 6	115.855 10	2.70 16	0.0067
ce-M- 5	126.20 20	19.1 5	0.0512

Radiation Type	Energy (keV)	Intensity (%)	Δ(g-rad/ μCi-h)
ce-K- 6	250.053 14	0.64 11	0.0034
ce-I- 6	314.568 14	0.27 5	0.0018
β- 1 max	453.7 20		
avg	133.4 7	0.330 20	0.0009
β- 2 max	628.9 20		
avg	193.4 7	0.160 20	0.0007
β- 3 max	739.1 20		
avg	233.0 8	0.55 3	0.0027
β- 4 max	771.8 20		
avg	244.9 8	0.63 4	0.0033
β- 5 max	983.9 20		
avg	324.7 8	1.76 8	0.0122
β- 6 max	1328.3 20		
avg	454.0 8	0.3	0.0029
β- 7 max	1629.0 20		
avg	583.9 9	1.34 13	0.0167
β- 8 max	1922.6 20		
avg	707.2 9	9.2 6	0.138
β- 9 max	2251.0 20		
avg	847.5 9	85.5 20	1.54
total β- avg	808.2 10	100.0 21	1.72

14 weak β's omitted ($\Sigma I\beta$ = 0.28%)

Radiation Type	Energy (keV)	Intensity (%)	Δ(g-rad/ μCi-h)
X-ray L	9.44	0.37 6	≈0
X-ray $K\alpha_2$	65.1220 20	0.24 3	0.0003
X-ray $K\alpha_1$	66.8320 20	0.40 5	0.0006
X-ray $K\beta$	75.7	0.176 23	0.0003
γ 4	293.541 14	2.5 4	0.0158
γ 5	300.741 14	0.35 6	0.0022
γ 6	328.448 14	13.0 21	0.0909
γ 10	589.179 17	0.139 22	0.0017
γ 14	621.971 19	0.33 6	0.0044
γ 15	645.146 20	1.16 19	0.0160

(CONTINUED ON NEXT COLUMN)

Radiation Type	Energy (keV)	Intensity (%)	Δ(g-rad/ μCi-h)
ce-M- 6	126.439 10	0.69 4	0.0019
ce-NOP- 5	128.78 20	6.48 21	0.0178
ce-NOP- 6	129.013 10	0.214 13	0.0006
X-ray L	9.44	74 9	0.0148
γ 3	30.876 6	2.56 17	0.0017
X-ray $K\alpha_2$	65.1220 20	22.4 12	0.0310
X-ray $K\alpha_1$	66.8320 20	38.4 20	0.0547
X-ray $K\beta$	75.7	16.7 9	0.0270
γ 4	98.857 10	11.3 6	0.0239
γ 6	129.735 10	2.81 14	0.0078

6 weak γ's omitted ($\Sigma I\gamma$ = 0.21%)

195AU EC DECAY (183 D 1) I(MIN) = 0.10%

Radiation Type	Energy (keV)	Intensity (%)	Δ(g-rad/ μCi-h)
Auger-L	7.24	103 7	0.0159
ce-L- 1	16.996 6	20.4 15	0.0074
ce-K- 2	20.462 10	62.9 25	0.0274
ce-M- 1	27.580 6	4.6 4	0.0027
ce-NOP- 1	30.154 6	1.20 10	0.0008
Auger-K	51	3.8 14	0.0041
ce-K- 3	51.340 10	0.381 19	0.0004
ce-L- 2	84.977 10	11.1 5	0.0200
ce-M- 2	95.561 10	2.50 8	0.0051
ce-NOP- 2	98.135 10	0.81 5	0.0017

(CONTINUED ON NEXT PAGE)

195AU IT DECAY (CONTINUED)

Radiation Type	Energy (keV)	Intensity (%)	Δ(g-rad/ μCi-h)
ce-L- 3	115.855 10	0.77 4	0.0019
ce-M- 3	126.439 10	0.198 10	0.0005
X-ray L	9.44	56 7	0.0112
γ 1	30.876 6	0.75 4	0.0005
X-ray $K\alpha_2$	65.1220 20	28.4 10	0.0394
X-ray $K\alpha_1$	66.8320 20	48.8 16	0.0694
X-ray Kβ	75.7	21.2 8	0.0342
γ 2	98.857 10	10.8 3	0.0228
γ 3	129.735 10	0.80 3	0.0022

2 weak γ's omitted (ΣIγ = 0.02%)

195AU IT DECAY (30.5 S 2) I(MIN) = 0.10%

Radiation Type	Energy (keV)	Intensity (%)	Δ(g-rad/ μCi-h)
Auger-L	7.42	62 4	0.0098
ce-L- 1	42.45 3	71.4 7	0.0646
ce-L- 2	47.087 20	1.56 10	0.0016
Auger-K	52.4	0.9 4	0.0010
ce-M- 1	53.38 3	21.2 5	0.0241
ce-NOP- 1	56.04 3	6.76 20	0.0081

197PT B- DECAY (CONTINUED)

Radiation Type	Energy (keV)	Intensity (%)	Δ(g-rad/ μCi-h)
β- 2 max	641.7 6		
avg		81.5	≈0
β- 3 max	450.3 6		
avg		7.9	≈0
total β- avg		100 3	≈0
γ 1	77.35 5	17.0 20	0.0280
γ 2	191.437 10	3.5 5	0.0142
γ 3	268.78 5	0.27 5	0.0016

197HG EC DECAY (64.1 H 1) I(MIN) = 0.10%

Radiation Type	Energy (keV)	Intensity (%)	Δ(g-rad/ μCi-h)
γ 1	77.3520 20	18 3	0.0297
γ 2	191.364 15	0.48 10	0.0020

1 weak γ's omitted (ΣIγ = 0.04%)

Radiation Type	Energy (keV)	Intensity (%)	Δ(g-rad/μCi-h)
ce-M- 2	58.015 20	0.388 24	0.0005
ce-NOP- 2	60.681 20	0.121 8	0.0002
ce-K- 3	119.65 3	0.276 16	0.0007
ce-K- 4	181.03 3	23.33 20	0.0900
ce-L- 3	186.02 3	0.253 15	0.0010
ce-K- 5	237.85 8	0.265 20	0.0013
ce-L- 4	247.40 3	4.40 5	0.0232
ce-M- 4	258.33 3	1.0403 14	0.0057
ce-NOP- 4	260.99 3	0.3247 5	0.0018
ce-L- 5	304.22 8	0.171 13	0.0011
X-ray L	9.7	35 4	0.0072
γ 2	61.440 20	0.17	0.0002
X-ray Kα2	66.9895 8	6.66 16	0.0095
X-ray Kα1	68.8037 8	11.38 25	0.0167
X-ray Kβ	78	4.98 13	0.0083
γ 3	200.37 3	1.63 10	0.0070
γ 4	261.75 3	68.1 6	0.380

2 weak γ's omitted (ΣIγ = 0.07%)

197PT B- DECAY (18.3 H 3) I(MIN) = 0.10%

Radiation Type	Energy (keV)	Intensity (%)	Δ(g-rad/μCi-h)
β- 1 max	719.0 6		
avg		11 3	≈0

(CONTINUED ON NEXT COLUMN)

198AU B- DECAY (2.696 D 2) I(MIN) = 0.10%

Radiation Type	Energy (keV)	Intensity (%)	Δ(g-rad/μCi-h)
Auger-L	7.6	2.10 16	0.0003
ce-K- 1	328.7021 14	2.87 3	0.0201
ce-L- 1	396.9651 15	1.031 20	0.0087
ce-M- 1	408.2428 16	0.267 20	0.0023
ce-NOP- 1	411.0041 15	0.101 7	0.0009
β- 1 max	285.3 6		
avg	79.60 20	1.30 10	0.0022
β- 2 max	961.2 6		
avg	314.80 20	98.70 10	0.662
total β-			
avg	311.78 21	100.02 15	0.664

1 weak β's omitted (ΣIβ = 0.03%)

Radiation Type	Energy (keV)	Intensity (%)	Δ(g-rad/μCi-h)
X-ray L	10	1.29 15	0.0003
X-ray Kα2	68.8950 20	0.812 22	0.0012
X-ray Kα1	70.8190 20	1.38 4	0.0021
X-ray Kβ	80.3	0.607 18	0.0010
γ 1	411.8044 11	95.505 19	0.838
γ 2	675.890 4	1.06 5	0.0153
γ 3	1087.692 4	0.23 6	0.0053

199AU B- DECAY (3.139 D 7) I(MIN) = 0.10%

Radiation Type	Energy (keV)	Intensity (%)	Δ(g-rad/ μCi-h)
Auger-L	7.6	20.9 15	0.0034
ce-L- 1	34.986 7	2.66 11	0.0020
ce-M- 1	46.263 7	0.75 4	0.0007
ce-NOP- 1	49.025 7	0.189 10	0.0002
Auger-K	53.8	0.5 4	0.0006
ce-K- 2	75.273 7	10.5 5	0.0168
ce-K- 3	125.099 7	5.5 3	0.0146
ce-I- 2	143.536 7	17.1 5	0.0522
ce-M- 2	154.813 7	4.5 4	0.0147
ce-NOP- 2	157.575 7	1.34 6	0.0045
ce-L- 3	193.362 7	1.02 4	0.0042
ce-M- 3	204.639 7	0.245 11	0.0011
β- 1 max	244.8 20		
avg	67.3 6	18.9 6	0.0271
β- 2 max	294.6 20		
avg	82.4 6	66.4 10	0.117
β- 3 max	453.0 20		
avg	132.9 6	14.7 5	0.0416
total β- avg	87.0 7	100.0 13	0.185
X-ray L	10	12.8 14	0.0027
γ 1	49.825 7	0.328 13	0.0003
X-ray Kα₂	68.8950 20	4.47 20	0.0066
X-ray Kα₁	70.8190 20	7.6 4	0.0115
X-ray Kβ	80.3	3.34 15	0.0057
γ 2	158.375 7	36.9 11	0.124
γ 3	208.201 7	8.4 3	0.0371

203HG B- DECAY (46.60 D 2) I(MIN) = 0.10%

Radiation Type	Energy (keV)	Intensity (%)	Δ(g-rad/ μCi-h)
Auger-L	7.78	8.8 7	0.0015
Auger-K	55.2	0.44 20	0.0005
ce-K- 1	193.659 5	13.37 21	0.0551
ce-L- 1	263.843 5	3.91 17	0.0220
β- 1 max	212.2 20		
avg	57.7 6	100	0.123
X-ray L	10.3	5.9 7	0.0013
X-ray Kα₂	70.8319 9	3.75 10	0.0057
X-ray Kα₁	72.8715 9	6.36 16	0.0099
X-ray Kβ	82.6	2.81 8	0.0049
γ 1	279.190 5	81.5 8	0.485

203PB EC DECAY (52.05 H 10) I(MIN) = 0.10%

Radiation Type	Energy (keV)	Intensity (%)	Δ(g-rad/ μCi-h)
Auger-L	7.78	54 4	0.0089
Auger-K	55.2	3.0 14	0.0035
ce-K- 1	193.659 5	13.14 21	0.0542
ce-K- 2	315.785 12	0.50 4	0.0034
X-ray L	10.3	36 4	0.0078
X-ray Kα₂	70.8319 9	25.2 6	0.0381

201TL EC DECAY (3.044 D 9) I(MIN) = 0.10%

Radiation Type	Energy (keV)	Intensity (%)	Δ(g-rad/ μCi-h)
ce-NOP- 1	0.78 4	38 22	0.0006
Auger-L	7.6	76 6	0.0123
ce-L- 2	15.76 3	11.4 6	0.0038
ce-L- 3	17.35 3	9.1 5	0.0033
ce-MNO- 2	27.04 3	3.63 16	0.0021
ce-MNO- 3	28.63 3	2.85 12	0.0017
ce-K- 4	52.24 4	7.5 4	0.0083
Auger-K	53.8	3.3 20	0.0038
ce-K- 5	82.78 7	0.29 4	0.0005
ce-K- 6	84.33 7	15.5 6	0.0278
ce-L- 4	120.50 4	1.27 7	0.0033
ce-MNO- 4	131.78 4	0.397 15	0.0011
ce-L- 6	152.59 7	2.62 9	0.0085
ce-MNO- 6	163.87 7	0.810 14	0.0028
X-ray L	10	47 6	0.0099
γ 2	30.60 3	0.310 13	0.0002
γ 3	32.19 3	0.285 12	0.0002
X-ray Kα2	68.8950 20	27.4 9	0.0402
X-ray Kα1	70.8190 20	46.6 14	0.0704
X-ray Kβ	80.3	20.5 7	0.0351
γ 4	135.34 4	2.65 10	0.0076
γ 5	165.88 7	0.180 20	0.0006
γ 6	167.43 7	10.00 17	0.0357

206BI EC DECAY (6.243 D 3) I(MIN) = 0.10%

Radiation Type	Energy (keV)	Intensity (%)	Δ(g-rad/ μCi-h)
Auger-L	8	71 5	0.0121
ce-I- 1	19.093 18	0.50 3	0.0002
ce-L- 2	28.249 18	0.111 13	≈0
ce-M- 1	31.103 18	0.118 7	≈0
Auger-K	56.7	3.7 15	0.0044
ce-K- 5	70.60 10	0.176 18	0.0003
ce-K- 6	96.02 3	22.2 8	0.0453
ce-I- 6	168.16 3	3.85 14	0.0138
ce-K- 12	174.71 5	1.58 6	0.0059
ce-M- 6	180.17 3	0.90 4	0.0035
ce-NOP- 6	183.13 3	0.296 11	0.0012
ce-L- 8	186.58 10	0.106 11	0.0004
ce-L- 12	246.85 5	0.272 10	0.0014
ce-K- 17	255.51 3	5.91 20	0.0322
ce-K- 21	310.00 3	1.83 6	0.0121
ce-L- 17	327.65 3	1.01 4	0.0070
ce-M- 17	339.66 3	0.234 8	0.0017
ce-L- 21	382.14 3	0.309 10	0.0025
ce-K- 28	409.06 4	1.44 5	0.0125

(CONTINUED ON NEXT PAGE)

206BI EC DECAY (CONTINUED)

Radiation Type		Energy (keV)		Intensity (%)		Δ(g-rad/ µCi-h)
ce-K- 29		428.18	4	1.94	7	0.0177
ce-K- 30		449.45	4	2.33	8	0.0223
ce-I- 28		481.20	4	0.242	8	0.0025
ce-L- 29		500.32	4	1.21	4	0.0129
ce-M- 29		512.33	4	0.314	10	0.0034
ce-NOP-29		515.29	4	0.106	4	0.0012
ce-L- 30		521.59	4	C.390	13	0.0043
ce-K- 35		532.48	5	0.299	10	0.0034
ce-K- 36		544.25	5	0.224	8	0.0026
ce-K- 43		715.10	5	0.800	24	0.0122
ce-K- 46		793.01	5	C.450	15	0.0076
ce-K- 47		807.12	5	0.318	10	0.0055
ce-K- 50		930.63	8	0.111	4	0.0022
X-ray	L	10.6		49	5	0.0112
X-ray	$K\alpha_2$	72.8042	9	32.2	8	0.0500
X-ray	$K\alpha_1$	74.9694	9	54.5	12	0.0870
X-ray	$K\beta$	84.9		24.2	6	0.0438
γ 6		184.02	3	15.8	3	0.0620
γ 10		234.26	7	0.241	12	0.0012
γ 12		262.71	5	3.02	5	0.0169
γ 15		313.67	7	0.359	10	0.0024
γ 17		343.51	3	23.4	3	0.172
γ 20		386.20	7	0.516	10	0.0042
γ 21		398.00	3	10.744	12	0.0911
γ 25		452.84	8	0.156	8	0.0015
γ 28		497.06	4	15.31	15	0.162
γ 29		516.18	4	40.8	4	0.448
γ 30		537.45	4	30.5	3	0.349
γ 33		576.36	10	0.112	10	0.0014
γ 34		581.97	8	0.485	25	0.0060
γ 35		620.48	5	5.76	6	0.0761

207BI EC DECAY (38 Y 3) I(MIN) = 0.10%

Radiation Type		Energy (keV)		Intensity (%)		Δ(g-rad/ µCi-h)
Auger-L		8		52	4	0.0088
Auger-K		56.7		2.5	11	0.0030
ce-K- 2		481.665	20	1.56	5	0.0160
ce-L- 2		553.809	20	C.430	13	0.0051
ce-K- 4		975.615	20	7.27	24	0.151
ce-L- 4		1047.759	20	1.80	6	0.0401
ce-MNO- 4		1059.769	20	0.599	9	0.0135

1 weak β's omitted (ΣIβ = 0.01%)

Radiation Type		Energy (keV)		Intensity (%)		Δ(g-rad/ µCi-h)
X-ray	L	10.6		36	4	0.0081
X-ray	$K\alpha_2$	72.8042	9	21.9	6	0.0339
X-ray	$K\alpha_1$	74.9694	9	36.9	9	0.0590
X-ray	$K\beta$	84.9		16.4	5	0.0297
γ 2		569.670	20	97.8	5	1.19
γ 3		897.70	10	0.147	10	0.0028
γ 4		1063.620	20	74.9	11	1.70
γ 5		1442.20	20	0.147	20	0.0045
γ 6		1770.23	4	6.85	20	0.258

γ 36	632.25 5	4.47 5	0.0602
γ 37	657.16 5	1.91 3	0.0267
γ 39	739.24 8	0.157 8	0.0025
γ 40	754.96 7	0.527 10	0.0085
γ 42	784.58 7	0.536 10	0.0090
γ 43	803.10 5	98.93 5	1.69
γ 45	841.28 7	0.186 9	0.0033
γ 46	881.01 5	66.2 7	1.24
γ 47	895.12 5	15.66 16	0.299
γ 50	1018.63 8	7.60 8	0.165
γ 56	1098.26 7	13.50 15	0.316
γ 57	1142.37 10	0.111 5	0.0027
γ 60	1194.69 8	0.277 15	0.0070
γ 61	1202.58 10	0.105 6	0.0027
γ 65	1332.33 10	0.282 15	0.0080
γ 67	1405.01 8	1.434 25	0.0429
γ 71	1496.18 8	0.176 10	0.0056
γ 72	1560.30 8	0.378 20	0.0126
γ 73	1565.34 8	0.304 15	0.0101
γ 75	1595.27 8	5.02 6	0.170
γ 76	1718.70 7	31.9 4	1.17
γ 77	1844.49 10	0.569 25	0.0223
γ 78	1878.65 8	2.01 4	0.0804
γ 79	1903.56 10	0.349 15	0.0142
γ 84	2599.60 20	0.130 10	0.0072

46 weak γ's omitted (ΣIγ = 1.25%)

226RA A DECAY (1600 Y 7) I(MIN) = 0.10%

Radiation Type	Energy (keV)	Intensity (%)	Δ(g-rad/ μCi-h)
α 1	4601.9 5	5.55 5	0.544
α 2	4784.50 25	94.45 5	9.63
γ 1	185.99 4	3.28 3	0.0130

222RN A DECAY (3.8235 D 3) I(MIN) = 0.10%

Radiation Type	Energy (keV)	Intensity (%)	Δ(g-rad/ μCi-h)
α 1	5489.7 3	99.920 10	11.68

2 weak α's omitted (ΣIα = 0.08%)

1 weak γ's omitted (ΣIγ = 0.08%)

218PO A DECAY (3.05 M) I(MIN) = 0.10%

Radiation Type	Energy (keV)	Intensity (%)	Δ(g-rad/ μCi-h)
α 1	6002.55 9	99.9789 20	12.78

214PB B- DECAY (26.8 M) I(MIN) = 0.10%

Radiation Type	Energy (keV)	Intensity (%)	Δ(g-rad/ μCi-h)
Auger-L	8.15	17.6 14	0.0031
ce-L- 1	36.838 14	10.1 6	0.0079
ce-M- 1	49.227 14	2.37 13	0.0025
ce-NOP- 1	52.288 14	0.80 5	0.0009
Auger-K	58.2	0.6 4	0.0008
ce-K- 2	151.455 11	4.88 20	0.0157
ce-K- 3	168.26 6	0.303 14	0.0011
ce-K- 5	204.687 11	6.9 4	0.0302
ce-L- 2	225.593 11	0.85 4	0.0041
ce-M- 2	237.982 11	0.198 8	0.0010
ce-K- 7	261.395 11	8.8 4	0.0488
ce-L- 5	278.825 11	1.24 4	0.0074
ce-M- 5	291.214 11	0.293 9	0.0018
ce-L- 7	335.533 11	1.51 6	0.0108
ce-M- 7	347.922 11	0.353 14	0.0026
ce-NOP- 7	350.983 11	0.118 5	0.0009

214BI B- DECAY (19.9 M 4) I(MIN) = 0.10%

Radiation Type	Energy (keV)	Intensity (%)	Δ(g-rad/ μCi-h)
Auger-L	8.33	0.45 4	≈0
ce-K- 30	516.207 11	0.64 3	0.0070
ce-K- 97	1322.695 4	0.353 16	0.0099
β- 1 max	541 12		
avg	162 4	0.389 19	0.0013
β- 2 max	551 12		
avg	165 4	0.181 13	0.0006
β- 3 max	575 12		
avg	173 5	0.111 8	0.0004
β- 4 max	762 12		
avg	239 5	0.10 3	0.0005
β- 5 max	764 12		
avg	240 5	0.18 3	0.0009
β- 6 max	788 12		
avg	248 5	0.97 5	0.0051
β- 7 max	822 12		
avg	261 5	2.55 11	0.0142
β- 8 max	977 12		
avg	318 5	0.520 20	0.0035
β- 9 max	1003 12		
avg	328 5	0.13 4	0.0009
β-10 max	1061 12		
avg	350 5	0.330 20	0.0025
β-11 max	1066 12		
avg	352 5	5.11 16	0.0383
β-12 max	1077 12		
avg	357 5	0.83 4	0.0063
β-13 max	1122 12		
avg	374 5	0.55 3	0.0044
β-14 max	1151 12		
avg	385 5	4.36 12	0.0358

Radiation		Energy		Intensity		
β- 1	max	185	12			
	avg	50	4	2.36	7	0.0025
β- 2	max	490	12			
	avg	145	4	0.93	9	0.0029
β- 3	max	672	12			
	avg	207	5	44.6	12	0.197
β- 4	max	729	12			
	avg	227	5	39.3	11	0.190
β- 5	max	1024	12			
	avg	337	5	12.8	22	0.0919
total β-						
	avg	227	6	100	3	0.484
X-ray	L	10.8		12.8	13	0.0029
γ 1		53.226	14	1.05	5	0.0012
X-ray	Kα2	74.8148	10	5.89	19	0.0094
X-ray	Kα1	77.1079	10	9.9	3	0.0163
X-ray	Kβ	87.3		4.42	15	0.0082
γ 2		241.981	11	6.84	18	0.0353
γ 3		258.79	6	0.511	18	0.0028
γ 4		274.9	5	0.30	4	0.0018
γ 5		295.213	11	17.9	5	0.113
γ 7		351.921	11	34.3	9	0.257
γ 8		480.42	8	0.281	15	0.0029
γ 9		487.08	8	0.396	13	0.0041
γ 10		533.69	8	0.170	10	0.0019
γ 13		580.15	4	0.341	12	0.0042
γ 14		785.910	20	1.02	4	0.0170
γ 15		839.025	15	0.550	22	0.0098

3 weak γ's omitted (ΣIγ = 0.13%)

Radiation		Energy		Intensity		
β-15	max	1181	12			
	avg	397	5	0.140	10	0.0012
β-16	max	1253	12			
	avg	425	5	2.36	8	0.0214
β-17	max	1259	12			
	avg	427	5	1.43	5	0.0130
β-18	max	1275	12			
	avg	434	5	1.15	4	0.0106
β-19	max	1380	12			
	avg	475	5	1.51	5	0.0153
β-20	max	1423	12			
	avg	492	5	7.72	20	0.0809
β-21	max	1505	12			
	avg	525	5	16.1	5	0.180
β-22	max	1527	12			
	avg	534	5	0.260	20	0.0030
β-23	max	1540	12			
	avg	539	5	16.4	5	0.188
β-24	max	1609	12			
	avg	567	5	0.92	10	0.0111
β-25	max	1727	12			
	avg	615	5	3.08	9	0.0403
β-26	max	1855	12			
	avg	668	5	1.00	5	0.0142
β-27	max	1892	12			
	avg	684	5	7.05	20	0.103
β-28	max	1995	12			
	avg	726	5	0.18	4	0.0028
β-29	max	3270	12			
	avg	1269	6	24.0	19	0.649
total β-						
	avg	681	7	100.0	21	1.45

18 weak β's omitted (ΣIβ = 0.35%)

(CONTINUED ON NEXT PAGE)

214BI B- DECAY (CONTINUED)

Radiation Type		Energy (keV)	Intensity (%)	Δ(g-rad/ µCi-h)
X-ray	L	11	0.34 4	≈0
X-ray	$K\alpha_2$	76.862 5	0.281 11	0.0005
X-ray	$K\alpha_1$	79.290 5	0.470 17	0.0008
X-ray	Kβ	89.8	0.211 8	0.0004
γ	2	273.7 4	0.162 22	0.0009
γ	11	387.0 4	0.290 15	0.0024
γ	12	389.1 4	0.375 16	0.0031
γ	15	405.74 3	0.158 9	0.0014
γ	16	454.77 12	0.273 10	0.0026
γ	17	461.80 10	0.200 18	0.0020
γ	18	469.69 12	0.124 9	0.0012
γ	22	511.00 10	0.69 3	0.0076
γ	30	609.312 10	42.6 11	0.553
γ	39	665.453 22	1.44 5	0.0203
γ	44	703.11 4	0.45 4	0.0068
γ	46	719.86 3	0.371 16	0.0057
γ	49	741.5 10	0.319 12	0.0050
γ	50	752.84 3	0.107 5	0.0017
γ	51	768.356 12	4.64 13	0.0760
γ	52	786.1 4	0.29 9	0.0049
γ	54	806.174 18	1.15 4	0.0198
γ	56	821.18 3	0.136 18	0.0024
γ	60	904.25 25	0.102 13	0.0020
γ	62	934.061 14	2.95 9	0.0587
γ	64	964.08 3	0.362 16	0.0074
γ	72	1051.96 3	0.298 19	0.0067
γ	74	1069.96 8	0.268 15	0.0061
γ	77	1120.287 12	13.9 4	0.332
γ	79	1133.66 3	0.256 15	0.0062
γ	80	1155.190 20	1.60 5	0.0393
γ	82	1207.68 3	0.439 17	0.0113
γ	85	1238.110 14	5.51 15	0.145

214PO A DECAY (164.3 US 20) I(MIN) = 0.10%

Radiation Type	Energy (keV)	Intensity (%)	Δ(g-rad/ µCi-h)
α 1	7687.09 6	99.9896 6	16.37

2 weak α's omitted (ΣIα = 0.01%)

2 weak γ's omitted (ΣIγ = 0.01%)

210PB B- DECAY (22.3 Y 2) I(MIN) = 0.10%

Radiation Type		Energy (keV)	Intensity (%)	Δ(g-rad/ µCi-h)
Auger-L		8.15	34 3	0.0058
ce-I- 1		30.116 15	57.9 21	0.0372
ce-M- 1		42.504 15	13.6 5	0.0123
ce-NOP- 1		45.565 15	4.54 17	0.0044
β- 1	max	16.5 5		
	avg	4.14 13	80 3	0.0071
β- 2	max	63.0 5		
	avg	16.13 14	20 3	0.0069
total β-				
	avg	6.54 18	100 5	0.0139
X-ray	L	10.8	24.3 25	0.0056
γ 1		46.503 15	4.05 8	0.0040

Radiation Type	Energy (keV)	Intensity (%)	Δ(g-rad/ μCi-h)
γ 86	1280.960 20	1.41 5	0.0385
γ 87	1303.76 8	0.101 9	0.0028
γ 92	1377.65 3	3.78 11	0.111
γ 93	1385.31 3	0.77 3	0.0228
γ 95	1401.50 4	1.33 5	0.0397
γ 96	1407.98 4	2.36 8	0.0706
γ 101	1509.228 17	2.04 7	0.0655
γ 102	1538.50 6	0.537 25	0.0176
γ 103	1543.32 6	0.298 15	0.0098
γ 104	1583.22 4	0.69 4	0.0234
γ 105	1594.73 8	0.290 15	0.0098
γ 106	1599.31 6	0.332 15	0.0113
γ 109	1661.28 6	1.07 4	0.0380
γ 110	1683.99 4	0.213 10	0.0076
γ 111	1729.595 17	2.68 8	0.0987
γ 112	1764.494 16	14.6 4	0.548
γ 114	1838.36 5	0.354 16	0.0138
γ 115	1847.42 3	1.93 6	0.0758
γ 116	1873.16 6	0.213 10	0.0085
γ 118	1896.3 3	0.16 3	0.0064
γ 129	2118.55 3	1.08 4	0.0486
γ 133	2204.19 4	4.59 14	0.215
γ 139	2293.36 12	0.307 12	0.0150
γ 148	2447.78 8	1.41 5	0.0737

125 weak γ's omitted ($\Sigma I\gamma$ = 3.33%)

210BI B- DECAY (5.012 D 9) I(MIN) = 0.10%

Radiation Type	Energy (keV)	Intensity (%)	Δ(g-rad/ μCi-h)
β- 1 max	1161.5 10		
avg	389.0 4	100	0.829

210PO A DECAY (138.38 D 1) I(MIN) = 0.10%

Radiation Type	Energy (keV)	Intensity (%)	Δ(g-rad/ μCi-h)
α 1	5304.51 7	100	11.30

239NP B- DECAY (2.355 D 4) I(MIN) = 0.10%

Radiation Type	Energy (keV)	Intensity (%)	Δ(g-rad/ μCi-h)
ce-M- 1	1.9171 14	6.6 6	0.0003
ce-NOP- 1	6.2914 8	2.19 19	0.0003

(CONTINUED ON NEXT PAGE)

239NP B- DECAY (CONTINUED)

Radiation Type	Energy (keV)	Intensity (%)	Δ(g-rad/ μCi-h)
Auger-L	10.3	40 7	0.0088
ce-L- 3	21.553 20	7.0 18	0.0032
ce-L- 4	26.313 20	9.5 21	0.0053
ce-L- 5	34.163 20	25 4	0.0180
ce-L- 7	38.383 5	0.33 5	0.0003
ce-M- 3	38.717 20	1.6 4	0.0013
ce-NOP- 3	43.091 20	0.43 11	0.0004
ce-M- 4	43.477 20	2.6 6	0.0024
ce-L- 8	44.78 3	6.6 17	0.0063
ce-NOP- 4	47.851 20	0.80 18	0.0008
ce-M- 5	51.327 20	6.9 10	0.0076
ce-MNO- 7	55.547 5	0.105 16	0.0001
ce-NOP- 5	55.701 20	2.6 4	0.0030
ce-K- 14	59.89 8	0.33 7	0.0004
ce-M- 8	61.95 3	1.8 5	0.0024
ce-NOP- 8	66.32 3	0.63 16	0.0009
Auger-K	76	0.9 10	0.0015
ce-L- 10	83.033 11	4.3 8	0.0076
ce-L- 11	83.37 4	0.42 8	0.0007
ce-K- 15	87.93 5	7.8 9	0.0146
ce-M- 10	100.197 10	1.14 20	0.0024
ce-MNO-11	100.54 4	0.139 25	0.0003
ce-NOP-10	104.571 10	0.39 8	0.0009
ce-K- 16	104.60 10	0.6 4	0.0013
ce-K- 17	106.37 5	20.9 20	0.0473
ce-K- 18	132.59 10	0.18 5	0.0005
ce-K- 19	151.02 9	0.116 22	0.0004
ce-K- 20	155.78 6	17.2 8	0.0571
ce-L- 15	186.653 11	1.62 18	0.0064
ce-L- 16	203.32 8	0.139 19	0.0006
ce-L- 17	205.093 11	4.4 5	0.0192
ce-MNO-17	222.257 10	1.50 9	0.0071

239NP B- DECAY (CONTINUED)

Radiation Type	Energy (keV)	Intensity (%)	Δ(g-rad/ μCi-h)
γ 22	315.88 4	1.59 11	0.0107
γ 23	334.30 5	2.03 18	0.0145

20 weak γ's omitted (ΣIγ = 0.37%)

242CM A DECAY (162.8 D 4) I(MIN) = 0.10%

Radiation Type	Energy (keV)	Intensity (%)	Δ(g-rad/ μCi-h)
Auger-L	10.3	7.6 12	0.0017
ce-L- 1	20.98 3	19.0 4	0.0085
ce-M- 1	38.15 3	5.25 15	0.0043
ce-NOP- 1	42.52 3	1.74 7	0.0016
α 1	6069.63 12	25.9 5	3.35
α 2	6112.92 8	74.1 5	9.65

6 weak α's omitted (ΣIα = 0.04%)

Radiation Type	Energy (keV)	Intensity (%)	Δ(g-rad/ μCi-h)
X-ray L	14.3	11.4 12	0.0035

9 weak γ's omitted (ΣIγ = 0.04%)

Radiation		Energy (keV)	Intensity (%)	Δ (g-rad/µCi-h)
ce-L- 20		254.50 3	3.52 15	0.0191
ce-MNO-20		271.67 3	1.13 4	0.0065
β− 1	max	211 3		
	avg	57.0 10	1.80 20	0.0022
β− 2	max	332 3		
	avg	93.0 10	33.0 20	0.0654
β− 3	max	393 3		
	avg	112.0 10	7 3	0.0167
β− 4	max	438 3		
	avg	126.0 10	53 5	0.142
β− 5	max	666 3		
	avg	201.0 10	2.0 10	0.0086
β− 6	max	715 3		
	avg	219.0 10	4.0 20	0.0187
total β−				
	avg	118.1 11	101 7	0.254

4 weak β's omitted (ΣIβ = 0.03%)

Radiation		Energy (keV)	Intensity (%)	Δ (g-rad/µCi-h)
X-ray	L	14.3	60 7	0.0183
γ 4		49.410 20	0.100 22	0.0001
γ 5		57.260 20	0.151 21	0.0002
γ 7		61.480 4	0.96 15	0.0013
X-ray	$K\alpha_2$	99.55 5	13.8 8	0.0292
X-ray	$K\alpha_1$	103.76 5	22.1 12	0.0489
γ 10		106.130 10	22.7 13	0.0513
X-ray	Kβ	117	10.4 6	0.0259
γ 14		181.71 6	0.111 15	0.0004
γ 15		209.750 10	3.24 25	0.0145
γ 16		226.42 8	0.34 5	0.0016
γ 17		228.190 10	10.7 7	0.0521
γ 18		254.41 8	0.100 18	0.0005
γ 20		277.60 3	14.1 4	0.0834
γ 21		285.41 3	0.78 8	0.0047

(CONTINUED ON NEXT COLUMN)

Appendix B

The Statistics of Radioactive Decay

B.1 The Experiments of Rutherford and Soddy

In the early experiments of Rutherford and Soddy, it was observed that, in the cases of all separated radioactive materials, the activity decreased in a geometric progression, which they expressed as

$$\frac{I_t}{I_0} = e^{-\lambda t}, \tag{B.1}$$

where I_0 was the ionization current, due to the radiations, observed at time $t = 0$, I_t was the ionization current observed at time t, and λ was a constant (Rutherford and Soddy, 1902).

On the assumption that each "ray or projected particle" would produce a "definite number of ions in its path", and that "each changing system gives rise to one ray", these authors concluded that

$$\frac{N_t}{N_0} = e^{-\lambda t}, \tag{B.2}$$

where N_t was the number of systems unchanged at time t and N_0 was the number initially present.

Treating N_t as a continuous function of t and differentiating, they deduced that "the rate of change of the system at any time", what would now be called the activity of a radionuclide at time t, "is always proportional to the amount remaining unchanged", namely

$$\frac{dN_t}{dt} = -\lambda N_t. \tag{B.3}$$

From this equation we see that λ, now known as the decay constant, is also the fractional rate (100λ is the percentage rate) at which N_t atoms at time t are decaying.

B.2 The Statistical Approach of von Schweidler

At the International Congress of Radiology held in Liège in 1905, von Schweidler (1906) pointed out that on the assumption that the probability of an atom disintegrating in time dt is λdt, then the probability of its surviving for time t would be $e^{-\lambda t}$, and its mean life would be $1/\lambda$. The proof of this result was given some years later in their book *Radioactivität*, by Stefan Meyer and E. von Schweidler (1927).

In this proof it was assumed that every atom was an entity on its own, that its decay was a statistical process governed by the laws of chance, and that its probability of decay is completely uninfluenced by the decay probability of any neighboring atoms. It was also assumed that the probability of decay of an individual atom was wholly independent of its previous history.

Meyer and von Schweidler then considered a small interval of time Δt and said that the probability of an atom decaying in this interval was proportional to the magnitude of that interval of time, and they set it equal to $\lambda \Delta t$, where λ was a constant characteristic of a given radioactive element. The probability p of an atom surviving this small interval of time is therefore

$$p = (1 - \lambda \Delta t). \tag{B.4}$$

Now consider the probability of an atom surviving the next interval of time Δt. As the atom knows nothing of its previous history, the probability of surviving the second interval is again equal to p and the probability of surviving both intervals is p^2. Considering k such intervals of time from $t = 0$ to time t such that $k\Delta t = t$, then in the limit, as k tends to become infinitely large, the probability of one atom surviving for a time t is given by

$$p = \lim_{k\to\infty} (1 - \lambda \Delta t)^k = \lim_{k\to\infty} \left(1 - \frac{\lambda t}{k}\right)^k = e^{-\lambda t}. \tag{B.5}$$

The probability of the complementary event of the atom disintegrating in the interval of time between $t = 0$ and time t is

$$w = 1 - e^{-\lambda t}. \tag{B.6}$$

If N_0 is the initial number of radioactive atoms at $t = 0$, then, because the probability, $e^{-\lambda t}$, of each atom surviving beyond time t is independent of that of all other atoms, then Eq. B.2, the empirically discovered law of Rutherford and Soddy, follows from the addition theorem for independent events. (By the addition theorem, the chance of getting either "heads" or "tails" is ½ + ½, in the toss of a coin.)

B.3 The Distribution of Observed Decays at a Given Time

B.3.1 *The Bernoulli or Binomial Distribution*

Suppose, now, that a series of experiments is conducted in which N_0 atoms of a radionuclide are initially present, and the number of decays n, in a fixed interval of time t, is determined. In actual practice such an experiment could be carried out by using a very long-lived substance such as ^{14}C, where the observation of n events, in a time t that is negligible compared with the halflife of ^{14}C, leaves N_0 essentially unchanged during a series of measurements of n.

Because radioactivity is a random, or stochastic, process, the measured value of n will vary from one measurement to the other. If, however, a sufficiently large number of measurements of n is carried out, the mean value, $\bar{n}$, will approach the product of the total number of atoms and the probability of decay in time t, namely, $N_0(1 - e^{-\lambda t})$.

But because the value of n, determined for a fixed time t, varies from one measurement to another, it is of interest to know how n varies, and in what manner it is distributed around $\bar{n}$.

Consider the probability p_n of exactly n of N_0 atoms decaying in time t. The probability of a single atom decaying in the interval $t = 0$ to time t is $1 - e^{-\lambda t}$ and the probability of a single atom surviving beyond time t is $e^{-\lambda t}$.

By way of example, consider the probability of exactly one atom decaying ($n = 1$). Suppose that a measurement starts with N_0 atoms at time $t = 0$ and is "frozen" at time t. At that point the decayed atoms are placed into a box A and undecayed atoms into a box B. Suppose, further, that the atoms are identified $1, 2, 3 \cdots N_0$.

During the time interval from zero to time t of each measurement, the probability of one given atom decaying is w, and of $(N_0 - 1)$ other atoms surviving is $(1 - w)^{N_0 - 1}$ (see Eq. B.6).

Therefore, on stopping the measurement at time t, the probability of finding one given atom in box A, and the rest in box B, is

$$p_1 = w(1 - w)^{N_0 - 1} = (1 - e^{-\lambda t})(e^{-\lambda t})^{N_0 - 1}. \quad \text{(B.7)}$$

But it is immaterial *which* of the N_0 atoms finds its way into box A, because we are interested only in the number distribution of atoms in box A and box B at time t, and not the order in which they are placed in the two boxes. In other words, the one atom in box A can be any one of $1, 2, 3 \cdots N_0$. Thus, the probability of finding *any* atom in box A is greater than that of finding a *given* atom in box A by a factor N_0, and is given by

$$p_1 = N_0 p_1 = N_0(1 - e^{-\lambda t})(e^{-\lambda t})^{N_0 - 1}. \tag{B.8}$$

Similarly, the probability of finding any two atoms in box A is

$$P_2 = \frac{N_0(N_0 - 1)}{2!}(1 - e^{-\lambda t})^2(e^{-\lambda t})^{N_0 - 2}, \tag{B.9}$$

for, again, we are only interested in the number of possible combinations of two atoms in box A (1 + 2, 1 + 3, ···· 2 + 3, 2 + 4, ···· etc.) and not in the order in which they arrive; i.e., not in the number of permutations.

Thus, in the general case, the probability of exactly n decays occurring in time t, P_n, is given by the general term of the binomial expansion, namely

$$P_n = \frac{N_0(N_0 - 1) \cdots\cdot (N_0 - n + 1)}{n!}(1 - e^{-\lambda t})^n(e^{-\lambda t})^{N_0 - n}. \tag{B.10}$$

This distribution is known as the Bernoulli or binomial distribution. It gives the distribution of the probabilities of decays, in time t, occurring for values of n equal to 1, 2, 3, ···· N_0. In general, the region of interest is in the values of n around average value of $\bar{n}$, e.g., $\bar{n} + 1$, $\bar{n} + 2$, ···· $\bar{n} - 1$, $\bar{n} - 2$, ··· . This distribution function was also derived by von Schweidler in 1903.

If the term $((1 - w) + w)^{N_0}$, where $w = 1 - e^{-\lambda t}$, is expanded by means of the binomial theorem, then

$$\begin{aligned}((1 - w) + w)^{N_0} &= (1 - w)^{N_0} + N_0(1 - w)^{N_0 - 1}w + \cdots\cdot \\ &+ \frac{N_0(N_0 - 1) \cdots\cdot (N_0 - n + 1)}{n!}(1 - w)^{N_0 - n}w^n + \cdots\cdot \\ &+ N_0(1 - w)w^{N_0 - 1} + w^{N_0}.\end{aligned} \tag{B.11}$$

The first and last terms on the right-hand side of Eq. B.11 correspond respectively to the probability of finding no atoms in box A and all N_0 atoms in box A.

The left-hand side of this equation is identically equal to unity, which simply means that the sum of all probabilities of every possible decay occurring between $n = 0$ and $n = N_0$ is equal to unity; i.e.,

$$\sum_{n=0}^{n=N_0} P_n = 1. \tag{B.12}$$

Suppose that, in a large number of experiments each carried out for time t, ν_n is the frequency of n, i.e., the number of times *exactly n* decays occur in time t. Then as we increase the number of experiments,

$\nu_n/\sum_0^{N_0} \nu_n$ approaches closer and closer to the probability P_n, so that for a large number of experiments

$$\nu_n/\sum_0^{N_0} \nu_n = P_n. \tag{B.13}$$

The mean value, or the expectation value $E(n)$ of n is given by

$$E(n) = \frac{\sum_0^{N_0} \nu_n n}{\sum_0^{N_0} \nu_n} = \sum_0^{N_0} P_n n\,. \tag{B.14}$$

Using the value for P_n given in Eq. B.10, the expectation value has been calculated (Friedlander *et al.*, 1964; Stevenson, 1966) to be

$$E(n) = wN_0 = N_0(1 - e^{-\lambda t}). \tag{B.15}$$

Consideration of the decay of, say, one microcurie of ^{14}C, for which N_0 is of the order of 10^{16} atoms, will give some appreciation for the orders of magnitude involved. The decay rate is 3.7×10^4 s^{-1}, and, for $t = 10$ min, the expectation value for the number of disintegrations, $E(n)$, will be of the order of 10^7. An observation time of $t = 10$ min is of the order of 10^{-9} of the half life (5730 y), and the probability of surviving 10 min, $1 - w = e^{-\lambda t}$, is less than unity by only 2 in 10^9.

The binomial distribution has been derived for actual decay rates, but the same considerations apply to *count* rates observed when using radiation detectors having efficiencies $\epsilon < 1$. The probability of recording a decay will be scaled down by ϵ so that the expectation value for the number of counts in time t will be $\epsilon N_0 (1 - e^{-\lambda t})$.

B.3.2 *The Poisson Distribution*

If, as was the case with ^{14}C, $N_0 \gg 1$, $n \ll N_0$, and $\lambda t \ll 1$, then the binomial or Bernoulli distribution can be greatly simplified. Equation B.10 can be written in the form

$$P_n = \frac{N_0(N_0 - 1) \cdots (N_0 - n + 1)}{n!} w^n(1 - w)^{N_0 - n}. \tag{B.16}$$

But, for $N_0 \gg 1$ and $n \ll N_0$,

$$N_0 \approx N_0 - 1 \approx N_0 - 2 \cdots$$

and

$$N_0(N_0 - 1) \cdots (N_0 - n + 1) \approx N_0^n.$$

Also, for $\lambda t \ll 1$, w is a very small quantity, and then,

$$
\begin{aligned}
(1-w)^{N_0-n} &\approx (1-w)^{N_0} \\
&= 1 - wN_0 + \frac{N_0(N_0-1)}{2!}w^2 - \cdots . \\
&\approx 1 - wN_0 + \frac{(wN_0)^2}{2!} - \cdots . \\
&\approx e^{-wN_0}.
\end{aligned}
$$

Substituting these results in equation B.16 gives

$$P_n = \frac{(wN_0)^n e^{-wN_0}}{n!}, \tag{B.17}$$

which is the Poisson distribution. This distribution also gives a mean value, or expectation value, of $E(n) = wN_0$. The Poisson distribution is not symmetrical about the mean value as it follows that P_{wN_0} is equal to P_{wN_0-1}, i.e.,

$$
\begin{aligned}
P_{wN_0} &= \frac{(wN_0)^{wN_0}\, e^{-wN_0}}{(wN_0)!} \\
&= \frac{(wN_0)^{wN_0-1}\, e^{-wN_0}}{(wN_0-1)!} \\
&= P_{wN_0-1}\, .
\end{aligned}
$$

B.3.3 *The Gaussian or Normal Distribution*

For larger values of *n*, greater than, say, 100 (Friedlander *et al.*, 1964), the Stirling formula can be used to give an approximate value of $E(n)!$, namely

$$E(n)! \approx E(n)^{E(n)} e^{-E(n)} \sqrt{2\pi E(n)}\, . \tag{B.18}$$

If we use $E(n)$ for wN_0 in Eq. B.17, then

$$P_n = \frac{[E(n)]^n e^{-E(n)}}{n!}\, . \tag{B.19}$$

Following Stevenson (1966), we can now transform the variable n to $x = [n - E(n)]$. This means that instead of taking n from 0 to N_0 we have x distributed, plus and minus, around the expected value of n, i.e., $E(n) \pm 1$, $E(n) \pm 2$, $\cdots$. Then,

$$P_x = \frac{[E(n)]^{E(n)+x}\, e^{-E(n)}}{[E(n)+x]!}$$

$$= \frac{[E(n)]^{E(n)}[E(n)]^x \, e^{-E(n)}}{[E(n)]! \, [E(n)+1] \, [E(n)+2] \cdots [E(n)+x]} \tag{B.20}$$
$$= \frac{[E(n)]^{E(n)} e^{-E(n)}}{[E(n)]!} \, \frac{[E(n)]^x}{[E(n)+1] \, [E(n)+2] \cdots [E(n)+x]},$$

whence, substituting Eq. B.18,

$$P_x = \frac{1}{\sqrt{2\pi E(n)}} \, \frac{1}{\left[1+\frac{1}{E(n)}\right]\left[1+\frac{2}{E(n)}\right]\cdots\left[1+\frac{x}{E(n)}\right]}. \tag{B.21}$$

If now we restrict consideration of the distribution to a range of values, x, around the expectation value, $E(n)$, such that $|E(n) - n| \ll E(n)$, i.e., that $|x| \ll E(n)$, then the following approximate relations hold:

$$e^{1/E(n)} \approx 1 + 1/E(n),$$

and

$$e^{2/E(n)} \approx 1 + 2/E(n), \; \cdots, \; e^{x/E(n)} \approx 1 + x/E(n). \tag{B.22}$$

Therefore,

$$[1 + 1/E(n)] \, [1 + 2/E(n)] \cdots [1 + x/E(n)]$$
$$\approx e^{(1+2+\cdots x)/E(n)} \approx e^{(x^2+x)/2E(n)}. \tag{B.23}$$

Substitution in Eq. B.21, neglecting x compared with x^2, gives

$$P_x = \frac{1}{\sqrt{2\pi \, E(n)}} \, e^{-x^2/2E(n)}, \tag{B.24}$$

which is the ordinate of the Gaussian or normal distribution at point x, and which is seen to be symmetrical about $E(n)$.

B.3.4 *Deviations from the Mean*

In determining the expectation value of a distribution $E(n)$ (designated also as $\langle n \rangle$), it is also important to be able to obtain a measure of the deviations, or the scatter, of measured values on either side of the expectation value or mean.

One indication of the scatter would be given by a plot of the distribution of $(n - \bar{n})$, but the mean value of all $(n - \bar{n})$'s, the mean deviation, of course, vanishes as, in taking the sum, values that are larger or smaller than the mean must cancel out by the very nature of the concept of the mean. A measure of the scatter would, however, be given by the mean of the absolute deviations, namely $\langle |n - \bar{n}| \rangle$.

Another measure is the mean of the squares of the deviations, $\langle (n - \bar{n})^2 \rangle$, known as the mean square deviation.

The mean square deviation is the mean value, for all values of n, of $(n - \bar{n})^2$, that is,

$$\begin{aligned} \langle (n - \bar{n})^2 \rangle &= \frac{\sum_n \nu_n (n - \bar{n})^2}{\sum_n \nu_n} \\ &= \sum_n P_n (n - \bar{n})^2 \\ &= \sum_n P_n n^2 - 2\sum_n P_n n\bar{n} + \sum_n P_n \bar{n}^2 \\ &= \langle n^2 \rangle - \langle n \rangle^2, \end{aligned} \tag{B.25}$$

or

$$\langle (n - \bar{n})^2 \rangle = E(n^2) - [E(n)]^2.$$

B.3.5 *Mean and Variance of "n" Based on a Poisson Distribution*

As an exercise, the expected value (mean) and the variance of n may be derived from Eq. B.17. The expected value $\mathrm{E}(n)$, remembering that $0! = 1$ by definition, is given by summing nP_n over all values of n, i.e.

$$\begin{aligned} \mathrm{E}(n) &= \sum_{n=0}^{N_0} n \frac{(wN_0)^n \mathrm{e}^{-wN_0}}{n!} \\ &= (wN_0) \sum_{n=1}^{N_0} \frac{(wN_0)^{n-1} \mathrm{e}^{-wN_0}}{(n-1)!} \\ &= wN_0 \;, \end{aligned} \tag{B.26}$$

since the sum term is equal to unity for large N_0.

For the variance, which is defined as

$$E(n^2) - [E(n)]^2,$$

the first term may be written as

$$\begin{aligned} E(n^2) &= E[n(n-1) + n] \\ &= \sum_{n=1}^{N_0} n(n-1) \frac{(wN_0)^n \mathrm{e}^{-wN_0}}{n!} + E(n) \\ &= (wN_0)^2 \sum_{n=2}^{N_0} \frac{(wN_0)^{n-2} \mathrm{e}^{-wN_0}}{(n-2)!} + wN_0 \end{aligned}$$

$$= (wN_0)^2 + wN_0. \qquad (B.27)$$

Hence, the variance of n, or

$$\begin{aligned} E(n^2) - [E(n)]^2 &= (wN_0)^2 + wN_0 - (wN_0)^2 \\ &= wN_0. \end{aligned} \qquad (B.28)$$

Thus, the mean and the variance of n are both equal to wN_0.

In actual measurements, the mean, $E(n)$, is estimated from k sets of data by

$$\bar{n} = \frac{1}{k} \sum_{i=1}^{k} n_i$$

and the variance, as a measure of scatter, can be estimated by the mean square deviation, or

$$\text{Var}\ (n) = \frac{1}{k-1} \sum_{i=1}^{k} (n_i - \bar{n})^2.$$

The square root of the variance, "the standard deviation of a single measurement", is the usual measure of the dispersion of the data and is in the same units as the data.

B.3.6 *Summary*

The Gaussian or normal distribution is symmetrical about $E(n)$, whereas for large n the Poisson distribution is nearly symmetrical about $E(n)$. The Bernoulli or binomial distribution is asymmetrical. As the Poisson and normal distributions are derived from the binomial by the assumption of large n, the binomial is more appropriate when considering the statistics of the measurement of activity at low or environmental levels. The derivations in this Appendix have been based on a consideration of actual numbers of disintegrations in a given time. In the counting of emitted radiations, in practice, the probability for detection is greatly reduced by the overall efficiency of the detecting system, but the statistics of counting will still conform to the distributions derived from a consideration of the decay of radioactive atoms. For a much fuller treatment of the subject matter in this Appendix, the reader should consult Evans (1955).

References

ADAMS, R. J. (1969). "Absolute standardization of ^{195}Au by β-γ coincidence counting," Int. J. Appl. Radiat. Isot. **20,** 808.

ADAMS, F. AND DAMS, R. (1970). *Applied Gamma-Ray Spectrometry,* 2nd ed. (Pergamon Press, Oxford and New York).

ADAMS, J. A. S. AND LOWDER, W. M., Eds. (1964). *The Natural Radiation Environment* (University of Chicago Press, Chicago).

ADAMS, R. J. AND BAERG, A. P. (1967). "Least squares polynomial fitting of coincidence counting data for the evaulation of absolute disintegration rates," page 123 in *Standardization of Radionuclides,* IAEA/STI/PUB/139 (International Atomic Energy Agency, Vienna).

ADAMS, R., JANSEN, C., GRAMES, G. M., AND ZIMMERMAN, C. D. (1974). "Deadtime of scintillation camera systems—definitions, measurement and applications," Med. Phys. **1,** 198.

ALLEN, R. A. (1957). "The standardization of electron capture isotopes," Int. J. Appl. Radiat. Isot. **1,** 289.

ALTSHULER, B. AND PASTERNACK, B. S. (1963). "Statistical measures of the lower limit of detection of a radioactivity counter," Health Phys. **9,** 293.

ANDERSON, E. C., LIBBY, W. F., WEINHOUSE, S., REID, A. F., KIRSHENBAUM, A. D., AND GROSSE, A. V. (1947). "Natural radiocarbon from cosmic radiation," Phys. Rev. **72,** 931.

ANDREWS, G. A., KNISELEY, R. M., AND WAGNER, H. N., Eds. (1966). *Radioactive Pharmaceuticals,* CONF-651111 (National Technical Information Service, Springfield, Virginia).

ANSI (1975). American National Standards Institute. *Calibration Techniques for the Calorimetric Assay of Plutonium-Bearing Solids Applied to Nuclear Materials Control.* Report N15.22-1975 (American National Standards Institute, New York).

ANSI (1976). American National Standards Institute. *Glossary of Terms in Nuclear Science and Technology.* Report ANSI N1.1-1976/ANS-9 (American National Standards Institute, New York).

ANSI (1977). American National Standards Institute. *Calibration and Usage of Germanium Detectors for Measurement of Gamma-Ray Emission Rates of Radionuclides.* Report N717, in preparation (American National Standards Institute, New York).

ARMANTROUT, G. A., BRADLEY, A. E., AND PHELPS, P. L. (1972). "Sensitivity problems in biological and environmental counting," IEEE Trans. Nucl. Sci. **NS-19,** 107.

ATTIX, F. H. AND RITZ, V. H. (1957). "A determination of the gamma-ray emission of radium," J. Res. Nat. Bur. Stand. **59,** 293.

AXTON, E. J. AND RYVES, T. B. (1963). "Dead-time corrections in the measurement of short-lived radionuclides," Int. J. Appl. Radiat. Isot. **14,** 159.

AYRES, R. L. (1977). Private communication. (Radioactivity Section, National Bureau of Standards, Washington).

BAERG, A. P. (1965). "Variation on the paired source method of measuring dead time," Metrologia **1,** 131.

BAERG, A. P. (1966). "Measurement of radioactive disintegration rate by the coincidence method," Metrologia **2,** 23.

BAERG, A. P. (1967). "Absolute measurement of radioactivity," Metrologia **3,** 105.

BAERG, A. P. (1971). *Statistical Analysis of Coincidence Counting Data,* National Research Council Report No. NRC 12114 (National Research Council, Ottawa).

BAERG, A. P. (1973a). "Dead time, decay and background effects in counting," Int. J. Appl. Radiat. Isot. **24,** 401.

BAERG, A. P. (1973b). "Pressurized proportional counters for coincidence measurements," Nucl. Instrum. Methods **112,** 95.

BAERG, A. P. (1973c). "The efficiency extrapolation method in coincidence counting," Nucl. Instrum. Methods **112,** 143.

BAERG, A. P. AND BOWES, G. C. (1971). "Standardization of ^{63}Ni by efficiency tracing," Int. J. Appl. Radiat. Isot. **22,** 781.

BAERG, A. P., MEGHIC, S., AND BOWES, G. C. (1964). "Extension of the efficiency tracing method for the calibration of pure β-emitters," Int. J. Appl. Radiat. Isot. **14,** 279.

BAERG, A. P., BOWES, G. C., AND ADAMS, R. J. (1967). "Calibration of radionuclides with a coincidence system using a pressurized proportional $4\pi\beta$-detector," page 91 in *Standardization of Radionuclides,* IAEA/STI/PUB/139. (International Atomic Energy Agency, Vienna).

BAERG, A. P., MUNZENMAYER, K., AND BOWES, G. C. (1976). "Live-timed anticoincidence counting with extending dead-time circuitry," Metrologia **12,** 77.

BAILLIE, L. A. (1960). "Determination of liquid scintillation counting efficiency by pulse height shift," Int. J. Appl. Radiat. Isot. **8,** 1.

BAMBYNEK, W. (1967). "Precise solid angle counting," page 373 in *Standardization of Radionuclides,* IAEA/STI/PUB/139. (International Atomic Energy Agency, Vienna).

BAMBYNEK, W. (1973). "On selected problems in the field of proportional counters," Nucl. Instrum. Methods **112,** 103.

BAMBYNEK, W., CRASEMANN, B., FINK, R. W., FREUND, H. U., MARK, H., SWIFT, C. D., PRICE, R. E., AND RAO, P. V. (1972). "X-ray fluorescence yields, Auger, and Coster-Kronig transition probabilities," Rev. Mod. Phys. **44,** 716.

BARNES, I. L., GARFINKEL, S. B., AND MANN, W. B. (1971). "Nickel-63: standardization, half-life, and neutron-capture cross-section," Int. J. Appl. Radiat. Isot. **22,** 777.

BARRALL, R. C. (1972). "Reduction of radioactive impurities in radiopharmaceuticals," J. Nucl. Med. **13,** 570.

BASSON, J. K. AND STEYN, J. (1954). "Absolute alpha standardization with liquid scintillators," Proc. Phys. Soc., London, Sect. A. **67,** 297.

BATEMAN, H. (1910). "Solution of a system of differential equations occurring in the theory of radioactive transformations," Proc. Cambridge Philos. Soc. **15,** 423.

BAY, Z., MANN, W. B., SELIGER, H. H., AND WYCKOFF, H. O. (1957). "Absolute measurement of W_{air} for sulfur-35 beta rays," Radiat. Res. **7,** 558.

BEARDEN, J. A. (1967). "X-ray wavelengths," Rev. Mod. Phys. **39,** 78.

BECKER, K. (1969). "Alpha particle registration in plastics and its applications for radon and neutron personnel dosimetry," Health Phys. **16,** 113.

BELCHER, E. H. AND VETTER, H., Eds. (1971). *Radioisotopes in Medical Diagnosis* (Butterworth and Co., London).

BELLEMARE, M., LACHANCE, Y., AND ROY, J. C. (1972). "The preparation of ^{241}Am sources by surface adsorption on different backings," Nucl. Instrum. Methods **104,** 615.

BENABIS, A. (1971). "A new gelifying agent in liquid scintillation counting," page 735 in *Organic Scintillators and Liquid Scintillation Counting,* Horrocks, D. L. and Peng, C. T., Eds. (Academic Press, New York).

BENSON, R. H. (1976). "Characteristics of solgel scintillators for liquid scintillation counting of aqueous and non-aqueous samples," Int. J. Appl. Radiat. Isot. **27,** 667.

BERLMAN, I. B. (1971). *Handbook of Fluorescent Spectra of Aromatic Molecules,* 2nd. ed. (Academic Press, New York).

BERTOLINI, G. AND COCHE, A., Eds. (1968). *Semiconductor Detectors* (North Holland Publishing Company, Amsterdam).

BILLINGHURST, M. W., GROOTHEDDE, M., AND PALSER, R. (1974). "Radiochemical purity of ^{99m}Tc-pertechnetate," J. Nucl. Med. **15,** 266.

BIPM (1964). Bureau International des Poids et Mesures. *Rapport Sur La Comparison Internationale de ^{241}Am* (Bureau International des Poids et Mesures, Sèvres).

BIPM (1973). Bureau International des Poids et Mesures. *Le Systeme International d'Unites (SI),* 2e ed. (Bureau International des Poids et Mesures, Sèvres).

BIPM (1976). Bureau International des Poids et Mesures. *The Detection and Estimation of Spurious Pulses,* Monographie BIPM-2 (Bureau International des Poids et Mesures, Sèvres).

BIRKS, J. B. (1964). *The Theory and Practice of Scintillation Counting* (Pergamon Press, Oxford).

BIRKS, J. B. (1970). "Physics of the liquid scintillation process," page 3 in *Current Status of Liquid Scintillation Counting,* Bransome, E. D., Ed. (Grune & Stratton, New York).

BIRKS, J. B. (1975). *Introduction to Liquid Scintillation Counting* (Koch-Light Laboratories, Ltd., Colnbrook, Bucks, England).

BOAG, J. W. (1966). "Ionization chambers," page 1 in *Radiation Dosimetry,* vol. 2 Attix, F. H., Roesch, W. C., and Tochilin, E., Eds. (Academic Press, New York and London).

BOLOTIN, H. H., STRAUSS, M. G., AND MCCLURE, D. A. (1970). "Simple technique for precise determinations of counting losses in nuclear pulse processing systems," Nucl. Instrum. Methods **83,** 1.

BORTNER, T. E. AND HURST, G. S. (1954). "Ionization of pure gases and mixtures of gases by 5-MeV alpha particles," Phys. Rev. **93,** 1236.

BOWEN, B. M. AND WOOD, D. E. (1972). Authors' reply, J. Nucl. Med. **13,** 570.

BOWES, G. C. AND BAERG, A. P. (1970). "Sampling and storage of radioactive solutions," Report No. NRC-11513 (National Research Council, Ottawa).

BRANSOME, E. D., Ed. (1970). *The Current Status of Liquid Scintillation Counting* (Grune & Stratton, New York).

BRAUER, F. P. AND KAYE, J. H. (1974). "Detection systems for the low level radiochemical analysis of iodine-131, iodine-129, and natural iodine in environmental samples," IEEE Trans. Nucl. Sci. **NS-21,** 496.

BRAUER, F. P., KAYE, J. H., AND CONNALLY, R. E. (1970). "X-ray and β-γ coincidence spectrometry applied to radiochemical analysis of environmental samples," page 231 in *Radionuclides in the Environment* (American Chemical Society, Washington).

BRAUER, F. P., FAGER, J. E., AND MITZLAFF, W. A. (1973). "Evaluation of a large Ge(Li) detector for low-level radionuclide analysis," IEEE Trans. Nucl. Sci. **NS-20,** 57.

BRINER, W. H. AND HARRIS, C. C. (1974). "Radionuclidic contamination of eluates from fission-product molybdenum-technetium generators," J. Nucl. Med. **15,** 466.

BRINKMAN, G. A. AND ATEN, A. H. W. (1963). "Absolute standardization with a NaI(Tl) crystal—III," Int. J. Appl. Radiat. Isot. **14,** 503.

BRINKMAN, G. A. AND ATEN, A. H. W. (1965). "Absolute standardization with a NaI(Tl) crystal—V," Int. J. Appl. Radiat. Isot. **16,** 177.

BRINKMAN, G. A., ATEN, A. H. W., AND VEENBOER, J. TH. (1963a). "Absolute standardization with a NaI(Tl) crystal—I," Int. J. Appl. Radiat. Isot. **14,** 153.

BRINKMAN, G. A., ATEN, A. H. W., AND VEENBOER, J. TH. (1963b). "Absolute standardization with a NaI(Tl) crystal—II," Int. J. Appl. Radiat. Isot. **14,** 433.

BRINKMAN, G. A., ATEN, A. H. W., AND VEENBOER, J. TH. (1965). "Absolute standardization with a NaI(Tl) crystal—IV," Int. J. Appl. Radiat. Isot. **16,** 15.

BROUNS, R. J. (1968). *Absolute Measurement of Alpha Emission and Spontaneous Fission,* Monograph NAS-NS 3112 (National Technical Information Service, Springfield, Virginia).

BROWNLEE, K. A. (1960). *Statistical Theory and Methodology in Science and Engineering* (John Wiley & Sons, New York).

BRUCER, M. (1970). "Standards and the calibration of a well counter," *Vignettes in Nuclear Medicine,* **2** [47], 1 (Mallinckrodt, Inc., St. Louis, Missouri).

BRYANT, J. (1962). "Anticoincidence counting method for standardizing radioactive materials," Int. J. Appl. Radiat. Isot. **13,** 273.

BRYANT, J. (1963). "Coincidence counting corrections for dead-time loss and accidental coincidences," Int. J. Appl. Radiat. Isot. **14,** 143.

BRYANT, J. (1967). "Advantage of anticoincidence counting for standardizing radionuclides emitting delayed gamma rays," page 129 in *Standardization of Radionuclides,* IAEA/STI/PUB/139 (International Atomic Energy Agency, Vienna).

BRYANT, J. AND MICHAELIS, M. (1952). "The measurement of micro-microcurie amounts of radon by a scintillation counting method," Radiochemical Centre Report R26 (The Radiochemical Centre, Amersham, United Kingdom).

BUDINGER, T. F. (1974). "Quantitative nuclear medicine imaging: application of computers to the gamma camera and whole-body scanner," in *Recent Advances in Nuclear Medicine,* Lawrence, J. H., Ed. (Grune & Stratton, New York).

BUNEMANN, O., CRANSHAW, T. E., AND HARVEY, J. A. (1949). "Design of grid ionisation chambers," Can. J. Res. **A27,** 191.

BUNTING, R. L. AND KRAUSHARR, J. J. (1974). "Short-lived radioactivity induced in Ge(Li) gamma-ray detectors by neutrons," Nucl. Instrum. Methods **118,** 565.

CAMERON, J. F. (1967). "Survey of systems for concentration and low background counting of tritium in water," page 543 in *Radioactive Dating and Methods of Low-Level Counting,* IAEA/STI/PUB/152 (International Atomic Energy Agency, Vienna).

CAMPION, P. J. (1959). "The standardization of radioisotopes by the beta-gamma coincidence method using high efficiency detectors," Int. J. Appl. Radiat. Isot. **4,** 232.

CAMPION, P. J. (1973). "Spurious pulses in proportional counters: a review," Nucl. Instrum. Methods **112,** 75.

CAMPION, P. J., TAYLOR, J. G. V., AND MERRITT, J. S. (1960). "The efficiency tracing technique for eliminating self-absorption errors in $4\pi\beta$-counting," Int. J. Appl. Radiat. Isot. **8,** 8.

CAMPION, P. J., DALE, J. W. G., AND WILLIAMS, A. (1964). "A study of weighing techniques used in radionuclide standardization," Nucl. Instrum. Methods **31,** 253.

CARSWELL, D. J. AND MILSTED, J. (1957). "A new method for the preparation of thin films of radioactive material," J. Nucl. Energy **4,** 51.

CAVALLO, L. M., COURSEY, B. M., GARFINKEL, S. B., HUTCHINSON, J. M. R., AND MANN, W. B. (1973). "Needs for radioactivity standards and measurements in different fields," Nucl. Instrum. Methods **112,** 5.

CEMBER, H. (1969). *Introduction to Health Physics* (Pergamon Press, Oxford and New York).

CHARALAMBUS, S. AND GOEBEL, K. (1963). "Low-level proportional counter for tritium," Nucl. Instrum. Methods **25,** 109.

CHASE, R. L. (1961). *Nuclear Pulse Spectrometry* (McGraw-Hill, New York).

CHASE, R. L. AND POULO, L. R. (1967). "High precision D.C. restorer," IEEE Trans. Nucl. Sci. **NS-14,** 83.

CHERRY, R. D. (1963). "The determination of thorium and uranium in geological samples by an alpha-counting technique," Geochim. Cosmochim. Acta **27,** 183.

CHEW, W. M., MCGEORGE, J. C., AND FINK, R. W. (1973). "A multiwire proportional counter technique for standardization of low energy photon sources," Nucl. Instrum. Methods **106,** 499.

CHEW, W. M., XENOULIS, A. C., FINK, R. W., AND PINAJIAN, J. J. (1974a). "Nuclear matrix elements from L/K electron capture ratios in ^{59}Ni decay," Nucl. Phys. **A218,** 372.

CHEW, W. M., XENOULIS, A. C., FINK, R. W., SCHIMA, F. J., AND MANN, W. B. (1974b). "The L/K electron capture ratio in first-forbidden ^{81g}Kr decay," Nucl. Phys. **A229,** 79.

CHYLINSKI, A., PASIEWICZ, K., RADOSZEWSKI, T., WOLSKI, D., AND WOJTOWICZ, S. (1972). "$4\pi\beta$-γ anticoincidence method and apparatus for radioactive concentration measurements of radioactive solutions," Nucl. Instrum. Methods **98,** 109.

CLARKE, R. K. (1950). "Absolute determination of the emission rate of beta rays," Rev. Sci. Instrum. **21,** 753.

CLINE, J. E. (1971). "Gamma rays emitted by the fissionable nuclides and associated isotopes," Idaho Nuclear Corporation Report IN-1448 (U.S. Atomic Energy Commission, Idaho Falls, Idaho).

CLINE, J. E., PUTNAM, M. H., AND HELMER, R. G. (1973). "Gauss VI, a computer program for automatic batch analysis of gamma-ray spectra from Ge(Li) spectrometers," AEC Report No. ANCR-1113 (National Technical Information Service, Springfield, Virginia).

CLULEY, H. J. (1962). "Suspension scintillation counting of carbon-14 barium carbonate," Analyst **87,** 170.

COCHRAN, J. A., SMITH, D. G., MAGNO, P. G., AND SHLEIEN, B. C. (1970). *An Investigation of Airborne Radioactive Effluent from an Operating Nuclear Fuel Reprocessing Plant* BRH/NERHL 70-3 (National Technical Information Service, Springfield, Virginia).

COHEN, E. J. (1974). "Live time and pile-up corrections for multichannel analyzer spectra," Nucl. Instrum. Methods **121,** 25.

COHN, C. E. (1966). "The effect of deadtime on counting errors," Nucl. Instrum. Methods **41,** 338.

COLEY, R. F. AND FRIGERIO, N. A. (1975). "Self absorption and geometric correction factors for reactor off-gas samples relative to NBS standards," page 666 in *Proceedings of the Noble Gases Symposium,* ERDA CONF-730915, Stanley, R. E. and Moghissi, A. A., Eds. (U.S. Energy Research and Development Administration, Washington).

COLLÉ, R. (1976). "AIF-NBS radioactivity measurements assurance program for the radiopharmaceutical industry," page 71 in *Proceedings of the NBS 75th Anniversary Symposium,* Nat. Bur. Stand. Spec. Publ. 456 (National Bureau of Standards, Washington).

COLLINSON, A. J. L., DEMETSOPOULLOS, I. C., DENNIS, J. A., AND ZARZYCKI, J. M. (1960). "Self-quenching Geiger counters containing mixtures of permanent gases," Nature **185,** 369.

CONDE, C. A. N., SANTOS, M. C. M., FERREIRA, M. F. A., AND SOUSA, C. A.

(1975). "An argon gas scintillation counter with uniform electric field," IEEE Trans. Nucl. Sci. **NS-22,** 104.

COOPER, J. A. (1970). "Factors determining the ultimate detection sensitivity of Ge(Li) gamma-ray spectrometers," Nucl. Instrum. Methods **82,** 273.

COOPER, J. A. AND PERKINS, R. W. (1971). *A Versatile Ge(Li)-NaI(Tl) Coincidence-Anti-Coincidence Gamma-Ray Spectrometer for Environmental and Biological Problems,* BNWL-SA-3972 (Battelle Pacific Northwest Laboratories, Richland, Washington).

COOPER, J. A., HOLLANDER, J. M., AND RASMUSSEN, J. O. (1965). "Effect of the chemical state on the lifetime of the 24-second isomer of Nb^{90*}," Phys. Rev. Letters **15,** 680.

CORSON, D. R. AND WILSON, R. R. (1948). "Particle and quantum counters," Rev. Sci. Instrum. **19,** 207.

COURSEY, B. M. (1976). "Use of NBS mixed-radionuclide gamma-ray standards for calibration of Ge(Li) detectors used in the assay of environmental radioactivity," page 173 in *Proceedings of a Symposium on Measurements for the Safe Use of Radiation,* NBS Spec. Publ. 456, Fivozinsky, S. P., Ed. (National Bureau of Standards, Washington).

COURSEY, B. M., NOYCE, J. R., AND HUTCHINSON, J. M. R. (1975). *Interlaboratory Intercomparisons of Radioactivity Measurements Using National Bureau of Standards Mixed Radionuclide Test Solutions,* Nat. Bur. Stand. Tech. Note 875 (National Bureau of Standards, Washington).

COX, D. R. AND ISHAM, V. (1977). "A bivariate point process connected with electronic counters," Proc. Roy. Soc. **A356,** 149.

CRADDUCK, T. D. (1973). "Fundamentals of scintillation counting," Semin. Nucl. Med. **3,** 205.

CRAMER, H. (1946). *Mathematical Methods of Statistics* (Princeton University Press, Princeton, New Jersey).

CRASEMAN, B. (1973). "Some aspects of atomic effects in nuclear transformations," Nucl. Instrum. Methods **112,** 33.

CRAWFORD, J. A. (1949). "Theoretical calculations concerning back-scattering of alpha particles," page 1307 in *The Transuranium Elements,* Pt. II, Seaborg, G. T., Katz, J. J., and Manning, W. M., Eds. (McGraw-Hill, New York).

CUDDIHY, R. G., MCCLELLAN, R. O., CHAPMAN, L. D., WAYLAND, J. R., AND DUGON, V. L. (1976). "Simulation modelling of environmental transport and health consequences of radioactive effluents from nuclear power stations," page 657 in *Transuranium Nuclides in the Environment,* IAEA/STI/PUB/410 (International Atomic Energy Agency, Vienna).

CULHANE, J. L. AND FABIAN, A. C. (1972). "Circuits for pulse rise time discrimination in proportional counters," IEEE Trans. Nucl. Sci. **NS-19,** 569.

CURRAN, S. C. (1950). "Counting and spectrometry with proportional tubes," Atomics **1,** 221.

CURRAN, S. C. AND WILSON, H. W. (1965). "Proportional counters and pulse ion chambers," page 303 in *Alpha-, Beta-, and Gamma-Ray Spectroscopy, Vol. I.,* Siegbahn, K., Ed. (North-Holland Publishing Co., Amsterdam).

CURRIE, L. A. (1968). "Limits for qualitative detection and quantitative determination. Applications to radiochemistry," Anal. Chem. **40,** 586.

CURRIE, L. A. (1972). "The measurement of environmental levels of rare gas nuclides and the treatment of very low-level counting data," IEEE Trans. Nucl. Sci. **NS-19,** 119.

CURRIE, L. A. AND LINDSTROM, R. M. (1975). "The NBS measurement system for natural argon-37," page 40 in *Proceedings of the Noble Gases Symposium,* ERDA CONF-730915, Stanley, R. E. and Moghissi, A. A., Eds. (U. S. Energy Research and Development Administration, Washington).

CURTIS, M. L. (1972). "Detection and measurement of tritium by bremsstrahlung counting," Int. J. Appl. Radiat. Isot. **23,** 17.

CURTISS, L. F. AND DAVIS, F. J. (1943). "A counting method for the determination of small amounts of radium and of radon," J. Res. Nat. Bur. Stand. **31,** 181.

DALE, J. W. G. (1961). "A beta-gamma ionization chamber for substandards of radioactivity—II," Int. J. Appl. Radiat. Isot. **10,** 72.

DALE, J. W. G., PERRY, W. E., AND PULFER, R. F. (1961). "A beta-gamma ionization chamber for substandards of radioactivity—I," Int. J. Appl. Radiat. Isot. **10,** 65.

DAMON, P. E. AND WINTERS, P. N. (1954). "Resolution losses in counters and trigger circuits," Nucleonics **12,** No. 12, 36.

DARRALL, K. G., RICHARDSON, P. J., AND TYLER, J. F. C. (1973). "An emanation method for determining radium using liquid scintillation counting," Analyst **98,** 610.

DAVIS, D. M. AND GUPTON, E. D. (1949). *Health Physics Instrument Manual,* ORNL-332, (TID-4500) (Oak Ridge National Laboratory, Oak Ridge, Tennessee).

DEBERTIN, K., SCHOTZIG, U., WALZ, K. F., AND WEISS, H. M. (1976). "Efficiency calibration of semiconductor spectrometers—techniques and accuracies," page 59 in *Proceedings of ERDA Symposium on X- and Gamma-Ray Sources and Applications,* Report CONF-760539 (National Technical Information Service, Springfield, Virginia).

DE BRUIN, M., THEN, S. S., BODE, P., AND KORTHOVEN, P. J. M. (1974). "A simple deadtime stabilizer for gamma-ray spectrometers," Nucl. Instrum. Methods **121,** 611.

DERUYTTER, A. J. (1962). "Evaluation of the absolute activity of alpha emitters and of the number of nuclei in thin alpha active layers," Nucl. Instrum. Methods **15,** 164.

DEVOE, J. R. (1961). *Radioactive Contamination of Materials Used in Scientific Research,* NAS-NRC Publication 895 (National Academy of Sciences, Washington).

DIAMOND, H., FIELDS, P. R., STEVENS, C. S., STUDIER, M. H., FIELD, S. M., INGRAHAM, M. G., HESS, D. C., PYLE, G. L., MECH, J. F., AND MANNING, W. M. (1960). "Heavy isotope abundances in Mike thermonuclear device," Phys. Rev. **119,** 2000.

DIETHORN, W. S. (1974). "Counter deadtime: a question of poor textbook advice," Int. J. Appl. Radiat. Isot. **25,** 55.

DIETZ, L. A. AND PACHUCKI, C. F. (1973). "^{137}Cs and ^{134}Cs half-lives determined by mass spectrometry," J. Inorg. Nucl. Chem. **35,** 1769.

DITMARS, D. A. (1976). "Measurement of the average total decay power of two plutonium heat sources in a Bunsen ice calorimeter," Int. J. Appl. Radiat. Isot. **27,** 469.

DIXON, W. J. AND MASSEY, F. J. (1957). *Introduction to Statistical Analysis* (McGraw-Hill, New York).

DOBRILOVIČ, L. AND SIMOVIČ, M. (1973). "Preparation of mono-molecular radioactive sources," Nucl. Instrum. Methods **112,** 359.

DONN, J. J. AND WOLKE, R. L. (1977). "The statistical interpretation of counting data from measurements of low-level radioactivity," Health Phys. **32,** 1.

DRAPER, N. H. AND SMITH, H. (1966). *Applied Regression Analysis* (John Wiley & Sons, New York).

DREVER, R. W. P., MOLJK, A., AND CURRAN, S. C. (1957). "A proportional counter system with small wall effect," Nucl. Instrum. Methods **1,** 41.

DUNWORTH, J. V. (1940). "The application of the method of coincidence counting to experiments in nuclear physics," Rev. Sci. Instrum. **11,** 167.

DUVAL, C. (1963). *Inorganic Thermogravimetric Analysis* (Elsevier, Amsterdam).

EASTHAM, J. F., WESTBROOK, H. L., AND GONZALES, D. (1962). "Liquid scintillation detection of tritium and other radioisotopes in insoluble or quenching organic samples," page 203 in *Tritium and the Physical and Biological Sciences,* vol. 1, IAEA/STI/PUB/39 (International Atomic Energy Agency, Vienna).

EASTWOOD, T. A., BROWN, F., AND CROCKER, I. H. (1964). "A krypton-81 half-life determination using a mass separator," Nucl. Phys. **58,** 328.

EDWARDS, R. R. (1962). "Iodine-129: its occurrence in nature and its utility as a tracer," Science **137,** 851.

EICHELBERGER, J. F., JORDAN, K. C., ORR, S. R., AND PARKS, J. R. (1954). "Calorimetric determination of the half-life of polonium-210," Phys. Rev. **96,** 719.

EISENHART, C. (1963). "Realistic evaluation of the precision and accuracy of instrument calibration systems," J. Res. Nat. Bur. Stand. **67C,** 161. (Reprinted in Ku, 1969.)

EISENHART, C. (1968). "Expression of the uncertainties of final results," Science **160,** 1201. (Reprinted in Ku, 1969.)

ELDRIDGE, J. S. AND CROWTHER, P. (1964). "Absolute determination of ^{125}I in clinical applications," Nucleonics, **22** No. 6, 56.

ELDRIDGE, J. S., O'KELLEY, G. D., NORTHCUTT, K. J., AND SCHONFELD, E. (1973). "Nondestructive determination of radionuclides in lunar samples using a large low-background gamma-ray spectrometer and a novel application of least squares fitting," Nucl. Instrum. Methods **112,** 319.

EMANUEL, C. F. (1973). "Delivery precision of micropipets," Anal. Chem. **45,** 1568.

ENGELKE, C. E. (1960). "High pressure gas scintillators," IRE Trans. Nucl. Sci. **NS-7,** 32.

ENGELKEMEIR, A. G. AND LIBBY, W. F. (1950). "End and wall corrections for absolute beta-counting in gas counters," Rev. Sci. Instrum. **21,** 550.

EPA (1973). Environmental Protection Agency. "Tritium surveillance system, April-June 1973," Radiat. Data Rep. **14,** 601 (Environmental Protection Agency, Washington).

EVANS, R. D. (1935). "Apparatus for the determination of minute quantities of radium, radon and thoron in solids, liquids and gases," Rev. Sci. Instrum. **6,** 99.

EVANS, R. D. (1937). "Radium poisoning: II. The quantitative determination of the radium content and radium elimination rate of living persons," Am. J. Roentgenol. Radium Ther. **37,** 368.

EVANS, R. D. (1955). *The Atomic Nucleus* (McGraw-Hill, New York).

EVANS, R. D. AND EVANS, R. O. (1948). "Studies of self-absorption in gamma-ray sources," Rev. Mod. Phys. **20,** 305.

FAIRSTEIN, E. AND HAHN, J. (1965). "Nuclear pulse amplifiers—fundamentals and design practice," Nucleonics **23** No. 7, 56; **23** No. 9, 81; **23** No. 11, 50; **24** No. 1, 54; **24** No. 3, 68.

FANO, U. (1947). "Ionization yield of radiations, II. Fluctuations of the number of ions," Phys. Rev. **72,** 26.

FEINGOLD, Q. M. AND FRANKEL, S. (1955). "Geometrical corrections in angular correlation instruments," Phys. Rev. **97,** 1025.

FELLER, W. (1948). "On probability problems in the theory of counters," page 105 in *Studies and Essays,* presented to R. Courant on his 60th Birthday, Jan. 8, 1948, Friedrichs, K. O., Neugebaur, O. E., and Stoker, J. J., Eds. (Interscience Publishers, Inc., New York).

FELLER, W. (1968). *An Introduction to Probability Theory and Its Applications,* Vol. I, 3rd ed. (John Wiley & Sons, New York).

FICKEN, V., ANDERSON, D., COX, P., AND COOTS, H. (1973). "Radionuclide contamination of fission generator eluate," J. Nucl. Med. Technol. **1,** 11.

FIELDS, P. F., DIAMOND, H., METTA, D. N., ROKOP, D. J., AND STEVENS, C. M. (1972). "^{237}Np, ^{236}U, and other actinides on the moon," Geochim. Cosmochim. Acta, Suppl. 3, **2,** 1637.

FINK, R. W. (1973). "Radioactive x-ray standards," Nucl. Instrum. Methods **112,** 243.

FLEISCHER, R. L., ALTER, H. W., FURMAN, S. C., PRICE, P. B., AND WALKER, R. M. (1972). "Particle track etching," Science **178,** 255.

FLYNN, K. F., GLENDENIN, L. E., AND PRODI, V. (1971). "Absolute counting of low energy beta emitters using liquid scintillation counting techniques," page 687 in *Organic Scintillators and Liquid Scintillation Counting,* Horrocks, D. L. and Peng, C. T., Eds. (Academic Press, New York).

FOGLIO PARA, A. AND BETTONI, M. MANDELLI (1970). "Statistical determination of half lives," Nucl. Instrum. Methods **79,** 125.

FOLSOM, T. R., YOUNG, D. R., JOHNSON, J. N., AND PILLAI, K. C. (1963). "Manganese-54 and zinc-65 in coastal organisms of California," Nature **200,** 327.

FOLSOM, T. R., PILLAI, K. C., AND FINNIN, L. E. (1965). "Assay of small traces of caesium-137 by internal conversion electrons," page 699 in *Radioisotopic*

Sample Measurement Techniques in Medicine and Biology. IAEA/STI/PUB/106 (International Atomic Energy Agency, Vienna).

FOX, B. W. (1968). "The application of Triton X-100 colloid scintillation counting in biochemistry," Int. J. Appl. Radiat. Isot. **19,** 717.

FRANCOIS, B. (1973). "Detection of hard beta-emitting radionuclides in aqueous solution using Čerenkov radiation," Int. J. Nucl. Med. Biol. **1,** 1.

FREEDMAN, G. S., KINSELLA, T., AND DWYER, A. (1972). "A correction method for high-count-rate quantitative radionuclide angiography," Radiology **104,** 713.

FRIEDLANDER, E. M. (1964). "Correlation counting of coincidence events," Nucl. Instrum. Methods **31,** 293.

FRIEDLANDER, G., KENNEDY, J. W., AND MILLER, J. M. (1964). *Nuclear and Radiochemistry,* 2nd ed. (John Wiley & Sons, New York).

FRIEDMAN, H. (1949). "Geiger counter tubes," Proceedings IRE **37,** 791.

GANDY, A. (1961). "Mesure absolue de l'activité des radionuclides par la méthode des coïncidence béta-gamma à l'aide de détecteurs de grand efficacité. Étude des coincïdences instrumentales," Int. J. Appl. Radiat. Isot. **11,** 75.

GANDY, A. (1962). "Mesure absolue de l'activité des radionuclides par la méthode des coïncidences béta-gamma à l'aide de detécteurs de grande efficacité: Corrections de temps morts," Int. J. Appl. Radiat. Isot. **13,** 501.

GARDNER, D. G. AND MEINKE, W. W. (1958). "β-ray spectroscopy using a hollow plastic scintillator," Int. J. Appl. Radiat. Isot. **3,** 232.

GARFINKEL, S. B. (1959). "Semiautomatic Townsend balance system," Rev. Sci. Instrum. **30,** 439.

GARFINKEL, S. B. AND HUTCHINSON, J. M. R. (1962). "Determination of source self-absorption in the standardization of electron-capturing radionuclides," Int. J. Appl. Radiat. Isot. **13,** 629.

GARFINKEL, S. B. AND HUTCHINSON, J. M. R. (1966). "The standardization of Cobalt-57," Int. J. Appl. Radiat. Isot. **17,** 587.

GARFINKEL, S. B. AND HUTCHINSON, J. M. R. (1970). "Recent activities of the NBS Radioactivity Section, with special reference to the standardization of thallium-208 in Radioactivity Calibration Standards," Nat. Bur. Stand. Spec. Publ. 331, Mann, W. B. and Garfinkel, S. B., Eds. (U.S. Government Printing Office, Washington).

GARFINKEL, S. B. AND HINE, G. J. (1973). *Dose Calibrator Pilot Study,* Nat. Bur. Stand. Tech. Note 791 (U.S. Government Printing Office, Washington).

GARFINKEL, S. B., MANN, W. B., MEDLOCK, R. W., AND YURA, O. (1965). "The calibration of the National Bureau of Standards tritiated-toluene standard of radioactivity," Int. J. Appl. Radiat. Isot. **16,** 27.

GARFINKEL, S. B., MANN, W. B., AND PARARAS, J. L. (1973a). "The National Bureau of Standards $4\pi\beta$-γ coincidence-counting and γ-ray comparator automatic sample changers," Nucl. Instrum. Methods **112,** 213.

GARFINKEL, S. B., MANN, W. B., SCHIMA, F. J., AND UNTERWEGER, M. P. (1973b). "Present status in the field of internal gas counting," Nucl. Instrum. Methods **112,** 59.

GEIGER, E. L., TSCHAECHA, A. N., AND WHITTAKER, E. L. (1961). "Simplified autoradiography technique for α-emitters," Health Phys. **11,** 1092.

GEIGER, H. AND WERNER, A. (1924). "Die Zahl der von Radium ausgesandten α-Teilchen," Z. Phys. **21,** 187.

GENERAL ELECTRIC COMPANY (1962). *Tables of the Individual and Cumulative Terms of Poisson Distribution.* (D. Van Nostrand Co., Inc., Princeton, New Jersey).

GENNA, S., WEBSTER, E. W., BRANTLEY, J. C., BROWNELL, G. L., CARDARELLI, J. A., CASTRONOVO, F. P., CHANDLER, H. L., AND MASSE, F. X. (1972). "A nuclear medicine quality control program," J. Nucl. Med. **13,** 285.

GERRARD, M. (1971). "Survey of the use of radionuclides in medicine—Phase II," Isot. Radiat. Technol. **8,** 236.

GIBSON, J. A. B. AND MARSHALL, M. (1972). "The counting efficiency for ^{55}Fe and other E.C. nuclides in liquid scintillator solutions," Int. J. Appl. Radiat. Isot. **23,** 321.

GIBSON, W. M., MILLER, G. L., AND DONOVAN, P. F. (1965). "Semiconductor particle spectrometers," page 345 in *Alpha-, Beta-, and Gamma-Ray Spectroscopy,* Vol. 1, Siegbahn, K., Ed. (North Holland Publishing Company, Amsterdam).

GOLD, R., ARMANI, R. J., AND ROBERTS, J. H. (1968). "Absolute fission rate measurements with solid-state track recorders," Nucl. Sci. Eng. **34,** 13.

GOLD, S., BARKHAU, H. W., SHLEIEN, B., AND KAHN, B. (1964). "Measurement of naturally occurring radionuclides in air," page 369 in *The Natural Radiation Environment,* Adams, J. A. S. and Lowder, W. M., Eds. (University of Chicago Press, Chicago).

GOLDANSKY, V. I. AND PODGORETSKY, M. I. (1955). "Possible use of a correlation function for the study of nuclear disintegration." Dok. Akad. Nauk. SSSR **100,** 237.

GOODIER, I. W., HUGHES, F. H., AND WILLIAMS, A. (1965). "Precise measurement of the activity of low-energy beta-emitters with calibrated ionization chambers," page 613 in *Radioisotope Sample Measurement Techniques in Medicine and Biology* IAEA/STI/PUB/106 (International Atomic Energy Agency, Vienna).

GOULDING, F. S. (1966). "Semiconductor detectors for nuclear spectrometry," Nucl. Instrum. Methods **43,** 1.

GRAY, L. H. (1949). "The experimental determination by ionization methods of the rate of emission of beta- and gamma-ray energy by radioactive substances," Br. J. Radiol. **22,** 677.

GREENE, R. E., PRESSLY, R. S., AND CASE, F. N. (1972). *A Review of Alpha Source Preparation Methods and Applications,* ORNL-4819 (Oak Ridge National Laboratory, Oak Ridge, Tennessee).

GRODSTEIN, G. W. (1957). *X-Ray Attenuation Coefficients from 10 keV to 100 MeV,* Nat. Bur. Stand. Circ. 583 (U.S. Government Printing Office, Washington).

GROSS, W. AND FAILLA, G. (1950). "A new method of radioactivity measurement," Phys. Rev. **79,** 209.

GROSS, W., WINGATE, C., AND FAILLA, G. (1957a). "The average energy lost by S^{35} beta rays per ion pair produced in air," Radiat. Res. **7,** 570.

GROSS, W., WINGATE, C., AND FAILLA, G. (1957b). "Determination of disintegration rate for gamma-emitting isotopes," Radiology **69,** 699.

GRUMMITT, W. E., BROWN, R. M., CRUIKSHANK, A. J., AND FOWLER, I. L. (1956). "Recent developments in low-background Geiger-Müller counters," Can. J. Chem. **34,** 206.

GRUN, A. E. AND SCHOPPER, E. (1951). "Über die Fluoreszenz von Gasen bei Anregung dürch α-Teilchen," Z. Naturf. **6a,** 698.

GUNN, S. R. (1964). "Radiometric calorimetry: a review," Nucl. Instrum. Methods **29,** 1.

GUNN, S. R. (1970). "Radiometric calorimetry: a review (1970 supplement)," Nucl. Instrum. Methods **85,** 285.

GUNNINK, R. AND NIDAY, J. B. (1976). "A working model for Ge(Li) detector counting efficiencies," page 55 in *Proceedings of ERDA Symposium on X- and Gamma-Ray Sources and Applications,* May 19–21, 1976, Ann Arbor, Michigan, CONF-760539 (National Technical Information Service, Springfield, Virginia).

GUNNINK, R., COLBY, L. J., JR., AND COBBLE, J. W. (1959). "Absolute beta standardization using 4π beta-gamma coincidence techniques," Anal. Chem. **31,** 796.

HAIGHT, F. A. (1967). *Handbook of the Poisson Distribution.* (John Wiley & Sons, New York).

HALLDEN, N. A. AND HARLEY, J. H. (1960). "An improved alpha-counting technique," Anal. Chem. **32,** 1861.

HALLIDAY, D. (1955). *Introductory Nuclear Physics,* 2d ed. (John Wiley & Sons, New York).

HANDLER, L. H. AND ROMBERGER, J. A. (1973). "The use of a solubilizer-scintillator mixture for counting aqueous systems containing high- and low-energy radionuclides," Int. J. Appl. Radiat. Isot. **24,** 129.

HANSEN, J. S. MCGEORGE, J. C., AND FINK, R. W. (1973). "Efficiency calibration of semiconductor detectors in the X-ray region," Nucl. Instrum. Methods **112,** 239.

HANSEN, W. L. AND HALLER, E. E. (1974). "A view of the present status and future prospects of high purity germanium," IEEE Trans. Nucl. Sci. **NS-21,** 251.

HARDING, J. E., SCHWEBEL, A., AND STOCKMANN, L. L. (1958). "Report on the determination of small amounts of radium in solution," J. Assoc. Off. Agri. Chem. **41,** 311.

HARDY, E. P., Ed. (1973). *Fallout Program Quarterly Summary Report,* USAEC Report HASL-274 (National Technical Information Service, Springfield, Virginia).

HARE, D. L., HENDEE, W. R., WHITNEY, W. P., AND CHANCY, E. L. (1974). "Accuracy of well ionization chamber isotope calibrators," J. Nucl. Med. **15,** 1138.

HARLEY, J. H., Ed. (1972) *HASL Procedures Manual,* USAEC Report HASL-300 (National Technical Information Service, Springfield, Virginia).

HARLEY, J. H., HALLDEN, N. A., AND FISENNE, I. M. (1962). "Beta scintillation counting with thin plastic phosphors," Nucleonics **20** No. 1, 59.

HARMS, J. (1967). "Automatic dead-time correction for multichannel pulse-height analyzers at variable counting rates," Nucl. Instrum. Methods **53,** 192.

HARPER, P. V., SIEMENS, W. D., LATHROP, K. A., AND ENDLICH, H. (1963). "Production and use of Iodine-125," J. Nucl. Med. **4,** 277.

HASL (1969). "Mechanical and electrical drawings for low level beta scintillation counters," HASL Technical Memorandum 62-7 (Internal Publication, ERDA Health and Safety Laboratory, New York).

HAUSER, W. (1974). "The 1972 nuclear medicine survey—radionuclide identification and activity measurement," Am. J. Clin. Pathol. **61,** 943.

HAUSER, W. AND CAVALLO, L. M. (1975). "Measurement and quality assurance of the amount of administered tracer," page 154 in *Quality Control in Nuclear Medicine* (C. V. Mosby Company, St. Louis).

HAWKINGS, R. C., MERRITT, W. F., AND CRAVEN, J. H. (1952). "The maintenance of radioactive standards with the 4 π proportional counter," page 35 in *Developments and Techniques in the Maintenance of Standards,* National Physical Laboratory, May 1951. (H.M. Stationery Office, London).

HAYWARD, R. W. (1961). "On the determination of disintegration rates by the coincidence method using high efficiency detectors," Int. J. Appl. Radiat. Isot. **12,** 148.

HAYWARD, R. W., HOPPES, D. D., AND MANN, W. B. (1955). "Branching ratio in the decay of polonium-210," J. Res. Nat. Bur. Stand. **54,** 47.

HEATH, R. L. (1957). *Scintillation Spectrometry Gamma-Ray Spectrum Catalogue,* page 13, USAEC Report, IDO-16408 (U.S. Atomic Energy Commission, Idaho Falls, Idaho).

HEATH, R. L. (1964). *Scintillation Spectrometry, Gamma-Ray Spectrum Catalogue,* 2nd. ed., USAEC Report IDO-16880-1 (U.S. Atomic Energy Commission, Idaho Falls, Idaho).

HEATH, R. L. (1966). "Computer techniques for the analysis of gamma-ray spectra obtained with NaI and lithium-ion drifted germanium detectors," Nucl. Instrum. Methods **43,** 209.

HEATH, R. L. (1969). "Gamma-ray spectrometry and automated data systems for activation analysis," page 959 in *Modern Trends in Activation Analysis,* Vol. II Nat. Bur. Stand. Spec. Publ. 312, DeVoe, J. R., Ed. (U.S. Government Printing Office, Washington).

HEATH, R. L. (1974). *Gamma-Ray Spectrum Catalogue, Ge(Li) and Si(Li) Spectrometry,* Vol. 2, Aerojet Nuclear Co. Report ANCR-1000-2 (U.S. Atomic Energy Commission, Idaho Falls, Idaho).

HEATH, R. L., HELMER, R. G., SCHMITTROTH, L. A., AND CAZIER, G. A. (1967). "A method for generating single gamma-ray shapes for the analysis of spectra," Nucl. Instrum. Methods **47,** 281.

HEIMBUCH, A. H. (1961). "Scintillating ion exchange resins," Trans. Am. Nucl. Soc. **4,** 251.

HELF, S., WHITE, C. G., AND SHELLEY, R. N. (1960). "Radioassay of finely divided solids by suspension in a gel scintillator," Anal. Chem. **32,** 238.

HELMER, R. G. AND PUTNAM, M. H. (1972). "Gauss V, a computer program for the analysis of gamma-ray spectra from Ge(Li) spectrometers," USAEC

Report No. ANCR-1043 (National Technical Information Service, Springfield, Virginia).

HELMER, R. G., HEATH, R. L., PUTNAM, M., AND GIPSON, D. H. (1967a). "Photopeak analysis program for photon energy and intensity determinations (Ge(Li) and NaI(Tl) spectrometers)," Nucl. Instrum. Methods **57,** 46.

HELMER, R. G., HEATH, R. L., SCHMITTROTH, L. A., JAYNE, G. A., AND WAGNER, L. M. (1967b). "Analysis of gamma-ray spectra from NaI(Tl) and Ge(Li) spectrometers. Computer programs," Nucl. Instrum. Methods **47,** 305.

HELMER, R. G., CLINE, J. E., AND GREENWOOD, R. C. (1975). "Gamma-ray energy and intensity measurements with Ge(Li) spectrometers," page 775 in *The Electromagnetic Interaction in Nuclear Spectroscopy,* Hamilton, W. D., Ed. (North Holland and American Elsevier Publishing Companies, Amsterdam and New York).

HENDEE, W. R. (1973). *Radioactive Isotopes in Biological Research* (John Wiley & Sons, New York).

HENDRICKS, R. W. (1972). "Gas amplification factor in xenon filled proportional counters," Nucl. Instrum. Methods **102,** 309.

HENSLEY, W. K., BASSETT, W. A., AND HIUZENGA, J. R. (1973). "Pressure dependence of the radioactive decay constant of beryllium-7," Science **181,** 1164.

HERCEG NOVI (1973). "Proceedings of the first international summer school on radionuclide metrology," Nucl. Instrum. Methods **112,** 1–398.

HERCULES, D. M. (1970). "Physical basis of chemiluminescence," page 315 in *The Current Status of Liquid Scintillation Counting,* Bransome, E. D., Ed. (Grune & Stratton, New York).

HIGASHIMURA, T., YAMADA, O., NOHARA, N., AND SHIDEI, T. (1962). "External standard method for the determination of the efficiency in liquid scintillation counting," Int. J. Appl. Radiat. Isot. **13,** 308.

HILL, C. R. (1961). "A method of alpha particle spectroscopy for materials of very low specific activity," Nucl. Instrum. Methods **12,** 299.

HINE, G. J. (1956). "Appendix," in *Radiation Dosimetry,* Hine, G. J. and Brownell, G. L., Eds. (Academic Press, New York).

HINE, G. J. (1967). "γ-ray sample counting," page 275 in *Instrumentation in Nuclear Medicine,* Vol. 1, Hine, G. J., Ed. (Academic Press, New York).

HINE, G. J. AND SORENSON, J. A., Eds. (1974). *Instrumentation in Nuclear Medicine,* Vol. 2 (Academic Press, New York).

HIRSHFELD, A. T., HOPPES, D. D., AND SCHIMA, R. J. (1976). "Germanium detector efficiency calibration with NBS standards," page 90 in *Proceedings of ERDA Symposium on X- and Gamma-Ray Sources and Applications,* Report CONF-760539 (National Technical Information Service, Springfield, Virginia).

HOFFER, P. B., BECK, R. N., AND GOTTSCHALK, A., Eds. (1971). *Semiconductor Detectors in the Future of Nuclear Medicine* (Society of Nuclear Medicine, New York).

HOFSTADTER, R. (1948). "Alkali halide scintillation counters," Phys. Rev. **74,** 100.

HOLM, E. AND PERSSON, R. B. R. (1976). "Transfer of fall-out plutonium in the food-chain lichen → reindeer → man," page 435 in *Transuranium Nuclides in the Environment,* IAEA/STI/PUB/410 (International Atomic Energy Agency, Vienna).

HORROCKS, D. L. (1966a). "Low-level alpha disintegration rate determinations with a one-multiplier phototube liquid scintillation spectrometer," Int. J. Appl. Radiat. Isot. **17,** 441.

HORROCKS, D. L., Ed. (1966b). *Organic Scintillators.* (Gordon and Breach Science Publishers, New York).

HORROCKS, D. L. (1968). "Direct measurement of $^{14}Co_2$ in a liquid scintillation counter," Int. J. Appl. Radiat. Isot. **19,** 859.

HORROCKS, D. L. (1971). "Obtaining the possible maximum 90 percent efficiency for counting of ^{55}Fe in liquid scintillator solutions," Int. J. Appl. Radiat. Isot. **22,** 258.

HORROCKS, D. L. (1974). *Applications of Liquid Scintillation Counting.* (Academic Press, New York).

HORROCKS, D. L. (1975). "Liquid scintillation counting in high-density polyethylene vials," Int. J. Appl. Radiat. Isot. **26,** 243.

HORROCKS, D. L. AND STUDIER, M. H. (1964). "Determination of radioactive noble gases with a liquid scintillator," Anal. Chem. **36,** 2077.

HORROCKS, D. L. AND PENG, C. T., Eds. (1971). *Organic Scintillators and Liquid Scintillation Counting* (Academic Press, New York).

HOUTERMANS, F. G. AND OESCHGER, H. (1958). "Proportionalzählrohr zur Messung schwacher Aktivitäten weicher β-Strahlung," Helv. Phys. Acta **31,** 117.

HOUTERMANS, H. (1973). "Probability of non-detection in liquid scintillation counting," Nucl. Instrum. Methods **112,** 121.

HOUTERMANS, H. AND MIGUEL, M. (1962). "4π-β-γ coincidence counting for the calibration of nuclides with complex decay schemes," Int. J. Appl. Radiat. Isot. **13,** 137.

HUBBELL, J. H. (1969). *Photon Cross Sections, Attenuation Coefficients, and Energy Absorption Coefficients from 10 keV to 100 GeV.* Report NSRDS-NBS 29 (U.S. Government Printing Office, Washington).

HUBBELL, J. H., VEIGELE, W. J., BRIGGS, E. A., BROWN, R. T., CROMER, D. T., AND HOWERTON, R. J. (1975). "Atomic form factors, incoherent scattering functions, and photon scattering cross sections," J. Phys. Chem. Ref. Data **4,** 471.

HUGHES, D. J. (1953). *Pile Neutron Research.* (Addison-Wesley, Cambridge, Massachusetts).

HUGHES, E. E. AND MANN, W. B. (1964). "The half-life of carbon-14; comments on the mass-spectrometric method," Int. J. Appl. Radiat. Isot. **15,** 97.

HUTCHINSON, J. M. R., NAAS, C. R., WALKER, D. H., AND MANN, W. B. (1968). "Backscattering of alpha particles from thick metal backings as a function of atomic weight," Int. J. Appl. Radiat. Isot. **19,** 517.

HUTCHINSON, J. M. R., LANTZ, J. L., MANN, W. B., MULLEN, P. A., AND RODRIGUEZ-PASQUES, R. H. (1972). "An anti-coincidence shielded NaI(Tl) system at NBS," IEEE Trans. Nucl. Sci. **NS-19,** 117.

HUTCHINSON, J. M. R., MANN, W. B., AND MULLEN, P. A. (1973a). "Sum-peak counting with two crystals," Nucl. Instrum. Methods **112,** 187.

HUTCHINSON, J. M. R., MANN, W. B., AND PERKINS, R. W. (1973b). "Low-level radioactivity measurements," Nucl. Instrum. Methods **112,** 305.

HUTCHINSON, J. M. R., LUCAS, L. L., AND MULLEN, P. A. (1976a). "Study of the scattering correction for thick uranium-oxide and other alpha-particle sources—Part II: Experimental," Int. J. Appl. Radiat. Isot. **27,** 43.

HUTCHINSON, J. M. R., MANN, W. B., AND MULLEN, P. A. (1976b). "Development of the National Bureau of Standards low-energy-photon-emission-rate radioactivity standards," page 25 in *Proceedings of ERDA Symposium on X- and Gamma-Ray Sources and Applications,* CONF 760539 (National Technical Information Service, Springfield, Virginia).

IAEA (1965). International Atomic Energy Agency. *Radioisotope Sample Measurement Techniques in Medicine and Biology,* IAEA/STI/PUB/106 (International Atomic Energy Agency, Vienna).

IAEA (1967a). International Atomic Energy Agency. *Standardization of Radionuclides,* IAEA/STI/PUB/139 (International Atomic Energy Agency, Vienna).

IAEA (1967b). International Atomic Energy Agency. *Radioactive Dating and Methods of Low-Level Counting,* IAEA/STI/PUB/152 (International Atomic Energy Agency, Vienna).

IAEA (1970). International Atomic Energy Agency. *Analytical Control of Radiopharmaceuticals,* IAEA/STI/PUB/253 (International Atomic Energy Agency, Vienna).

IAEA (1971). International Atomic Energy Agency. *Radioisotope Production and Quality Control,* IAEA/STI/DOC/10/128 (International Atomic Energy Agency, Vienna).

IAEA (1973). International Atomic Energy Agency. *New Developments in Radiopharmaceuticals and Labelled Compounds,* IAEA/STI/PUB/344 (International Atomic Energy Agency, Vienna).

ICRP (1975). International Commission on Radiological Protection. *Report of the Task Group on Reference Man,* ICRP Publication 23 (Pergamon Press, New York).

ICRU (1961). International Commission on Radiological Units and Measurements (1959), published as NBS Handbook 78 (International Commission on Radiation Units and Measurements, Washington).

ICRU (1963). International Commission on Radiological Units and Measurements. *Radioactivity,* ICRU Report 10c, issued as National Bureau of Standards Handbook 86 (International Commission on Radiation Units and Measurements, Washington).

ICRU (1964). International Commission on Radiological Units and Measurement. *Physical Aspects of Irradiation,* published as National Bureau of Standards Handbook 85 (International Commission on Radiation Units and Measurements, Washington).

ICRU (1968). International Commission on Radiation Units and Measurements. *Certification of Standardized Radioactive Sources,* ICRU Report 12 (International Commission on Radiation Units and Measurements, Washington).

ICRU (1970). International Commission on Radiation Units and Measurements. *Radiation Dosimetry: X Rays Generated at Potentials of 5 to 150 kV,* ICRU Report 17 (International Commission on Radiation Units and Measurements, Washington).

ICRU (1971). International Commission on Radiation Units and Measurements. *Radiation Quantities and Units,* ICRU Report 19. (International Commission on Radiation Units and Measurements, Washington).

ICRU (1972). International Commission on Radiation Units and Measurements. *Measurement of Low-Level Radioactivity,* ICRU Report 22 (International Commission on Radiation Units and Measurements, Washington).

ICRU (1978). International Commission on Radiation Units and Measurements. *Average Energy Required to Produce an Ion Pair* (in preparation).

ISRAEL, H. I., LIER, D. W., AND STORM, E. (1971). "Comparison of detectors used in measurement of 10 to 300 keV x-ray spectra," Nucl. Instrum. Methods **91,** 141.

IVANOV, V. B. AND SHIPILOV, V. I. (1974). "A summing Ge(Li)-Compton spectrometer," Nucl. Instrum. Methods **119,** 313.

JAFFEY, A. H. (1954). "Radiochemical assay by alpha and fission measurements," page 596 in *The Actinide Elements,* Seaborg, G. T. and Katz, J. J., Eds. (McGraw-Hill, New York, Toronto, and London).

JARRETT, A. A. (1946). *Statistical Methods Used in the Measurement of Radioactivity with Some Useful Graphs and Nomographs,* USAEC Report AECU-262 (Department of Energy, Technical Information Center, Oak Ridge, Tennessee).

JEFFAY, H. AND ALVAREZ, J. (1961). "Liquid scintillation counting of carbon 14; use of ethanolamine-ethylene glycol monomethyl ether-toluene," Anal. Chem. **33,** 612.

JELLEY, J. V. (1958). *Čerenkov Radiation and Its Applications* (Pergamon Press, Oxford and New York).

JESSE, W. P. AND SADAUSKIS, J. (1953). "Alpha-particle ionization in pure gases and the average energy to make an ion pair," Phys. Rev. **90,** 1120.

JESSE, W. P. AND SADAUSKIS, J. (1955). "Ionization in pure gases and the average energy to make an ion pair for alpha and beta particles," Phys. Rev. **99,** 1668.

JOHNS, F. B., Ed. (1975). *Handbook of Radiochemical Analytical Methods.* EPA-680/4-75-001 (National Technical Information Service, Springfield, Virginia).

JOHNS, H. E. AND CUNNINGHAM, J. R. (1969). Table A-8 in *The Physics of Radiology,* 3rd ed. (C. C. Thomas, Springfield).

JOHNSON, N. R., EICHLER, E., AND O'KELLEY, G. D. (1963). *Nuclear Chemistry,* vol. 2 of *Techniques of Inorganic Chemistry,* Jonassen, H. B., Ed. (Interscience Publishers, New York).

JONES, D. G. AND MCNAIR, A. (1970). "Standardization of radionuclides by efficiency-tracing methods using a liquid-scintillation counter," page 37 in *Radioactivity Calibration Standards,* Nat. Bur. Stand. Spec. Publ. 331, Mann, W. B. and Garfinkel, S. B., Eds. (U.S. Government Printing Office, Washington).

JONES, W. M. (1955). "Half-life of tritium," Phys. Rev. **100,** 124.

JORDAN, K. C. (1973). Private communication. (Monsanto Mound Laboratory,

Miamisburg, Ohio).

JORDAN, K. C. AND BLANKE, B. C. (1967). "Half-lives of actinium-227 and thorium-227 measured calorimetrically," page 567 in *Standardization of Radionuclides,* IAEA/STI/PUB/139 (International Atomic Energy Agency, Vienna).

KAHN, B., BLANCHARD, R. L., KRIEGER, H. Z., KOLDE, H. E., SMITH, D. B., MARTIN, A., GOLD, S., AVERETT, W. J., BRINCK, W. L., AND KARCHES, G. J. (1971a). *Radiological Surveillance Studies at a Boiling Water Nuclear Power Reactor,* PHS Report BRH/DER 70-1. (National Technical Information Service, Springfield, Virginia).

KAHN, B., BLANCHARD, R. L., KOLDE, H. E., KRIEGER, H. L., GOLD, S., BRINCK, W. L., AVERETT, W. J., SMITH, D. B., AND MARTIN, A. (1971b). *Radiological Surveillance Studies at a Pressurized Water Nuclear Power Reactor,* EPA Report RD-71-1 (National Technical Information Service, Springfield, Virginia).

KALBHEN, D. A. (1970). "Chemiluminescence as a problem in liquid scintillation counting," page 337 in *The Current Status of Liquid Scintillation Counting,* Bransome, E. D., Ed. (Grune & Stratton, New York).

KALLMANN, H. (1947). "Der elementar Prozess der Lichtanregung in Leuchtstoffen. Über eine neue Methode der Zählung und Energiemessung geladener Teilchen," Natur Tech. (July).

KATCOFF, S., SCHAEFFER, O. A., AND HASTINGS, J. M. (1951). "Half-life of I^{129} and the age of the elements," Phys. Rev. **82,** 688.

KEENE, J. P. (1950). "An absolute method for measuring the activity of radioactive isotopes," Nature **166,** 601.

KEISCH, B., KOCH, R. C., AND LEVINE, A. S. (1965). "Determination of biospheric levels of I-129 by neutron activation analysis," page 284 in *Modern Trends in Activation Analysis.* (Center for Trace Characterization, Texas A & M University, College Station, Texas).

KEITH, H. D. AND LOOMIS, T. C. (1976). "Calibration and use of lithium-drifted silicon detector for accurate analysis of x-ray spectra," X-Ray Spectrometry, **5,** 93.

KEITH, R. L. G. AND WATT, D. E. (1966). "A routine procedure for 4π proportional counting weak beta emitters from solid sources," Nucl. Instrum. Methods **42,** 68.

KEITHLEY (1977). *Electrometer Measurements* revised 2nd ed. (Keithley Instruments, Inc., Cleveland, Ohio).

KLOTS, C. E. (1968). "Energy deposition mechanisms," page 1 in *Fundamental Processes in Radiation Chemistry,* Ausloos, P. J., Ed. (John Wiley & Sons, New York).

KOBAYASHI, Y. AND MAUDSLEY, D. V. (1970). "Practical aspects of double isotope counting," page 76 in *The Current Status of Liquid Scintillation Counting,* Bransome, E. D., Ed. (Grune & Stratton, New York).

KOCH, L. (1958). "La scintillation dans les gaz nobles et leurs mélanges," page 151 in *Nuclear Electronics,* IAEA/STI/PUB/2, vol. 1 (International Atomic Energy Agency, Vienna).

KOCH, R. C. AND KEISCH, B. (1964). "Neutron activation analysis studies of ^{129}I in the biosphere," page 47 in *L'Analyse Par Radioactivation* (Presses

Universitaires de France, Vendome).

KOHMAN, T. P. (1945). "A general method for determining coincidence corrections of counting instruments," page 1655 in *The Transuranium Elements, Part II,* Seaborg, G. T., Katz, J. J., and Manning, W. M., Eds. (McGraw-Hill, New York).

KONDO, S. AND RANDOLPH, M. L. (1960). "Effect of finite size of ionization chambers on measurements of small photon sources," Radiat. Res. **13,** 37.

KORENMAN, I. M. (1966). Analytical Chemistry of Low Concentrations. (Israel Program for Scientific Translations Limited, Jerusalem).

KOSTROUN, V. O., CHEN, M. S., AND CRASEMANN, B. (1971). "Atomic radiation transition probabilities to the 1*s* state and theoretical *K*-shell fluorescence yields," Phys. Rev. **A3,** 533.

KOWALSKI, E. (1965). "Absolutmessung der radioactiven Quellstarke nach der allgemeinen Koinzidenmethode; Einfluss der Richtungskorrelation, Streukoinzidenzen und Summenimpulse," Nucl. Instrum. Methods **33,** 29.

KOWALSKI, E., ANLIKER, R., AND SCHMID, K. (1967). "Criteria for the selection of solutes in liquid scintillation counting: new efficient solutes with high solubility," Int. J. Appl. Radiat. Isot. **18,** 307.

KRANER, H. W., SCHROEDER, G. L., LEWIS, A. R., AND EVANS, R. D. (1964). "Large volume scintillation chamber for radon counting," Rev. Sci. Instrum. **35,** 1259.

KREY, P. W. AND HARDY, E. P. (1970). "Plutonium in soil around the Rocky Flats plant," USAEC Report HASL-235 (National Technical Information Service, Springfield, Virginia).

KRUGERS, J. (1973). *Instrumentation in Applied Nuclear Chemistry.* (Plenum Press, New York).

KU, H. H. (1966). "Notes on the use of propagation of error formulas," J. Res. Nat. Bur. Stand. **70C,** 263. (Reprinted in Ku, 1969).

KU, H. H., Ed. (1969). *Precision Measurement and Calibration—Statistical Concepts and Procedures,* Nat. Bur. Stand. Publ. **300,** vol. 1 (U.S. Government Printing Office, Washington).

KUCHTA, D. R., RUNDO, J., AND HOLTZMAN, R. B. (1976). "Low-level determination of skeletal ^{228}Ra and ^{228}Th in the presence of gross amounts of ^{226}Ra," page 183 in *Biological and Environmental Effects of Low-Level Radiation,* Vol. II, IAEA/STI/PUB/409 (International Atomic Energy Agency, Vienna).

LACHANCE, Y. AND ROY, J. C. (1972). "Method for vacuum evaporation of SiO on thin plastic films," Rev. Sci. Instrum. **43,** 1382.

LAICHTER, Y., NOTEA, A., AND SHAFRIR, N. H. (1973). "A general approach to the calibration of a Ge(Li) gamma spectrometry system for measurement of environmental and biological samples," Nucl. Instrum. Methods **113,** 61.

LAPP, R. E. AND ANDREWS, H. L. (1964). *Nuclear Radiation Physics.* (Prentice-Hall, Englewood Cliffs, New Jersey).

LATNER, N. AND SANDERSON, C. G. (1972). "The HASL Ge(Li)-NaI(Tl) low-level counting system," IEEE Trans. Nucl. Sci. **NS-19,** 141.

LAVI, N. (1973). "^{166m}Ho as an efficiency calibration standard for Ge(Li) detectors," Nucl. Instrum. Methods **109,** 265.

LEDERER, C. M., HOLLANDER, J. M., AND PERLMAN, I. (1967). *Table of Isotopes,* 6th ed. (John Wiley & Sons, New York).

LEGRAND, J. (1973). "Calibration of γ-spectrometers," Nucl. Instrum. Methods **112,** 229.

LEGRAND, J., BLONDEL, M., AND MAGNIER, P. (1973). "High-pressure 4 π proportional counter for internal conversion electron measurements (^{139}Ce, ^{109}Cd, ^{99m}Tc)," Nucl. Instrum. Methods **112,** 101.

LEWIS, V. E., WOODS, M. J., AND GOODIER, I. W. (1972). "The calibration of the 1383A ionization chamber for ^{67}Ga," Int. J. Appl. Radiat. Isot. **23,** 279.

LEWIS, V. E., SMITH, D., AND WILLIAMS, A. (1973). "Correlation counting applied to the determination of absolute disintegration rates for nuclides with delayed states," Metrologia **9,** 14.

LIEBERMAN, R. AND MOGHISSI, A. A. (1970). "Low-level counting by liquid scintillation, II. Applications of emulsions in tritium counting," Int. J. Appl. Radiat. Isot. **21,** 319.

LIEBSON, S. H. AND FRIEDMAN, H. (1948). "Self-quenching halogen-filled cavities," Rev. Sci. Instrum. **19,** 303.

LINDENBAUM, A. AND SMYTH, M. A. (1971). "Determination of ^{239}Pu and ^{241}Am in animal tissues by liquid scintillation spectrometry," page 951 in *Organic Scintillators and Liquid Scintillation Counting,* Horrocks, D. L. and Peng, C. T., Eds. (Academic Press, New York).

LLOYD, R. D., MAYS, C. W., AND ATHERTON, D. R. (1967). "Knothole, a new side-well gamma ray detector," Nucl. Instrum. Methods **49,** 109.

LMRI (1975). Laboratoire de Metrologie de Rayonnements Ionisants. *Annual Report of the Laboratoire de Metrologie des Rayonnements Ionisants.* (Centre d'Études Nucleaires, Saclay).

LOEVINGER, R. AND BERMAN, M. (1951). "Efficiency criteria in radioactivity counting," Nucleonics **9** No. 1, 26.

LOEVINGER, R. AND BERMAN, M. (1976). *A Revised Schema for Calculating the Absorbed Dose from Biologically Distributed Radionuclides,* MIRD Pamphlet No. 1, Revised (Society of Nuclear Medicine, New York).

LOFTUS, T. P. AND WEAVER, J. T. (1974). "Standardization of ^{60}Co and ^{137}Cs gamma-ray beams in terms of exposure," J. Res. Nat. Bur. Stand. **78A,** 465.

LOOSLI, H., OESCHGER, H., STUDER, R., WAHLEN, M., AND WIEST, W. (1975). "Argon-37 activity in air samples and its atmospheric mixing," page 240 in *Proceedings of the Noble Gases Symposium,* Stanley, R. E. and Moghissi, A. A., Eds. ERDA CONF-730915 (U.S. Energy Research and Development Administration, Washington).

LOWENTHAL, G. C. (1969). "Secondary standard instruments for the activity measurement of pure β emitters—a review," Int. J. Appl. Radiat. Isot. **20,** 559.

LOWENTHAL, G. C. AND SMITH, A. M. (1964). "Use of Au-20% Pd for metallising thin source supports for 4π proportional gas flow counters," Nucl. Instrum. Methods **30,** 363.

LOWENTHAL, G. C. AND PAGE, V. (1970). "A syringe pycnometer for the accurate weighing of milligram quantities of aqueous solutions," Anal. Chem. **42,** 815.

LOWENTHAL, G. C. AND WYLLIE, H. A. (1973a). "Special methods of source preparation," Nucl. Instrum. Methods **112,** 353.

LOWENTHAL, G. C. AND WYLLIE, H. A. (1973b). "Thin 4π sources on electrosprayed ion exchange resin for radioactivity standardization," Int. J. Appl. Radiat. Isot. **24,** 415.

LOWENTHAL, G. C., PAGE, V., AND WYLLIE, H. A. (1973). "The use of sources made on electrosprayed pads of ion exchange resins for efficiency tracer measurements," Nucl. Instrum. Methods **112,** 197.

LUCAS, H. F. (1957). "Improved low-level alpha-scintillation counter for radon," Rev. Sci. Instrum. **28,** 680.

LUCAS, H. F. (1964). *The Low-Level Gamma Counting Room: Radon Removal and Control,* ANL Report ANL-6938 (Argonne National Laboratory, Argonne, Illinois).

LUCAS, H. F. (1965). *A Radon Removal System for the NASA Lunar Sample Laboratory: Design and Discussion,* ANL Report ANL-7060 (Argonne National Laboratory, Argonne, Illinois).

LUCAS, L. L. AND HUTCHINSON, J. M. R. (1976). "Study of the scattering correction for thick uranium-oxide and other alpha-particle sources—Part I; Theoretical," Int. J. Appl. Radiat. Isot. **27,** 35.

LUCAS, L. L. AND MANN, W. B., Eds. (1978). "The half life of ^{239}Pu: Measurements organized by the ERDA Half-Life Evaluation Committee," Int. J. Appl. Radiat. Isot. **29,** 479.

MAGNO, P., REAVEY, T., AND APIDANAKIS, J. (1970). *Liquid Waste Effluents from a Nuclear Fuel Reprocessing Plant,* PHS Report BRH/NERHL 70-2 (National Technical Information Service, Springfield, Virginia).

MAKOWER, W. (1909). "On the number and the absorption by matter of the β particles emitted by radium," Philos. Mag. 17 [6], 171.

MALM, H. L., RAUDORF, T. W., MARTINI, M., AND ZANIO, K. R. (1973). "Gamma-ray efficiency comparisons for Si(Li), Ge, CdTe, and HgI_2 detectors," IEEE Trans. Nucl. Sci. **NS-20,** 500.

MANLEY, J. H., AGNEW, H. M., BARSCHALL, H. H., BRIGHT, W. C., COON, J. H., GRAVES, E. R., JORENSON, T., AND WALDMAN, B. (1946). "Elastic backscattering of d-d neutrons," Phys. Rev. **70,** 602.

MANN, H. M., BILGER, H. R., AND SHERMAN, I. S. (1966). "Observations on the energy resolution of germanium detectors for 0.1 to 10 MeV gamma rays," IEEE Trans. Nucl. Sci. **NS-13,** 252.

MANN, W. B. (1954a). "Use of Callendar's 'Radio-Balance' for the measurement of the energy emission from radioactive sources," J. Res. Nat. Bur. Stand. **52,** 177.

MANN, W. B. (1954b). "A radiation balance for the calorimetric comparison of four national radium standards," J. Res. Nat. Bur. Stand. **53,** 277.

MANN, W. B. (1962). "Calorimetric measurements of radioactivity," page 538 in *Encyclopaedic Dictionary of Physics,* Vol. 1, Thewlis, J., Ed-in-Chief (Pergamon Press, Oxford and MacMillan Co., New York).

MANN, W. B. (1973). "Radioactive calorimetry: a review of the work at the National Bureau of Standards," Nucl. Instrum. Methods **112,** 273.

MANN, W. B. AND PARKINSON, G. B. (1949). "A Geiger-Müller counting unit and external quenching equipment for the estimation of C^{14} in carbon dioxide," Rev. Sci. Instrum. **20,** 41.

MANN, W. B. AND SELIGER, H. H. (1958). "Preparation, maintenance, and application of standards of radioactivity," Nat. Bur. Stand. Circ. **594** (U.S.

Government Printing Office, Washington).
MANN, W. B. AND GARFINKEL, S. B. (1966). *Radioactivity and Its Measurement* (D. Van Nostrand Co., Inc., Princeton, New Jersey). (See also Mann, Ayres, and Garfinkel, 1979).
MANN, W. B. AND HUTCHINSON, J. M. R. (1976). "The standardization of iodine-129 by beta-photon coincidence counting," Int. J. Appl. Radiat. Isot. **27,** 187.
MANN, W. B. AND WEISS, H. M. (1976). Private communication (National Bureau of Standards, Washington and Physikalishe-Tecknische Bundesanstalt, Braunschweig).
MANN, W. B., STOCKMANN, L. L., YOUDEN, W. J., SCHWEBEL, A., MULLEN, P. A., AND GARFINKEL, S. B. (1959). "The preparation of new solution standards of radium," J. Res. Nat. Bur. Stand. **62,** 27.
MANN, W. B., MARLOW, W. F., AND HUGHES, E. E. (1961). "The half-life of carbon-14," Int. J. Appl. Radiat. Isot. **11,** 57.
MANN, W. B., MEDLOCK, R. W., AND YURA, O. (1964). "A recalibration of the National Bureau of Standards tritiated water standards by gas counting," Int. J. Appl. Radiat. Isot. **15,** 351.
MANN, W. B., AYRES, R. L., AND GARFINKEL, S. B. (1979). *Radioactivity and Its Measurement,* 2nd ed. (Pergamon Press, Oxford and New York).
MARINENKO, G. AND TAYLOR, J. K. (1966). "Coulometric calibration of microvolumetric apparatus," J. Res. Nat. Bur. Stand. **70C,** 1.
MARLOW, W. F. AND MEDLOCK, R. W. (1960). "Preparation and standardization of a carbon-14 beta-ray standard: benzoic-acid-7-C^{14} in toluene," J. Res. Nat. Bur. Stand. **64A,** 143.
MARTIN, G. R. (1961). "The estimation of the resolving time of a counting apparatus," Nucl. Instrum. Methods **13,** 263.
MARTIN, M. J. (1973). *Radioactive Atoms—Supplement I.* Report ORNL-4923 (Oak Ridge National Laboratory, Oak Ridge, Tennessee).
MARTIN, M. J. (1976). *Nuclear Decay Data for Selected Radionuclides,* Report ORNL-5114 (Oak Ridge National Laboratory, Oak Ridge, Tennessee).
MARTIN, M. J. (1978). Personal communication. (Oak Ridge National Laboratory, Oak Ridge, Tennessee).
MARTIN, M. J. AND BLICHERT-TOFT, P. H. (1970). *Radioactive Atoms,* Nucl. Data Tables, **8,** No. 1-2. (Oak Ridge National Laboratory, Oak Ridge, Tennessee).
MASTERS, C. F. AND EAST, L. V. (1970). "An investigation into the Harms dead time correction procedure using Monte Carlo modelling techniques," IEEE Trans. Nucl. Sci. **NS-17,** 383.
MAY, H. A. AND MARINELLI, L. D. (1960). "Factors determining the ultimate detector sensitivity of gamma scintillation spectrometers," page 71 in Semi-Annual Report, Jan-June, Argonne National Laboratory Report ANL-6199 (National Technical Information Service, Springfield, Virginia).
MAZAKI, H., NAGATOMO, T., AND SHIMIZU, S. (1972). "Effect of pressure on the decay constant of ^{99m}Tc," Phys. Rev. **C5,** 1718.
MCBETH, G. W., WINYARD, R. A., AND LUTKIN, J. E. (1971). *Pulse Shape Discrimination with Organic Scintillators* (Koch-Light Laboratories, Ltd.,

Colnbrook, Bucks, England).

McCallum, G. J. and Coote, G. E. (1975). "Influence of source-detector distance on relative intensity and angular correlation measurements with Ge(Li) spectrometers," Nucl. Instrum. Methods **30,** 189.

McDaniel, E. W., Schaefer, H. J., and Colehour, J. K. (1956). "Dual proportional counter for low-level measurement of alpha activity of biological materials," Rev. Sci. Instrum. **27,** 864.

McDowell, W. J. (1971). "Liquid scintillation counting techniques for the higher actinides," page 937 in *Organic Scintillators and Liquid Scintillation Counting,* Horrocks, D. L. and Peng, C. T., Eds. (Academic Press, New York).

McDowell, W. J. (1975). "High resolution liquid scintillation method for the analytical determination of alpha emitters in environmental samples," IEEE Trans. Nucl. Sci. **NS-22,** 649.

McKay, K. (1951). "Electron-hole production in germanium by alpha particles," Phys. Rev. **84,** 829.

McNelles, L. A. and Campbell, J. L. (1973). "Absolute efficiency calibration of coaxial Ge(Li) detectors for the energy range 160 to 1330 keV," Nucl. Instrum. Methods **109,** 241.

McNish, A. G. (1958). "Classification and nomenclature for standards of measurement," IRE Trans. Instrum. **I 7,** 371.

McNish, A. G. (1960). "Nomenclature for standards of radioactivity," Int. J. Appl. Radiat. Isot. **8,** 145.

Merritt, J. S. (1973). "Present status in quantitative source preparation," Nucl. Instrum. Methods **112,** 325.

Merritt, J. S. and Taylor, J. G. V. (1967a). *Gravimetric Sampling in the Standardization of Solutions of Radionuclides.* Report No. AECL-2679 (Atomic Energy of Canada, Limited, Chalk River).

Merritt, J. S. and Taylor, J. G. V. (1967b). "Response of 4π proportional counters to γ rays," page 147 in *Standardization of Radionuclides,* IAEA/STI/PUB/139 (International Atomic Energy Agency, Vienna).

Merritt, J. S. and Taylor, J. G. V. (1971). "The standardization of ^{63}Ni and the decay of ^{65}Ni," Int. J. Appl. Radiat. Isot. **22,** 783.

Merritt, J. S., Taylor, J. G. V., and Campion, P. J. (1959). "Self-absorption in sources prepared for 4π beta counting," Can. J. Chem. **37,** 1109.

Merritt, J. S., Taylor, J. G. V., Merritt, W. F., and Campion, P. J. (1960). "The absolute counting of sulfur-35," Anal. Chem. **32,** 310.

Meyer, K. P., Schmid, P., and Huber, P. (1959). "Absolut-Messung radioaktiver quellstarken mit hilfe einer neugestaltung der koinzidenzmethode," Helv. Phys. Acta **32,** 423.

Meyer, R. A., Tirsell, K. G., and Armantrout, G. A. (1976). "Current research relevant to the improvement of γ-ray spectroscopy as an analytical tool," page 40 in *Proceedings of ERDA Symposium on X- and Gamma-Ray Sources and Applications,* CONF-760539 (National Technical Information Service, Springfield, Virginia).

Meyer-Schützmeister, L. and Vincent, D. H. (1952). "Absoluteichungen Energiearmer β Strahler mit dem 4π-Zählrohr," Z. Phys. **134,** 9.

Meyer, S. and von Schweidler, E. (1927). *Radioaktivitat,* ed. 2 (B. G.

Teubner, Berlin).

MILLER, W. W. (1947). "High-efficiency counting of long-lived radioactive carbon as CO_2," Science **105,** 123.

MOGHISSI, A. A. (1970). "Low-level liquid scintillation counting of α- and β-emitting nuclides," page 86 in *The Current Status of Liquid Scintillation Counting,* Bransome, E. D., Ed. (Grune & Stratton, New York).

MOGHISSI, A. A. AND CARTER, M. W. (1968). "Internal standard with identical system properties for determination of liquid scintillation efficiency," Anal. Chem. **40,** 812.

MOGHISSI, A. A. AND CARTER, M. W., Ed. (1973). *Tritium* (Messenger Graphics, Phoenix, Arizona).

MOGHISSI, A. A., MCNELIS, D. N., PLOTT, W. F., AND CARTER, M. W. (1971). "A rapid liquid scintillation technique for monitoring ^{14}C in air," page 391 in *Rapid Methods for Measuring Radioactivity in the Environment,* IAEA/STI/PUB/289 (International Atomic Energy Agency, Vienna).

MORAN, J. J. (1967). "Measurement of volume: liquid," page 4279 in *Treatise on Analytical Chemistry,* Part 1, vol. 7, Kolthoff, I. M., Elving, P. J., and Sandell, E. B., Eds. (John Wiley, New York).

MOSELEY, H. G. J. (1912). "The number of β particles emitted in the transformation of radium," Proc. R. Soc. London Ser. A **87,** 230.

MUGHABGHAB, S. F. AND GARBER, D. I. (1973). "Neutron cross sections," in *Vol. 1, Resonance Parameters,* Report BNL 325, 3rd ed. (also as Report TID-4500) (National Technical Information Service, Springfield, Virginia).

MÜLLER, J. W. (1973) "Dead-time problems," Nucl. Instrum. Methods **112,** 47.

MÜLLER, J. W. (1974). "Some formulae for a dead time distorted Poisson process," Nucl. Instrum. Methods **117,** 401.

MÜLLER, J. W. (1975a). "Principles of correlation counting," Report BIPM-75/5 (Bureau International des Poids et Mesures, Sévres).

MÜLLER, J. W. (1975b). "Bibliography on dead-time effects," Report BIPM-75/6 (Bureau International des Poids et Mesures, Sévres).

MÜLLER, J. W. (1976a). "On the evaluation of the correction factor $\mu(\rho',\tau)$ for the periodic pulse method," Report BIPM-76/3 (Bureau International des Poids et Mesures, Sévres).

MÜLLER, J. W. (1976b). "The source-pulser method revisited," Report BIPM-76/5 (Bureau International des Poids et Mesures, Sévres).

MUNZENMAYER, K. AND BAERG, A. P. (1969). "Delay matching of beta-gamma coincident pulses," Nucl. Instrum. Methods **70,** 157.

NAS-NRC (1974). National Academy of Sciences-National Research Council. *Users' Guides for Radioactivity Standards,* National Academy of Sciences Nuclear Science Series Report No. 3115 (National Academy of Sciences, Washington).

NATRELLA, M. G. (1963). *Experimental Statistics,* Nat. Bur. Stand. Handbook 91 (U.S. Government Printing Office, Washington). (Reprinted with corrections 1966).

NBS (1977). National Bureau of Standards. *The International System of Units (SI),* Nat. Bur. Stand. Spec. Publ. **330,** 3rd ed. (U.S. Government Printing Office, Washington).

NCRP (1961). National Committee on Radiation Protection and Measurements. *A Manual of Radioactivity Procedures*, NCRP Report No. 28, published as Nat. Bur. Stand. Handbook 80 (U.S. Government Printing Office, Washington).

NCRP (1975). National Council on Radiation Protection and Measurements. *Natural Background Radiation in the United States,* NCRP Report No. 45 (National Council on Radiation Protection and Measurements, Washington).

NCRP (1976a). National Council on Radiation Protection and Measurements. *Tritium Measurement Techniques for Laboratory and Environmental Use,* NCRP Report No. 47 (National Council on Radiation Protection and Measurements, Washington).

NCRP (1976b). National Council on Radiation Protection and Measurements. *Environmental Radiation Measurements,* NCRP Report No. 50 (National Council on Radiation Protection and Measurements, Washington).

NERVIK, W. E. AND STEVENSON, P. C. (1952). "Self-scattering and self-absorption of betas by moderately thick samples," Nucleonics **10** No. 3, 18.

NEYMAN AND PEARSON (1936). "Contribution to the theory of testing statistical hypotheses," page 1 in *Statistical Research Memoirs,* Vol. 1, Neyman, J. and Pearson, E. S., Eds. (The University Press, Cambridge, England).

NICHOLSON, P. W. (1974). *Nuclear Electronics* (John Wiley & Sons, New York).

NIELSEN, J. M. (1972). "Gamma-ray spectrometry," page 678 in *Physical Methods of Chemistry,* vol. 1, Weissberger, A. and Rossiter, B. W., Eds. (John Wiley & Sons, New York).

NIELSEN, J. M. AND PERKINS, R. W. (1967). "Anti-coincidence shielded multidimensional gamma-ray spectrometers for low-level counting," page 687 in *Radioactive Dating and Methods of Low-Level Counting,* IAEA/STI/PUB/152 (International Atomic Energy Agency, Vienna).

NIH (1973). National Institutes of Health. *The National Institutes of Health Radiation Safety Guide,* DHEW Publication NIH-73-18 (Department of Health, Education and Welfare, Washington).

NOAKES, J. E., NEARY, M. P., AND SPAULDING, J. D. (1973). "Tritium measurements with a new liquid scintillation counter," Nucl. Instrum. Methods **109,** 177.

NOBLES, R. A. (1956). "Detection of charged particles with gas scintillation counters," Rev. Sci. Instrum. **27,** 280.

NORTHRUP, J. A. AND NOBLES, R. A. (1956). "Some aspects of gas scintillation counters," IRE Trans. Nucl. Sci. **NS-3,** 59.

NOWLIN, C. H. AND BLANKENSHIP, C. L. (1965). "Elimination of undesirable undershoot in the operation and testing of nuclear pulse amplifiers," Rev. Sci. Instrum. **36,** 1830.

O'KELLEY, G. D. (1962). "Detection and measurement of nuclear radiation," NAS-NS 3105 (National Academy of Sciences, Washington).

OLLER, W. L. AND PLATO, P. (1972). "Beta spectrum analysis: a new method to analyze mixtures of beta-emitting radionuclides by liquid scintillation techniques," Int. J. Appl. Radiat. Isot. **23,** 481.

Olsson, I. U., Karlen, I., Turnbull, A. H., and Prosser, N. J. D. (1962). "A determination of the half-life of C^{14} with a proportional counter," Ark. Fys. **22,** 237.

Osborne, R. V. and Hill, C. R. (1964). "High resolution alpha particle spectroscopy at very low specific activities," Nucl. Instrum. Methods **29,** 101.

Ott, D. G., Richmond, C. R., Trujillo, T. T., and Foreman, H. (1959). "Cab-O-Sil suspensions for liquid-scintillation counting," Nucleonics **17** No. 9, 106.

Ouseph, T. J. (1975). *Introduction to Nuclear Radiation Detectors* (Plenum Publishing Company, New York).

Painter, K. (1974). "Choice of counting vials for liquid scintillation: A review," page 431 in *Recent Developments in Liquid Scintillation Counting,* Stanley, P. and Scoggins, B., Eds. (Academic Press, New York).

Palmer, H. E. (1975). "Recent advances in gas scintillation proportional counters," IEEE Trans. Nucl. Sci. **NS-22,** 100.

Palmer, H. E. and Beasley, T. M. (1967). "Iron-55 in man and the biosphere," Health Phys. **13,** 889.

Parker, R. P. and Elrick, R. H. (1970). "Čerenkov counting as a means of assaying β-emitting radionuclides," page 110 in *The Current Status of Liquid Scintillation Counting,* Bransome, E. D., Ed. (Grune & Stratton, New York).

Parmentier, J. H. and Ten Haaf, F. E. L. (1969). "Developments in liquid scintillation counting since 1968," Int. J. Appl. Radiat. Isot. **20,** 305.

Parzen, E. (1960). *Modern Probability Theory and Its Applications* (John Wiley & Sons, New York).

Pasternack, B. S. and Harley, N. H. (1971). "Detection limits for radionuclides in the analysis of multi-component gamma ray spectrometer data," Nucl. Instrum. Methods **91,** 533.

Pate, B. D. and Yaffe, L. (1955a). "A new material and technique for the fabrication and measurement of very thin films for use in 4π-counting," Can. J. Chem. **33,** 15.

Pate, B. D. and Yaffe, L. (1955b). "Disintegration-rate determination by 4π-counting, Pt. I," Can. J. Chem. **33,** 610.

Pate, B. D. and Yaffe, L. (1955c). "Disintegration-rate determination by 4π-counting, Pt. II. Source-mount absorption correction," Can. J. Chem. **33,** 929.

Pate, B. D. and Yaffe, L. (1955d). "Disintegration-rate determination by 4π-counting, Pt. III. Absorption and scattering of radiation," Can. J. Chem. **33,** 1656.

Pate, B. D. and Yaffe, L. (1956). "Disintegration-rate determination by 4π-counting, Pt. IV. Self-absorption correction: general method and application to Ni^{63} β-radiation," Can. J. Chem. **34,** 265.

Patterson, M. S. and Greene, R. C. (1965). "Measurement of low energy beta emitters in aqueous solution by liquid scintillation counting of emulsions," Anal. Chem. **37,** 854.

Peelle, R. W. and Maienschein, F. C. (1960). "Gamma-gamma coincidence standardization of sources, with application to the B^{11} (p,γ) C^{12} reaction," page 96 in *Neutron Physics Annual Progress Report,* ORNL-3016 (Oak Ridge National Laboratory, Oak Ridge, Tennessee).

PELL, E. M. (1960). "Ion drift in an n-p junction," J. Appl. Phys. **31,** 291.

PENG, C. T. (1975). "Liquid scintillation and Čerenkov counting," page 79 in *Radiochemical Methods in Analysis,* Coomber, D. I., Ed. (Plenum Press, New York).

PERKINS, R. W. (1965). "An anticoincidence shielded-multidimensional gamma-ray spectrometer," Nucl. Instrum. Methods **33,** 71.

PERKINS, R. W., NIELSEN, J. M., AND DIEBEL, R. N. (1960). "Total absorption gamma-ray spectrometers utilizing anticoincidence shielding." Rev. Sci. Instrum. **31,** 1344.

PERKINS, R. W., RANCITELLI, L. A., COOPER, J. A., KAYE, J. H., AND WOGMAN, N. A. (1970). "Cosmogenic and primordial radionuclides in lunar samples by nondestructive gamma-ray spectrometry," Science **167,** 577.

PETEL, M. AND HOUTERMANS, H. (1967). "Methodes d'indicateur pour l'etalonnage du fer-55," page 301 in Standardization of Radionuclides, IAEA/STI/PUB/139 (International Atomic Energy Agency, Vienna).

PETREE, B., HUMPHREYS, J. C., LAMPERTI, P. J., LOFTUS, T. P., AND WEAVER, J. T. (1975). "Comparison of Cobalt-60 exposure determinations by calorimetry and by ionization chamber techniques," Metrologia **11,** 11.

PETROFF, M. D. AND DOGGETT, W. O. (1956). "Application of a statistical correlation function for determining daughter half-lives in the range 10^{-4} to 10^{-1} second," Rev. Sci. Instrum. **27,** 838.

PHELPS, P. L. AND HAMBY, K. O. (1972). "Experience in the use of an anticoincidence shielded Ge(Li) gamma-ray spectrometer for low-level environmental radionuclide analysis," IEEE Trans. Nucl. Sci. **NS-19,** 155.

PICCIOTTO, E. AND WILGAIN, S. (1954). "Thorium determination in deep-sea sediments," Nature **173,** 632.

PILTINGSRUD, H. V. AND STENCEL, J. R. (1972). "Determination of ^{90}Y, ^{90}Sr, and ^{89}Sr in samples by use of liquid scintillation beta spectroscopy," Health Phys. **23,** 121.

POLICARPO, A. J. P. L., CONDE, C. A. N., AND ALVES, M. A. F. (1968). "Large light output gas proportional scintillation counters," Nucl. Instrum. Methods **58,** 151.

POLICARPO, A. J. P. L., ALVES, M. A. F., SANTOS, M. C. M., AND CARVALHO, M. J. T. (1972). "Improved resolution for low energies with gas proportional scintillation counters," Nucl. Instrum. Methods **102,** 337.

PRICE, W. J. (1958). *Nuclear Radiation Detection* (McGraw-Hill, New York, Toronto, London).

PRICE, W. J. (1964). *Nuclear Radiation Detection,* 2nd ed. (McGraw-Hill, New York).

PUTMAN, J. L. (1953). "Limitations and extensions of the coincidence method for measuring the activity of beta-gamma emitters," in AERE Report 1/M-26 (Atomic Energy Research Establishment, Harwell, United Kingdom).

RAGAINI, R. C. (1976). "Methods in environmental sampling for radionuclides," IEEE Trans. Nucl. Sci. **NS-23,** 1202.

RAMTHUN, H. (1973). "Recent developments in calorimetric measurements of radioactivity," Nucl. Instrum. Methods **112,** 265.

RANDOLPH, R. B. (1975). "Determination of strontium-90 and strontium-89 by Čerenkov and liquid scintillation counting," Int. J. Appl. Radiat. Isot. **26,** 9.

RAO, P. V., CHEN, M. H., AND CRASEMANN, B. (1972). "Atomic vacancy distributions produced by inner-shell ionization," Phys. Rev. **A5,** 997.

RAPKIN, E. (1967). "Preparation of samples for liquid scintillation counting," page 181 in *Instrumentation in Nuclear Medicine,* vol. 1, Hine, G. J., Ed. (Academic Press, New York).

RIECK, H. G., MYERS, I. T., AND PALMER, R. F. (1956). "Tritiated water standard," Radiat. Res. **4,** 451.

ROBERTS, P. B. AND DAVIES, B. L. (1966). "A transistorized radon measuring equipment," J. Sci. Instrum. **43,** 32.

ROBINSON, B. M. W. (1960). "Routine standardization of radionuclides in the UK," page 41 in *Metrology of Radionuclides,* IAEA/STI/PUB/6 (International Atomic Energy Agency, Vienna).

ROBINSON, C. V. (1967). "Geiger-Müller and proportional counters," page 57 in *Instrumentation in Nuclear Medicine,* vol. 1, Hine, G. J., Ed. (Academic Press, New York).

ROBINSON, H. P. (1960). "A new design for a high precision, high geometry alpha counter," page 147 in *Metrology of Radionuclides,* IAEA/STI/PUB/6 (International Atomic Energy Agency, Vienna).

RODRIGUEZ-PASQUÉS, R. H. (1971). "Microactivity radiochemistry," page 278 in *Proceedings of the Secunda Conferencia Interamericana de Radioquimica,* CONF-680423 (Organization of American States, Washington).

RODRIGUEZ-PASQUÉS, R. H., STEINBERG, H. L., HARDING, J. E., MULLEN, P. A., HUTCHINSON, J. M. R., AND MANN, W. B. (1972). "Low-level radioactivity measurements on aluminum, steel, and copper," Int. J. Appl. Radiat. Isot. **23,** 445.

ROLLE, R. (1970). "Determination of radon daughters on filters by simple liquid scintillation technique," Am. Ind. Hyg. Assoc. J. **31,** 718.

ROSE, M. E. AND KORFF, S. A. (1941). "An investigation of proportional counters I," Phys. Rev. **59,** 850.

ROSS, H. H. (1970). "Čerenkov radiation: photon yield application to carbon-14 assay," page 123 in *Current Status of Liquid Scintillation Counting,* Bransome, E. D., Ed. (Grune & Stratton, New York).

ROSS, H. H. (1971). "Performance parameters of selected waveshifting compounds for Čerenkov counting," page 757 in *Organic Scintillators and Liquid Scintillation Counting,* Horrocks, D. L. and Peng, C. T., Eds. (Academic Press, New York).

ROSSI, B. (1952). *High Energy Particles* (Prentice Hall, Inc., Englewood Cliffs, New Jersey).

ROUX, A. M. (1974). "Comparison of activity and power measurements of a 1 Ci ^{60}Co source," Metrologia **10,** 103.

ROY, J. C. AND RYTZ, A. (1965). "Analyse de resultats de la comparaison internationale de ^{35}S, Juin 1962," BIPM Report (Bureau International des Poids et Mesures, Sèvres, France).

Roy, J. C. and Turcotte, J. (1967). "Use of ionic bombardment to increase the conductivity of metalised thin plastic films for making source mounts," Can. J. Chem. **65,** 1379.

Rutherford, E. (1905). "Charge carried by the α and β rays of radium," Philos. Mag. **10** [6], 193.

Rutherford, E. and Soddy, F. (1902). "The cause and nature of radioactivity: Part I Growth and decay," Philos. Mag. **4** [6], 370.

Rutherford, E. and Soddy, F. (1903). "Radioactive change," Philos. Mag. **5** [6], 576.

Rutherford, E. and Barnes, H. T. (1904). "Heating effects of the radium emanation," Philos. Mag. **7** [6], 202.

Rutherford, E. and Robinson, H. (1913). "Heating effect of radium and its emanation," Philos. Mag. **25** [6], 312.

Rutherford, E. Chadwick, J., and Ellis, C. D. (1931). *Radiations From Radioactive Substances* (Cambridge University Press, Cambridge).

Rytz, A. (1964). *Rapport sur la comparaison international de ^{241}Am, Juillet 1963,* (Bureau International des Poids et Mesures Report, Sèvres).

Salmon, L. (1961). "Analysis of gamma-ray scintillation spectra by the method of least squares," Nucl. Instrum. Methods **14,** 193.

Sayres, A. and Wu, C. S. (1957). "Gas scintillation counter," Rev. Sci. Instrum. **28,** 758.

Schell, W. R. (1970). "An internal gas proportional counter for measuring low level environmental radionuclides," page 173 in *Radionuclides in the Environment* (American Chemical Society, Washington).

Schonfeld, E., Kibby, A. H., and Davis, W. (1966). "Determination of nuclide concentrations in solutions containing low levels of radioactivity by least-squares resolution of the gamma-ray spectra," Nucl. Instrum. Methods **45,** 1.

Schopper, H. (1966). *Weak Interactions and Nuclear Beta Decay* (North Holland Publishing Co., Amsterdam).

Schwebel, A., Isbell, H. S., and Karabinos, J. V. (1951). "A rapid method for measurement of carbon-14 in formamide solution," Science **113,** 465.

Schwebel, A., Isbell, H. S., and Moyer, J. D. (1954). "Determination of carbon-14 in solutions of carbon-14 labeled materials by means of a proportional counter," J. Res. Nat. Bur. Stand. **53,** 221.

Seliger, H. H. (1960). "Liquid scintillation counting of α particles and energy resolution of the liquid scintillator for α- and β-particles," Int. J. Appl. Radiat. Isot. **8,** 29.

Seliger, H. H. and Schwebel, A. (1954). "Standardization of beta-emitting nuclides," Nucleonics **12,** No. 7, 54.

Seliger, H. H., Mann, W. B., and Cavallo, L. M. (1958). "The average energy of sulfur-35 beta decay," J. Res. Nat. Bur. Stand. **60,** 447.

Shalek, R. J. and Stoval, M. (1969). "Dosimetry in implant therapy," page 743 in *Radiation Dosimetry,* vol. 3, 2nd ed., Attix, F. H. and Tochilin, E., Eds. (Academic Press, New York).

Shapiro, J. (1972). *Radiation Protection* (Harvard University Press, Cambridge, Massachusetts).

SHARPE, J. AND WADE, F. (1951). "T.P.A. Mk II ionization chamber," AERE EL/R806 (U.K. Atomic Energy Research Establishment, Harwell).

SHER, D. W., KAARTINEN, N., EVERETT, L. J., AND JUSTES, V. (1971). "Preparing samples for liquid scintillation counting with the Packard sample oxidizer," page 849 in *Organic Scintillators and Liquid Scintillation Counting,* Horrocks, D. L. and Peng, C. T., Eds. (Academic Press, New York).

SHONKA, F. R. AND STEPHENSON, R. J. (1949). "Chicago pressure ionization chamber for gamma ray measurement," Univ. Chicago Metallurgical Laboratory Report MUC-RJS-2, AECD-2463 (U.S. Atomic Energy Commission, Washington).

SHORE, L. G. (1949). "Long-lived self-quenching counter filling," Rev. Sci. Instrum. **20,** 956.

SHUPING, R. E., PHILLIPS, C. R., AND MOGHISSI, A. A. (1969). "Low-level counting of environmental krypton-85 by liquid scintillation," Anal. Chem. **41,** 2082.

SIEGBAHN, K., Ed. (1965). *Alpha-, Beta-, and Gamma-Ray Spectroscopy* (North Holland Publishing Co., Amsterdam).

SILL, C. W. AND OLSON, D. G. (1970). "Sources and prevention of recoil contamination of solid-state alpha detectors," Anal. Chem. **42,** 1596.

SIMMONS, G. H. (1970). *A Training Manual for Nuclear Medicine Technologists,* BRH/DMRE 70-3 (National Technical Information Service, Springfield, Virginia).

SINCLAIR, W. K., TROTT, N. G., AND BELCHER, E. H. (1954). "The measurement of radioactive samples for clinical use," Br. J. Radiol. **27,** 565.

SMITH, C. M. H. (1965). *Textbook of Nuclear Physics,* Chapter 11 (Pergamon Press, Oxford; distributed by MacMillan, New York).

SMITH, D. (1975). "An improved method of data collection for $4\pi\beta$-γ coincidence measurements," Metrologia **11,** 73.

SMITH, D. AND WILLIAMS, A. (1971). "A $4\pi\beta$-γ coincidence method for mixed electron-capture/positron emitters: The absolute measurement of ^{68}Ga," Int. J. Appl. Radiat. Isot. **22,** 615.

SMITH, D. AND STUART, L. E. H. (1975). "An extension of the $4\pi\beta$-γ coincidence technique: Two dimensional extrapolation," Metrologia **11,** 67.

SMITH, K. F. AND CLINE, J. E. (1966). "Low-noise charge sensitive preamplifier for semiconductor detectors using paralleled field-effect-transistors," IEEE Trans. Nucl. Sci. **NS-13,** 468.

SNELL, A. H., Ed. (1962). *Nuclear Instruments and Their Uses,* vol. 1. (John Wiley & Sons, New York).

SPERNOL, A. AND DENECKE, B. (1964a). "High-precision absolute gas counting of tritium. I—preparation of hydrogen and counting gas," Int. J. Appl. Radiat. Isot. **15,** 139.

SPERNOL, A. AND DENECKE, B. (1964b). "High-precision absolute gas counting of tritium. II—characteristics and construction of internal gas counters, especially for hydrogen-methane mixtures," Int. J. Appl. Radiat. Isot. **15,** 195.

SPERNOL, A. AND DENECKE, B. (1964c). "High precision absolute gas counting of tritium. III—absolute counting of tritium in the internal gas counter," Int. J. Appl. Radiat. Isot. **15,** 241.

SPERNOL, A., DAROOST, E., AND MUTTERER, M. (1973). "Problems and possibilities of bremsstrahlung counting," Nucl. Instrum. Methods **112,** 169.

STANLEY, R. E. AND MOGHISSI, A. A., Eds. (1975). *Noble Gases,* ERDA Report CONF-730915 (National Technical Information Service, Springfield, Virginia).

STAUB, H. H. (1953). "Detection methods," page 1 in *Experimental Nuclear Physics,* Segré, E., Ed. (John Wiley, New York and Chapman and Hall, London).

STEVENSON, P. C. (1966). *Processing of Counting Data,* National Academy of Sciences-National Research Council Nuclear Science Series, Report NAS-NS 3109 (National Technical Information Service, Springfield, Virginia).

STEYN, J. (1967). "Internal liquid scintillation counting applied to the absolute disintegration-rate measurement of electron capture nuclides," page 35 in *Standardization of Radionuclides,* IAEA/STI/PUB/139 (International Atomic Energy Agency, Vienna).

STEYN, J. (1973). "Special problems involved in liquid scintillation counting," Nucl. Instrum. Methods **112,** 137.

STEYN, J. AND HAASBROEK, F. J. (1958). "The application of internal liquid scintillation counting to a 4π beta-gamma coincidence method for the absolute standardization of radioactive nuclides," page 95 in *Peaceful Uses of Atomic Energy,* vol. 21 (United Nations, Geneva).

STOCKMANN, L. L. (1955). "Radium assay of ores and sludges by the radon method," Unpublished manuscript (Radioactivity Section, National Bureau of Standards, Washington).

SUBRAMANIAN, G., RHODES, B. A., COOPER, J. F., AND SODD, V. J., Eds. (1975). *Radiopharmaceuticals* (The Society of Nuclear Medicine, Inc., New York).

SUDAR, S., VAS, L., AND BIRO, T. (1973). "Pulse-shape discrimination in the proportional counting of tritium betas," Nucl. Instrum. Methods **112,** 399.

SUTHERLAND, L. G. AND BUCHANAN, J. D. (1967). "Error in the absolute determination of disintegration rates of extended sources by coincidence counting with a single detector—Application to I-125 and Co-60," Int. J. Appl. Radiat. Isot. **18,** 786.

SWIERKOWSKI, S. P., ARMANTROUT, G. A., AND WICHNER, R. (1974). "Recent advances with HgI_2 x-ray detectors," IEEE Trans. Nucl. Sci. **NS-21,** 302.

TAMERS, M. BIRBRON, R., AND DELIBRIAS, G. (1962). "A new method for measuring low-level tritium using a benzene liquid scintillator," page 303 in *Tritium in the Physical and Biological Sciences,* vol. 1 IAEA/STI/PUB/39 (International Atomic Energy Agency, Vienna).

TANAKA, E. (1961). "A low background beta-ray scintillation spectrometer using a coincidence method with a Geiger counter," Nucl. Instrum. Methods **13,** 43.

TANAKA, E., ITOH, S., HIRAMOTO, T., AND IINUMA, T. A. (1965). "Cosmic ray contribution to the background of NaI scintillation spectrometers," Japanese J. Appl. Phys. **4,** 785.

TANAKA, E., ITOH, S., MARUYAMA, T., KAWAMURA, S., AND HIRAMOTO, T. (1967). "Low-level beta-spectroscopy of solid samples by means of a coincidence-type scintillation spectrometer combined with a logarithmic amplifier," Int. J. Appl. Radiat. Isot. **18,** 161.

TANAKA, S., SAKAMOTO, K., AND TAKAGI, J. (1967). "An extremely low-level gamma-ray spectrometer," Nucl. Instrum. Methods **56,** 319.

TAVENDALE, A. J. (1964). "Semiconductor lithium-ion drift diodes as high-resolution gamma-ray pair spectrometers," IEEE Trans. Nucl. Sci. **NS-11,** 191.

TAVENDALE, A. J. AND EWAN, G. T. (1963). "A high-resolution lithium-ion drifted germanium gamma-ray spectrometer," Nucl. Instrum. Methods **25,** 185.

TAYLOR, D. (1962). "Ionization chambers," in *Encyclopaedic Dictionary of Physics,* Thewlis, J., Ed.-in-Chief, **4,** 50 (MacMillan, New York; Pergamon, Oxford and London).

TAYLOR, J. G. V. (1962). "The total internal conversion coefficient of the 279-keV transition following the decay of Hg^{203} as measured by a new coincidence method," Can. J. Phys. **40,** 383.

TAYLOR, J. G. V. (1967). "X-ray-x-ray coincidence counting methods for the standardization of ^{125}I and ^{197}Hg," Page 341 in *Standardization of Radionuclides,* IAEA/STI/PUB/139 (International Atomic Energy Agency, Vienna).

TAYLOR, J. G. V. AND MERRITT, J. S. (1962). "Branching ratio in the decay of Be^7," Can. J. Phys. **40,** 926.

TAYLOR, J. M. (1963). *Semiconductor Particle Detectors* (Butterworths, Inc., Washington).

THIESS, P. E. AND MILEY, G. H. (1974). "New near-infrared and ultraviolet gas-proportional scintillation counters," IEEE Trans. Nucl. Sci. **NS-21,** 125.

THOMAS, J. (1963). "Determination of count-rate dependent corrections by the substitution method," Risø Report No. 57 (Danish Atomic Energy Commission, Risø).

THORNGATE, J. H., MCDOWELL, W. J., AND CHRISTIAN, D. C. (1974). "An application of pulse shape discrimination to liquid scintillation alpha spectroscopy," Health Phys. **27,** 123.

TOLBERT, B. M. (1956). *Ionization Chamber Assay of Radioactive Gases,* USAEC Report UCRL-3499 (National Technical Information Service, Springfield, Virginia).

TOLBERT, B. M. (1963). "Tritium measurements using ionization chambers," page 167 in *Advances in Tracer Methodology,* Rothchild, S., Ed. (Plenum Press, New York).

TOWNSEND, J. S. (1903). "The genesis of ions by the motion of positive ions in a gas and a theory of the sparking potential," Philos. Mag. **6** [6], 598.

TROMBKA, J. I. AND SCHMADEBECK, R. L. (1968). "A method for the analysis of pulse-height spectra containing gain-shift and zero-drift compensating," Nucl. Instrum. Methods **62,** 253.

TROUGHTON, M. E. C. (1966). "The absolute standardization of cobalt-57," Int. J. Appl. Radiat. Isot. **17,** 145.

TROUGHTON, M. E. C. (1967). "Coincidence method of standardization for the pure electron capture nuclide caesium-131," page 323 in *Standardization of Radionuclides* IAEA/STI/PUB/139 (International Atomic Energy Agency, Vienna).

TROUGHTON, M. E. C. (1977). "A method for measuring the nuclear transformation rate of ^{109}Cd," Int. J. Appl. Radiat. Isot. **28,** 773.

TUBIS, M. AND WOLF, W., Eds. (1976). *Radiopharmacy* (John Wiley & Sons, New York).

TURNER, R. C., RADLEY, J. M., AND MAYNEORD, W. V. (1958). "The alpha-ray activity of human tissues," Br. J. Radiol. **31,** 397.

URQUHART, D. F. (1967). "New method of measuring the gamma counting efficiency of 4π beta gas counters," page 167 in *Standardization of Radionuclides* IAEA/STI/PUB/139 (International Atomic Energy Agency, Vienna).

USAEC (1965). *Radiocarbon and Tritium Dating,* USAEC Report CONF-650652 (National Technical Information Service, Springfield, Virginia).

USP (1975). *The United States Pharmacopeia,* Nineteenth Revision (United States Pharmacopeial Convention, Inc., Rockville, Maryland).

VALENTINE, J. M. AND CURRAN, S. C. (1958). "Average energy expenditure per ion pair in gases and gas mixtures," Rep. Progr. Phys. **21,** 1.

VAN DER EIJK, W., OLDENHOF, W., AND ZEHNER, W. (1973). "Preparation of thin sources," Nucl. Instrum. Methods **112,** 343.

VANINBROUKX, R. AND SPERNOL, A. (1965). "High-precision 4π liquid scintillation counting," Int. J. Appl. Radiat. Isot. **16,** 289.

VANINBROUKX, R. AND STANEF, I. (1973). "Present status in the field of precision liquid scintillation counting," Nucl. Instrum. Methods **112,** 111.

VATAI, E. (1970). "Correction of electron capture ratios measured by multiwire proportional counter," Acta Phys. Acad. Sci. Hung. **28,** 103.

VENVERLOO, L. A. J. (1971). *Practical Measuring Techniques for Beta Radiation* (MacMillan, London).

VINCENT, C. H. (1973). "Random pulse trains, their measurement and statistical properties," IEE Monograph Series 13 (Peter Peregrinus, London).

VON SCHWEIDLER, E. (1906). "Über Schwankungen der radioaktiven Umwandlung," page 1 in *The Comptes Rendus du Premier Congrès International Pour L'Étude de la Radiologie et de l'Ionisation, Liége, 1905* (Durrod, H. and Pinot, E., Eds. Paris).

WAGNER, H. N., Ed. (1968). *Principles of Nuclear Medicine* (W. B. Saunders Co., Philadelphia, Pennsylvania).

WALKER, D. H. (1965). "An experimental study of the backscattering of 5.3 MeV alpha particles from platinum and monel metal," Int. J. Appl. Radiat. Isot. **16,** 183.

WALTER, F. J. AND DABBS, J. W. T. (1958). *Broad-Area Germanium p-n Junction Counters,* USAEC Report ORNL-2501 (Oak Ridge National Laboratory, Oak Ridge, Tennessee).

WALZ, K. F. AND WEISS, H. M. (1970). "Messung der halbwertszeiten von ^{60}Co, ^{137}Cs, und ^{133}Ba," Naturforscher **25a,** 921.

WANG, C. H. AND WILLIS, D. L. (1975). *Radiotracer Methodology in the Biological, Environmental, and Physical Sciences* (Prentice-Hall, Englewood Cliffs, New Jersey).

WASHTELL, C. C. H. (1959). *An Introduction to Radiation Counters and Detectors* (Philosophical Library, New York).

WATT, D. E. AND RAMSDEN, D. (1964). *High Sensitivity Counting Techniques* (MacMillan, New York).

WATT, D. E., RAMSDEN, D., AND WILSON, H. W. (1961). "The half-life of carbon-14," Int. J. Appl. Radiat. Isot. **11,** 68.

WEBB, P. P. AND WILLIAMS, R. L. (1963). "Gamma-ray spectroscopy using a germanium lithium-drifted diode," Nucl. Instrum. Methods **22,** 361.

WEISS, H. M., Ed. (1968). *Compte Rendu de la Comparaison Internationale d'Une Solution de ^{54}Mn, BIPM Report* (Bureau International des Poids et Mesures, Sèvres).

WEISS, H. M. (1973). "$4\pi\gamma$-ionization chamber measurements," Nucl. Instrum. Methods **112,** 291.

WEISS, J. AND BERNSTEIN, W. (1955). "Energy required to produce one ion pair for several gases," Phys. Rev. **98,** 1828.

WELLER, R. I. (1964). "Low-level radioactive contamination," page 567 in *The Natural Radiation Environment,* Adams, J. A. S. and Lowder, W. M., Eds. (University of Chicago Press, Chicago).

WHITE, P. H. (1970). "Alpha and fission counting of thin foils of fissile material," Nucl. Instrum. Methods **79,** 1.

WIEN, W. (1903). "Über die Selbstelektrisierung des Radiums und die Intensität der von ihm ausgesandten Strahlen," Phys. Z. **4,** 624.

WIERNIK, M. (1971a). "Comparison of several methods proposed for correction of dead-time losses in the gamma-ray spectrometry of very short-lived nuclides," Nucl. Instrum. Methods **95,** 13.

WIERNIK, M. (1971b). "Normal and random pulse generators for the correction of dead-time losses in nuclear spectrometry," Nucl. Instrum. Methods **96,** 325.

WILKINSON, D. (1950). "A stable ninety-nine channel pulse amplitude analyzer for slow counting," Proc. Cambridge Philos. Soc. **46** [3], 508.

WILKINSON, D. H. (1950). *Ionization Chambers and Counters* (Cambridge University Press).

WILLIAMS, A. (1964). "An estimation of the inherent accuracy of the tracer technique for measurements of disintegration rate," Int. J. Appl. Radiat. Isot. **15,** 709.

WILLIAMS, A. AND BIRDSEYE, R. A. (1966). "Measurement of the decay scheme correction in the absolute standardization of cobalt-57," Int. J. Appl. Radiat. Isot. **17,** 366.

WILLIAMS, A. AND CAMPION, P. J. (1963). "Measurement of the γ-sensitivity of a $4\pi\beta$-counter," Int. J. Appl. Radiat. Isot. **14,** 533.

WILLIAMS, A. AND CAMPION, P. J. (1965). "On the relative time distribution of pulses in $4\pi\beta$-γ coincidence technique," Int. J. Appl. Radiat. Isot. **16,** 555.

WILLIAMS, A. AND SARA, R. I. (1968). "Correlation counting applied to the determination of absolute disintegration rates," Nucl. Instrum. Methods **60,** 189.

WILLIAMS, A., LEWIS, V. E., AND SMITH, D. (1973). "Correlation techniques in radioactivity measurement," Nucl. Instrum. Methods **112,** 285.

WILLIAMS, A. AND SMITH, D. (1973). "Afterpulses in liquid scintillation counters," Nucl. Instrum. Methods **112,** 131.

WILLIAMS, K. D. AND SUTTON, J. D. (1969). "Survey of the use of radionuclides in medicine," Isot. Radiat. Technol. **6,** 415.

WILZBACH, K. E., VAN DYKEN, A. R., AND KAPLAN, L. (1954). "Determination of tritium by ion current measurement," Anal. Chem. **26,** 880.

WOELLER, F. H. (1961). "Liquid scintillation counting of $C^{14}O_2$ with phenethylamine," Anal. Biochem. **2,** 508.

WOGMAN, N. A. (1974). "Comparison of Ge(Li) and anti-Compton systems for measurement of environmental samples," IEEE Trans. Nucl. Sci. **NS-21,** 526.

WOGMAN, N. A. AND BRODZINSKI, R. L. (1973). "The development and application of a beta-gamma-gamma multidimensional spectrometer," IEEE Trans. Nucl. Sci. **NS-20,** 73.

WOGMAN, N. A., PERKINS, R. W., AND KAYE, J. H. (1969). "An all sodium iodide anticoincidence shielded multidimensional gamma-ray spectrometer for low-activity samples," Nucl. Instrum. Methods **74,** 197.

WOOD, D. E. AND BOWEN, B. M. (1971). "^{95}Zr and ^{124}Sb in ^{99m}Tc generators," J. Nucl. Med. **12,** 307.

WOOD, R. E. AND PALMS, J. M. (1974). "A gamma-ray spectrum analysis technique for low-level environmental radionuclides," IEEE Trans. Nucl. Sci. **NS-21,** 536.

WOODS, M. J. AND LUCAS, S. E. M. (1975). "Calibration of the 1383A ionization chamber for ^{125}I," Int. J. Appl. Radiat. Isot. **26,** 488.

WOODS, M. J. (1970). "Calibration figures for the type 1383A ionization chamber," Int. J. Appl. Radiat. Isot. **21,** 752.

WOODS, M. J., GOODIER, I. W., AND LUCAS, S. E. M. (1975). "The half-life of ^{133}Xe and its calibration factor for the 1383A ionization chamber," Int. J. Appl. Radiat. Isot. **26,** 485.

YAFFE, L. (1962). "Preparation of thin films, sources, and targets," Annu. Rev. Nucl. Sci. **12,** 153.

YAFFE, L. AND FISHMAN, J. B. (1960). "Self-absorption studies with a $4\pi\beta$-proportional flow counter," page 185 in *Metrology of Radionuclides* IAEA/STI/PUB/6 (International Atomic Energy Agency, Vienna).

YAGODA, H. J. (1949). *Radioactive Measurements with Nuclear Emulsions* (John Wiley & Sons, New York).

YAMADA, S. (1962). "The detection of low activity in aqueous solution using Čerenkov radiation," J. Phys. Soc. Japan **17,** 865.

YOUDEN, W. J. (1951). *Statistical Methods for Chemists* (John Wiley & Sons, New York).

YOUDEN, W. J. (1963). "Statistics in chemical analysis," page 14-1 in *Handbook of Analytical Chemistry,* Meites, L., Ed., (McGraw-Hill, New York).

YOUNG, J. Z. AND ROBERTS, F. (1951). "A flying-spot microscope," Nature **167,** 231.

ZSDANSZKY, K. (1973). "Precise measurement of small currents," Nucl. Instrum. Methods **112,** 299.

ZUMWALT, L. R. (1950). *Absolute Beta Counting Using End-Window Geiger-Muller Counters and Experimental Data on Beta-Particle Scattering Effects,* AECU-567 (Oak Ridge National Laboratory, Oak Ridge, Tennessee).

The NCRP

The National Council on Radiation Protection and Measurements is a nonprofit corporation chartered by Congress in 1964 to:

1. Collect, analyze, develop, and disseminate in the public interest information and recommendations about (a) protection against radiation and (b) radiation measurements, quantities, and units, particularly those concerned with radiation protection;
2. Provide a means by which organizations concerned with the scientific and related aspects of radiation protection and of radiation quantities, units, and measurements may cooperate for effective utilization of their combined resources, and to stimulate the work of such organizations;
3. Develop basic concepts about radiation quantities, units, and measurements, about the application of these concepts, and about radiation protection;
4. Cooperate with the International Commission on Radiological Protection, the International Commission on Radiation Units and Measurements, and other national and international organizations, governmental and private, concerned with radiation quantities, units, and measurements and with radiation protection.

The Council is the successor to the unincorporated association of scientists known as the National Committee on Radiation Protection and Measurements and was formed to carry on the work begun by the Committee.

The Council is made up of the members and the participants who serve on the fifty-eight Scientific Committees of the Council. The Scientific Committees, composed of experts having detailed knowledge and competence in the particular area of the Committee's interest, draft proposed recommendations. These are then submitted to the full membership of the Council for careful review and approval before being published.

The following comprise the current officers and membership of the Council:

Officers

President	Warren K. Sinclair
Vice President	Hymer L. Friedell
Secretary and Treasurer	W. Roger Ney
Assistant Secretary	Eugene R. Fidell
Assistant Treasurer	Harold O. Wyckoff

Members

Seymour Abrahamson
S. James Adelstein
Roy E. Albert
Frank H. Attix
John A. Auxier
William J. Bair
Victor P. Bond
Harold S. Boyne
Robert L. Brent
A. Bertrand Brill
Reynold F. Brown
William W. Burr, Jr.
Melvin W. Carter
George W. Casarett
Randall S. Caswell
Arthur B. Chilton
Stephen F. Cleary
Cyril L. Comar
Patricia W. Durbin
Merril Eisenbud
Thomas S. Ely
Benjamin G. Ferris
Asher J. Finkel
Donald C. Fleckenstein
Richard F. Foster
Hymer L. Friedell
Arthur H. Gladstein
Robert A. Goepp
Barry B. Goldberg
Marvin Goldman
Robert O. Gorson
Douglas Grahn
Arthur W. Guy
Ellis M. Hall
John H. Harley
John W. Healy
Louis H. Hempelmann, Jr.
John M. Heslep
George B. Hutchison
Marylou Ingram
Seymour Jablon
Jacob Kastner
Edward B. Lewis
Thomas A. Lincoln
Charles W. Mays
Roger O. McClellan
James E. McLaughlin
Charles B. Meinhold
Mortimer M. Mendelsohn
Dade W. Moeller
Paul E. Morrow
Robert D. Moseley, Jr.
James V. Neel
Robert J. Nelsen
Frank Parker
Andrew K. Poznanski
William C. Reinig
Chester R. Richmond
Harald H. Rossi
Robert E. Rowland
Eugene L. Saenger
Harry F. Schulte
Raymond Seltser
Warren K. Sinclair
J. Newell Stannard
Chauncey Starr
John B. Storer
Roy C. Thompson
James D. Turner
John C. Villforth
George L. Voelz
Niel Wald
Edward W. Webster
George M. Wilkening
McDonald E. Wrenn

Honorary Members

LAURISTON S. TAYLOR, *Honorary President*

EDGAR C. BARNES	PAUL C. HODGES	EDITH H. QUIMBY
CARL B. BRAESTRUP	GEORGE V. LEROY	JOHN H. RUST
AUSTIN M. BRUES	KARL Z. MORGAN	SHIELDS WARREN
FREDERICK P. COWAN	RUSSELL H. MORGAN	HAROLD O. WYCKOFF
ROBLEY D. EVANS	HERBERT M. PARKER	

Currently, the following Scientific Committees are actively engaged in formulating recommendations:

SC-1: Basic Radiation Protection Criteria
SC-3: Medical X- and Gamma-Ray Protection up to 10 MeV (Equipment Performance and Use)
SC-11: Incineration of Radioactive Waste
SC-16: X-Ray Protection in Dental Offices
SC-18: Standards and Measurements of Radioactivity for Radiological Use
SC-23: Radiation Hazards Resulting from the Release of Radionuclides into the Environment
SC-24: Radionuclides and Labeled Organic Compounds Incorporated in Genetic Material
SC-25: Radiation Protection in the Use of Small Neutron Generators
SC-26: High Energy X-Ray Dosimetry
SC-30: Physical and Biological Properties of Radionuclides
SC-31: Selected Occupational Exposure Problems Arising from Internal Emitters
SC-32: Administered Radioactivity
SC-33: Dose Calculations
SC-34: Maximum Permissible Concentrations for Occupational and Non-Occupational Exposures
SC-37: Procedures for the Management of Contaminated Persons
SC-38: Waste Disposal
SC-39: Microwaves
SC-40: Biological Aspects of Radiation Protection Criteria
SC-41: Radiation Resulting from Nuclear Power Generation
SC-42: Industrial Applications of X Rays and Sealed Sources
SC-44: Radiation Associated with Medical Examinations
SC-45: Radiation Received by Radiation Employees
SC-46: Operational Radiation Safety
SC-47: Instrumentation for the Determination of Dose Equivalent
SC-48: Apportionment of Radiation Exposure
SC-50: Surface Contamination
SC-51: Radiation Protection in Pediatric Radiology and Nuclear Medicine Applied to Children
SC-52: Conceptual Basis of Calculations of Dose Distributions
SC-53: Biological Effects and Exposure Criteria for Radiofrequency Electromagnetic Radiation
SC-54: Bioassay for Assessment of Control of Intake of Radionuclides
SC-55: Experimental Verification of Internal Dosimetry Calculations
SC-56: Mammography
SC-57: Internal Emitter Standards
SC-58: Radioactivity in Water

In recognition of its responsibility to facilitate and stimulate cooperation among organizations concerned with the scientific and related aspects of radiation protection and measurement, the Council has created a category of NCRP Collaborating Organizations. Organizations or groups of organizations that are national or international in scope and are concerned with scientific problems involving radiation quantities, units, measurements and effects, or radiation protection may be admitted to collaborating status by the Council. The present Collaborating Organizations with which the NCRP maintains liaison are as follows:

American Academy of Dermatology
American Association of Physicists in Medicine
American College of Radiology
American Dental Association
American Industrial Hygiene Association
American Insurance Association
American Medical Association
American Nuclear Society
American Occupational Medical Association
American Podiatry Association
American Public Health Association
American Radium Society
American Roentgen Ray Society
American Society of Radiologic Technologists
Association of University Radiologists
Atomic Industrial Forum
College of American Pathologists
Defense Civil Preparedness Agency
Genetics Society of America
Health Physics Society
National Bureau of Standards
National Electrical Manufacturers Association
Radiation Research Society
Radiological Society of North America
Society of Nuclear Medicine
United States Air Force
United States Army
United States Department of Energy
United States Environmental Protection Agency
United States Navy
United States Nuclear Regulatory Commission
United States Public Health Service

The NCRP has found its relationships with these organizations to be extremely valuable to continued progress in its program.

The Council's activities are made possible by the voluntary contribution of the time and effort of its members and participants and the generous support of the following organizations:

Alfred P. Sloan Foundation
Alliance of American Insurers
American Academy of Dental Radiology
American Academy of Dermatology
American Association of Physicists in Medicine
American College of Radiology
American College of Radiology Foundation
American Dental Association
American Industrial Hygiene Association
American Insurance Association
American Medical Association
American Nuclear Society
American Occupational Medical Association
American Osteopathic College of Radiology
American Podiatry Association
American Public Health Association
American Radium Society
American Roentgen Ray Society
American Society of Radiologic Technologists
American Veterinary Medical Association
American Veterinary Radiology Society
Association of University Radiologists
Atomic Industrial Forum
Battelle Memorial Institute
College of American Pathologists
Defense Civil Preparedness Agency
Edward Mallinckrodt, Jr. Foundation
Electric Power Research Institute
Genetics Society of America
Health Physics Society
James Picker Foundation
National Association of Photographic Manufacturers
National Bureau of Standards
National Electrical Manufacturers Association
Radiation Research Society
Radiological Society of North America
Society of Nuclear Medicine
United States Department of Energy
United States Environmental Protection Agency
United States Navy
United States Nuclear Regulatory Commission
United States Public Health Service

To all of these organizations the Council expresses its profound appreciation for their support.

Initial funds for publication of NCRP reports were provided by a grant from the James Picker Foundation and for this the Council wishes to express its deep appreciation.

The NCRP seeks to promulgate information and recommendations based on leading scientific judgment on matters of radiation protection and measurement and to foster cooperation among organizations concerned with these matters. These efforts are intended to serve the public interest and the Council welcomes comments and suggestions on its reports or activities from those interested in its work.

NCRP Publications

NCRP publications are distributed by the NCRP Publications' office. Information on prices and how to order may be obtained by directing an inquiry to:

NCRP Publications
P.O. Box 30175
Washington, D.C. 20014

The extant publications are listed below.

Lauriston S. Taylor Lectures

No.	Title and Author
1	*The Squares of the Natural Numbers in Radiation Protection* by Herbert M. Parker
2	*Why Be Quantitative about Radiation Risk Estimates?* by Sir Edward Pochin

NCRP Reports

No.	Title
8	*Control and Removal of Radioactive Contamination in Laboratories (1951)*
9	*Recommendations for Waste Disposal of Phosphorus-32 and Iodine-131 for Medical Users (1951)*
12	*Recommendations for the Disposal of Carbon-14 Wastes (1953)*
16	*Radioactive Waste Disposal in the Ocean (1954)*
22	*Maximum Permissible Body Burdens and Maximum Permissible Concentrations of Radionuclides in Air and in Water for Occupational Exposure (1959) [Includes Addendum 1 issued in August 1963]*
23	*Measurement of Neutron Flux and Spectra for Physical and Biological Applications (1960)*
25	*Measurement of Absorbed Dose of Neutrons and of Mixtures of Neutrons and Gamma Rays (1961)*
27	*Stopping Powers for Use with Cavity Chambers (1961)*
30	*Safe Handling of Radioactive Materials (1964)*
32	*Radiation Protection in Educational Institutions (1966)*

33	*Medical X-Ray and Gamma-Ray Protection for Energies Up to 10 MeV—Equipment Design and Use (1968)*
35	*Dental Protection in Veterinary Medicine (1970)*
36	*Radiation Protection in Veterinary Medicine (1970)*
37	*Precautions in the Management of Patients Who Have Received Therapeutic Amounts of Radionuclides (1970)*
38	*Protection Against Neutron Radiation (1971)*
39	*Basic Radiation Protection Criteria (1971)*
40	*Protection Against Radiation From Brachytherapy Sources (1972)*
41	*Specification of Gamma-Ray Brachytherapy Sources (1974)*
42	*Radiological Factors Affecting Decision-Making in a Nuclear Attack (1974)*
43	*Review of the Current State of Radiation Protection Philosophy (1975)*
44	*Krypton-85 in the Atmosphere—Accumulation, Biological Significance, and Control Technology (1975)*
45	*Natural Background Radiation in the United States (1975)*
46	*Alpha-Emitting Particles in Lungs (1975)*
47	*Tritium Measurement Techniques (1976)*
48	*Radiation Protection for Medical and Allied Health Personnel (1976)*
49	*Structural Shielding Design and Evaluation for Medical Use of X Rays and Gamma Rays of Energies Up to 10 MeV (1976)*
50	*Environmental Radiation Measurements (1976)*
51	*Radiation Protection Design Guidelines for 0.1–100 MeV Particle Accelerator Facilities (1977)*
52	*Cesium-137 From the Environment to Man: Metabolism and Dose (1977)*
53	*Review of NCRP Radiation Dose Limit for Embryo and Fetus in Occupationally-Exposed Women (1977)*
54	*Medical Radiation Exposure of Pregnant and Potentially Pregnant Women (1977)*
55	*Protection of the Thyroid Gland in the Event of Releases of Radioiodine (1977)*
56	*Radiation Exposure From Consumer Products and Miscellaneous Sources (1977)*
57	*Instrumentation and Monitoring Methods for Radiation Protection (1978)*
58	*A Handbook of Radioactivity Measurements Procedures (1978)*

Binders for NCRP Reports are available. Two sizes make it possible to collect into small binders the "old series" of reports (NCRP Reports Nos. 8–31) and into large binders the more recent publications (NCRP Reports Nos. 32–58). Each binder will accommodate from five to seven reports. The binders carry the identification "NCRP Reports" and come with label holders which permit the user to attach labels showing the reports contained in each binder.

The following bound sets of NCRP Reports are also available:

Volume I.	NCRP Reports Nos. 8, 9, 12, 16, 22
Volume II.	NCRP Reports Nos. 23, 25, 27, 30
Volume III.	NCRP Reports Nos. 32, 33, 35, 36, 37
Volume IV.	NCRP Reports Nos. 38, 39, 40, 41
Volume V.	NCRP Reports Nos. 42, 43, 44, 45, 46
Volume VI.	NCRP Reports Nos. 47, 48, 49, 50, 51

(Titles of the individual reports contained in each volume are given above.)

The following NCRP reports are now superseded and/or out of print:

NCRP Report No.	Title
1	*X-Ray Protection (1931) [Superseded by NCRP Report No. 3]*
2	*Radium Protection (1934). [Superseded by NCRP Report No. 4]*
3	*X-Ray Protection (1936). [Superseded by NCRP Report No. 6]*
4	*Radium Protection (1938). [Superseded by NCRP Report No. 13]*
5	*Safe Handling of Radioactive Luminous Compounds (1941). [Out of print]*
6	*Medical X-Ray Protection up to Two Million Volts (1949). [Superseded by NCRP Report No. 18]*
7	*Safe Handling of Radioactive Isotopes (1949). [Superseded by NCRP Report No. 30]*
10	*Radiological Monitoring Methods and Instruments (1952). [Superseded by NCRP Report No. 57]*
11	*Maximum Permissible Amounts of Radioisotopes in the Human Body and Maximum Permissible Concentrations in Air and in Water (1953). [Superseded by NCRP Report No. 22]*
13	*Protection Against Radiations from Radium, Cobalt-60 and Cesium-137 (1954). [Superseded by NCRP Report No. 24]*
14	*Protection Against Betatron—Synchrotron Radiations Up to 100 Million Electron Volts (1954). [Superseded by NCRP Report No. 51].*
15	*Safe Handling of Cadavers Containing Radioactive Isotopes (1953). [Superseded by NCRP Report No. 21]*

17 *Permissible Dose from External Sources of Ionizing Radiation (1954) including Maximum Permissible Exposure to Man, Addendum to National Bureau of Standards Handbook 59 (1958). [Superseded by NCRP Report No. 39]*

18 *X-Ray Protection (1955). [Superseded by NCRP Report No. 26]*

19 *Regulation of Radiation Exposure by Legislative Means (1955). [Out of print]*

20 *Protection Against Neutron Radiation Up to 30 Million Electron Volts (1957). [Superseded by NCRP Report No. 38]*

21 *Safe Handling of Bodies Containing Radioactive Isotopes (1958) [Superseded by NCRP Report No. 37]*

24 *Protection Against Radiations from Sealed Gamma Sources (1960). [Superseded by NCRP Reports Nos. 33, 34, and 40]*

26 *Medical X-Ray Protection Up to Three Million Volts (1961). [Superseded by NCRP Reports Nos. 33, 34, 35, and 36]*

28 *A Manual of Radioactivity Procedures (1961) [Superseded by NCRP Report No. 58]*

29 *Exposure to Radiation in an Emergency (1962). [Superseded by NCRP Report No. 42]*

31 *Shielding for High-Energy Electron Accelerator Installations (1964). [Superseded by NCRP Report No. 51]*

34 *Medical X-Ray and Gamma-Ray Protection for Energies Up to 10 MeV—Structural Shielding Design and Evaluation (1970). [Superseded by NCRP Report No. 49]*

Statements

The following statements of the NCRP were published outside of the NCRP Report series:

"Blood Counts, Statement of the National Committee on Radiation Protection," Radiology 63, 428 (1954)

"Statements on Maximum Permissible Dose from Television Receivers and Maximum Permissible Dose to the Skin of the Whole Body," Am. J. Roentgenol., Radium Ther. and Nucl. Med. 84, 152 (1960) and Radiology 75, 122 (1960)

X-Ray Protection Standards for Home Television Receivers, Interim Statement of the National Council on Radiation Protection and Measurements (National Council on Radiation Protection and Measurements, Washington, 1968)

Specification of Units of Natural Uranium and Natural Thorium (National Council on Radiation Protection and Measurements, Washington, 1973)

Copies of the statements published in journals may be consulted in libraries. A limited number of copies of the last two statements listed above are available for distribution by NCRP Publications.

Index

Note: For individual radionuclides, see List of Radionuclides (Appendix A.2) for which nuclear-decay data are given in Appendix A.3

Note: For individual radionuclides, see List of Radionuclides (Appendix A.2) for which nuclear-decay data are given in Appendix A.3

Note: For individual radionuclides, see List of Radionuclides (Appendix A.2) for which nuclear-decay data are given in Appendix A.3

Note: For individual radionuclides, see List of Radionuclides (Appendix A.2) for which nuclear-decay data are given in Appendix A.3

Note: For individual radionuclides, see List of Radionuclides (Appendix A.2) for which nuclear-decay data are given in Appendix A.3

Note: For individual radionuclides, see List of Radionuclides (Appendix A.2) for which nuclear-decay data are given in Appendix A.3

Note: For individual radionuclides, see List of Radionuclides (Appendix A.2) for which nuclear-decay data are given in Appendix A.3

Note: For individual radionuclides, see List of Radionuclides (Appendix A.2) for which nuclear-decay data are given in Appendix A.3

Note: For individual radionuclides, see List of Radionuclides (Appendix A.2) for which nuclear-decay data are given in Appendix A.3

Note: For individual radionuclides, see List of Radionuclides (Appendix A.2) for which nuclear-decay data are given in Appendix A.3